N-Oxidation of Drugs

This volume is dedicated to Professor Dan Ziegler in recognition of his extensive contributions to the enzymology of *N*-oxidation.

N-Oxidation of Drugs
Biochemistry, pharmacology, toxicology

Edited by

P. Hlavica
Full Professor of Pharmacology and Toxicology,
Ludwig-Maximilians-Universität München, Munich, Germany

L.A. Damani
Lecturer in Biopharmacy, King's College, London, UK

CHAPMAN & HALL
London · New York · Tokyo · Melbourne · Madras

UK	Chapman & Hall, 2–6 Boundary Row, London SE1 8HN
USA	Chapman & Hall, 29 West 35th Street, New York NY10001
JAPAN	Chapman & Hall Japan, Thomson Publishing Japan, Hirakawacho Nemoto Building, 7F, 1-7-11 Hirakawa-cho, Chiyoda-ku, Tokyo 102
AUSTRALIA	Chapman & Hall Australia, Thomas Nelson Australia, 102 Dodds Street, South Melbourne, Victoria 3205
INDIA	Chapman & Hall India, R. Seshadri, 32 Second Main Road, CIT East, Madras 600 035

First edition 1991

© 1991 Chapman & Hall

Typeset in 10/12pt Times by EJS Chemical Composition,
Midsomer Norton, Bath, Avon
Printed in Great Britain by St Edmundsbury Press,
Bury St Edmunds, Suffolk

ISBN 0 412 36030 6

British Library Cataloguing in Publication Data

N-oxidation of drugs: Biochemistry, pharmacology, toxicology.
I. Hlavica, P.   II. Damani, L.A.
615

ISBN 0–412–36030–6

Library of Congress Cataloging-in-Publication Data

Available

Contents

Contributors

L.-O. Andersson

Varian AG, Zug, Switzerland

M. Aravagiri

Brentwood V.A. Medical Center and UCLA, California, USA

G.M.J. Beijersbergen van Henegouwen

Department of Medicinal Chemistry, Leiden University, The Netherlands

R.G. Böcker

Department of Biochemistry and Center in Molecular Toxicology, Vanderbilt University School of Medicine, Tennessee, USA

A. Bondon

Department of Biochemistry and Center in Molecular Toxicology, Vanderbilt University School of Medicine, Tennessee, USA

A.R. Boobis

Department of Clinical Pharmacology, Royal Postgraduate Medical School, London, UK

A.K. Cho

Department of Pharmacology and Jonsson Comprehensive Cancer UCLA, Center for Health Sciences, California, USA

S. Cholerton

Department of Pharmacology and Toxicology, St Mary's Hospital Medical School, London, UK

B. Clement

Institut für Pharmazeutische Chemie, Philipps-Universität, Marburg, Germany

J.F. Curtis

Laboratory of Molecular Biophysics, National Institute of Environmental, Health Sciences, North Carolina, USA

L.A. Damani   Chelsea Department of Pharmacy,
                        King's College, University of London, UK

D.S. Davies   Department of Clinical Pharmacology,
                        Royal Postgraduate Medical School,
                        London, UK

H. Diekmann   Institut für Mikrobiologie, Universität
                        Hannover, Germany

J.D. Duncan   Department of Pharmacology and Jonsson
                        Comprehensive Cancer UCLA, Center for
                        Health Sciences, California, USA

R.J. Edwards   Department of Clinical Pharmacology,
                        Royal Postgraduate Medical School,
                        London, UK

T.E. Eling   Laboratory of Molecular Biophysics,
                        National Institute of Environmental
                        Health Sciences, North Carolina, USA

P. Eyer   Walther Straub-Institut für Pharmakologie und
                        Toxikologie der Universität, München,
                        Germany

H. Fukuta   Division of Analytical Biochemistry,
                        Hokkaido University, Japan

J.M. Fukuto   Department of Pharmacology and Jonsson
                        Comprehensive Cancer UCLA, Center of
                        Health Sciences, California, USA

V. Glover   Department of Chemical Pathology,
                        Queen Charlotte's and Chelsea Hospital,
                        London, UK

I. Golly   Walther Straub-Institut für Pharmakologie
                        und Toxikologie der Universität, München,
                        Germany

J.W. Gorrod   Chelsea Department of Pharmacy,
                        King's College, University of London, UK

F.P. Guengerich   Department of Biochemistry and Center in
                            Molecular Toxicology, Vanderbilt University
                            School of Medicine, Tennessee, USA

E.M. Hawes   College of Pharmacy, University of
                            Saskatchewan, Canada

P. Hlavica   Walther Straub-Institut für Pharmakologie
                            und Toxikologie der Universität, München,
                            Germany

J.-Y. Hong   Department of Chemical Biology and
                            Pharmacognosy, Rutgers University,
                            New Jersey, USA

J.W. Hubbard   College of Pharmacy, University of
                            Saskatchewan, Canada

M. Immel   Institut für Pharmazeutische Chemie,
                            Philipps-Universität, Marburg, Germany

H. Ishizaki   Department of Chemical Biology and
                            Pharmacognosy, Rutgers University,
                            New Jersey, USA

S. Jatoe   Faculty of Pharmacy, University of Toronto,
                            Ontario, Canada

T.J. Jaworski   College of Pharmacy, University of
                            Saskatchewan, Canada

F.F. Kadlubar   Office of Research (HFT-100), National
                            Center for Toxicological Research, Arkansas,
                            USA

T. Kamataki   Division of Analytical Biochemistry,
                            Hokkaido University, Japan

Th. Kämpchen   Institut für Pharmazeutische Chemie,
                            Philipps-Universität, Marburg, Germany

S. Khan   Faculty of Pharmacy, University of Toronto,
                            Ontario, Canada

O. Kikuchi	Division of Analytical Biochemistry, Hokkaido University, Japan
C.M. King	Department of Chemical Carcinogenesis, Michigan Cancer Foundation, USA
M. Kitada	Division of Analytical Biochemistry, Hokkaido University, Japan
M. Komori	Division of Analytical Biochemistry, Hokkaido University, Japan
E.D. Korchinski	College of Medicine, University of Saskatchewan, Canada
S.P. Lam	Chelsea Department of Pharmacy, King's College, University of London, UK
S.J. Land	Department of Chemical Carcinogenesis, Michigan Cancer Foundation, USA
W. Lenk	Walther Straub-Institut für Pharmakologie und Toxikologie der Universität, München, Germany
P.D. Lotlikar	Fels Institute for Cancer Research and Molecular Biology, Temple University, School of Medicine, Pennsylvania, USA
I.L. Macdonald	Department of Chemistry, University of Virginia, USA
L.G. McGirr	Faculty of Pharmacy, University of Toronto, Ontario, Canada
G. McKay	College of Pharmacy, University of Saskatchewan, Canada
S.R. Marder	Brentwood V.A. Medical Center and UCLA, California, USA
R.P. Mason	Laboratory of Molecular Biophysics, National Institute of Environmental Health Sciences, North Carolina, USA

K.K. Midha	College of Pharmacy, University of Saskatchewan, Canada
B.P. Murray	Department of Clinical Pharmacology, Royal Postgraduate Medical School, London, UK
P.J. O'Brien	Faculty of Pharmacy, University of Toronto, Ontario, Canada
K. Ohta	Department of Analytical Biochemistry, Hokkaido University, Japan
H. Pfundner	Institut für Pharmazeutische Chemie, Philipps-Universität, Marburg, Germany
H.J. Plattner	Institut für Mikrobiologie, Universität Hannover, Germany
M. Riedl	Walther Straub-Institut für Pharmakologie und Toxikologie der Universität, München, Germany
M. Sandler	Department of Chemical Pathology, Queen Charlotte's and Chelsea Hospital, London, UK
S. Schmitt	Institut für Pharmazeutische Chemie, Philipps-Universität, Marburg, Germany
D. Sesardic	Department of Clinical Pharmacology, Royal Postgraduate Medical School, London, UK
R.L. Smith	Department of Pharmacology and Toxicology, St Mary's Hospital Medical School, London, UK
T. Smith	Department of Chemical Biology and Pharmacognosy, Rutgers University, New Jersey, USA
G. Talaska	Office of Research (HFT-100), National Center for Toxicological Research, Arkansas, USA

M. Taneda

Division of Analytical Biochemistry,
Hokkaido University, Japan

D.C. Thompson

Laboratory of Molecular Biophysics,
National Institute of Environmental Health
Sciences, North Carolina, USA

T. Uchida

Division of Analytical Biochemistry,
Hokkaido University, Japan

J. Uetrecht

Faculties of Pharmacy and Medicine,
University of Toronto, Ontario, Canada

E. Williams

Department of Food Science and Technology
and Toxicology Program, Oregon State
University, USA

C.S. Yang

Department of Chemical Biology and
Pharmacognosy, Rutgers University,
New Jersey, USA

J.S.H. Yoo

Department of Chemical Biology and
Pharmacognosy, Rutgers University,
New Jersey, USA

J. van der Zee

Laboratory of Molecular Biophysics,
National Institute of Environmental Health
Sciences, North Carolina, USA

D.M. Ziegler

Clayton Foundation Biochemical Institute
and Department of Chemistry, The University
of Texas at Austin, USA

M. Zimmermann

Institut für Pharmazeutische Chemie,
Philipps-Universität, Marburg, Germany

Preface

The metabolic *N*-oxidation of nitrogenous xenobiotics has been reported to occur in many biological systems, in addition to mammalian tissues, and the mechanisms appear to differ in many respects from those involved in oxidative attack at carbon centres. The extensive use of nitrogen-containing compounds as pharmaceuticals and chemical intermediates can lead to exposure to a large number of these agents under widely varying conditions. Biotransformation of these xenobiotics by *N*-oxidative pathways can effect detoxication, but equally well can induce formation of cytotoxic metabolites or potential promutagens and procarcinogens. The substantial progress, in recent years, in our understanding of the biochemistry and toxicology of *N*-oxidation of nitrogenous structures has created a need for a synthesis of current knowledge.

This book provides a wide-ranging review of the state-of-the-art in nitrogen xenobiochemistry divided into four parts. The introductory chapter discusses recent developments in trace analysis of radical intermediates and other *N*-oxygenated products by physical and immunochemical techniques. Special attention is given in Part Two to the enzymology of *N*-oxidation. Thus, detailed account is given of the mechanism and substrate specificity of the flavin-containing mono-oxygenase and factors regulating its activity are addressed. A separate chapter outlines the polymorphic expression of flavoprotein-dependent reactions. Similarly, the mechanistic background and inducibility of cytochrome *P*-450-catalysed turnover of specific types of nitrogenous compounds is highlighted. Data are also compiled describing the role of peroxidative *N*-oxidation of xenobiotics in extrahepatic tissues lacking significant amounts of cytochrome *P*-450.

Part Three summarizes reductions and conjugations of *N*-oxygenated compounds; some of these pathways are known to confer unusual reactivity of the products formed towards cellular macromolecules. Reviews on toxicological aspects of *N*-oxidation are grouped together in Part Four of this book. These include accounts on food-derived mutagenic heterocycles, selective destruction of nigrostriatal dopaminergic neurons in the CNS by

MPTP metabolites, phototoxic effects associated with the intake of certain nitrogenous xenobiotics, and drug hypersensitivity reactions.

Written by leading authorities in the area of *N*-oxidation, this book is intended to serve as a reference for graduates and research workers in the biochemical, biopharmaceutical and toxicological fields and for professionals in the drug, food and chemical industries. Thus, this compilation of data might contribute to focus attention on aspects of *N*-oxidation that still require more intensive investigation.

P. Hlavica and L.A. Damani

PART ONE

Analysis of *N*-Oxidized Products

1

Formation of aromatic amine free radicals by prostaglandin hydroperoxidase and peroxyl radicals: analysis by ESR and stable end products

T.E. Eling, J.F. Curtis, D.C. Thompson, J. Van der Zee and R.P. Mason

Laboratory of Molecular Biophysics
National Institute of Environmental Health Sciences
National Institutes of Health
Research Triangle Park, North Carolina 27709

1. Aromatic amines are excellent substrates for peroxidases and are metabolized to free radicals as detected by analysis of the stable end-products and ESR.
2. *N*-substituted aromatic amines are oxidatively cleaved by peroxidases by a mechanism that involves nitrogen centered free radicals, with carbon centered free radicals being formed in certain cases.
3. For aromatic amines that are not good substrates for peroxidase, metabolism appears to be mediated by peroxyl radicals which form *C*-oxygenated metabolites.

1.1 INTRODUCTION

Arachidonic acid (AA) is converted into a number of biologically active metabolites, including prostaglandins, prostacyclin, thromboxane and leukotrienes via either the cyclo-oxygenase or lipoxygenase pathways (Pace-Asciak *et al.*, 1983). Prostaglandin H synthase (PHS) is the initial enzyme in the cyclo-oxygenase pathway which commits AA to the formation of prostaglandins (Fig. 1.1). Two catalytic activities copurify with PHS, cyclo-oxygenase and peroxidase (Miyamoto *et al.*, 1976). The cyclo-oxygenase catalyses the addition of two moles of oxygen to one mole of AA forming a cyclic endoperoxide hydroperoxide, prostaglandin (PG) G_2. The peroxidase subsequently reduces the hydroperoxide to the corresponding alcohol, PGH_2. PGH_2 is then converted into a variety of other prostanoid metabolites, including thromboxanes, prostacyclin, and PGs E_2 and $F_{2\alpha}$.

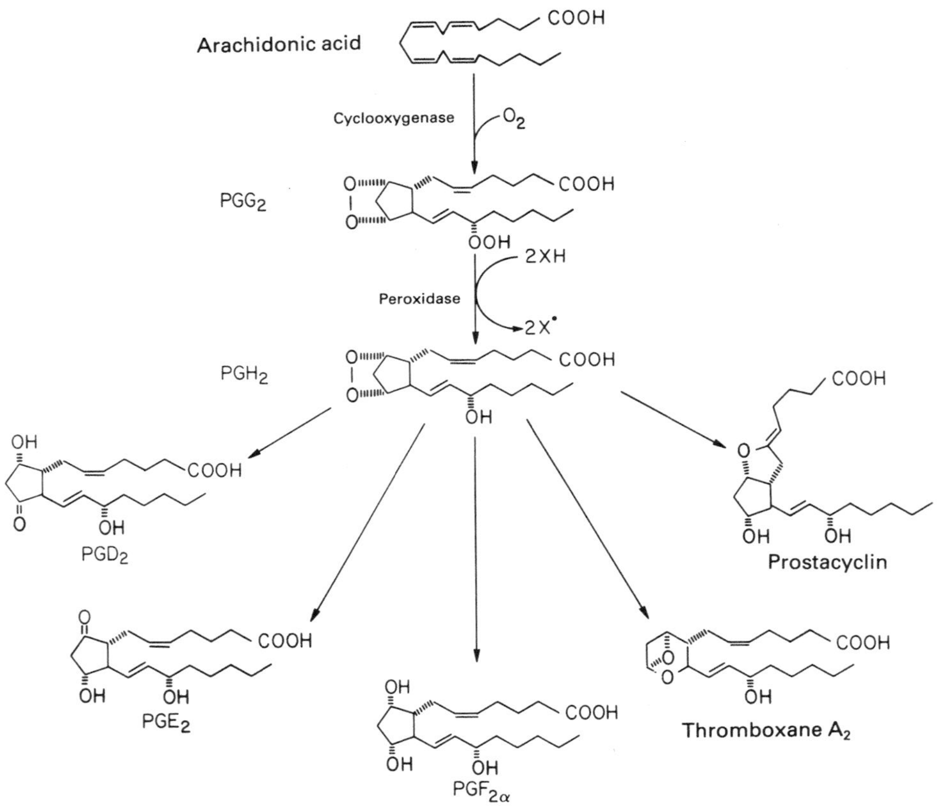

Figure 1.1 Metabolism of arachidonic acid by prostaglandin H synthase.

A number of oxidizing species are generated during prostaglandin biosynthesis. PHS peroxidase is in itself an oxidizing species as it can metabolize many chemicals, including aromatic amines, by one-electron oxidation. Peroxyl radicals are also significant oxidizing agents which are generated as intermediates during PG biosynthesis. Existence of these peroxyl radicals has been demonstrated in studies of the epoxidation of the carcinogen benzo(a)pyrene 7,8-dihydrodiol.

Ruf and his colleagues (Karthein *et al.*, 1988; Dietz *et al.*, 1988) have extensively investigated the higher oxidation states of PHS using rapid spectroscopic methods and low temperature electron spin resonance (ESR). Two intermediates were detected and characterized (Fig. 1.2). Intermediate I was proposed as [(protoporphyrin IV)$^{\bullet+}$Fe^{IV}O] analogous to HRP compound I. Intermediate II resembles compound II of HRP, but contains a tyrosyl radical with the proposed structure [(protoporphyrin IX)Fe^{IV}O]Tyr$^{\bullet}$ which is formed by an electron transfer from a tyrosine residue to the porphyrin cation radical. Intermediate II resembles the ES complex of cytochrome *c* peroxidase. Thus, intermediate II of PHS is different from compound II of HRP. Intermediate II is reduced to compound II at the expense of an electron donor (cosubstrate) which undergoes a one-electron oxidation. Compound II is reduced to the resting enzyme (Fe(III)) at the expense of a second electron donor molecule. In some cases, compound I can directly transfer the oxygen atom to an acceptor molecule and is reduced

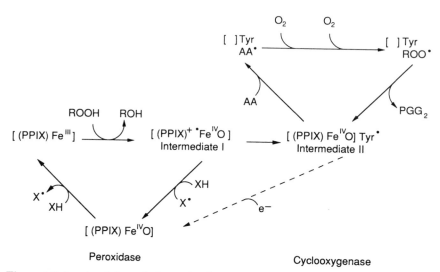

Figure 1.2 A potential catalytic mechanism for PHS cyclo-oxygenase and peroxidase adapted from the proposal by Dietz *et al.* (1988).

to the resting enzyme. Thus, electron or oxygen transfer can occur. Since the electron donor transfers a single electron to the peroxidase, a free radical metabolite of the electron donor is generated.

One of the unique features of PHS is that two enzymatic activities, the cyclo-oxygenase and the peroxidase, reside within a single protein. Ruf and co-workers (Karthein *et al.*, 1988; Dietz *et al.*, 1988) have also proposed that the tyrosyl radical of intermediate II participates in the cyclo-oxygenase reaction as shown in Fig. 1.2. By this proposed mechanism, the tyrosyl radical initiates the cyclo-oxygenase by abstracting a hydrogen at C-13 of AA to form a carbon-centered radical. This radical rearranges and reacts with molecular oxygen to form a cyclic endoperoxide peroxyl radical which reoxidizes the tyrosine yielding the tyrosyl radical and PGG_2. Thus, the haem Fe does not change oxidation state during the initiation of the cyclo-oxygenase reaction, which is in contrast to previous proposals.

1.2 METABOLISM OF AROMATIC AMINES

Aromatic amines are metabolized by PHS peroxidase via one-electron oxidation to nitrogen-centered free radicals. The ease of oxidation is altered by substituents which affect the electron density on the nitrogen. For example, benzidine is an excellent cosubstrate for PHS peroxidase, while the less easily oxidized 2-aminofluorene (2-AF) (Bull, 1987) is a poorer cosubstrate. Acetylation of the amine reduces the ease of oxidation by PHS and other peroxidases. Acetylbenzidine, e.g., is a much poorer cosubstrate than benzidine for PHS (Josephy *et al.*, 1983a,b). The addition of electron withdrawing groups to the aromatic ring of *N*-methyl anilines decreases the metabolism of these compounds by PHS peroxidase (Sivarajah *et al.*, 1982).

Evidence for a one-electron oxidation of aromatic amines by PHS peroxidase to free radical metabolites comes from a number of investigations. The stable, isolated metabolites produced by PHS peroxidase-catalysed co-oxidation of aromatic amines indicate free radical formation, and the chemical nature of the free radical dictates the nature of the stable metabolites. Tertiary and secondary aromatic amines undergo *N*-dealkylation. For example, aminopyrine and *N*-methylaniline are *N*-demethylated (Sivarajah *et al.*, 1982). For primary aromatic amines, the free radicals undergo nitrogen-to-nitrogen or nitrogen-to-carbon coupling reactions. The carcinogen 2-AF is oxidized by PHS peroxidase and HRP to 2,2'-azobisfluorene, 2-aminodifluorenyl amine and polymeric material (Boyd and Eling, 1984) as shown in Fig. 1.3. Benzidine is also oxidized to azobenzidine by peroxidases, but a complete characterization of the stable metabolites that are formed by PHS peroxidase or HRP is not published. Thus, indirect evidence obtained by analysis of the stable end products

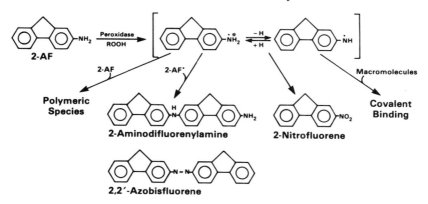

Figure 1.3 The oxidation of 2-aminofluorene by PHS.

suggests a free radical mechanism for the oxidation of aromatic amines by PHS.

Direct evidence for the formation of aromatic amine free radicals is obtained from studies using ESR to detect the free radical. The use of ESR to detect amine free radicals is compromised by the sensitivity of the method relative to the inherent instability of the aromatic amine free radicals. Spin trapping techniques are not suitable since spin traps do not react with amine free radicals. The stability problem can be overcome, in some cases, by flowing the incubation mixture past the detector rather than by the normal static detection technique.

Aminopyrine is oxidized by PHS peroxidase to an aminopyrine cation free radical that was detected by ESR and characterized by UV/visible spectroscopy (Lasker *et al.*, 1981; Eling *et al.*, 1985). Benzidine and 3,3′,5,5′-tetramethylbenzidine undergo a one-electron oxidation by PHS peroxidase or HRP to cation radicals which can undergo a second one-electron oxidation to the diimine (Josephy *et al.*, 1982a, 1983a,b). The one-electron oxidation products were detected by ESR techniques and the diimine or other two-electron oxidation products were characterized by spectroscopic methods. In addition, the formation of several free radicals has been detected during the oxidative N-dealkylation of N-substituted aromatic amines catalysed by horseradish peroxidase (HRP) which will be discussed below. In this case the free radicals are relatively stable and can be detected. However, we were unsuccessful in attempts to detect the free radical formed during the oxidation of 2-AF, 2-NA, and the heterocyclic aromatic amine carcinogens using both static and flow techniques.

Indirect ESR evidence was also used to support the hypothesis for a one-electron oxidation of aromatic amines by PHS. Petry *et al.* (1986)

investigated the metabolism of carcinogenic heterocyclic aromatic amines derived from the pyrolysis of amino acids and proteins. Again, free radicals were not observed by ESR, but the detection of a glutathionyl radical (GS) formed by reduction of a free radical by glutathione indicated that these amines are also undergoing a one-electron oxidation. Thus, characterization of the stable isolated metabolites of aromatic amines and both direct and indirect ESR data indicate that aromatic amines undergo one-electron oxidation to free radical metabolites.

1.3 MECHANISM FOR THE *N*-DEALKYLATION OF *N*-SUBSTITUTED AMINES

A study on the *N*-dealkylation of aromatic amines illustrates the use of ESR and analysis of stable end products as a powerful approach to understanding the mechanism of peroxidase-catalysed oxidations. It has been shown (Lasker *et al.*, 1981; Eling *et al.*, 1985) that aminopyrine is oxidized to a nitrogen-centered radical cation by peroxidases. This radical then reacts with a second aminopyrine cation radical (disproportionates) to yield an iminium cation and aminopyrine. The iminium cation reacts with water to give formaldehyde and monomethyl amine (Fig. 1.4; pathway D). It was proposed (Griffin, 1978; Griffin and Ting, 1978) that the nitrogen-centered aromatic amine free radicals are enzymatically oxidized to form the iminium cation (pathway C), whereas Galliani *et al.* (1978) proposed that the nitrogen-centered free radical deprotonates to form a carbon-centered free radical (pathway A/B). This carbon-centered free radical is either oxidized to the iminium cation or reacts with molecular oxygen to form a peroxyl radical which decomposes to formaldehyde and the monomethylamine (pathway A and B). In contrast, other investigators (Kedderis and Hollenberg, 1983, 1984; Kedderis *et al.*, 1986; Miwa *et al.*, 1983) proposed an initial hydrogen abstraction to form a carbon-centered aromatic amine free radical, which is subsequently oxidized to the iminium cation eventually yielding formaldehyde (pathway E). We have studied the oxidation of the following *N*-substituted aromatic amines: BAPTA, 5,5-methyl BAPTA, *N*,*N*-dimethylaniline, half-dimethyl BAPTA, and *N*,*N*-dimethyl-*p*-toluidine by HRP in an attempt to understand more clearly the mechanism for *N*-dealkylation of aromatic amines (Van der Zee *et al.*, 1989).

By using either static or flow conditions, we were able to detect directly by ESR the nitrogen-centered free radicals formed by peroxidase-catalysed oxidation of the *N*-methyl aromatic amines; *N*,*N*-dimethyl-*p*-toluidine, *N*,*N*-dimethylaniline and the *N*-CH$_2$-COOH-substituted aromatic amines half-dimethyl BAPTA, and 5,5-dimethyl BAPTA the spectra of which are shown in Fig. 1.5. Computer simulations using Fourier transformation

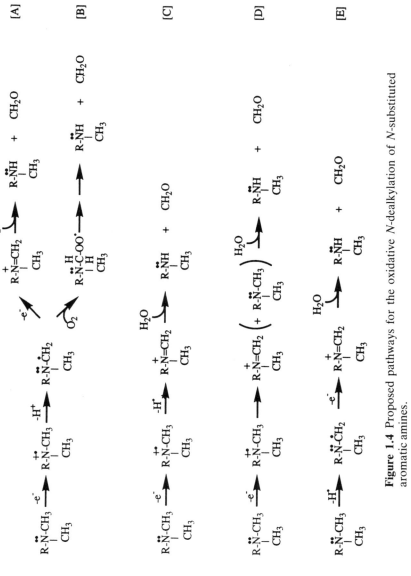

Figure 1.4 Proposed pathways for the oxidative *N*-dealkylation of *N*-substituted aromatic amines.

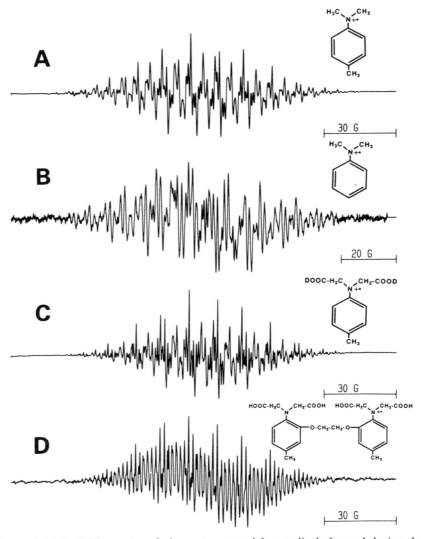

Figure 1.5 The ESR spectra of nitrogen-centered free radicals formed during the *N*-dealkylation of *N*-substituted aromatic amines.

confirmed the structure of the nitrogen-centered free radicals. Furthermore, we used the spin traps DMPO and *t*-NB to trap any carbon-centered free radicals also formed from these amines by the peroxidase. Carbon-centered radicals were only observed with 5,5-dimethyl BAPTA and half-dimethyl BAPTA (Fig. 1.6), but not with *N*-methyl amines under a

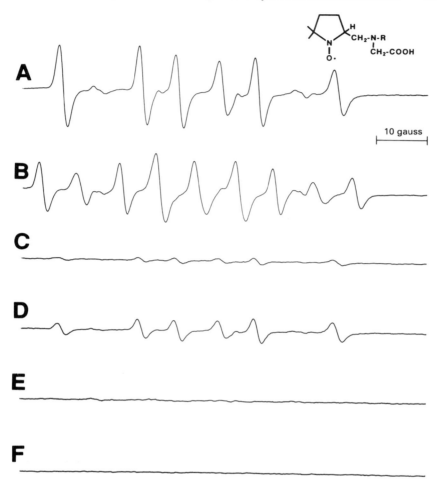

Figure 1.6 The ESR spectra of carbon-centered free radicals observed during the N-dealkylation of 5,5'-dimethyl BAPTA.

variety of incubation conditions. These results suggest that the N-methyl amines were oxidized by HRP to aromatically delocalized nitrogen-cation free radicals with little or no formation of carbon-centered radicals.

Previous studies with N-methyl aromatic amines and dimethyl BAPTAs indicated oxidative cleavage of the methyl group or N-CH_2COOH groups. Therefore, oxygen consumption and formaldehyde formation were measured under conditions similar to those used for ESR studies. Carbon-centered free radicals, but not the delocalized aromatic amine free radicals,

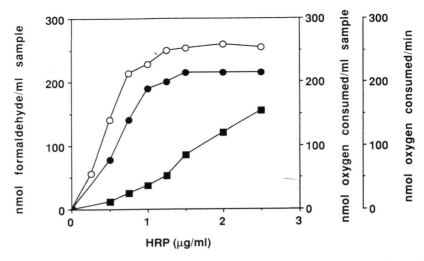

Figure 1.7 The formation of formaldehyde and oxygen consumption during the oxidation of 5,5'-dimethyl-BAPTA. ○ = formaldehyde formed (nmol/ml); oxygen consumed: ● = nmol/ml; ■ = nmol/min.

react with molecular oxygen. Extensive formaldehyde formation and oxygen consumption occurred with the oxidation of the dimethyl BAPTAs (Fig. 1.7), while less formaldehyde and little or no oxygen consumption occurred during the oxidation of the *N*-methylaniline derivatives or aminopyrine as shown in Fig. 1.8. The addition of DMPO, which reacts with carbon-centered but not delocalized free radicals, inhibited oxygen consumption and formaldehyde formation from the dimethyl BAPTAs but had little or no effect on formaldehyde formation from the *N*-methylanilines or aminopyrine (data not shown).

These results clearly demonstrate that nitrogen-centered free radicals are detected during the peroxidase oxidation of both *N*-methyl and CH_2COOH-substituted amines. Carbon-centered free radicals were detected only with the CH_2COOH-substituted amines. There was general agreement between formation of the carbon-centered free radicals, oxygen consumption and inhibition of formaldehyde production by spin traps. The data suggest that aromatic amines are initially oxidized by HRP to a nitrogen-centered free radical, not a carbon-centered free radical, and indicate that pathway E is an unlikely mechanism for the oxidation of the amines. The results also indicate that the mechanism greatly depends on the structure of the nitrogen-centered free radical. For the CH_2COOH-substituted amines, the nitrogen-centered free radical undergoes bond breakage through a rearrangement. A

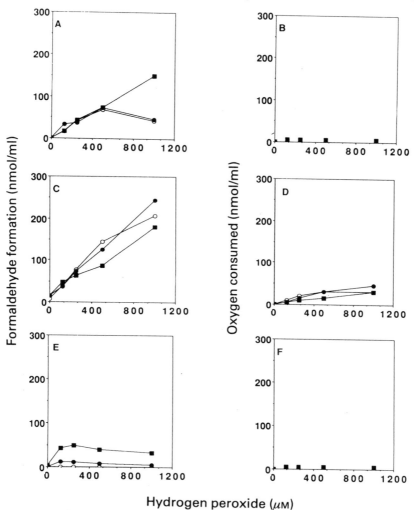

Figure 1.8 Formaldehyde formation and oxygen consumption during the oxidation of *N*-substituted aromatic amines. A/B = *N,N'*-dimethylaniline; C/D = *N,N'*-dimethyl-*p*-toluidine; E/F = aminopyrine; ○ = 0.5 µg/ml HRP; ● = 1.0 µg/ml HRP; ■ = 5.0 µg/ml HRP.

carbon-centered free radical is formed due to the loss of CO_2 which reacts with molecular oxygen to form a peroxyl radical as shown in Fig. 1.9. The iminium cation is formed upon decomposition of the peroxyl radical with the formation of superoxide anion radical, which was also detected. The

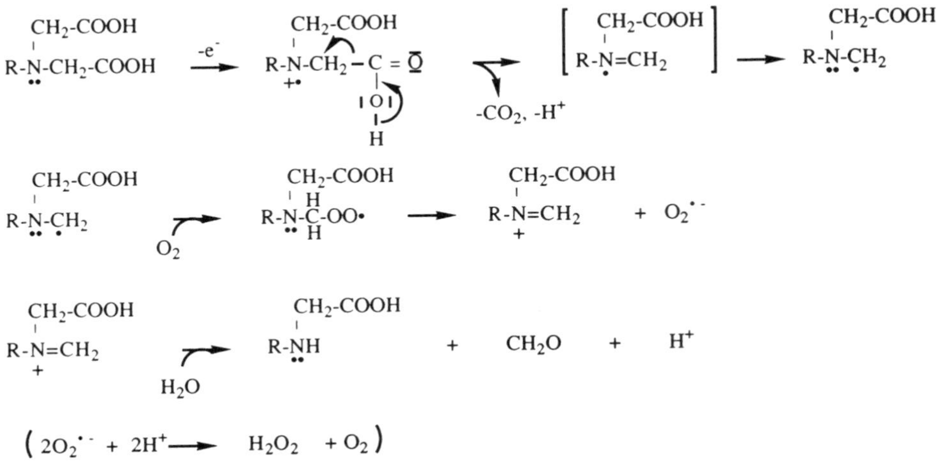

$$CH_2\text{-}COOH$$
$$\underset{\bullet\bullet}{R}\text{-}N\text{-}CH_2\text{-}COOH \xrightarrow{-e^-}$$

$$CH_2\text{-}COOH$$
$$R\text{-}\underset{+\bullet}{N}\text{-}CH_2\text{-}\overset{}{C}=\overset{\bullet\bullet}{O}$$
$$\overset{|}{\underset{\quad H}{\text{I O I }}} \xrightarrow{-CO_2,\,-H^+}$$

$$\left[\begin{array}{c} CH_2\text{-}COOH \\ R\text{-}N{=}CH_2 \end{array} \right] \longrightarrow$$

$$CH_2\text{-}COOH$$
$$\underset{\bullet\bullet}{R}\text{-}\underset{\bullet}{N}\text{-}CH_2$$

$$CH_2\text{-}COOH$$
$$\underset{\bullet\bullet}{R}\text{-}\underset{\bullet}{N}\text{-}CH_2 \xrightarrow{O_2}$$

$$CH_2\text{-}COOH$$
$$\underset{\bullet\bullet}{R}\text{-}N\text{-}\overset{H}{\underset{H}{C}}\text{-}OO\bullet \longrightarrow$$

$$CH_2\text{-}COOH$$
$$\underset{+}{R}\text{-}N{=}CH_2 \quad + \quad O_2^{\bullet-}$$

$$CH_2\text{-}COOH$$
$$\underset{+}{R}\text{-}N{=}CH_2 \xrightarrow{H_2O}$$

$$CH_2\text{-}COOH$$
$$\underset{\bullet\bullet}{R}\text{-}NH \quad + \quad CH_2O \quad + \quad H^+$$

$$\left(2O_2^{\bullet-} + 2H^+ \longrightarrow H_2O_2 + O_2 \right)$$

Figure 1.9 A mechanism for the *N*-dealkylation of *N*-CH$_2$-COOH-substituted aromatic amines.

iminium cation is also unstable and hydrolyses to formaldehyde and monomethyl amine. For monomethyl amines, rearrangement of the nitrogen-centered free radical is unlikely since there is no good leaving group. There are several possible pathways for the oxidation of the nitrogen-centered free radical, two of which are outlined in Fig. 1.4. Further studies are required to fully elucidate this mechanism.

1.4 A NEW MECHANISM FOR THE OXIDATION OF AMINES

Many investigators studying the oxidation of chemicals by PHS used the assumption, supported by data, that PHS and HRP operated by similar mechanisms. Indeed, for many studies this assumption appears to be valid and identical metabolites are formed by oxidations catalysed by HRP and PHS. However, the ability of chemicals to serve as a reducing cosubstrate is not identical for HRP and PHS, suggesting differences do exist (Markey *et al.*, 1987). Boyd and Eling (1987) investigated the metabolism of the carcinogen 2-naphthylamine (2-NA) by HRP and PHS. HRP, under a variety of incubation conditions, oxidizes 2-NA to exclusively nitrogen-to-nitrogen and nitrogen-to-carbon coupling products, as expected from a nitrogen-centered free radical. Figure 1.10 compares the HPLC profile obtained from incubating 2-NA with HRP or PHS. The products from these

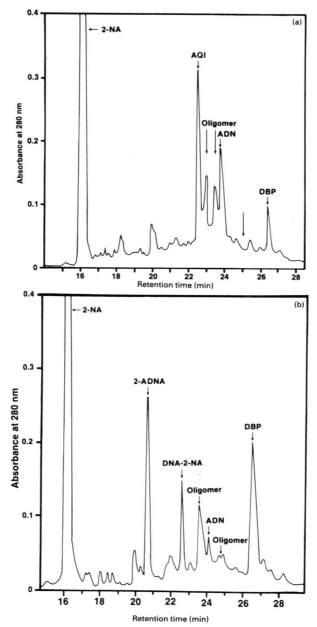

Figure 1.10 The oxidation of 2-naphthylamine by PHS (a) and horseradish peroxidase (b); HPLC analysis of products.

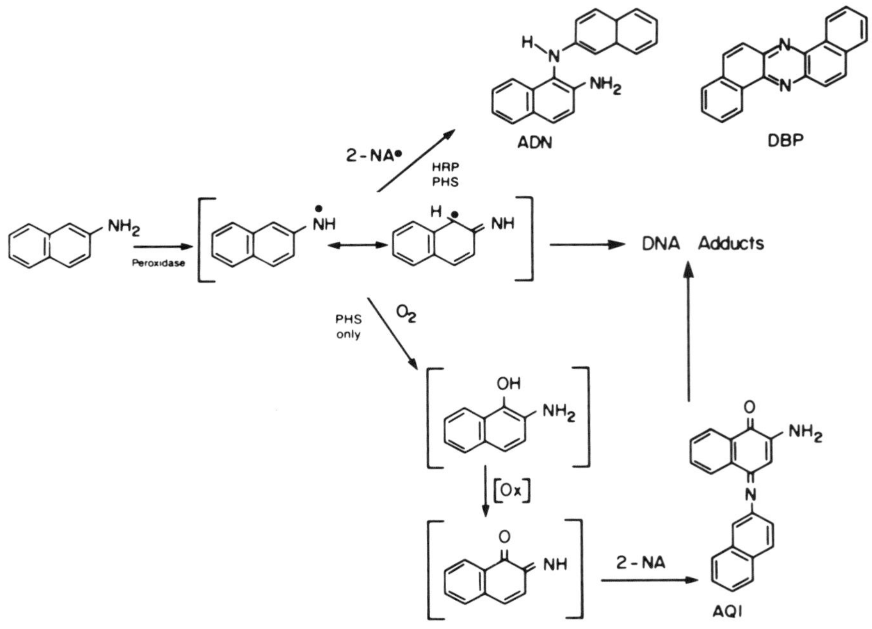

Figure 1.11 Proposal for the oxidation of 2-NA by HRP and PHS.

incubations were isolated and characterized by UV-visible spectrometry and mass spectrometry, the structures of which are shown in Fig. 1.11. With PHS, the major metabolites were the oxygenated rather than the coupling products observed with HRP. We have also investigated the metabolism of the amines *o*- and *p*-anisidine. For *p*-anisidine, an apparent difference was observed in the metabolic profile obtained from incubation of the aromatic amine with HRP and PHS (data not shown). The data clearly indicate that differences do exist between PHS and HRP and indicate that caution should be used in using HRP as a model enzyme for PHS.

A possible explanation for this difference may be the presence of peroxyl radicals in incubations with PHS, particularly in incubations using ram seminal vesicle microsomes as the source of PHS. To investigate this further, 2-NA was incubated with a peroxyl radical-generating system (15-HPETE and haematin). Under these incubation conditions, the oxygenated metabolites of 2-NA were formed, as were seen with the PHS/arachidonic acid system (data not shown). The nitrogen-to-nitrogen and nitrogen-to-carbon-coupling products were also detected in small amounts in these incubations, as also seen with the PHS system. These preliminary studies suggest that PHS and HRP may operate by similar, but distinct mechanisms, and that with PHS the peroxyl radicals formed during arachidonic acid oxidation may contribute to the oxidation of aromatic amines by PHS. Clearly, further studies are required to understand fully the mechanism by which PHS oxidizes chemicals.

1.5 CONCLUSIONS

Metabolism of aromatic amines by PHS is certainly not completely understood. Recent data indicate that an additional mechanism exists with PHS that results in the formation of oxygenated metabolites. Some data suggest a role for peroxyl radicals in this process, but further studies are clearly required. These investigators may require the use of ESR techniques to detect and characterize unstable free radical intermediates. The use of ESR coupled with analysis of the stable end-products illustrated by studies on the *N*-dealkylation of *N*-substituted amines is a powerful approach that can be used in investigations of the mechanisms responsible for the oxidation of aromatic amines by peroxidases.

REFERENCES

Boyd, J.A. and Eling, T.E. (1984) Evidence for a one-electron mechanism of 2-aminofluorene oxidation by prostaglandin H synthase and horseradish peroxidase. *J. Biol. Chem.*, **259**, 13885–96.

Boyd, J.A. and Eling, T.E. (1987) Prostaglandin H synthase-catalyzed metabolism and DNA binding of 2-naphthylamine, *Cancer Res.*, **47**, 4007–14.

Bull, A.W. (1987) Reducing substrate activity of some aromatic amines for prostaglandin H synthase. *Carcinogenesis*, **8**, 387–90.

Dietz, R., Nastaincczyk, W. and Ruf, H.H. (1988) Higher oxidation states of prostaglandin H synthase. Rapid electronic spectroscopy detected two spectral intermediates during the peroxidase reaction with prostaglandin G_2. *Eur. J. Biochem.*, **171**, 321–8.

Eling, T.E., Mason, R.P. and Sivarajah, K. (1985) The formation of aminopyrine cation radical by the peroxidase activity of prostaglandin H synthase and subsequent reactions of the radical. *J. Biol. Chem.*, **260**, 1601–7.

Galliani, G., Rindone, B. and Marchesini, A. (1978) Horseradish peroxidase-catalyzed oxidation of aromatic tertiary amines with hydrogen peroxide. *J. Chem. Soc. Perkin Trans.*, **1**, 456–60.

Griffin, B.W. (1978) Evidence for a free radical mechanism of *N*-demethylation of *N,N*-dimethylaniline and an analog by hemeprotein-H_2O_2 systems. *Arch. Biochem. Biophys.*, **190**, 850–3.

Griffin, B.W. and Ting, P.L. (1978) Mechanism of *N*-demethylation of aminopyrine by hydrogen peroxide catalyzed by horseradish peroxidase, metmyoglobin, and protohemin. *Biochemistry*, **17**, 2206–11.

Josephy, P.D., Eling, T. and Mason, R.P. (1982a) The horseradish peroxidase-catalyzed oxidation of 3,5,3',5'-tetramethylbenzidine. Free radical and charge-transfer complex intermediates. *J. Biol. Chem.*, **257**, 3669–75.

Josephy, P.D., Mason, R.P. and Eling, T. (1982b) Cooxidation of the clinical reagent 3,5,3',5'-tetramethylbenzidine by prostaglandin synthase. *Cancer Res.*, **42**, 2567–70.

Josephy, P.D., Eling, T.E. and Mason, R.P. (1983a) Cooxidation of benzidine by prostaglandin synthase and comparison with the action of horseradish peroxidase. *J. Biol. Chem.*, **258**, 5561–9.

Josephy, P.D., Eling, T.E. and Mason, R.P. (1983b) An electron spin resonance study of the activation of benzidine by peroxidases. *Mol. Pharmacol.*, **23**, 766–70.

Karthein, R., Dietz, R., Nastainczyk, W. and Ruf, H.H. (1988) Higher oxidation states of prostaglandin H synthase. EPR study of a transient tyrosyl radical in the enzyme during the peroxidase reaction. *Eur. J. Biochem.*, **171**, 313–20.

Kedderis, G.L. and Hollenberg, P.F. (1983) Characterization of the N-demethylation reactions catalyzed by horseradish peroxidase. *J. Biol. Chem.*, **258**, 8129–38.

Kedderis, G.L. and Hollenberg, P.F. (1984) Peroxidase-catalyzed N-demethylation reactions. *J. Biol. Chem.*, **259**, 3663–8.

Kedderis, G.L., Rickert, D.E., Pandey, R.N. and Hollenberg, P.F. (1986) ^{18}O studies of the peroxidase-catalyzed oxidation of N-methylcarbazole. *J. Biol. Chem.*, **261**, 15910–14.

Lasker, J.M., Sivarajah, K., Mason, R.P. *et al.* (1981) A free radical mechanism of prostaglandin synthase-dependent aminopyrine demethylation. *J. Biol. Chem.*, **256**, 7764–7.

Markey, C.M., Alward, A., Weller, P.E. and Marnett, L.J. (1987) Quantitative studies of hydroperoxide reduction by prostaglandin H synthase. Reducing substrate specificity and the relationship of peroxidase to cyclooxygenase activities. *J. Biol. Chem.*, **262**, 6266–79.

Miwa, G.T., Walsh, J.S., Kedderis, G.L. and Hollenberg, P.F. (1983) The use of intramolecular isotope effects to distinguish between deprotonation and hydrogen atom abstraction mechanisms in cytochrome *P*-450- and peroxidase-catalyzed N-demethylation reactions. *J. Biol. Chem.*, **258**, 14445–9.

Miyamoto, T., Ogino, N., Yamamoto, S. and Hayaishi, O. (1976) Purification of prostaglandin endoperoxide synthetase from bovine vesicular gland microsomes. *J. Biol. Chem.*, **251**, 2629–36.

Pace-Asciak, C.R. and Smith, W.L. (1983) Enzymes in the biosynthesis and catabolism of the eicosanoids: prostaglandins, thromboxanes, leukotrienes and hydroxy fatty acids, in *The Enzymes* (ed. P.D. Boyer), Vol. 16, Academic Press, New York, pp. 543–603.

Petry, T.W., Krauss, R.S. and Eling, T.E. (1986) Prostaglandin H synthase-mediated bioactivation of the amino acid pyrolysate product Trp P-2. *Carcinogenesis*, 7, 1397–400.

Sivarajah, K., Lasker, J.M., Eling, T.E. and Abou-Donia, M.B. (1982) Metabolism of N-alkyl compounds during the biosynthesis of prostaglandins. N-dealkylation during prostaglandin biosynthesis. *Mol. Pharmacol.*, **21**, 133–41.

Van der Zee, J., Duling, R.D., Mason, R.P. and Eling, T.E. (1989) The oxidation of N-substituted aromatic amines by horseradish peroxidase. *J. Biol. Chem.*, **264**, 19 828–36.

2

The application of ^{15}N-NMR in the analysis of N-oxygenated amidines and guanidines

B. Clement and Th. Kämpchen

Institut für Pharmazeutische Chemie, Philipps-Universität, D-3550 Marburg, FRG

1. Nitrogen is considered with regard to the suitability of its isotopes for nuclear magnetic resonance (NMR) spectroscopy.
2. Measurement conditions and the characteristics of ^{15}N-NMR spectra are described briefly.
3. ^{15}N-NMR spectra of N-hydroxyguanidines, amidoximes and N-hydroxyamidines were used to demonstrate the suitability of ^{15}N-NMR spectroscopy for structural analysis of N-oxygenated compounds.
4. N-hydroxyguanidines and amidoximes both exist in the oxime-type tautomeric form. In any case, protonation occurs at the double-bonded nitrogen atom. In the case of the N-oxygenated derivative of N-methylbenzamidine, the α-aminonitrone form clearly predominates over the N-hydroxyamidine form.
5. In the cases of N-monosubstituted and N,N-disubstituted N-hydroxyguanidines, rapid isomerization takes place at room temperature, whereas E/Z-isomers of N,N'-disubstituted derivatives can be detected at room temperature. N,N-unsubstituted and N-monosubstituted amidoximes exist exclusively in the Z-configuration, whereas in the case of N,N-disubstituted amidoximes the E-isomers are thermodynamically more stable.

2.1 INTRODUCTION

With regard to their magnetic moments, both nitrogen isotopes, ^{14}N and ^{15}N, would appear to be suitable for nuclear magnetic resonance spectroscopy (NMR). Unfortunately, however, the ^{14}N nucleus (natural abundance 99.63%) with a nuclear spin quantum number of I = 1 possesses a quadrupole moment (Table 2.1) and is thus not appropriate for high resolution studies. As a result of quadrupole relaxation, the line widths of the signals are very broad and relatively sharp resonance lines can be observed only in special cases.

The stable isotope ^{15}N, like ^{1}H and ^{13}C, has the nuclear spin quantum number I = 1/2 (Table 2.1), so that this isotope is suitable for high-resolution experiments. Unfortunately, ^{15}N has a natural abundance of 0.37% (Table 2.1) and is, therefore, not only a very rare isotope, but is also characterized by a very small magnetogyric ratio (Table 2.1) as compared to ^{1}H and ^{13}C and is negative in sign (for reviews on the properties of the nitrogen NMR probe see Blomberg and Rüterjans, 1983; Kalinowski *et al.*, 1984; Levy and Lichter, 1979; Lichter, 1983; Breitmaier, 1983).

The problem of the low natural abundance of ^{15}N can be solved by enrichment with this isotope of the samples to be investigated. Unfortunately, the synthesis of ^{15}N-labelled compounds are, in most cases, laborious and expensive. On the other hand, measurement conditions have recently been attained which permit recording of spectra of compounds containing ^{15}N in natural abundance within a relatively short period of time. One prerequisite for this was the development of the pulse Fourier transform technique (PFT). Commercial introduction of high-field superconducting magnets, together with the existence of polarization transfer pulse sequences (INEPT, DEPT), have resulted in a breakthrough in ^{15}N-NMR spectroscopy and allowed recording of spectra of non-enriched samples within practicable time periods. As a result ^{15}N has effectively

Table 2.1 Comparison of nuclear properties

Isotope	Nuclear spin quantum number I	Natural abundance (%)	Magnetogyric ratio γ ($\times 10^7$ rad. $T^{-1} S^{-1}$)	Quadrupole moment Q ($\times 10^{-28} m^2$)
^{1}H	½	99.98	26.751	—
^{13}C	½	1.108	6.726	—
^{14}N	1	99.63	1.932	1.6×10^{-2}
^{15}N	½	0.37	−2.711	—

supplanted ^{14}N in most studies on molecular structure by high resolution nitrogen NMR (Levy and Lichter, 1979; Lichter, 1983).

As an example, the characteristics of the ^{15}N-NMR spectra of N-hydroxy-guanidines, amidoximes and N-hydroxyamidines are interpreted in detail. The chemical shifts δ (^{15}N) and coupling constants ^{1}J (^{15}N, ^{1}H) can be used to elucidate the constitution, configuration, position of tautomeric equilibrium, and the site of protonation. The present review summarizes the results of ^{15}N-NMR studies on the N-oxygenated compounds mentioned above. Analogous data on the parent compounds (amidines, guanidines) are not included (for relevant information see Clement, 1986a; Clement and Kämpchen, 1986). However, it has to be noted that ^{15}N-NMR spectroscopy cannot be readily applied to the identification of metabolites in biological specimens without prior enrichment with ^{15}N-nuclides, since, despite the progress in experimental techniques described above, the substance requirements for ^{15}N nuclear magnetic resonance spectroscopy with material naturally abundant in ^{15}N is still large. Hence, ^{15}N nuclear magnetic resonance spectroscopy with material of natural abundance in ^{15}N is usually only employed for the characterization of compounds prepared synthetically in larger amounts for comparative purposes, and the precise structures of metabolites analysed by other methods are deduced on the basis of these NMR data.

2.2 TECHNICAL ASPECTS OF MEASURING ^{15}N-NMR SPECTRA

2.2.1 Measurement conditions

Typical pulsed FT-NMR methods may be used for ^{15}N-spectroscopy, with some experimental modifications, to account for the small-valued and negative ^{15}N magnetogyric ratio (Levy and Lichter, 1979; Lichter, 1983).

Conventional ^{1}H-noise-decoupled ^{15}N-PFT-NMR spectra can be recorded, but spectra obtained by polarization transfer pulse sequences, such as the INEPT ('insensitive nuclei enhanced by polarization transfer') pulse sequence, have certain advantages (Morris and Freeman, 1979; Morris, 1980). The increase in sensitivity for the detection of nitrogen nuclei bonded to protons is considerable. As a consequence of the pulse sequence, which forms the basis for INEPT processes, the zero intensity of the central line of a multiplet, containing an odd number of lines and the inversion of one half of each multiplet, is typical for simple INEPT-NMR spectra with spin-spin coupling (multiplet patterns for NH: $-1,+1$; NH$_2$: $-1,0,+1$; NH$_3$: $-1,-1,+1,+1$). The integrated total intensity of a multiplet in an INEPT spectrum is thus zero (Morris, 1980; Benn and Günther, 1983).

^{15}N Chemical shifts are influenced by the solvent (Levy and Lichter, 1979). Thus, if possible, identical solvents should be used for measurements with a series of analogous molecules. In this way, chemical shifts can be easily compared. If some of the information on tautomeric forms is to be used in the discussion of biotransformation studies, water would be an ideal solvent. However, in some cases solubility of the substances in water is limited. On the other hand, most of the substances (salts and free bases) have a sufficient solubility in dimethyl sulphoxide (DMSO). Thus, DMSO has been used in ^{15}N-NMR experiments with *N*-hydroxyguanidines, amidoximes and *N*-hydroxyamidines (Clement, 1986a). In the case of water-soluble compounds, additional spectra in aqueous solution can be recorded for comparison. As can be expected, aberrations of a few ppm were indeed observed, but a change in the preference for a particular tautomeric form was not detected with *N*-hydroxyguanidines, amidoximes and *N*-hydroxyamidines (Clement, 1986a).

2.2.2 Overview of the ^{15}N-spectral characteristics

^{15}N-Spectra are generally simple first order spectra as a result of the spin of 1/2 of the ^{15}N nucleus. Chemical shifts (ppm) are generally referred to the NH$_3$ signal with $\delta = 0$ ppm. Values of ppm then are usually positive in sign as the nitrogen atoms are in a more deshielded environment when compared with the ammonia nitrogen (Breitmaier, 1983). Chemical shifts, as reported in this chapter, also refer to the ammonia standard. All literature data referring to other standards have been appropriately corrected (Levy and Lichter, 1979).

The chemical shift range is very large and covers more than 900 ppm. Nitroso compounds contain the most highly deshielded nitrogen atoms (900–800 ppm), followed by azo (550–500 ppm) and nitro (400–340 ppm) groups. Imine or imine-type nitrogen atoms appear in a middle-deshielded region (380–150 ppm), whereas amine nitrogen atoms are only slightly more deshielded than the ammonia nitrogen (Breitmaier, 1983).

Protonation effects are of considerable importance. When a double-bonded nitrogen atom is protonated it will become more shielded. For example, protonation of *trans*-azobenzene or diphenylketimine decreases the ppm values by 150 and 140 ppm, respectively (Levy and Lichter, 1979). This striking difference allows identification of the site of protonation in unclear cases. A reverse, but much smaller effect can be observed for ammonia and amines. Thus, the ammonium ion is deshielded by + 25 ppm as compared to ammonia (Levy and Lichter, 1979).

As regards the coupling constant, this chapter focuses on one-bond couplings. There is a good correlation between the s proportion and the magnitude of the $^1J(^{15}N\text{-}^1H)$ coupling constant. When the s proportion increases, the coupling constant is expected to increase, too. For sp^3 nitrogen atoms with a low s proportion the coupling constant is in the range of 60 to 80 Hz; for sp^2 nitrogen atoms values between 80 and 100 Hz have been found, and for protonated nitriles, i.e. sp nitrogen atoms, the values range from 130 to 140 Hz (Breitmaier, 1983).

2.3 ANALYSIS OF VARIOUS CLASSES OF N-OXIDIZED COMPOUNDS

2.3.1 N-Hydroxyguanidines*

N-Hydroxyguanidines have been synthesized in the course of studies on the N-oxidative metabolism of guanidines (Clement, 1986b) and ^{15}N-NMR analyses have been performed (Clement and Kämpchen, 1985; Clement, 1986a). Hydroxyguanidines of the type (1) should exist in the oxime-type structure I and not in the tautomeric hydroxylamine-type form II containing an imine-type nitrogen atom (Fig. 2.1), as has been discussed on the basis of IR- and ^{1}H-NMR spectroscopic data (Belzecki *et al.*, 1970). ^{15}N-Nuclear magnetic resonance spectroscopic investigations were able to shed more light on this subject, since the spectra provided more information than the ^{1}H-NMR spectra (Clement, 1986a).

The ^{15}N chemical shifts and coupling constants $^1J(^{15}N^1H)$ of compounds (**1a–d**) are summarized in Table 2.2. Unambiguous NH-coupled spectra were obtained for all N-hydroxyguanidines in DMSO solution. Signal broadening as a result of rapid proton exchange could not be detected, thus

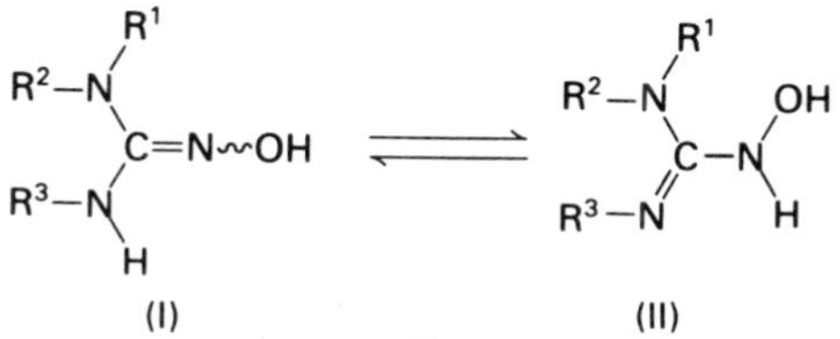

Figure 2.1 Tautomerism of N-hydroxyguanidines (**1**) (for R see Table 2.2).

* Although not in complete accordance with the IUPAC rules, for better understanding primes are used throughout to assign substitution at the nitrogen atom (see the Chemical Abstracts). Thus, in guanidines and related compounds N and N' refer to single-bonded and N'' to double-bonded nitrogen. Similarly, for amidines and related compounds N and N' (for double-bonded nitrogen) are used.

Table 2.2 ^{15}N-Chemical shifts[a] and coupling constants (Hz) of N''-hydroxyguanidines (**1**) (solvent DMSO). For structure see Figure 2.1

No.	R^1	R^2	R^3	$\delta(^{15}N, ppm)$			$^1J(^{15}N, {}^1H)$ (Hz)		
				N	N'	N''	N–H	N'–H	N''–H
1a	CH₃–C– with CH₃ above and CH₃ below	H	H	85.1 (d)	57.6 (t)	248.4 (s)	84.2	85.5	—
1b	CH₃–C–CH₂–C– with CH₃, CH₃ above and CH₃, CH₃ below	H	H	84.2 (d)	56.7 (t)	246.7 (s)	84.2	86.7	—
1b HCl	CH₃–C–CH₂–C– with CH₃, CH₃ above and CH₃, CH₃ below	H	H	100.9 (d)	77.1 (t)	136.8 (d)	91.6	92.8	98.9
1c	Tetrahydro-isoquinoline		H	60.3 (s)	57.2 (t)	254.4 (s)	—	83.1	—
1c HCl	Tetrahydro-isoquinoline		H	77.4 (s)	82.5 (t)	142.4 (d)	—	92.0	99.0
1d	C₆H₅	H	C₆H₅	90.2[b] (d)	84.6[b] (d)	273.5 (s)	89.7[b]	90.3[b]	—

[a] Chemical shifts are given in ppm relative to NH₃.
[b] Signals for two topomers.
From Clement and Kämpchen (1985), with permission of VCH Verlagsgesellschaft.

demonstrating that the measured chemical shifts do not represent the average values of two tautomeric forms (Clement, 1986a).

The spectrum of *N*-hydroxydebrisoquine (**1c**) (Fig. 2.2) can only be interpreted in terms of an oxime-type structure. The spectrum of a hydroxylamine-type tautomer with an imine-type nitrogen atom (see general structure **II**, $R^3 = H$, Fig. 2.1) would be expected to consist of two doublets and one singlet.

The spectrum of the hydrochloride of (**1c**) (Fig. 2.2*b*) is also as expected. In the case of the unsaturated nitrogen atoms of oximes, it is known that protonation of these structures results in the usual (see Section 2.2.2) high-field shift of the ^{15}N resonances (Allen and Roberts, 1980). The difference in

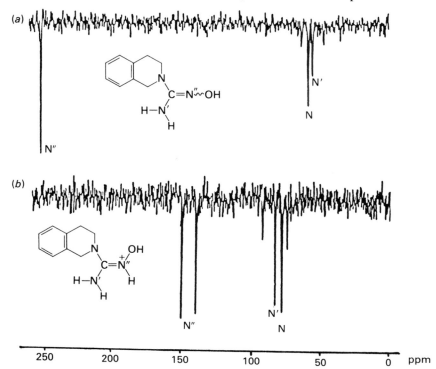

Figure 2.2 ^{15}N-NMR spectra of *N*-hydroxydebrisoquine (**1c**) in DMSO (gated decoupling); (*a*) base; (*b*) **1c** · HCl.

the chemical shift of N″ in **1c** amounts to 112.0 ppm, and is thus comparable with the values found for oximes (Allen and Roberts, 1980). An analogous result can be deduced from the spectrum of the hydrochloride of compound **1b** (Fig. 2.3*c,d*). The chemical shifts of N″ in the hydrochlorides of **1b** and **1c** and the splitting of the signals into doublets provide unequivocal evidence that the double-bonded nitrogen atom had been protonated, as can be expected by assuming mesomerism.

The spectra of the *N*-monosubstituted derivatives **1a** and **1b** are very similar to those of **1c** (Figs 2.2 and 2.3). However, in these cases, the tautomer **II** would possess the same signal multiplicities as the tautomer **I**.

In addition to the analogies with the spectra of **1c**, the chemical shifts are also in favour of the tautomer **I**. If structure **II** were present, the resonances at 84.2 or 85.1 ppm, respectively, for a nitrogen atom bearing a proton would have to be assigned to the hydroxylamine-type nitrogen atom of **II**.

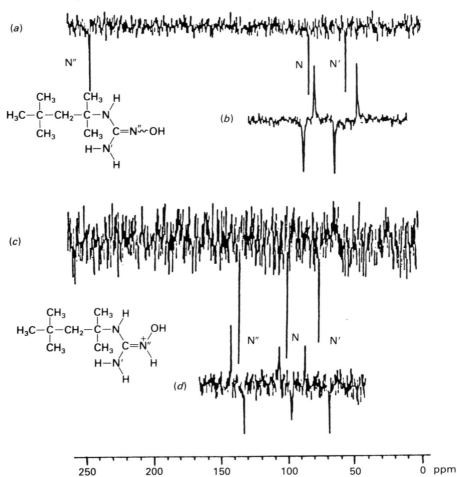

Figure 2.3 ^{15}N-NMR spectra of **1b** in DMSO; (*a*) base, proton-decoupled spectrum; (*b*) base, proton-coupled spectrum (INEPT); (*c*) **1b** · HCl, proton-decoupled spectrum, (*d*) **1b** · HCl, proton-coupled spectrum (INEPT). (From Clement and Kämpchen (1985), with permission of VCH Verlagsgesellschaft.)

Nitrogen atoms bearing a hydroxy group in structures where the free electron pair of the nitrogen is additionally in conjugation with a double bond, should resonate at higher ppm values. The chemical shift values for substituted hydroxylamines, for example, are all in excess of 100 ppm (Gouesnard and Dorie, 1982). Furthermore, the values of 246.7 or

248.4 ppm, respectively, for N″ are larger than comparable data for imine-type nitrogen atoms within a guanidine function (values of 193.2 and 204.3 ppm have been recorded for pentamethyleneguanidine and *N,N,N′,N′*-tetramethyl-*N″*-phenylguanidine, respectively (Witanowski *et al.*, 1976; Naulet and Martin, 1979)).

The ^{15}N-NMR spectra of **1a–c** do not provide any evidence for the presence of geometric isomers (*E,Z*) at the carbon–nitrogen double bond. This observation can be explained either by assuming identity of the

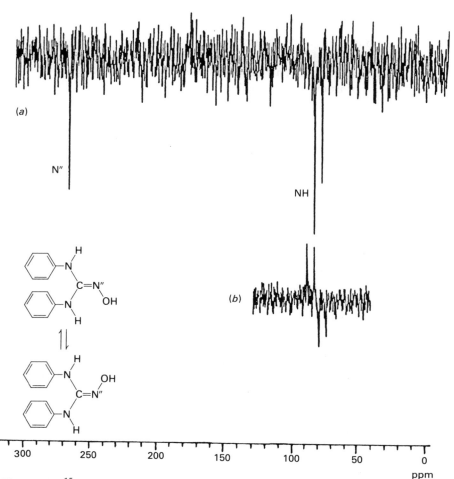

Figure 2.4 ^{15}N-NMR spectra of **1d** in DMSO; (*a*) proton-decoupled spectrum; (*b*) proton-coupled spectrum (INEPT).

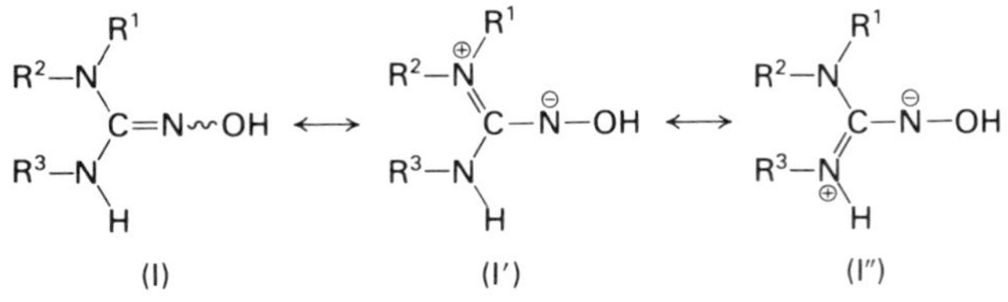

Figure 2.5 Delocalization of electrons in N-hydroxyguanidines (**1**) (for R see Table 2.2).

chemical shifts, the existence of only one isomer, or rapid isomerization. Investigations on N,N'-disubstituted and cyclic hydroxyguanidines indicated that rapid interconversion is likely to be the most probable explanation. The energy barriers determined are very low, and the possible mechanisms of isomerization of this type have been discussed (Piotrowska *et al.*, 1973; Trojnar *et al.*, 1975).

In the case of the N,N'-diphenyl derivative **1d**, the chemical shift values are again only in agreement with an oxime-type structure. However, two doublets for NH protons are observed and attributed to the existence of geometrical isomers (topomers) (Fig. 2.4). This has been demonstrated by ^{1}H-NMR-spectroscopic investigations of structurally similar derivatives such as N,N'-dibenzyl-N''-hydroxyguanidine (Piotrowska *et al.*, 1973; Trojnar *et al.*, 1975). In contrast to **1a–c**, rapid isomerization of **1d** does not occur at room temperature, instead coalescence of the NH-protons of **1d** was observed in the ^{1}H-NMR spectrum recorded at 58°C (Clement, 1986a).

For all the investigated compounds (**1**) (Table 2.2), the ^{15}N chemical shifts of the double-bonded nitrogen atoms N'' are smaller than those of oximes (values of about 340 ppm; Botto *et al.*, 1978). On the other hand, the chemical shifts of the nitrogen atoms N or N', respectively, are larger than those of amines (Levy and Lichter, 1979). These facts are indicative of a delocalization of the electrons with participation of the resonance structures I' and I'' (Fig. 2.5). A prerequisite for this state of affairs is the sp^2 hybridization of all nitrogen atoms. The magnitudes of the measured $^1J(^{15}N,^1H)$ coupling constants (83–93 Hz, see Table 2.2) are in the range of those for sp^2 nitrogen atoms (see section 2.2.2).

2.3.2 Amidoximes

The N-hydroxylated compounds (**2**) (Fig. 2.6) have been detected as N-oxidative biotransformation products of amidines (Clement, 1983; Clement and Zimmermann, 1987).

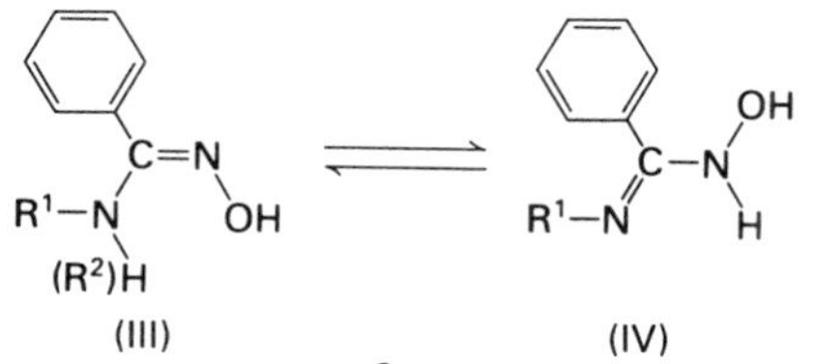

Figure 2.6 Tautomerism of amidoximes (**2**) (for R see Table 2.3).

Amidines hydroxylated at N' (**2**) were designated as amidoximes, since all of them, including those capable to undergo tautomerism, should exist in the oxime-type structure **III** and not in the hydroxylamine-type structure **IV** (Fig. 2.6) (Dignam *et al.*, 1980). These previous investigations (Dignam *et al.*, 1980) were further substantiated by ^{15}N-NMR spectroscopic studies (Clement and Kämpchen, 1985; Clement 1986a). In the case of N-unsubstituted representatives, this is demonstrated not only by the magnitudes of the chemical shifts, but also by the multiplicities of the signals (see Table 2.3 and Fig. 2.7). In the case of N-monosubstituted derivatives, the same splittings should be observed for the tautomers **III** and **IV**, but,

Table 2.3 ^{15}N-chemical shifts[a] and coupling constants (Hz) of amidoximes (**2**) (solvent DMSO). For structure see Figure 2.6

No.	Configuration	R^1	R^2	$\delta(^{15}N, ppm)$		$^1J(^{15}N, {}^1H)$ (Hz)	
				N	N'	N–H	N'–H
2a	Z	H	H	63.2 (t)	287.7 (s)	87.0	—
2a HCl		H	H	101.9	168.6	ND[b]	
2b (*p*-Nitro)	Z	H	H	63.8 (t)	295.8 (s)	87.3	—
2c	Z	C_6H_5	H	92.8 (d)	308.7 (s)	91.6	—
2c HCl		C_6H_5	H	118.0	188.0	ND[b]	
2d	Z	CH_3	H	57.9 (d)	289.0 (s)	87.9	—
2e	E	CH_3	CH_3	55.6	302.1	—	—
2f	Z	$-(CH_2)_5-$		71.4	309.1	—	—
2f	E	$-(CH_2)_5-$		78.7	305.0	—	—

[a] Chemical shifts are given in ppm relative to NH_3.
[b] ND = Not determined.
From Clement and Kämpchen (1985), with permission of VCH Verlagsgesellschaft.

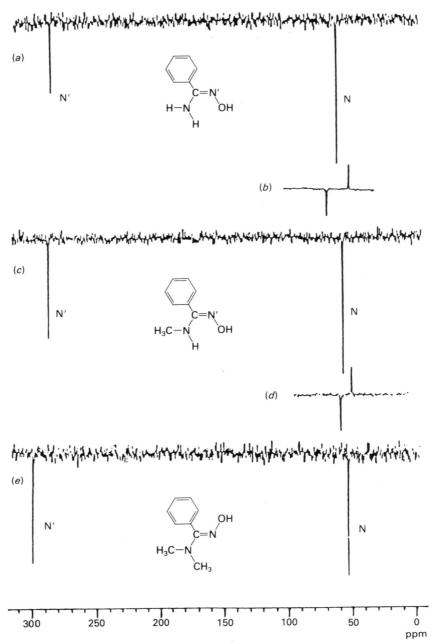

Figure 2.7 ^{15}N-NMR spectra of amidoximes (**2**) in DMSO; (*a*) **2a**, proton-decoupled spectrum; (*b*) **2a**, proton-coupled spectrum (INEPT); (*c*) **2d**, proton-decoupled spectrum; (*d*) **2d**, proton-coupled spectrum (INEPT); (*e*) **2e**, proton-decoupled spectrum. (From Clement and Kämpchen (1985), with permission of VCH Verlagsgesellschaft.)

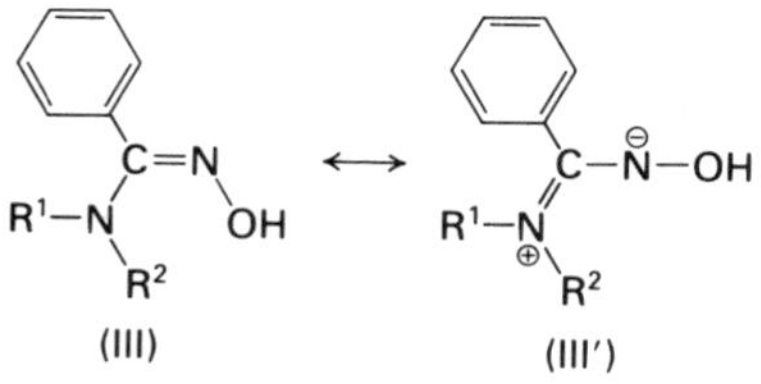

Figure 2.8 Delocalization of electrons in amidoximes (2) (for R see Table 2.3).

again, the magnitudes of the chemical shifts, as discussed on a basis similar to that for the spectra of *N*-monosubstituted hydroxyguanidines, can only be accounted for by structure **III**. Furthermore, the chemical shifts of tautomeric compounds are only slightly different from those, where the possibility of tautomerization is blocked through replacement of the NH protons by alkyl groups. This is especially well illustrated by comparison of the shift values for the unsubstituted (**2a**) and mono-methyl-substituted derivative (**2d**) with those for *N*,*N*-dimethylbenzamidoxime (**2e**) (Fig. 2.7).

Although the magnitudes of the ^{15}N chemical shifts of the doubly bonded nitrogen atoms in amidoximes are larger than those of the *N″*-hydroxyguanidines, they are still markedly smaller than those of oximes. This can also be explained in terms of the resonance structure **III′** (Fig. 2.8).

The magnitudes of the $^{1}J(^{15}N,{}^{1}H)$ coupling constants (87–92 Hz, Table 2.3) in the amidoximes are again indicative of the sp^2 hybridization of the nitrogen atoms which, in turn, makes electron delocalization possible. Since, as a result of the absence of a nitrogen atom, delocalization is smaller than that in *N″*-hydroxyguanidines the magnitudes of the chemical shifts of the double-bonded nitrogens are between those for *N″*-hydroxyguanidines and those for oximes.

In accordance with previously reported results (Dignam *et al.*, 1980), the ^{15}N-NMR spectra of unsubstituted and *N*-monosubstituted derivatives (Table 2.3 and Fig. 2.7) do not provide any evidence of the existence of geometrical isomers (E,Z) at the carbon–nitrogen double bond. Unlike the situation with some *N″*-hydroxyguanidines, this is not the result of rapid isomerization, but rather arises from the presence of only one isomer, namely the *Z*-isomer; the formation of intramolecular hydrogen bonds contributes to increased stability of this *Z*-form. It has as yet not been possible chemically to synthesize the *E*-isomer of any of these compounds (Dignam *et al.*, 1980).

However, in the case of the *N*,*N*-disubstituted representatives (**2e, 2f**) the *Z*-isomers, prepared by the 1,3-dipolar addition of amines to nitrile oxides, undergo complete conversion to the thermodynamically more stable *E*-isomers on heating (Clement, 1986a). The influence of protonation on the chemical shifts of nitrogen resonances is comparable to that observed for the N''-hydroxyguanidines, although with the hydrochlorides of **2a** and **2c** clear NH-coupled spectra could not be obtained (Clement, 1986a). This is indicative of a rapid proton exchange. On the other hand, the high-field shifts of the resonances of the double-bonded nitrogen atoms in the salts of **2a** and **2c** again suggest that this is the site of protonation (Table 2.3).

In summary, it may be concluded that the oxime-type structure is preferred in amidoximes and N''-hydroxyguanidines. This may be explained by the fact that mesomeric stabilization of both systems (see **I'**, **I''** and **III'**, respectively) is particularly favoured when the electron-withdrawing hydroxy group is positioned at the double-bonded nitrogen atom.

2.3.3 *N*-Hydroxyamidines/α-aminonitrones

Compounds of the type **3** (Fig. 2.9) can also be formed from N-mono-substituted benzamidines by metabolic N-oxidation (Clement, 1986a). ^{15}N-NMR investigations on the structure of this class of compounds have been undertaken with the N-methylderivative (**3**) (Clement and Kämpchen, 1987). The ^{15}N-NMR spectra of **3** – both as the free base and the salt – are shown in Fig. 2.9.

The spectrum of the hydrochloride of **3** in DMSO is quite usual (Fig. 2.9b). Two resonances occur at 163.4 ppm and 110.1 ppm, and the latter is split into a triplet ($^1J(^{15}N,^1H) = 91.6$ Hz) in the [1]H-coupled spectrum.

In all the compounds (free bases) discussed up to now, NH couplings were observed in the ^{15}N-nuclear magnetic resonance spectra. However, in the case of the free base of **3**, it has not been possible to establish a coupling constant (Clement, 1986a). This may be explained in terms of a rapid proton exchange. A tautomeric equilibrium exists between the N-hydroxyamidine structure **V** and the aminonitrone **VI** in which the positive charge is stabilized through resonance (Fig. 2.9).

Protonation of either form results in the formation of the same cation. The measured chemical shifts of the nitrogen atoms in the base represent average values. Since the magnitudes of the chemical shifts (196.9 ppm vs. 85.5 ppm) of the base differ widely, the two tautomers cannot be present in about the same proportion. The average value for the imine- or amine-type nitrogen should be approximately 147 ppm (Clement and Kämpchen, 1986) and that for the hydroxylamine- or nitrone-type nitrogen should be of a similar magnitude.

On the basis of the magnitudes of the chemical shifts of the base, considering the unambiguously coupled spectrum and the influence of protonation on chemical shift values, it may be assumed that the aminonitrone **IV** clearly predominates. For a compound of type **V**, a shift value higher than 196.9 ppm would be expected with respect to the imine

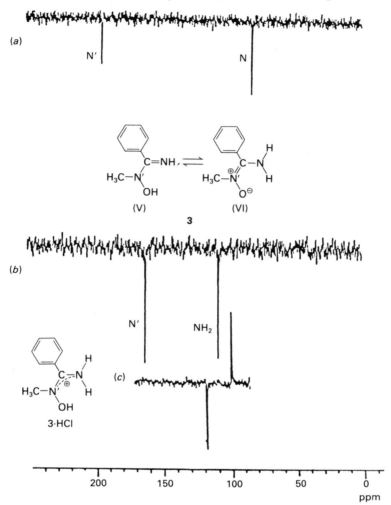

Figure 2.9 ^{15}N-NMR spectra of **3** in DMSO; (*a*) base, proton-decoupled spectrum; (*b*) **3** · HCl, proton-decoupled spectrum; (*c*) **3** · HCl, proton-coupled spectrum (INEPT). (From Clement and Kämpchen (1987), with permission of VCH Verlagsgesellschaft.)

nitrogen (Clement and Kämpchen, 1986); the shift change of 33.5 ppm resulting from the influence of protonation is too small (see section 2.2.2). The chemical shift of 85.5 ppm is too small to account for the hydroxylamine-type nitrogen in **V**, since all values for substituted hydroxylamines are in excess of 100 ppm (Gouesnard and Dorie, 1982). On the other hand, the value of 196.9 ppm for a nitrone-nitrogen atom, being shifted by 33.5 ppm after protonation at the oxygen, as well as the value of 85.5 ppm for a NH_2-group in conjugation with a double bond are in good agreement with structure **VI**.

2.4 CONCLUSIONS

In summary, ^{15}N-NMR spectroscopy is highly suitable for elucidation of the structural features of *N*-oxygenated compounds. Furthermore, *N*-oxygenated products can be readily distinguished from the parent compounds. Both in *N*-hydroxyguanidines and amidoximes the nitrogen atom bonded to oxygen is deshielded by about 50–70 ppm as compared to the corresponding nitrogen centre in guanidines and amidines (Clement, 1986a).

In many cases, ^{15}N-NMR spectra of nitrogen-containing compounds provide additional information not available from ^{1}H- or ^{13}C-NMR data.

REFERENCES

Allen, M. and Roberts, J.D. (1980) Effects of protonation and hydrogen bonding on nitrogen-15 chemical shifts of compounds containing the C≡N-group. *J. Org. Chem.*, **45**, 130–5.

Belzecki, C., Hintze, B. and Kwiatkowska, S. (1970) Synthesis of hydroxyguanidine derivatives from disubstituted cyanamides. *Bull. Acad. Pol. Sci.*, *Ser. Sci. Chim.*, **18**, 375–8.

Benn, R. and Günther, H. (1983) Moderne Pulsfolgen in der hochauflösenden NMR-Spektroskopie. *Angew. Chem.*, **95**, 381–411.

Blomberg, F. and Rüterjans, H. (1983) Nitrogen-15 NMR in biological systems. *Biol. Magn. Reson.*, **5**, 21–73.

Botto, R.E., Westerman, P.W. and Robert, J.D. (1978) Nitrogen-15 nuclear magnetic resonance spectroscopy. Natural-abundance nitrogen-15 spectra of aliphatic oximes. *Org. Magn. Reson.*, **11**, 510–15.

Breitmaier, E. (1983) Die Stickstoff-15-Kernresonanz-Grenzen und Möglichkeiten. *Pharmazie in unserer Zeit*, **12**, 161–80.

Clement, B. (1983) The *N*-oxidation of benzamidines *in vitro*. *Xenobiotica*, **13**, 467–73.

Clement, B. (1986a) in *Biotransformation stark basischer funktioneller Gruppen durch Stickstoffoxidation*, Hochschulverlag, Freiburg, pp. 83–114.

Clement, B. (1986b) *In vitro*-Untersuchungen zur mikrosomalen N-Oxidation einiger Guanidine. *Arch. Pharm.* (*Weinheim*), **319**, 961–8.

Clement, B. and Kämpchen, T. (1985) [15]*N*-NMR-Studien an 2-Hydroxyguanidinen und Amidoximen. *Chem. Ber.*, **118**, 3481–91.

Clement, B. and Kämpchen, T. (1986) [15]*N*-NMR-Studien der Tautomerie in *N*-monosubstituierten Amidinen und in *N*,*N*′-Diphenylguanidin, *Chem. Ber.*, **119**, 1101–4.

Clement, B. and Kämpchen, T. (1987) ^{15}N-NMR-Studien einer α-Aminonitron/ N-Hydroxyamidin-Tautomerie. *Arch. Pharm. (Weinheim)*, **320**, 566–9.

Clement, B. and Zimmermann, M. (1987) Characteristics of the microsomal N-hydroxylation of benzamidine to benzamidoxime. *Xenobiotica*, **17**, 659–67.

Dignam, K.J., Hegarty, A.F. and Begley, M.J. (1980) Structural studies on isolable (*E*)-and (*Z*)-*N*,*N*-disubstituted-amidoximes. Crystal and molecular structure of (*E*)-morpholino-*p*-nitro-benzamidoxime. *J. Chem. Soc. Perkin Trans.*, **2**, 704–9.

Gouesnard, J.P. and Dorie, J. (1982) Nitrogen-15 NMR study on *N*,*N*-di-tert.-butylamine compounds; steric and electronic effects. *Nouv. J. Chim.*, **6**, 143–7.

Kalinowski, H.-O., Berger, S. and Braun, S. (1984) in ^{13}C-NMR-Spektroskopie, Georg Thieme Verlag, Stuttgart, p. 4.

Levy, G.C. and Lichter, R.L. (1979) in *Nitrogen-15 Nuclear Magnetic Resonance Spectroscopy*. John Wiley, New York, pp. 1–33.

Lichter, R.L. (1983) Nitrogen nuclear magnetic resonance spectroscopy, in *The Multinuclear Approach to NMR Spectroscopy* (ed. J.B. Lambert and F.G. Riddel), D. Reidel, Netherlands, pp. 207–44.

Morris, G.A. (1980) Sensitivity enhancement in ^{15}N-NMR: polarization transfer using the INEPT pulse sequence. *J. Am. Chem. Soc.*, **102**, 428–9.

Morris, G.A. and Freeman, R. (1979) Enhancement of nuclear magnetic resonance signals by polarization transfer. *J. Am. Chem. Soc.*, **101**, 760–2.

Naulet, N. and Martin, G.J. (1979) Application de la resonance de l'azote 15 et du carbone 13 a l'étude de la transmission des effets électroniques dans les systèmes C≡N. *Tetrahedron Lett.*, 1493–6.

Piotrowska, K., Trojnar, J. and Belzecki, C. (1973) Geometrical isomerism of hydroxyguanidines. *Bull. Acad. Pol. Sci., Ser. Sci. Chim.*, **21**, 265–70.

Trojnar, J., Piotrowska, K. and Belzecki, C. (1975) Hindered rotation in hydroxyguanidines. *Bull. Acad. Pol. Sci., Ser. Sci. Chim.*, **23**, 125–8.

Witanowski, M., Stefaniak, L., Szymánski, S. and Webb, G.A. (1976) A ^{14}N NMR study of the isomeric structures of urea and its analogues. *Tetrahedron*, **32**, 2127–9.

3

Radioimmunoassay and other methods for trace analysis of N-oxide compounds

K.K. Midha, T.J. Jaworski, E.M. Hawes, G. McKay, J.W. Hubbard, M. Aravagiri* and S.R. Marder*

College of Pharmacy, University of Saskatchewan, S7N 0W0, Canada and
*Brentwood V.A. Medical Center and UCLA, Los Angeles, CA 90073, USA

1. The difficulties encountered in the trace analysis of N-oxides and the methods required for their intact analysis are discussed.
2. Emphasis in the discussion is focused on the tertiary aliphatic amines and the tertiary arylalkylamines referred to here as the aliphatic tertiary amines.
3. The techniques which fulfil the requirements necessary for the analysis of intact N-oxides include high performance liquid chromatography (HPLC), mass spectrometry (MS) (with soft ionization techniques) and radioimmunoassay (RIA).
4. A thorough understanding of the difficulties encountered in the analysis of N-oxides will enable the correct application of these modern methods of analysis to determine better the metabolic importance of aliphatic tertiary amine N-oxides.

3.1 INTRODUCTION

When undertaking development of methodology to analyse a compound containing an N-oxide, it is important to consider whether the compound is the N-oxide of a tertiary aliphatic amine such as trimethylamine, or tertiary arylalkylamine such as dimethylphenylamine, or heteroaromatic amine such as pyridine or quinoline. The first two types are referred to here as aliphatic tertiary amines. It is established that there are chemical and metabolic differences between these two types of amine oxides. Generally the N-oxides of heteroaromatic amines are more stable both chemically and metabolically than the N-oxides of aliphatic amines (Ochiai, 1967; Bickel, 1969; Jenner, 1971; Lindsay, 1979; Damani, 1985). A comparison of the stability of the two types of N-oxides to heat and some reducing agents is given in Table 3.1.

In view of their greater lability, this chapter will essentially only discuss methods of analysis for the aliphatic tertiary amine type of N-oxide in order that the problems associated with the development of methods can be brought into focus. Such lability has resulted in difficulties associated with the development of suitable analytical techniques for the isolation, identification and quantitation of the intact aliphatic tertiary amine N-oxide compounds from biological media. The difficulties will be discussed later in this chapter. In addition to these difficulties, the polarity of N-oxides makes their isolation from biological fluids more difficult. The earlier reported

Table 3.1 Comparison of stability between aliphatic tertiary amine oxides and aromatic heterocyclic N-oxides

Condition	Aliphatic	Aromatic
Reduction with titanous chloride	Generally easily reduced at room temperature	Generally easily reduced at room temperature
Reduction with sulphurous acid or sodium dithionite or sodium metabisulphite	Generally easily reduced at room temperature	Generally resistant to reduction at room temperature
Application of heat	Tendency to undergo decomposition by deoxygenation, loss of aldehyde to form secondary amine, Cope elimination	Tendency to undergo deoxygenation to tertiary amine

methods for the qualitative and quantitative analysis of aliphatic tertiary amine *N*-oxides were often indirect methods because of difficulties associated with their isolation. Consequently, their identification and quantitation was often based on measurement of either reduced or thermally decomposed products of the *N*-oxides (Beckett *et al.*, 1971; Ames and Powis, 1978; Gruenke *et al.*, 1985; Jacob *et al.*, 1986).

The current trend in method development is that the intact *N*-oxide molecule is analysed directly. Thus the analytical techniques which have met the challenges posed by the labile nature of the aliphatic tertiary amine *N*-oxides include methods based on high performance liquid chromatography (HPLC), mass spectrometry (MS) (with soft ionization techniques) and radioimmunoassay (RIA). Two important features of HPLC are its operation at ambient temperature and its suitability for the chromatography of polar compounds. Both these features increase the chances of successful application to the analysis of intact *N*-oxides. The availability of soft modes of ionization in MS makes these techniques suitable for the analysis of labile compounds. Moreover the use of combined HPLC–MS by means of interfaces such as plasmaspray or thermospray has been an important development. RIA, although initially involving a large investment of time and effort, can reduce the complexity of sample preparation and results in high sample turnover with an overall decrease in analysis time. This chapter will focus on the use of these modern techniques for the qualitative and quantitative analysis of aliphatic tertiary amine *N*-oxides. The examples taken from the literature include some based on the experience of this research group with the *N*-oxide metabolites of various psychotropic drugs, drugs where the aliphatic tertiary amine group is commonly encountered. The difficulties involved in the analysis of *N*-oxide compounds will be discussed first. It is only through an understanding of these difficulties that the modern methods of analysis can be correctly applied so that true understanding of the metabolic importance of aliphatic tertiary amine *N*-oxides is fully realized.

3.2 DEFINING THE PROBLEM

Early literature on the *N*-oxides was reviewed authoritatively by Bickel (1969), an author who has himself contributed to our understanding of these compounds. In the pharmaceutical world, although there are few drugs marketed as *N*-oxides *per se*, an enormous number of drugs are nitrogenous compounds which can give rise to *N*-oxide metabolites. It is only in recent years, however, that analytical techniques have advanced to the point where it is possible to measure trace levels of intact *N*-oxides in biological matrices.

As a result, the true significance of the pharmacokinetics and pharmaco-dynamics of N-oxides as drug metabolites, drugs or prodrugs is only beginning to be realized.

3.2.1 Problems which have created challenges in the analysis of the aliphatic tertiary amine N-oxides

(a) Polarity of compounds

There are marked physicochemical differences between an aliphatic tertiary amine and its corresponding N-oxide (Jenner, 1971). Formation of an N-oxide generally leads to a marked increase in polarity (Beckett and Triggs, 1967) and a decrease in pK_a (Bickel, 1969). Solvents commonly used for the extraction of lipophilic drugs from aqueous biological media (hexane, pentane or diethyl ether, etc.) do not possess enough polar character to extract aliphatic tertiary amine N-oxides. More polar solvents such as dichloromethane, ethyl acetate, 2-propanol or acetone are required. The use of these more polar solvents results in the extraction of extraneous materials from the biological matrix resulting in chromatographic problems. Clean up procedures required to avoid chromatographic problems often lead to poor overall extraction recoveries.

(b) Thermal decomposition

There have been many cases reported in the literature where N-oxides have been shown to undergo thermal decomposition in the injection port of the gas chromatograph or in the mass spectrometer. Aliphatic tertiary amine N-oxides which have an N-methyl group characteristically decompose by loss of oxygen to yield the corresponding tertiary amine, loss of formaldehyde to yield the corresponding secondary amine and Cope elimination to yield an alkene. Among others these observations have been reported for chlorpromazine N-oxide (Gudzinowicz *et al.*, 1964), doxylamine N-oxide (Ganes *et al.*, 1986), meperidine N-oxide (Linberg and Bogentoft, 1975) and tamoxifen N-oxide (Foster *et al.*, 1980).

(c) Decomposition during biological sample preparation for analysis

A dramatic example of the decomposition of an N-oxide during sample preparation causing major errors in analysis has been reported (Hubbard *et al.*, 1985). Plasma samples from patients medicated with chlorpromazine appeared to contain much higher concentrations of chlorpromazine after extraction from aliquots of plasma alkalinized with sodium hydroxide (pH 13.6), compared with aliquots of the same plasma alkalinized with sodium carbonate (pH 11.7). Krieglstein *et al.*, (1979) had previously suggested the possibility of chlorpromazine *N*-oxide being converted into chlorpromazine and chlorpromazine sulphoxide during an extraction procedure which involved adjustment of the biological sample to pH 14. Hubbard *et al.* (1985) showed that reduction of the tertiary amine *N*-oxide occurred as a result of chemical reaction between chlorpromazine *N*-oxide and plasma constituents which were modified under strongly alkaline conditions. It was postulated that reactive groups which were produced by, for example, cleavage of the disulphide bonds of albumin (Field, 1977) to yield thiolate anions, are capable of reducing susceptible functional groups such as *N*-oxides. It was confirmed with further experiments that on addition of chlorpromazine *N*-oxide to albumin solutions buffered at pH 14, the presence of chlorpromazine was detected. When a similar experiment was conducted where no albumin was added to a solution buffered at pH 14, reduction of chlorpromazine *N*-oxide was not detected.

The significance of these observations to the measurement of chlorpromazine and chlorpromazine *N*-oxide in plasma samples obtained from patients is obvious. In fact, in an investigation of plasma samples ($n = 10$) obtained from patients under chronic oral treatment with chlorpromazine, it was shown that the apparent chlorpromazine concentrations were up to 343% higher when sodium hydroxide was used for alkalinization rather than sodium carbonate. A survey of the extraction procedures reported in the literature where measurements are made of plasma or serum concentrations of chlorpromazine in patients indicated that almost invariably alkali is added and that sodium hydroxide is a very frequent choice. However, it was also shown that the extent of conversion of chlorpromazine *N*-oxide into chlorpromazine depends not only on the number of added equivalents of sodium hydroxide but also other aspects of the extraction procedure employed. For example, when plasma samples that contain added chlorpromazine *N*-oxide and sodium hydroxide were extracted by otherwise similar procedures of either single (10 ml) or double aliquots (2×5 ml) of the same extraction solvent, 1.4% and 23.9%, respectively, was recovered as chlorpromazine. The reason for this vast difference in recovery of chlorpromazine was probably the contact time between alkalinized plasma and the extraction solvent in that the double-aliquot procedure took considerably longer to perform (Hubbard *et al.*, 1985).

In the case of the analysis of chlorpromazine in whole blood it is essential that alkali should not be used at all in the extraction procedure since it was found that both sodium carbonate and sodium hydroxide rapidly converted chlorpromazine *N*-oxide into chlorpromazine, and a portion (10–14%) of chlorpromazine was converted into chlorpromazine sulphoxide (McKay *et al.*, 1985). Finally it is a paradox that chlorpromazine, chlorpromazine sulphoxide and chlorpromazine *N*-oxide are stable under physiological conditions in whole blood, and they can be efficiently extracted at physiological pH after the appropriate separation of plasma and red blood cells (Hawes *et al.*, 1986).

These observations regarding the stability of chlorpromazine *N*-oxide under alkaline conditions probably apply to some extent to other *N*-oxides of aliphatic tertiary amines. In fact, it was clearly shown that the *N*-oxides of certain tricyclic antidepressant drugs, namely amitriptyline, doxepin and imipramine were substantially *N*-deoxygenated when plasma was alkalinized with sodium hydroxide, but not sodium carbonate (Hubbard *et al.*, 1986).

It should also be considered that the use of oxidizing or reducing agents at any stage of analysis can potentially cause artifactual interconversion between an *N*-oxide and the corresponding tertiary amine. For example, it has been demonstrated that the addition of antioxidants to prevent decomposition of plasma samples during storage is not wise when an *N*-oxide is present. Thus the addition of sodium metabisulphite or ascorbic acid before storage of plasma samples at 4°C or − 18°C led to the partial formation of chlorpromazine from chlorpromazine *N*-oxide (Whelpton, 1978).

(d) Decomposition during chemical derivatization procedures

The chemical derivatization of organic compounds is generally carried out to facilitate their chromatographic behaviour and/or to enhance their sensitivity to detection. Acid chlorides and acid anhydrides are two examples of agents which are commonly used for this purpose. Unfortunately, aliphatic tertiary amine *N*-oxides have been shown to decompose as a result of these derivatization procedures. For example, a well-known reaction which occurs with aliphatic tertiary amine *N*-oxides and acid anhydrides is the Polonovski reaction, which is illustrated in Fig. 3.1. Because of the ease with which these reactions occur, it is best to avoid acid anhydrides or acid chlorides at all times when working with samples likely to contain *N*-oxides.

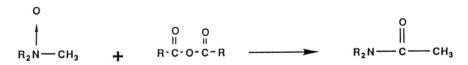

Figure 3.1 Polonovski reaction with aliphatic tertiary amine *N*-oxides.

3.3 TECHNIQUES AVAILABLE FOR THE QUALITATIVE AND QUANTITATIVE ANALYSIS OF INTACT N-OXIDES

As mentioned previously, the earlier methods of quantitation almost invariably involved the use of indirect methods of analysis because of the instabilities associated with N-oxides. A widely applied method was based on the use of a reducing agent to convert the N-oxide back into the tertiary amine. Titanous chloride (Beckett *et al.*, 1971), sodium metabisulphite (Curry and Evans, 1976) and sodium dithionite (Aravagiri *et al.*, 1990) are examples of agents that have been used for this purpose. The latter two reagents are preferred, since, unlike titanous chloride, they selectively reduce the N-oxide group in the presence of the sulphoxide group (Curry and Evans, 1976; Aravagiri *et al.*, 1990).

New techniques which are suitable for the analysis of the intact N-oxides include RIA, HPLC and MS with the aid of soft ionization modes. These techniques will be illustrated below with a number of examples.

3.3.1 Qualitative analysis

(a) High performance liquid chromatography

HPLC has been used qualitatively to investigate the presence of N-oxide metabolites in biological samples. For example, HPLC methods were used in the investigation of the metabolites of chlorpromazine N-oxide in urine and faeces of rat, dog and man after oral administration of the N-oxide to each of the three species (Jaworski *et al.*, 1988, 1990). The administered compound and metabolites were isolated by collection of eluates from HPLC and their identities confirmed with the aid of MS. The advantage of this procedure was that it was able to isolate and identify the intact N-oxide compounds from biological matrices. A sample chromatogram is shown in Fig. 3.2.

(b) Mass spectrometry

Electron impact (EI) mass spectra generally provide insufficient evidence for identification of most aliphatic tertiary amine N-oxides because of their labile nature. Therefore, methods such as chemical ionization (CI) and fast atom bombardment (FAB) have been used for the identification of intact molecules present in samples of biological and non-biological origin since these methods yield information about pseudomolecular ions.

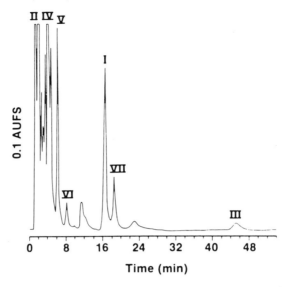

Figure 3.2 HPLC chromatogram of an extract from urine from a dog dosed orally with chlorpromazine *N*-oxide. I, chlorpromazine *N*-oxide; II, chlorpromazine; III, chlorpromazine *N,S*-dioxide; IV, 7-hydroxychlorpromazine; V, chlorpromazine sulphoxide; VI, *N*-desmethylchlorpromazine; VII, *N*-desmethylchlorpromazine sulphoxide.

McKay *et al.* (1986) have examined the use of three different modes of ionization, EI, CI and FAB. In the case of EI or CI, samples were analysed via a solid direct probe inlet system, whereas with FAB samples these were applied to a stainless-steel target. Samples which were run in EI mode either gave no molecular ion or one of very weak intensity. On the other hand, although the relative abundance varied from compound to compound CI spectra invariably showed pseudomolecular ions. FAB analysis gave abundant pseudomolecular ions which in many cases constituted the base peak.

(c) *Radioimmunoassay*

Immunoassay techniques can provide a very simple method to carry out a rapid qualitative screen for the presence of *N*-oxide metabolites in biological fluids. Interaction of a ligand with the antibody may be taken as good evidence for the presence of an *N*-oxide metabolite in the sample, which should be investigated and confirmed by more rigorous methods.

3.3.2 Quantitative analysis

(a) *High performance liquid chromatography*

Some advantages of using HPLC for the quantitation of aliphatic tertiary amine *N*-oxides can be exemplified by an analytical method developed for the measurement of chlorpromazine and metabolites, including the *N*-oxide metabolite, in plasma (Midha *et al.*, 1987). The development of this method was a great challenge since chlorpromazine is extensively metabolized to numerous products, many of which are very difficult to extract and separate in a single chromatographic run. The method eventually developed was able to separate chlorpromazine plus eight metabolites in a single chromatographic run. In addition, the compounds were detected by UV spectrophotometry which avoids much of the interference from endogenous plasma constituents encountered with electrochemical detectors.

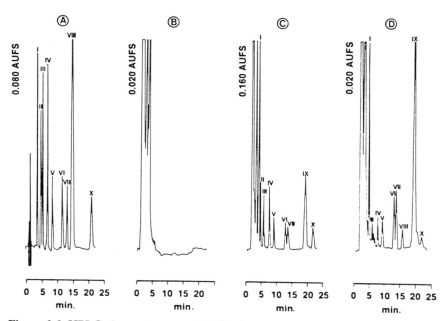

Figure 3.3 HPLC chromatograms of (A) reference standards in organic solution, (B) an extract of blank plasma, (C) an extract of plasma 'spiked' with reference standards, and (D) an extract of plasma from a volunteer dosed with chlor-promazine. I, chlorpromazine; II, 7-hydroxychlorpromazine; III, *N,N*-didesmethyl-chlorpromazine; IV, *N*-desmethylchlorpromazine; V, 7-hydroxy-*N*-desmethyl-chlorpromazine; VI, chlorpromazine sulphoxide; VII, chlorpromazine *N*-oxide; VIII, *N,N*-didesmethylchlorpromazine sulphoxide; IX, mesoridazine (internal standard); X, *N*-desmethylchlorpromazine sulphoxide.

A further advantage of this method is the simple extraction procedure which is much improved over a previously reported method (Gruenke *et al.*, 1985). The procedure involves a single extraction with an organic solvent mixture consisting of dichloromethane : pentane : 2-propanol (49 : 46 : 5 by vol.); the solvent mixture allows for extraction of analytes with a wide range of polarities. One of the reasons that this method was able to separate at least nine related components was due to a mixed cationic buffer system (ammonium and sodium) contained in the mobile phase, since

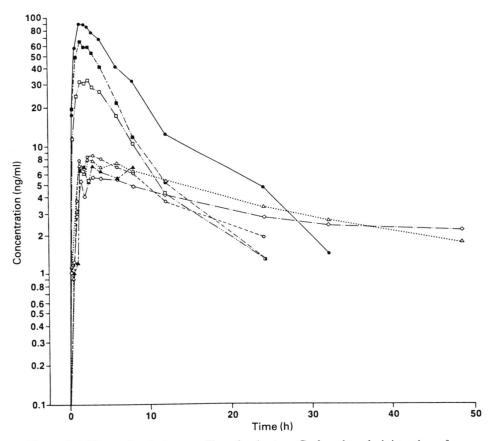

Figure 3.4 Plasma level–time profiles of volunteer C after the administration of a single 100 mg oral dose of chlorpromazine hydrochloride. Plasma concentrations were determined by an HPLC method. ●, Chlorpromazine; ○, 7-hydroxy-chlorpromazine; ◇, *N*-desmethylchlorpromazine; △, 7-hydroxy-*N*-desmethyl-chlorpromazine; □, chlorpromazine sulphoxide; ■, chlorpromazine *N*-oxide; ▲, *N*-desmethylchlorpromazine sulphoxide.

the addition of only one of the cations was insufficient for separation of all nine components. The separation of chlorpromazine and its metabolites from plasma is shown in Fig. 3.3. As mentioned earlier, the use of polar extraction solvents can lead to chromatographic problems through the co-extraction of endogenous constituents from the biological matrix. Difficulties of this type were in fact encountered in the chromatography of chlorpromazine in extracts of plasma from some individuals. The spuriously high values for chlorpromazine were detected by analysis of aliquots of the same plasma samples by RIA. This illustrates the fact that it is possible for an immunoassay to be more specific than a chromatographic method in a multicomponent analysis. No chromatographic problems of this type were encountered, however, in the analysis of chlorpromazine N-oxide.

The method can quantitate chlorpromazine metabolites over a concentration range of 2.5–50 ng/ml with an overall coefficient of variation in the range 3.8–7.7%. The method has been successfully applied to the analysis of plasma samples collected over 48 h from four healthy volunteers each given 100 mg single oral doses of chlorpromazine. An example of a plasma level–time profile is shown in Fig. 3.4.

(b) Mass spectrometry (fast atom bombardment)

Until recently, examination of intact N-oxides by MS has been restricted to qualitative analysis. With the advances made in soft ionization techniques, however, the development of quantitative methods for the intact N-oxides has been made possible. One example of a method recently reported was the use of FAB tandem MS for the analysis of a substituted benzazepine N-oxide compound, SK&F 102102 shown in Fig. 3.5 (Straub and Levandoski, 1985). The procedure was carried out on a Finnigan MAT triple stage quadrupole MS with xenon as the primary ionizing beam for the FAB ionization process.

The method can be applied to both plasma or urine samples. After the addition of a pentadeuterated internal standard, extraction of the N-oxide from the biological fluid was accomplished rapidly with the aid of solid-phase extraction cartridges. An aliquot of the crude extract eluted from the cartridges was then applied to the FAB probe tip and inserted into the mass

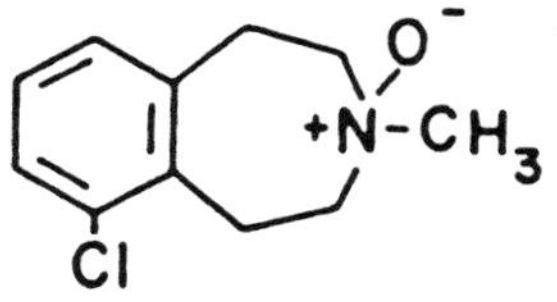

Figure 3.5 The structure of SK&F 102102.

spectrometer. Quantitation was carried out with selected reaction monitoring such that only the protonated molecular ion $[M + H]^+$ of the sample or internal standard was allowed to pass through quadrupole 1. Quadrupole 2 was used as a collision cell for the generation of collision-activated dissociation (CAD) daughter ions. Quadrupole 3 allowed the passage of the daughter ion generated in quadrupole 2 for each specific parent ion. Measurements of the daughter ion current of m/z 180 and 185 were then carried out and the ratio of the daughter ion signal intensities from the sample and internal standard, respectively, were plotted versus concentration to generate a standard curve. The method was validated by using quality control samples of three different concentrations and standard curves which contained 10 data points/curve. Cross validation was carried out with the aid of an independent assay involving GC analysis of the N-oxide after reduction to the tertiary amine. The advantage of this method is its utility for the detection and quantitation of a labile compound within a complex sample matrix. In addition, sample preparation is simple, and assay time per sample is very rapid (usually less than 5 minutes).

(c) Radioimmunoassay

Over the past few years RIAs have been successfully developed and applied to the measurement of N-oxide metabolites of some phenothiazine antipsychotic agents. For example, chlorpromazine N-oxide has been shown to be a major metabolite in patients medicated with chlorpromazine. In order to develop an RIA procedure for chlorpromazine N-oxide, the first step was the production of a suitable antiserum. Since compounds with low molecular weight are not immunogenic, they must be linked covalently to a macromolecule in order to stimulate an immunogenic response. A commonly used macromolecule for this purpose is the protein bovine serum albumin (BSA) which has a molecular weight of approximately 65 000. Yeung *et al.* (1987) showed that the most specific antibodies are formed when the protein is attached to the molecule at a site most distal from the most characteristic functional group. With the N-oxide of chlorpromazine, for example, the most distinguishable functional group is the tertiary amine N-oxide moiety located at the dimethylaminopropyl side chain. Therefore, the best position for attachment to the protein would be on the phenothiazine ring system. Since a reactive functional group is not available on the ring system for direct coupling to a protein, a group must be attached for this purpose. In this case a functional group was attached by means of a Friedel–Crafts acylation reaction. The purity and integrity of this product, commonly referred to as a hapten, was established by means of TLC, NMR, UV and MS because impurities could result in the production of antibodies for the impurities and, therefore, compromise specificity of a hapten for

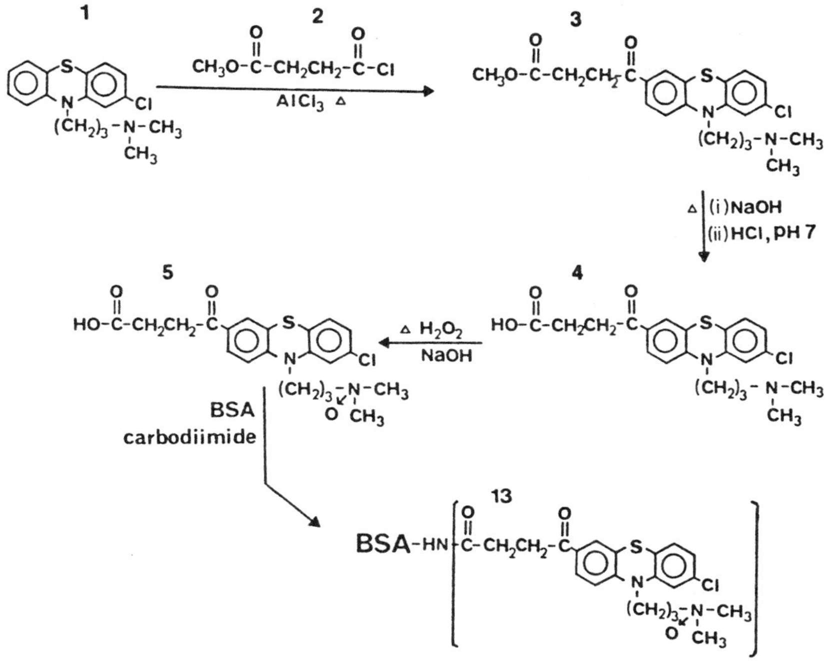

Figure 3.6 Synthesis of a hapten for chlorpromazine *N*-oxide and its bovine serum albumin (BSA) conjugate. 1, Chlorpromazine; 2, 3-methoxycarbonylpropionyl chloride; 3, 7-(3-methoxycarbonylpropionyl)-chlorpromazine; 4, 7-(3-carboxy-propionyl) chlorpromazine; 5, 7-(3-carboxypropionyl) chlorpromazine *N*-oxide (hapten for chlorpromazine *N*-oxide).

chlorpromazine *N*-oxide. The synthesis of the hapten for chlorpromazine *N*-oxide is shown in Fig. 3.6.

After production of the antiserum it is essential to determine its cross-reactivity profile. In the past, the method of Abraham (1969) has been widely adopted and such data are shown for chlorpromazine *N*-oxide in Table 3.2. (Yeung *et al.*, 1987). Subsequent work, however, has shown that the criteria of Abraham do not define the specificity of the antisera adequately. A test which is more rigorous than Abraham's method is the cross-reactants challenge (Aravagiri *et al.*, 1984). In this procedure, the analyte is added to a blank biological matrix containing, in 2–5 times excess, a number of potential cross-reactants which are likely to be present in test samples. The apparent concentration of the test analyte is then determined experimentally and compared with the measurements made in the absence of cross-reactants. From the result of the cross-reactants challenge, it was decided that direct RIA for chlorpromazine *N*-oxide gave inflated values.

Table 3.2 Cross-reactions of the chlorpromazine N-oxide anti-serum[a]

Compound	Cross-reaction (%)
Chlorpromazine N-oxide	100
Chlorpromazine	< 1
7-Hydroxychlorpromazine	< 1
Chlorpromazine sulphoxide	< 1
N-Desmethylchlorpromazine	< 1
N,N-Didesmethylchlorpromazine	< 1
Chlorpromazine N,S-dioxide	< 1
7-Hydroxychlorpromazine-O-glucuronide	< 1

[a] Values are expressed as percentage cross-reaction which was calculated by the method of Abraham (1969).

Therefore, a procedure was developed to extract the N-oxide from plasma and RIA was then carried out on the extracts (Table 3.3). The extraction RIA was shown to be useful over the concentration range 0.25–250 ng/ml and was capable of monitoring chlorpromazine N-oxide levels in plasma for up to 48 h after administration of a single oral dose of 50 mg chlorpromazine hydrochloride. An example of a plasma level–time profile is shown in Fig. 3.7 (Yeung *et al.*, 1987).

An RIA procedure was also developed for the measurement of trifluoperazine $N-4'$-oxide concentrations in plasma (Aravagiri *et al.*, 1986).

Table 3.3 Determination of plasma chlorpromazine N-oxide concentrations in the absence and presence of chlorpromazine and other major metabolites

	Concentration added as CPZNO, 1 ng/ml	Concentration added as CPZNO (CPZ + CPSO + 7-OHCPZ), 1 ng/ml; 5 ng/ml each	Difference of means (t test)
Direct RIA[a] (n = 3)	1.03 ± 0.12^c	1.33 ± 0.05	$P < 0.05$
Extraction RIA[b] (n = 6)	0.97 ± 0.09	0.98 ± 0.04	$P > 0.05$

[a] CPZNO was measured directly in plasma without prior extraction.
[b] CPZNO was measured after selective extraction from plasma, as described in Yeung *et al.* (1987).
[c] Mean ± SD of the assayed plasma concentration. CPZ, chlorpromazine; CPZNO, chlorpromazine N-oxide; CPZSO, chlorpromazine sulphoxide; 7-OHCPZ, 7-hydroxychlorpromazine.

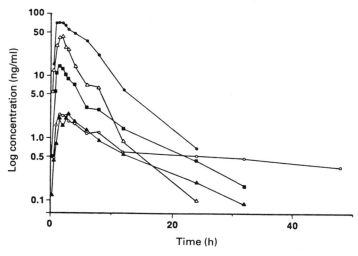

Figure 3.7 Plasma concentrations of chlorpromazine and metabolites in a healthy volunteer after a single 50 mg oral dose of chlorpromazine hydrochloride. Plasma concentrations were determined by RIA subsequent to extraction. ▲, Chlorpromazine; △, chlorpromazine *N*-oxide; ●, conjugated 7-hydroxychlorpromazine; ○, free 7-hydroxychlorpromazine; ■, chlorpromazine sulphoxide.

Until this procedure was developed, the levels of the *N*-oxide metabolite in plasma were largely unknown. As in the development of the RIA procedure for chlorpromazine *N*-oxide it was necessary to prepare a hapten. The Friedel–Crafts acylation failed, however, probably because of the presence of the strongly electronegative 2-trifluoromethyl group of the phenothiazine

Table 3.4 Cross-reactions of the antiserum for the $N^{4'}$-oxide metabolites of piperazine type of phenothiazine antipsychotic agents

Compound	Cross-reaction (%)
Trifluoperazine $N^{4'}$-oxide	100
Fluphenazine $N^{4'}$-oxide	67
Prochlorperazine $N^{4'}$-oxide	100
Trifluoperazine	< 1
Fluphenazine	< 1
Prochlorperazine	< 1

Values are expressed as percentage cross-reaction which was calculated by the method of Abraham (1969).

ring. Fortunately, an antiserum generated to a particular phenothiazine derivative will generally cross-react with an analogue differing in structure only at the ring 2-position. Thus a hapten was prepared by Friedel–Crafts acylation of the 2-chloro analogue (prochlorperazine). The antiserum subsequently raised to the hapten–BSA conjugate cross-reacted 100% with trifluoperazine $N^{4'}$-oxide and cross-reacted significantly with fluphenazine $N^{4'}$-oxide (Table 3.4). It was then decided to use this antisera for the development of RIA procedures for trifluoperazine $N^{4'}$-oxide and fluphenazine $N^{4'}$-oxide.

The assay is capable of quantitating trifluoperazine $N^{4'}$-oxide from 0.1–10 ng/ml of plasma and, therefore, can measure trifluoperazine $N^{4'}$-oxide in plasma samples collected as late as 72 h after administration of a single oral dose of trifluoperazine dihydrochloride (5 mg). Figure 3.8 shows the mean plasma concentrations of trifluoperazine $N^{4'}$-oxide in the plasma of six healthy volunteers after each had received a single oral dose (5 mg) of trifluoperazine dihydrochloride.

An RIA has also been developed for the $N^{4'}$-oxide of fluphenazine (Aravagiri *et al.*, 1990). This method was also based on an antiserum raised to the prochlorperazine $N^{4'}$-oxide immunogen. The antiserum displayed an insignificant cross-reactivity with other major metabolites of fluphenazine such as 7-hydroxyfluphenazine, fluphenazine sulphoxide or N-deshydroxy-ethylfluphenazine. Moreover, the cross-reactants challenge test showed

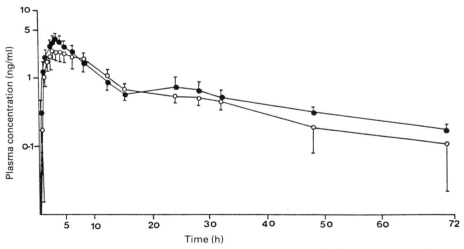

Figure 3.8 Mean plasma concentrations of trifluoperazine (○) and trifluoperazine N-oxide (●). Concentrations in six healthy volunteers after a single 5 mg oral dose of trifluoperazine dihydrochloride were determined by RIA procedures. Each point represents the mean ± SD.

Table 3.5 Plasma concentrations (mean ± SD) of FLU and FLUNO in patients ($n = 29$) under chronic treatment with biweekly intra-muscular doses of 5 mg of FLU decanoate

Length of treatment (weeks)	n	Plasma concentrations (ng/ml)	
		FLU	*FLUNO*
4	26	0.32 ± 0.16 (26)	0.28 ± 0.46 (10)
8	26	0.32 ± 0.18 (26)	0.25 ± 0.31 (11)
26	17	0.31 ± 0.32 (16)	0.29 ± 0.26 (7)

Both FLU (McKay *et al.*, 1989) and FLUNO (Aravagiri *et al.*, 1990) were determined by the respective extraction RIA procedure. The numbers in parentheses are the number of patient samples where the analytes could be quantitated ($\geqslant 0.05$ ng/ml). FLU, fluphenazine; FLUNO, fluphenazine $N^{4'}$-oxide.

that extraction was necessary prior to quantitation of the $N^{4'}$-oxide. The RIA was capable of quantitating 0.1–5 ng/ml of fluphenazine $N^{4'}$-oxide in plasma. It could accurately determine fluphenazine $N^{4'}$-oxide in most patients receiving daily oral doses of 5–20 mg of fluphenazine dihydro-chloride in whom the levels of the $N^{4'}$-oxide ranged from 13–378% of the observed fluphenazine concentrations. The method has also been applied to measurement of fluphenazine $N^{4'}$-oxide in plasma samples of some patients under chronic treatment with biweekly intramuscular injections of 5 mg of fluphenazine decanoate (Table 3.5).

3.4 CONCLUSIONS

The measurement of intact N-oxide metabolites has been promoted and enhanced by a number of developments. The advent of HPLC has permitted the separation of trace levels of intact N-oxides from other components of various biological matrices. Moreover, the development of soft ionization techniques in MS has greatly assisted in the identification and measurement of numerous polar and unstable analytes including N-oxides. The combination of HPLC with MS by modern interface devices such as plasmaspray or thermospray will become widely used techniques in the analysis of N-oxides in the next decade. Greater understanding of the physicochemical properties of N-oxides and their behaviour during manipulation of biological samples has also contributed to their successful analysis. It has also been demonstrated that immunoassay techniques can be developed and applied successfully to the analysis of N-oxides or to provide a rapid qualitative screen for their detection.

REFERENCES

Abraham, G.E. (1969) Solid phase radioimmunoassay of estradiol-17β. *J. Clin. Endocrinol. Metab.* **29**, 866–70.

Ames, M.M. and Powis, G. (1978) Determination of indicine *N*-oxide and indicine in plasma and urine by electron-capture gas-liquid chromatography. *J. Chromatogr.*, **166**, 519–26.

Aravagiri, M., Hawes, E.M. and Midha, K.K. (1986) Radioimmunoassay for the sulfoxide metabolite of trifluoperazine and its application to a kinetic study in humans. *J. Pharm. Sci.*, **73**, 1383–7.

Aravagiri, M., Hawes, E.M. and Midha, K.K. (1986) Development and application of a specific radioimmunoassay for trifluoperazine $N^{4'}$-oxide to a kinetic study in humans. *J. Pharmacol. Exp. Ther.*, **237**, 615–22.

Aravagiri, M., Marder, S.R., Van Putten, T., *et al.* (1990) Therapeutic monitoring of steady-state levels of $N^{4'}$-oxide metabolite of fluphenazine in chronically treated schizophrenic patients determined by a specific and sensitive radioimmunoassay. *Ther. Drug Monit.*, **12**, 268–76.

Beckett, A.H., Gorrod, J.W. and Jenner, P. (1971) The analysis of nicotine-1'-*N*-oxide in urine, in the presence of nicotine and cotinine, and its application to the study of *in vivo* nicotine metabolism in man. *J. Pharm. Pharmacol.*, **23**, 55S–61S.

Beckett, A.H. and Triggs, E.J. (1967) Buccal absorption of basic drugs and its application as an *in vivo* model of passive drug transfer through lipid membranes. *J. Pharm. Pharmacol.*, **19**, 31S–41S.

Bickel, M.H. (1969) The pharmacology and biochemistry of *N*-oxides. *Pharmacol. Rev.*, **21**, 325–55.

Curry, S.H. and Evans, S. (1976) A note of the assay of chlorpromazine *N*-oxide and its sulphoxide in plasma and urine. *J. Pharm. Pharmacol.* **28**, 467–8.

Damani, L.A. (1985) Oxidation of tertiary heteroaromatic amines in *Biological Oxidation of Nitrogen in Organic Molecules* (eds J.W. Gorrod and L.A. Damani), Ellis Horwood, Chichester, pp. 205–18.

Field, L. (1977) Disulfides and polysulfide, in *The Organic Chemistry of Sulfur* (ed. S. Owe), Plenum Press, New York, pp. 303–82.

Foster, A.B., Griggs, L.J., Jarman, M. *et al.* (1980) Metabolism of tamoxifen by rat liver microsomes: formation of the *N*-oxide, a new metabolite. *Biochem. Pharmacol.* **29**, 1977–9.

Ganes, D.A., Hindmarsh, K.W. and Midha, K.K. (1986) Doxylamine metabolism in rat and monkey. *Xenobiotica*, **16**, 781–94.

Gruenke, L.D., Craig, J.C., Klein, F.D. *et al.* (1985) Determination of chlorpromazine and its major metabolites by gas chromatography/mass spectrometry: application to biological fluids. *Biomed. Mass Spectrom.* **12**, 707–13.

Gudzinowicz, B.J., Martin, H.F. and Driscoll, J.L. (1964) Gas chromatographic analysis of thermal decomposition products of chlorpromazine, chlorpromazine S-oxide and chlorpromazine N-oxide. *J. Gas Chromatogr.*, **2**, 265–9.

Hawes, E.M., Hubbard, J.W., Martin, M. *et al.* (1986) Therapeutic monitoring of chlorpromazine III: minimal interconversion between chlorpromazine and metabolites in human blood. *Ther. Drug Monit.*, **8**, 37–41.

Hubbard, J.W., Cooper, J.K., Hawes, E.M. *et al.* (1985) Therapeutic monitoring of chlorpromazine I: pitfalls in plasma analysis. *Ther. Drug Monit.*, **7**, 222–8.

Hubbard, J.W., Cooper, J.K., Sanghvi, S. *et al.* (1986) Elevation of tricyclic antidepressants in plasma due to decomposition of their N-oxide metabolites. *Acta Pharmacol. Toxicol.* **59** (Suppl. V, Abstr. II), 236.

Jacob, P., Benowitz, N.L., Yu, L. and Shulgin, A.T. (1986) Determination of nicotine N-oxide by gas chromatography following thermal conversion to 2-methyl-6-(3-pyridyl)tetrahydro-1,2-oxazine. *Anal. Chem.* **58**, 2218–21.

Jaworski, T.J., Hawes, E.M., McKay, G. and Midha, K.K. (1988) The metabolism of chlorpromazine N-oxide in the rat. *Xenobiotica*, **18**, 1439–47.

Jaworski, T.J., Hawes, E.M., McKay, G. and Midha, K.K. (1990) The metabolism of chlorpromazine N-oxide in man and dog. *Xenobiotica, **20**, 107–15.

Jenner, P. (1971) The role of nitrogen oxidation in the excretion of drugs and foreign compounds. *Xenobiotica*, **1**, 399–418.

Krieglstein, J., Rieger, H. and Schütz, H. (1979) Effects of chlorpromazine and some of its metabolites on the EEG and on dopamine metabolism of the isolated perfused rat brain. *Eur. J. Pharmacol.* **56**, 363–70.

Linberg, C. and Bogentoft, C. (1975) Mass spectrometric and gas chromatographic properties of pethidine N-oxide. *Acta Pharm. Suec.*, **12**, 507–10.

Lindsay, R.J. (1979) Aromatic amines, in *Comprehensive Organic Chemistry – The synthesis and reactions of organic compounds.* Vol. 2 (ed. I.O. Sutherland), Pergamon Press, Oxford.

McKay, G., Cooper, J.K., Hawes, E.M. *et al.* (1985) Therapeutic monitoring of chlorpromazine II: pitfalls in whole blood analysis. *Ther. Drug Monit.*, **7**, 472–7.

McKay, G., Edom, R., Hawes, E.M. *et al.* (1986) Mass spectral analysis of phenothiazine N-oxides. Presented at the 34th Annual Conference on Mass Spectrometry and Allied Topics, Cincinnati, OH, 8–13 June.

McKay, G., Steeves, T., Cooper, J.K. *et al.* (1990) Development and application of a radioimmunoassay for fluphenazine based on monoclonal antibodies and its comparison with alternate assay methods. *J. Pharm. Sci.*, **79**, 240–3.

Midha, K.K., Hubbard, J.W., Cooper, J.K. *et al.* (1987) Therapeutic monitoring of chlorpromazine IV: comparison of a new high-performance liquid chromatographic method with radioimmunoassays for parent drug and some of its major metabolites. *Ther. Drug Monit.*, **9**, 358–65.

Ochiai, E. (1967) in *Aromatic Amine Oxides*. Elsevier, Amsterdam, pp. 6–18.

Straub, K.M. and Levandoski, P. (1985) Quantitative analysis of an N-oxide metabolite by fast atom bombardment tandem mass spectrometry. *Biomed. Mass Spectrom.*, **12**, 338–43.

Whelpton, R. (1978) What happens when we freeze and thaw plasma? *Acta Pharm. Suec.*, **15**, 458.

Yeung, P.K.F., Hubbard, J.W., Korchinski, E.D. and Midha, K.K. (1987) Radioimmunoassay for the N-oxide metabolite of chlorpromazine in human plasma and its application to a pharmacokinetic study in healthy humans. *J. Pharm. Sci.*, **76**, 803–8.

PART TWO
Enzymology of *N*-Oxidation

4

Mechanism, multiple forms and substrate specificities of flavin-containing mono-oxygenases

D.M. Ziegler

Clayton Foundation Biochemical Institute and Department of Chemistry,
The University of Texas at Austin, Austin, TX 78712, USA

1. Multisubstrate flavin-containing mono-oxygenases (FMOs) with similar catalytic mechanisms have been isolated from both eukaryotes and prokaryotes.
2. These enzymes differ from all other mono-oxygenases requiring an external reductant in that an oxygenatable substrate is not required for formation of the enzyme-bound oxygenating intermediate.
3. Any soft nucleophile gaining access to this intermediate is oxidized and substrate specificity is largely determined by steric factors that limit access.
4. FMO's apparently discriminate between different nucleophiles by excluding non-substrates rather than selectively binding substrates.
5. Evidence collected with different nucleophilic amines suggest that differences in dimensions of enzyme substrate channels may be responsible for many of the differences in substrate specificities exhibited by tissue specific forms of FMO.

4.1 INTRODUCTION

Enzymes bearing flavin prosthetic groups are quite common in all eukaryotic and prokaryotic organisms. The tricyclic ring of isoalloxizine is an exceptionally versatile molecule that in different microenvironments carries out a variety of different biological redox functions. Most of these are well known and the present discussion will be restricted to a subgroup of flavin-dependent mono-oxygenases that catalyse the oxidation of an exceptionally broad range of substrates *via* a mechanism distinctly different from all other mono-oxygenases or oxidases.

In vertebrates these flavoproteins have apparently evolved to catalyse the detoxication of soft nucleophiles such as alkaloids and other nutritionally useless compounds so abundant in our plant foods (Liener, 1980). These flavoproteins are ideally adapted for this function and unique steps in their catalytic mechanism that permit such exceptionally loose specificity without seriously compromising catalytic efficiency will be the principal focus of this chapter.

However, before taking up this and a few other properties of these flavoproteins, some comments on nomenclature appear in order. Systematic nomenclature of multisubstrate enzymes with distinct but overlapping specificities is a recurring problem in the field of drug metabolism and the flavin-dependent drug oxygenases are no exception. It has become common practice to refer to enzymes similar in composition and mechanism to the extensively studied porcine liver microsomal enzyme as flavin-containing mono-oxygenases – often abbreviated simply as FMO. This designation, although useful, fails to address species and tissue variants and both should be specified for accuracy. The following discussions will follow the current practice of referring to this type of flavoprotein from vertebrates as FMO but, as necessary for clarity, the abbreviation will be preceded by the species and tissue.

4.2 MULTIPLE FORMS

Flavoprotein catalysing N-oxygenation of N,N-dimethylaniline and a variety of other amines have been isolated from eukaryotes and prokaryotes as indicated by representative examples listed in Table 4.1. The enzymes present in mammalian liver microsomes are quite similar in composition and catalytic properties although some qualitative species differences have been described (Ziegler, 1980, 1988a,b). Indirect evidence based on differences in thermal stability and optimal pH also suggest that some rodents may express at least two structurally different forms of FMO in liver (Dixit and Roche, 1984). Whether the observed differences in rodent liver are due to

Table 4.1 Multisubstrate flavin-containing mono-oxygenases purified to homogeneity

Species (tissue)	Reference
Porcine (liver)	Ziegler and Mitchell (1972)
Mouse (liver)	Sabourin *et al.* (1984)
Mouse (lung)	Tynes *et al.* (1985)
Rat (liver)	Kimura *et al.* (1983)
Rabbit (lung)	Williams *et al.* (1984a, b)
	Tynes *et al.* (1985)
Trypanosomi cruzi (cytosol)	Kuwahara *et al.* (1985)
Acinetobacter (cytosol)	Ryerson *et al.* (1982)
Beneckea harvieyi (cytosol)	Hastings and Balny (1975)

different gene products or to post-translation modification of a single protein are not known. Either possibility (or even localization in different phospholipid environments) could account for most, if not all, the apparent structural differences in rodent liver FMO.

In addition to liver, immunochemical analysis of microsomal proteins (Tynes and Philpot, 1987), suggests that an enzyme quite similar to the liver enzyme is also expressed in kidney and lung of all species examined except rabbit (Williams *et al.*, 1984b; Tynes and Philpot, 1987). Lung tissue from the latter species (and perhaps from sheep, Williams, personal communication) contain a form of FMO catalytically and immunologically distinct from the rabbit liver enzyme.

However, tissue-specific forms of FMO cannot be detected in all species. For instance, analysis of liver genomic DNA (Gassner *et al.*, 1990) suggests that in the pig, a single gene encodes for FMO and the same protein is also expressed in lung and kidney in this species. Thus it would appear that, at least in the pig, the loose substrate specificity of FMO in liver, lung and kidney is due to activity of a single protein and not to a family of flavoproteins almost identical in physical properties. Although the existence of different forms of FMO in other tissues cannot be ruled out, the evidence currently available suggests that in the major organs of entry in the pig only a single gene encodes for FMO. But as indicated above this is certainly not true for all mammalian species and expression of tissue-specific forms of FMO may well be species dependent.

In addition to major organs of entry, activities characteristic of FMO have been detected in virtually all nucleated mammalian cells examined (Ziegler, 1988a). Although amounts of enzyme estimated from activity or immunochemical measurements are usually much lower, activity in some tissues can reach levels present in liver. For example Heinze *et al.* (1970)

described a dramatic increase in *N,N*-dimethylaniline *N*-oxygenation in microsomes isolated from corpus lutea during the latter stages of the oestrus cycle in pigs. Fivefold induction of FMO in the placenta of mice has also been reported (Omitz and Kulkarni, 1982) and the concentration of FMO in rabbit lung increases at least fivefold and becomes the major microsomal protein in lung tissue from pregnant rabbits (Williams *et al.*, 1984a). The available evidence suggests that FMO is widely distributed in tissues from vertebrates but tissue distribution and presence of tissue-specific forms appears to vary with species.

Flavoproteins similar in catalytic properties to the mammalian enzymes have also been isolated from protozoa and bacteria and the more thoroughly characterized are listed in Table 4.1. The flavoprotein isolated from *T. cruzi* by Kuwakara *et al.* (1985) appears similar in composition and substrate specificity to liver FMO. The enzyme is apparently overexpressed in antihelmintic tolerant organisms and Agosin and Ankley (1987) suggest that the increased concentration of FMO may be related to drug resistance.

The prokaryotic mono-oxygenases listed in Table 4.1 possess the loose specificity characteristic of FMO, but each catalyses reactions not shared by the mammalian enzymes. For instance Baeyer-Villiger reactions carried out by cyclohexanone oxygenase (Ryerson *et al.*, 1982) cannot be demonstrated with the pig liver enzyme (unpublished observations, this laboratory). The oxygenation of long-chain aldehydes by luciferases from luminescent marine bacteria also yields a photon of blue–green light in addition to the acid, but emission of light by the mammalian enzymes has not been reported. The luciferases from marine organisms also differ from the other mono-oxygenases listed in Table 4.1 in composition and subunit structure. Active preparations of luciferase require separate reductases for electron transfer from NADH or NADPH to the flavin-dependent mono-oxygenase and the latter specifically requires FMN for activity (Hastings *et al.*, 1985).

The preceding summary indicates that flavoproteins similar in mechanism to the liver microsomal enzyme are widely distributed in nature. While species and tissue-specific forms exhibit distinct substrate specificities, each flavoprotein in this category catalyses NAD(P)H- and oxygen-dependent oxygenation of an exceptionally broad range of different compounds. Steps in the catalytic cycle, common to all these enzymes, responsible for such loose specificity are summarized in the following section.

4.3 MECHANISM

Unlike all other oxidases or mono-oxygenases, oxygenatable substrate is not required for flavin reduction or its reoxidation by molecular oxygen in FMO. This property distinguishes FMOs from all other flavin, haem, or other metalloprotein oxygenases. This difference is largely responsible for the exceptionally loose substrate specificities of this class of flavoproteins and the following discussion focuses on advantages and disadvantages of this unusual mechanism on reactions catalysed by these enzymes. Since most of the past studies on biological *N*-oxygenations have focused on mammalian systems, the mechanism of the porcine liver FMO will be used to illustrate the major steps in the catalytic cycle.

The studies on mechanism (Poulsen and Ziegler, 1979; Beaty and Ballou, 1981a,b) demonstrated unequivocally that FMO exists within the hepatocyte in the 4a-hydroperoxyflavin form. The catalytic cycle illustrated in Fig. 4.1, although similar to ones depicted previously (Ziegler, 1980, 1988), emphasizes this aspect by starting with the oxygenation of substrate as the first step in the cycle. Any nucleophile susceptible to oxidation by an organic peroxide is a potential substrate if it can gain access to the enzyme-bound flavin hydroperoxide.

Nitrogen compounds that penetrate to this oxidant apparently react via a nucleophilic displacement of the nitrogen atom on the terminal oxygen of the hydroperoxide. This, followed by heterolytic cleavage of the peroxide produces the oxygenated product and the flavin pseudobase. The nature of the products formed appear identical to those produced by chemical oxidation of the substrates with an organic peroxide. Without exception

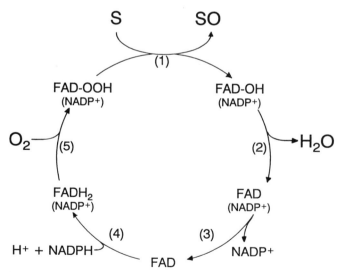

Figure 4.1 Major steps in the catalytic cycle of mammalian microsomal flavin-containing mono-oxygenases. The cycle is based on data taken from Poulsen and Ziegler (1979), Ballou (1981a, b) and Williams *et al.* (1984 a, b).

tertiary amines are oxidized to *N*-oxides, whereas primary and secondary amines are oxidized to oximes and nitrones respectively through intermediate hydroxylamines (see previous reviews for references to the original literature, Ziegler, 1980, 1988). The oxygenated product diffuses from the active site and apparently does not interact with any of the subsequent steps in the cycle. Steps 2–5 (Fig. 4.1) simply regenerate, with endogenous substrates (i.e. NADPH, H^+ and O_2), the oxygenating form of the enzyme from the flavin psuedobase.

There is no evidence that binding of substrate to the enzyme lowers the energy of activation. As a result, precise fit of substrate to the active site does not appear essential. As stated earlier, any compound readily oxidized by an organic hydroperoxide that can gain access to the flavin-hydroperoxide will be oxidized. It is this feature that is primarily responsible for the exceptionally loose substrate specificity of these flavoproteins.

On the other hand, lack of substrate activation to facilitate oxygen transfer from the hydroperoxide to substrate severely limits the types of compounds that can be oxygenated by FMO. Although substrates for FMO have essentially no structural features in common, all are soft nucleophiles (Ziegler, 1988), and only compounds bearing a polarizable, electron rich centre are potential substrates. However, the disadvantages due to lack of tight substrate binding are more than offset by the exceptionally loose specificity of FMO for such compounds. Furthermore, the nature of the enzyme-bound oxidant is ideally suited to the oxidation of soft nucleophiles. Pearson's principle (Pearson and Songstad, 1967; also Anders, 1988) states that soft nucleophiles preferentially react with soft electrophiles and the hydroperoxyflavin is a relatively soft electrophile. Thus it would appear that functional groups bearing nitrogen classified as soft nucleophiles would be oxidized primarily by FMO. Among different functional groups bearing nitrogen only amines, hydroxylamines and hydrazines would fall into this class and *N*-oxygenations catalysed by FMO are, for the most part restricted to these types of organic compounds bearing nitrogen.

However, not even all these are substrates since steric elements around the active site prevent entry of nucleophiles that resemble physiologically essential cellular metabolites. Liver FMO and other flavoproteins of this type appear to discriminate between essential and foreign nucleophiles by excluding the former rather than selectively binding the latter. These enzymes are unusual in that specificity is determined by *exclusion* rather than *binding* and differences in specificity among different forms of FMO must be due to variations in functional groups around the active site that prevent entry of potential substrates. Although essentially nothing is known about the tertiary structure of the flavoproteins from vertebrates, some empirically determined differences in specificities of the porcine liver and rabbit lung FMOs are summarized in the following section.

4.4 SPECIFICITY

While only amines and hydrazines are sufficiently nucleophilic to be readily oxidized by FMO, the number of xenobiotic amines and hydrazines that fit the substrate requirements is extensive and lists of both synthetic and naturally occurring xenobiotics oxygenated by liver FMO have been described in detail (Ziegler, 1980, 1985, 1988a,b; Ziegler *et al.*, 1980). The present discussion will, therefore, be limited to some unexpected anomalies and differences between the substrate specificities of two well-characterized mammalian FMOs – porcine liver and rabbit lung FMO.

The distinct substrate specificities of FMO from rabbit lung and liver, first detected with microsomal preparations (Ohmiya and Mehendale, 1982, 1983, 1984) have been confirmed and extended on studies with the purified flavoproteins (Williams *et al.*, 1984a; Tynes *et al.*, 1985; Poulsen *et al.*, 1986). Although the catalytic cycles of both the rabbit lung and porcine liver FMOs are similar, some striking differences in the *N*-oxidation of amines have been described. For instance, in all species tested, hepatic FMO will not catalyse the oxidation of primary alkylamines, whereas the lung flavoprotein readily catalyses their oxidation. Although both enzymes share common substrates in *N,N*-dimethylaniline and trifluoperazine, the pulmonary FMO does not catalyse the oxidation of many of the better tertiary amine substrates for the liver enzyme (Williams *et al.*, 1984a; Poulsen *et al.*, 1986).

These differences are undoubtedly due to subtle differences in structural elements that control access of nucleophiles into the active sites of the respective enzymes. While as mentioned earlier nothing is known about the tertiary structure of these enzymes, elements on substrates known to prevent activity with either or both are summarized in Table 4.2.

Despite a mechanism designed for loose specificity these enzymes must possess some methods for discriminating between essential and xenobiotic nucleophiles. The number and position of ionic groups appears to be the principal method used to discriminate among different nucleophiles as suggested from the representative examples listed in Table 4.2. Almost without exception, essential metabolites bearing nucleophilic sulphur, nitrogen, or selenium atoms contain one or more charged groups that prevent entry of the metabolite into the active site. However, differences in charge are not the only consideration since unlike lung FMO, the liver enzyme has some mechanism that permits it to exclude lipophilic primary alkylamines. This is apparently related to the fact that liver FMO will not accept nucleophiles bearing two hydrogen atoms covalently linked to the heteroatom. How the hepatic FMO can discriminate between amine groups bearing one or two hydrogen atoms is not known.

A recent report (Nagata *et al.*, 1990) suggests that many of the other differences in specificities between rabbit lung and porcine liver FMO can be

Table 4.2 Steric factors that exclude nucleophilic amines from liver and lung FMO: similarities and differences

Parameter	Examples	Lung	Liver
Charge			
Two or more cationic groups	Putrescine	NA[a]	NA
	Cadaverine	NA	NA
One or more anionic groups	Lysine	NA	NA
	Taurine	NA	NA
No. of hydrogen atoms covalently linked to nitrogen			
None	N,N-Dimethylaniline	+	+
	N,N-Dimethyloctylamine	+	+
One	N-Methyloctylamine	+	+
	Octylhydroxylamine	+	+
Two	Octylamine	+	NA
	Dodecylamine	+	NA
Position of large substituents relative to the amine group			
Within 0.3–0.4 nm	Benzphetamine	NA	+
	Imipramine	NA	+
	Chlorpromazine	NA	+
	Chlorcyclizine	NA	+
More distal than 0.6 nm	Prochlorperazine	+	+

[a] NA: No detectable activity.
Summary based on data from Williams *et al.* (1984), Tynes *et al.* (1985), Poulsen *et al.* (1986).

attributed to differences in the dimensions of their respective substrate channels. Data collected with tricyclics bearing alkylamine side chains varying in length from C_2 to C_7 suggest that compounds bearing substituents much larger than a six membered ring within 0.3–0.4 nm of the terminal amine group are completely excluded from the enzyme-bound oxidant in the rabbit lung enzyme. The channel leading the 4a-flavin hydroperoxide in hepatic FMO is considerably more open and this apparent structural difference appears responsible for many of the distinct substrate specificities of these enzymes.

4.5 INHIBITORS

At present there are no selective mechanism-based inhibitors available for either liver or lung FMO. Approaches generally used to design selective enzyme inhibitors have been unsuccessful with FMO for reasons that are apparent from the mechanism. First, substrate is oxidized by an ionic rather than a radical mechanism. Therefore formation of reactive intermediate substrate radicals does not occur, and methods used for designing mechanism based inhibitors for flavoproteins that catalyse the oxidation of amines by sequential one electron oxidations (Silverman and Yamasaki, 1984) cannot be used. Second, xenobiotic substrates for FMO apparently require only a single point of contact with the enzyme and do not form an extended, tightly bound transition state complex. FMO is also remarkably immune to inactivation by enzyme-generated electrophiles (Taylor and Ziegler, 1987) and selective inactivation of the enzyme by reactive inter-mediates generated from the oxidation of dithio acids and thiocarbamides cannot be detected.

The lack of specific inhibitors or other means of manipulating activity has severely hampered studies on the role of FMO in the metabolism of amines *in vivo*. While the oxidation of amine drugs can be inhibited by sulphur-containing substrates for FMO and methimazole has been used for this purpose, this can lead to results that are difficult to interpret. Methimazole (and a host of related organic sulphur compounds) are oxidized to reactive intermediates that lead to the loss of GSH and oxidation of protein thiols (Poulsen *et al.*, 1979; Krieter *et al.*, 1984). Methimazole-dependent damage to the cellular metabolic machinery may well affect the metabolism of amines but the effects may be indirect.

Alternate substrates that are relatively non-toxic and specific for FMO are few in number. Of the few potential candidates dimethylsulphoxide (DMSO) appears to be one of the more promising. DMSO is a known substrate for FMO, although the K_m is around 3–4 mM. The rather high concentrations required to block the oxidation of other substrates is an obvious disadvantage although animals are relatively tolerant of high doses of DMSO. In addition DMSO does not appear to affect the oxidation of amines known to be catalysed by P-450-dependent mono-oxygenases or other oxidases present in hepatic microsomes. Preliminary studies carried out by Kaderlik and Poulsen indicate that the ratio of trimethylamine oxide to trimethylamine decreases dramatically in urine from rats injected with DMSO. DMSO also blocks methimazole-dependent oxidation of liver GSH in live hamsters which again indicates that DMSO may be a useful alternate substrate for inhibiting the oxidation of amines by FMO *in vivo*. Methods for manipulating activity of FMO in live animals would greatly facilitate studies on the role of FMO in the metabolism of amine drugs *in vivo*. This remains one of the major unsolved problems and could, therefore, become a productive area for future work.

REFERENCES

Agosin, M. and Ankley, G.T. (1987) Conversion of *N,N*-dimethylaniline to *N,N*-dimethylaniline-*N*-oxide by a cytosolic flavin-containing enzyme from *Trypanosoma cruzi. Drug Metab. Dispos.* **15**, 200–3.

Anders, M.W. (1988) Bioactivation mechanisms and hepatocellular damage, in *The Liver: Biology and Pathobiology*, 2nd edn (eds J.M. Arias, H. Popper, W.B. Jakoby, D.S. Schachter and D.A. Shafritz) Raven Press, New York, pp. 389–400.

Beaty, N.S. and Ballou, D.P. (1981a) The reductive half-reaction of liver microsomal FAD-containing monooxygenase. *J. Biol. Chem.*, **256**, 4611–18.

Beaty, N.S. and Ballou, D.P. (1981b) The oxidative half-reaction of liver microsomal FAD-containing monooxygenase. *J. Biol. Chem.*, **256**, 4619–25.

Dixit, A. and Roche, T.E. (1984) Spectrophotometric assay of the flavin-containing monooxygenase and changes in activity in female mouse liver with nutritional and diurnal conditions. *Arch. Biochem. Biophys.*, **233**, 50–63.

Gassner, R., Tynes, R.E., Lawton, M.P. *et al.* (1990) The flavin-containing monooxygenase expressed in pig liver: primary sequence, distribution and evidence for a single gene. *Biochemistry*, **29**, 119–24.

Hasting, J.W. and Balny, C. (1975) The oxygenated bacterial luciferase–flavin intermediate: reaction products via the light and dark pathways. *J. Biol. Chem.*, **250**, 7288–93.

Hastings, J.W., Potrikus, C.J., Gupta, S.C. *et al.* (1985) Biochemistry and physiology of bioluminescent bacteria. *Adv. Microb. Physiol.*, **26**, 235–91.

Heinze, E., Hlavica, P., Kiese, M. and Lipowsky, G. (1970) *N*-oxygenation of arylamines in microsomes prepared from corpora lutea of the cycle and other tissues of the pig. *Biochem. Pharmacol. Ther.*, **15**, 32–8.

Kimura, T., Kodama, M. and Nagata, C. (1983) Purification of mixed-function amine oxidase from rat liver microsomes. *Biochem. Biophys. Res. Commun.*, **110**, 640–5.

Kuwakahara, T., White, R.A. Jr and Agosin, M. (1985) A cytosolic FAD-containing enzyme catalyzing cytochrome *c* reduction in *Trypanosoma cruzi*. I. Purification and properties. *Arch. Biochem. Biophys.*, **239**, 18–25.

Krieter, P.A., Ziegler, D.M., Hill, K.A. and Burk, R.F. (1984) Increased biliary GSSG efflux from rat livers perfused with thiocarbamide substrates for the flavin-containing monooxygenase. *Mol. Pharmacol.*, **26**, 122–7.

Liener, J.E. (1980) *Toxic Constituents of Plant Foodstuff*, 2nd edn, Academic Press, New York.

Nagata, T., Williams, D.E. and Ziegler, D.M. (1990) Substrate specificities of rabbit lung and porcine liver flavin-containing monooxygenases: differences due to substrate size, *Chem. Res. Toxicol.*, **3**, 372–6.

Ohmiya, Y. and Mehendale, H.M. (1982) Metabolism of chlorpromazine by pulmonary microsomal enzymes in the rat and rabbit. *Biochem. Pharmacol.*, **31**, 157–62.

Ohmiya, Y. and Mehendale, H.M. (1983) *N*-Oxidation of *N,N*-dimethylaniline in the rabbit and rat lung. *Biochem. Pharmacol.*, **32**, 1281–5.

Ohmiya, Y. and Mehendale, H.M. (1984) Effect of mercury on accumulation and metabolism of chlorpromazine and imipramine in rat lungs. *Drug Metab. Dispos.*, **12**, 376–8.

Omitz, T.G. and Kulkarni, A.P. (1982) Oxidative metabolism of xenobiotics during pregnancy: significance of microsomal flavin-containing monooxygenase. *Biochem. Biophys. Res. Commun.*, **109**, 1164–71.

Pearson, R.G. and Songstad, J. (1967) Applications of the principles of hard and soft acids and bases to organic chemistry. *J. Am. Chem. Soc.*, **89**, 1827–36.

Poulsen, L.L., Hyslop, R.M. and Ziegler, D.M. (1979) *S*-oxygenation of *N*-substituted thioureas catalyzed by the liver microsomal FAD-containing monooxygenase. *Arch. Biochem. Biophys.*, **198**, 78–98.

Poulsen, L.L., Taylor, K., Williams, D.E. *et al.* (1986) Substrate specificity of the rabbit lung flavin-containing monooxygenase for amines: oxidation products of primary alkylamines. *Mol. Pharmacol.*, **30**, 680–5.

Poulsen, L.L. and Ziegler, D.M. (1979) The liver microsomal FAD-containing monooxygenase: spectral characterization and kinetic studies. *J. Biol. Chem.*, **254**, 6449–55.

Ryerson, C.C., Ballou, D.P. and Walsh, C.Z. (1982) Mechanistic studies on cyclohexanone oxygenase. *Biochemistry*, **21**, 2644–55.

Sabourin, P.J., Smyser, B.P. and Hodgson, E. (1984) Purification of the flavin-containing monooxygenase from mouse and pig liver microsomes. *Int. J. Biochem.*, **16**, 713–20.

Silverman, R.B. and Yamasaki, R.B. (1984) Mechanism-based inactivation of mitochondrial monoamine oxidase by *N*-(1-methylcyclopropyl) benzylamine. *Biochemistry*, **23**, 1322–38.

Taylor, K.L. and Ziegler, D.M. (1987) Studies on substrate specificity of the hog liver flavin-containing monooxygenase: anionic organic sulfur compounds. *Biochem. Pharmacol.*, **36**, 141–6.

Tynes, R.E. and Philpot, R.M. (1987) Tissue- and species-dependent expression of multiple forms of mammalian microsomal flavin-containing monooxygenase. *Mol. Pharmacol.*, **31**, 569–74.

Tynes, R.E., Sabourin, P.J. and Hodgson, E. (1985) Identification of distinct hepatic and pulmonary forms of microsomal flavin-containing monooxygenase in the mouse and rabbit. *Biochem. Biophy. Res. Commun.*, **126**, 1069–75.

Williams, D.E., Hale, S.E., Meurhoff, A.S. and Masters, B.S.S. (1984a) Rabbit lung flavin-containing monooxygenase: purification, characterization and induction during pregnancy. *Mol. Pharmacol.*, **28**, 381–90.

Williams, D.E., Ziegler, D.M., Nordin, D.J. *et al.* (1984b) Rabbit lung flavin-containing monooxygenase is immunochemically and catalytically distinct from the liver enzyme. *Biochem. Biophys. Res. Commun.*, **125**, 116–22.

Ziegler, D.M. (1980) Microsomal flavin-containing monooxygenase: oxygenation of nucleophilic nitrogen and sulfur compounds in *Enzymatic Basis of Detoxication*, Vol. 1 (ed. W.B. Jakoby), Academic Press, New York, pp. 201–25.

Ziegler, D.M. (1985) Molecular basis for *N*-oxygenation of *sec*- and *tert*-amines, in *Biological Oxidation of Nitrogen in Organic Molecules* (eds J.W. Gorrod and L.A. Damani), Ellis Horwood, Chichester, pp. 43–52.

Ziegler, D.M. (1988a) Flavin-containing monooxygenases: catalytic mechanism and substrate specificities. *Drug Metab. Rev.*, **19**, 1–32.

Ziegler, D.M. (1988b) Functional groups activated via flavin-containing mono-oxygenases, in *Microsomes and Drug Oxidations* (eds J. Miners, D.J. Birkett, R. Drews and M. McManus), Taylor and Francis, London, New York and Philadelphia, pp. 297–304.

Ziegler, D.M. and Mitchell, C.H. (1972) Microsomal oxidase. IV. Properties of a mixed-function amine oxidase isolated from pig liver microsomes. *Arch. Biochem. Biophys.*, **150**, 116–25.

Ziegler, D.M., Poulsen, L.L. and Duffel, M.W. (1980) Kinetic studies on ·mechanism and substrate specificity of the microsomal flavin-containing monooxygenase, in *Microsomes, Drug Oxidations and Chemical Carcinogenesis* (eds M.J. Coon, A.H. Conney, R.W. Estabrook, H.V. Gelboin, J.R. Gillette and P.J. O'Brien), Academic Press, New York, pp. 637–46.

5

On the genetic polymorphism of the flavin-containing mono-oxygenase

P. Hlavica and I. Golly

Walther-Straub-Institut für Pharmakologie und Toxikologie der Universität,
Nussbaumstrasse 26, D-8000 München 2, FRG

1. All available evidence from kinetic and structural studies suggests that the flavin-containing mono-oxygenase represents a family of polymorphically expressed isoenzymes characterized by distinct substrate specificity.
2. In view of this, attention is drawn to the question of how the qualitative as well as quantitative variances in the individual FMO forms will modulate metabolism of drugs and other foreign compounds, including mutagens and carcinogens.

5.1 INTRODUCTION

The mammalian flavin-containing mono-oxygenase (FMO; EC 1.14.13.8) provides a significant route for the NADPH- and oxygen-dependent biotransformation of a wide variety of nitrogen- and sulphur-containing compounds that possess a nucleophilic heteroatom. These include secondary and tertiary aliphatic amines, arylamines, hydrazines, sulphides, thiols, thioamides and thioureas. Depending on the nature of the parent compound and the reaction conditions employed, oxidative attack on nucleophilic centres within these structures can give rise to multiple end products. While FMO activity can be detected in virtually all nucleated cells, the properties of this enzyme in various tissues from different species were largely inferred from the known characteristics of the purified porcine liver mono-oxygenase. The subject has been comprehensively reviewed (Ziegler, 1980, 1988).

It was originally believed that substrate turnover is catalysed by basically the same type of FMO in tissues from all vertebrates (Ziegler, 1980). However, studies with highly purified FMO preparations, which are now available from tissues of mice, rats, guinea pigs and rabbits (Kimura *et al.*, 1983; Tynes *et al.*, 1986; Brodfuehrer and Zannoni, 1987), suggest that there exist tissue and species differences in this enzyme system. Moreover, both liver and lung microsomal fractions have been shown to contain multiple forms of the hepatic and pulmonary FMO (Tynes and Philpot, 1987; Ozols, 1989). In addition, protozoal (Agosin and Ankley, 1987) and bacterial (Hofsteenge *et al.*, 1983) flavin-containing oxygenases have been isolated, which may be considered to represent analogues of the mammalian enzymes characterized by more restricted substrate specificity. It is thus concluded that the FMO represents a family of polymorphically expressed variants. This chapter focuses primarily on the description of factors that serve to substantiate this hypothesis.

5.2 FACTORS PROVIDING INDIRECT EVIDENCE OF THE EXISTENCE OF MULTIPLE FMO FORMS

5.2.1 Differences in the response of FMO activity to steroidal hormones

Devereux and Fouts (1975) reported that FMO activity in lung microsomes from female Dutch Belt rabbits increases about two-fold during late pregnancy, peak activity being observable on day 20 of gestation. However, differences in FMO activity were not seen in corresponding hepatic microsomal preparations at any stage of pregnancy. Using female New

Zealand White rabbits, Williams *et al.* (1985) have furnished evidence by activity measurements and immunoquantitation that there is a parallel induction of FMO activity and enzyme content in lung microsomes on day 28 of gestation, whereas the level of liver FMO is not significantly altered. Although the precise nature of the inducer remains unclear, deoxy-corticosterone, which is generated from progesterone and increases in plasma during late pregnancy, might account for the observed effect, since treatment of non-pregnant rabbits with the mineralocorticoid mimics the effect of pregnancy on the pulmonary and hepatic FMO (Devereux and Fouts, 1975). In contrast to the rabbit, FMO-associated activity does not increase during gestation in the lung of female CD-1 mice, but induction occurs in placental microsomes from this species (Osimitz and Kulkarni, 1982).

Testosterone-treatment of female CF-1 mice depresses specific FMO activity in the liver without affecting this parameter in kidney tissue (Duffel *et al.*, 1981). Administration of dexamethasone to male Sprague–Dawley rats decreases the level of hepatic FMO by 98%, as determined by immunochemical techniques, but does not produce a change in the specific FMO content in the livers of female rats (Williams *et al.*, 1989). These observations are suggestive of a species-, sex- and tissue-specific induction or repression of selective enzyme forms by steroidal hormones.

5.2.2 Differences in substrate turnover

(a) *Differences in substrate specificity*

Although, in general, the lung and liver FMOs catalyse oxidation of the same classes of compounds, there are some noticeable differences. Thus, chlorpromazine and imipramine were reported to be readily *N*-oxidized by pulmonary microsomes from rats but not from rabbits (Ohmiya and Mehendale, 1981, 1982). The lack of activity towards these substrates is not due to the absence of enzyme in the latter tissue preparations, since immunochemical staining for FMO in pulmonary microsomes from male and female rabbits reveals the presence of significant amounts of the flavoprotein (Dannan and Guengerich, 1982). Moreover, lung microsomes from the rabbit rapidly *N*-oxidize other substrates of the FMO system (Ohmiya and Mehendale, 1983). Hence, the failure of the rabbit lung FMO to attack nitrogen centres in the chlorpromazine and imipramine molecule appears to reflect an inherent property of this enzyme form, as is also documented by experiments with the highly purified pulmonary mono-oxygenase (Williams *et al.*, 1984). In contrast, the isolated pig liver enzyme readily metabolizes both amines. The latter enzyme species also *N*-oxidizes the pyrrolizidine alkaloid senecionine, whereas this compound does not appear to serve as a substrate for the purified rabbit lung oxidase (Williams *et al.*, 1989). The mouse and rabbit lung FMO enzymes, but not the liver enzymes, efficiently utilize certain primary alkylamines (Tynes *et al.*, 1985, 1986; Poulsen *et al.*, 1986) which prove to be poor substrates for the hog liver mono-oxygenase (Ziegler, 1988). The reverse applies to the metabolic transformation of some primary arylamines (Ziegler, 1988). Finally, liver microsomes from the rat but not from the rabbit catalyse *N*-oxide formation from the tertiary amine tamoxifen (Ruenitz *et al.*, 1984). These observations suggest the species- and tissue-dependent existence of catalytically distinct variants of the FMO.

Apart from these qualitative differences, kinetic analysis of the purified and membrane-bound FMO from the liver and lung of various animal species shows these enzymes to be characterized by widely divergent substrate affinities. Table 5.1 summarizes the apparent K_m values for interaction of a series of representative compounds. Inspection of the data for *N*,*N*-dimethylaniline, which has been frequently used as a marker for FMO-dependent activity, reveals that the tertiary amine has a very low K_m value for the porcine hepatic flavoprotein, whereas the hepatic forms from other species, such as the rabbit, rat and mouse, exhibit 8- to 10-fold lower reactivity towards this compound. There is also considerable discrepancy in the affinity of the rabbit liver and lung enzyme for this dialkylarylamine. These findings lend support to the notion of the occurrence of species- and tissue-specific subforms of the FMO, as is also suggested by comparison of the kinetic data for other substrates listed in Table 5.1.

Although data on oxygen affinity of the enzyme system are scarce, all evidence available hints at species-dependent differences in the reactivity of the FMO towards O_2. Thus, at pH 8.4, the apparent K_m values for oxygen of the hepatic enzymes from mice, rabbits and pigs were determined to be 40 μM, 158 μM and 250 μM, respectively (Hlavica and Kehl, 1974; Poulsen and Ziegler, 1979; Tynes *et al.*, 1986). The rabbit pulmonary form has a K_m value of 100 μM (Tynes *et al.*, 1986). Most interestingly, at physiological pH the membrane-bound rabbit liver mono-oxygenase exhibits sigmoidal kinetics of oxygen saturation during *N*-oxidation of *N*,*N*-dimethylaniline, suggesting cooperativity in O_2 binding (Hlavica, 1972; Hlavica and Kehl, 1974). This is not observed with the porcine hepatic enzyme (Poulsen and Ziegler, 1979).

(b) Differences in stereoselective substrate metabolism

Recently, attention has been drawn to stereoselective phenomena in the biotransformation of certain amine substrates. Thus, hog liver microsomes *N*-oxygenate the *Z*-isomers of zimeldine and homozimeldine in marked preference to the *E*-isomers ($E/Z = 0.4$), whereas rat liver microsomes

Table 5.1 K_m Values for interaction of various substrates with the FMO

| | Mouse | | Rat | | Hamster | Rabbit | | Pig | T. cruzi | |
Substrate	Liver	Lung	Liver	Lung	liver	Liver	Lung	liver	cytosol	Reference
										K_m (μM)
n-Octylamine		1900					12000			Tynes et al. (1986), Poulsen et al. (1986),
N-Methyloctylamine							120	400		Poulsen et al. (1986), Ziegler (1988)
Trimethylamine	2340							617		Sabourin and Hodgson (1984)
Chlorpromazine				41[a]				9		Ohmiya and Mehendale (1982), Ziegler (1988)
Benzphetamine	1670							74		Sabourin and Hodgson (1984)
N-Methylaniline	1060					1200[a]	8000[a]	343		Uehleke (1973), Sabourin and Hodgson (1984)
N-Benzylaniline	2030[a]		1940[a]		5620[a]	420[a]			56	Gorrod and Gooderham (1987)
N,N-Dimethylaniline	105		93[a]	99[a]		89	330	11		Willi and Bickel (1973), Hlavica and Kehl, (1977), Ohmiya and Mehendale (1983), Sabourin and Hodgson (1984), Poulsen et al. (1986), Agosin and Ankley (1987)
Perazine						200[a]	1500[a]	8000		Uehleke (1973), Ziegler (1988)
Ethylmorphine	1650							284		Sabourin and Hodgson (1984)

[a] Kinetic data for membrane-bound FMO.

metabolize the E-isomers to a greater extent than the Z-isomers ($E/Z = 2$) (Cashman *et al.*, 1988). The purified FMO from rat and hog liver gives an S/R ratio for N-oxygenation of (R)- and (S)-verapamil of 6.6 and 10.1, respectively (Cashman, 1989). Finally, the *trans/cis* ratio for N-oxide formation from (S)-($-$)-nicotine, as measured with guinea-pig liver microsomes and the porcine liver FMO, has a value of 0.27 and 1.0, respectively. (R)-($+$)-nicotine gives more *trans*- than *cis*-N-oxide in guinea-pig liver, whereas no *cis*-product can be detected with the hog liver enzyme (Damani *et al.*, 1988). In contrast, rabbit liver preparations produce more *cis*-N-oxide (Jenner *et al.*, 1973). These observations favour the concept of the existence of species-dependent structural specificities in the FMO active site, which may prohibit binding of certain conformers of a given substrate through steric hindrance.

5.2.3 Differences in pH optima for enzyme activity

At saturating NADPH and O_2 concentration, oxygenations catalysed by the hog liver microsomal FMO exhibit a distinct pH optimum at 8.3–8.4 (Ziegler, 1980). Similar pH activity optima were observed with the hepatic microsomal enzymes from rats (Stefek *et al.*, 1989), hamsters (Gorrod and Gooderham, 1987), guinea pigs (Gorrod and Patterson, 1983) and rabbits (Hlavica, 1970; Tynes and Hodgson, 1985; Tynes *et al.*, 1985). The mouse liver mono-oxygenase has an activity peak at pH 8.8–9.2, which is slightly higher than that of the pig FMO (Sabourin and Hodgson, 1984; Tynes and Hodgson, 1985; Tynes *et al.*, 1985). There is, however, a drastic change in the response to H^+ concentration of the pulmonary FMO from mice and rabbits, in that these enzyme species display activity optima around pH 10 (Tynes and Hodgson, 1985; Tynes *et al.*, 1985; Williams *et al.*, 1985). On the other hand, the rat lung FMO has a pH optimum at 7–8 (Ohmiya and Mehendale, 1982). This behaviour reflects pronounced differences in alkaline stability of the individual enzyme forms, most likely resulting from variations in protein conformation.

5.2.4 Differences in the response to activators

Oxygenations catalysed by the porcine liver FMO have been recognized to be stimulated by the presence of alkyl guanidines or primary alkylamines, such as n-octylamine or DPEA (Ziegler *et al.*, 1973). This has been attributed to interaction of the lipophilic compounds with a regulatory site on the oxidase (Ziegler *et al.*, 1971) to enhance the rate of breakdown of the flavin pseudobase; this reaction is considered to constitute the rate-limiting step in the catalytic cycle (Beaty and Ballou, 1981). Stimulation by n-octylamine has also been observed with the hepatic enzymes from hamsters, guinea pigs and humans (Poulsen *et al.*, 1974; McManus *et al.*, 1987; Stefek *et al.*, 1989), whereas the liver FMO from mice, rats and rabbits is relatively refractory towards the stimulatory effect of n-octylamine (Poulsen *et al.*, 1974; Hlavica and Kehl, 1977; Dixit and Roche, 1984; Sabourin and Hodgson, 1984; Tynes and Hodgson, 1985; Stefek *et al.*, 1989). Similarly, the rabbit lung enzyme is not prone to activation by the primary alkylamine (Williams *et al.*, 1985). The aberrant sensitivity towards n-octylamine has been tentatively interpreted to mean that the rate-limiting step in substrate turnover varies with certain prototypes of the FMO (Ziegler, 1988). Moreover, primary alkylamines serve as substrates for the rabbit lung enzyme (Poulsen *et al.*, 1986) and thus would be expected to inhibit rather than to stimulate oxygenation of other compounds by this FMO species. The latter explanation is in contrast to results obtained with the lung enzyme from mice. This FMO form, too, metabolizes a series of primary alkylamines but shows allosteric activation by n-octylamine (Tynes *et al.*, 1986). It has been proposed that the different mono-oxygenases probably have a conserved alkylamine-binding site of defined specificity (Tynes *et al.*, 1986). With respect to this it seems noteworthy that n-octylamine, although not hydroxylated, enhances enzyme activity in the rat lung (Ohmiya and Mehendale, 1982).

5.2.5 Differences in the response to divalent metal ions, chelators and detergents

The response of the FMOs to divalent metal ions has been used to further characterize the individual variants of this enzyme system. Devereux and Fouts (1974) reported stimulation of the rabbit lung mono-oxygenase by 100 μM-Hg^{2+}, but depression of enzyme activity in the rabbit liver. Stimulation of the pulmonary form is dependent on substrate concentration (Devereux *et al.*, 1977). Mercurials, such as *p*-chloromercuribenzoate and phenylmercuriacetate, affect *N*-oxidase activity in the two rabbit tissues analogous to inorganic mercury (Hlavica and Kiese, 1969; Devereux and Fouts, 1974). Most interestingly, a reverse situation is met in rat tissues, where 100 μM-Hg^{2+} inhibits pulmonary enzyme activity, but stimulates *N*-oxygenation in the liver (Arrhenius, 1969; Ohmiya and Mehendale, 1982). Similarly, *p*-chloromercuribenzoate activates rat liver FMO, but inhibits the hepatic enzyme from guinea pigs (Beckett *et al.*, 1971; Willi and Bickel, 1973).

Other SH-blocking agents, such as *N*-ethylmaleimide or iodoacetamide, do not affect the liver and lung FMO from rabbits (Devereux and Fouts, 1974), but enhance *N*-oxidase activity in hepatic microsomes from rats (Arrhenius, 1969; Willi and Bickel, 1973). However, catalytic capacity of the porcine liver mono-oxygenase is diminished (Heinze *et al.*, 1970).

High concentrations of Mg^{2+} decrease enzyme activity in the livers of rats (Nakazawa, 1970; Gigon and Bickel, 1971), rabbits (Devereux and Fouts, 1974) and pigs (Gigon and Bickel, 1971), but stimulate turnover in the rabbit lung (Devereux and Fouts, 1974). In contrast, the rat pulmonary enzyme is blocked (Ohmiya and Mehendale, 1982).

Chelators, such as EDTA, increase *N*-oxidase activity in rat liver microsomes, but decrease it in hog liver microsomal preparations (Gigon and Bickel, 1971).

Of special interest is the influence of anionic detergents on the catalytic properties of the FMO. Low concentrations of bile salts, such as cholate, deoxycholate or glycocholate, have been reported to enhance metabolic rates with both the hepatic enzyme from mice (Dixit and Roche, 1984) and rats (Ziegler and Pettit, 1964; Arrhenius, 1969; Willi and Bickel, 1973) and the pulmonary form from rabbits (Williams *et al.*, 1985). On the other hand, these agents perturb activity of the rabbit liver (Gorrod *et al.*, 1975) and pig liver (Williams *et al.*, 1985) oxidase.

5.2.6 Differences in thermostability of the various FMO forms

The porcine liver FMO has been reported to exhibit unusual thermal lability above 30°C in the absence of $NADP^+$ or NADPH (Ziegler, 1980). When exposed to 45°C, about 90% of activity is lost during a time period of 2 min (Williams *et al.*, 1985). Thermal treatment between 30° and 40°C has been demonstrated to be associated with irreversible changes in protein secondary structure (Kitchell *et al.*, 1978). High sensitivity to heat is also shared by hepatic microsomal FMO forms from mice (Dixit and Roche, 1984; Rouer *et al.*, 1987), rats (Cashman *et al.*, 1988; Cashman, 1989) and humans (McManus *et al.*, 1987).

However, extreme thermolability is not an intrinsic property of all FMO species. Hlavica and Kehl (1977) have shown that more than 90% of activity is recovered when the partially purified rabbit liver enzyme is preincubated at 37°C in the absence of NADPH for 10 min. Similarly, the rabbit lung FMO is stable at 45°C in the absence of cofactor for at least 10 min (Williams *et al.*, 1985). Such differences in thermostability of the individual FMO forms suggest differences in enzyme structure responsible for variances in backbone unfolding.

5.2.7 Differences in immunochemical behaviour

Ouchterlony double-diffusion analysis, using polyclonal antibodies directed against the liver and lung FMO from various animal species, reveals that the hepatic and pulmonary mono-oxygenases from rabbits represent immuno-chemically distinct proteins (Williams *et al.*, 1984, 1985; Tynes *et al.*, 1985). This also applies to the mouse liver and lung enzymes (Tynes *et al.*, 1985). It also would appear that there exist species-specific variants of the hepatic flavoprotein, in that anti-mouse FMO sera do not recognize the purified hepatic enzyme from rabbits and pigs (Tynes *et al.*, 1985), and antibodies raised to the porcine liver oxidase fail to cross-react with the mouse liver FMO (Sabourin and Hodgson, 1984). Immunochemical differences have also been detected between the hog liver enzyme and mono-oxygenase isolated from rabbit lung (Williams *et al.*, 1984). Western-blot analysis reveals only faint precipitin bands, when antibodies to the rabbit pulmonary enzyme are allowed to react with renal microsomal preparations from mice (Tynes and Philpot, 1987). Analogous experiments with microsomal fractions from various tissues of mice, rats, rabbits, dogs and humans demonstrate that not all antigenic determinants of the hog liver FMO are present in the enzymes found in the animal species tested (Dannan and Guengerich, 1982).

5.3 FACTORS PROVIDING DIRECT EVIDENCE OF THE EXISTENCE OF MULTIPLE FMO FORMS

5.3.1 Differences in physical properties of the various FMO species

(a) Differences in subunit size

The FMO has been purified to apparent homogeneity from various tissues of mice (Sabourin *et al.*, 1984; Tynes *et al.*, 1986), rats (Kimura *et al.*, 1983; Rouer *et al.*, 1988), guinea pigs (Brodfuehrer and Zannoni, 1987), rabbits (Hlavica and Hülsmann, 1979; Williams *et al.*, 1985; Ozols, 1989) and pigs (Poulsen and Ziegler, 1979; Sabourin *et al.*, 1984) and subjected to sodium dodecyl sulphate–polyacrylamide gel electrophoresis (SDS-PAGE) analysis. Figure 5.1 shows the electrophoretic profiles of a series of representative enzyme forms. The apparent subunit molecular masses, as calculated from electrophoretic mobilities, are summarized in Table 5.2. As can be seen, there exist marked species and tissue differences in subunit size. It is noteworthy that Western blotting of microsomal preparations from various animal species and tissues reveals immunoreactive bands at M_r

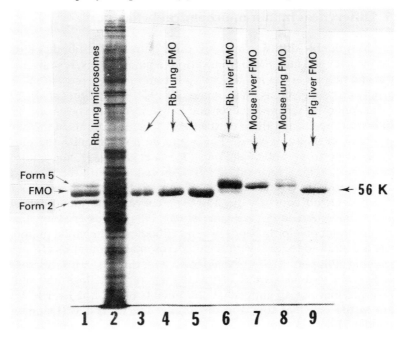

Figure 5.1 SDS-PAGE analysis of flavin-containing mono-oxygenases purified from various animal species and tissues. For comparison, the electrophoretic profiles of cytochrome *P*-450 isoenzymes 2 and 5 from rabbit (Rb) liver are shown in lane 1 (Tynes *et al.*, 1986, with permission).

55 to 59 kDa (Dannan and Guengerich, 1982; Tynes and Philpot, 1987), indicating that the highly purified FMO isoenzymes migrate similarly to the nascent proteins.

Table 5.2 also includes data for fungal, protozoal and bacterial flavin-containing mono-oxygenases belonging to the EC 1.14.13 enzyme family. These FMOs may be considered analogues of the eukaryotic forms characterized by more restricted substrate specificity. Thus, 4-amino-benzoate hydroxylase from the mushroom *Agaricus bisporus* oxygenates a series of aniline derivatives (Tsuji *et al.*, 1986). It should be noted that mammalian FMO also metabolizes a selected number of primary aromatic amines (Ziegler, 1988). A cytosolic flavoprotein from *Trypanosoma cruzi* attacks certain secondary and tertiary arylamines as well as sulphur-containing compounds, but does not hydroxylate aniline (Kuwahara *et al.*, 1985; Agosin and Ankley, 1987). The non-substrate 6-hydroxy-nicotinate, a heteroaromatic amine, acts as an allosteric effector of *p*-hydroxybenzoate hydroxylase from *Pseudomonas fluorescens* (Howell

Table 5.2 Apparent subunit molecular masses of FMO species from various sources

Source of enzyme	Apparent M_r (kDa)[a]	Reference
Mouse liver	56; 58	Sabourin *et al.*, (1984)
Mouse lung	59	Tynes *et al.* (1986)
Rat liver		
Normal	50; 59	Kimura *et al.* (1983)
		Rouer *et al.* (1988)
Diabetic	49	Rouer *et al.* (1988)
Guinea-pig liver	56	Brodfuehrer and Zannoni (1987)
Rabbit liver	59	Hlavica and Hülsmann (1979)
	58; 58.5	Ozols (1989)
Rabbit lung	59	Williams *et al.* (1985)
Hog liver	56	Poulsen and Ziegler (1979)
		Sabourin *et al.* (1984)
A. bisporus	50	Tsuji *et al.* (1986)
T. cruzi	52	Kuwahara *et al.* (1985)
P. fluorescens	43	Müller *et al.* (1979)
Acinetobacter	56	Cummings Ryerson *et al.* (1982)

[a] Apparent molecular masses are based on electrophoretic mobility in polyacrylamide gels in the presence of SDS.

and Massey, 1970). This resembles activation of certain eukaryotic FMO forms by primary amines, which do not undergo oxygenation (Ziegler *et al.*, 1973). Cyclohexanone oxygenase from *Acinetobacter* NCIB 9871 carries out *S*-oxidation of pentamethylene sulphide (Cummings Ryerson *et al.*, 1982). Apparent subunit size of these FMO variants appears to be distinctly lower than that of the mammalian isoenzymes.

(b) Differences in spectral properties

Although the visible absorption spectra of the diverse FMO variants are in accord with those of many other flavoproteins, the precise position of the absorbance bands changes somewhat with the enzyme species investigated. Table 5.3 summarizes the optical data. As can be seen, in the oxidized form the shorter-wavelength peak is located at 374–384 nm with the mammalian enzymes, but is shifted to 362–375 nm with the non-eukaryotic subforms; the longer-wavelength band is positioned at 437–455 nm. Reduction bleaches flavin absorbance, but reoxygenation gives rise to the formation of a peroxy species with a single absorbance maximum at 360–388 nm. The variance in peak position of the oxidized enzymes and the reduced oxygen derivatives clearly indicates specific modulation of the spectral properties of the FAD moiety on combination with apoenzyme of distinct conformation.

Table 5.3 Spectral properties of FMO species from various sources

Source of enzyme	Oxidized (λ_{max})	Reduced + O_2 (λ_{max})	Reference
Mouse liver	374; 452	ND	Sabourin *et al.*, (1984)
Rabbit lung	378; 447	388	Williams *et al.* (1985)
Hog liver	384; 450	375	Poulsen and Ziegler (1979)
A. bisporus	362; 450	ND	Tsuji *et al.* (1986)
T. cruzi	374; 455	ND	Kuwahara *et al.* (1985)
P. fluorescens	372; 447	383	Entsch *et al.* (1976)
Acinetobacter	375; 437	360	Cummings Ryerson *et al.* (1982)

ND = not determined.

5.3.2 Differences in enzyme structure

(a) Differences in amino acid composition

Inspection of Table 5.4 reveals that amino acid composition of the rabbit liver FMO is similar to but not identical with that of the rabbit pulmonary enzyme. Differences among the analyses for the two flavoproteins occur in several amino acids, most notably glutamine/glutamate and phenylalanine. Comparison of these data with the previously published amino acid composition of the porcine liver FMO results in a few apparently significant variances. Thus, both rabbit isoenzymes contain a higher percentage of glutamine/glutamate and glycine and a lower amount of leucine than does the hog liver enzyme. As calculated on a molar basis, there also appears to be a striking degree of similarity in amino acid composition between the fungal and bacterial variants and the mammalian enzyme forms.

In all flavoproteins listed, amino acids bearing a hydrophobic side chain consistently amount to 40–46% of the total amino acids present, as is typical of globular proteins.

(b) Differences in amino acid sequence

Ozols (1989) reported isolation from the livers of male rabbits of two FMO species designated form 1 and form 2. Sequence alignment of tryptic fragments from these subforms shows a high degree of homology, suggesting the existence of structurally similar but not identical hepatic isoenzymes. Most interestingly, the NH_2-terminus of form 2 enzyme exhibits some 40% identity to residues 1 to 33 in the NH_2-terminal segment of *p*-hydroxy-benzoate hydroxylase from *Pseudomonas fluorescens*.

Comparative structural analysis has also been performed with FMO from rabbit liver and lung (Hlavica *et al.*, 1990). HPLC profiles of tryptic peptides

Table 5.4 Amino acid composition of FMO species from various sources

	Rabbit lung[a]		Rabbit liver[a]		Hog liver[b]		A. bisporus[c]		P. fluorescens[d]		Acinetobacter[e]	
	Count	Mol. %	Count	Mol. %	Count	Mol. %	Count	Mol. %	Count	Mol. %	Count	Mol. %
Asx	50	9.6	53	10.2	52	9.1	43	9.5	23	5.9	56	12.1
Thr	28	5.4	27	5.2	33	5.8	21	4.8	19	4.9	24	5.3
Ser	38	7.3	38	7.3	38	6.7	26	5.8	18	4.6	21	4.5
Glx	70	13.5	60	11.4	52	9.1	38	8.5	49	12.5	60	12.9
Pro	37	7.1	31	5.9	42	7.4	29	6.5	17	4.4	18	3.9
Gly	48	9.2	47	9.0	37	6.5	37	8.3	34	8.7	34	7.4
Ala	25	4.8	27	5.3	31	5.4	38	8.5	36	9.3	40	8.6
Cys	4	0.8	6	1.1	9	1.6	5	1.1	5	1.3	3	0.6
Val	37	7.2	33	6.4	44	7.7	28	6.3	28	7.2	33	7.2
Met	11	2.1	13	2.5	10	1.8	10	2.1	6	1.5	10	2.2
Ile	21	4.0	19	3.6	29	5.1	30	6.8	17	4.4	25	5.4
Leu	44	8.5	48	9.2	58	10.2	48	10.7	45	11.5	34	7.3
Tyr	12	2.3	19	3.6	19	3.3	14	3.2	15	3.8	20	4.3
Phe	25	4.8	35	6.7	37	6.4	11	2.6	11	2.8	20	4.3
Lys	33	6.3	30	5.7	39	6.8	23	5.1	12	3.1	36	7.8
His	10	1.9	13	2.5	10	1.8	19	4.2	9	2.3	7	1.5
Arg	27	5.2	23	4.4	21	3.7	20	4.4	36	9.2	13	2.8
Trp	ND	—	ND	—	9	1.6	7	1.6	10	2.6	9	1.9
Total	520	100	522	100	570	100	447	100	390	100	463	100

[a] Hlavica et al. (1990)
[b] Poulsen and Ziegler (1979)
[c] Tsuji et al. (1986)
[d] Müller et al. (1979)
[e] Donoghue et al. (1976)

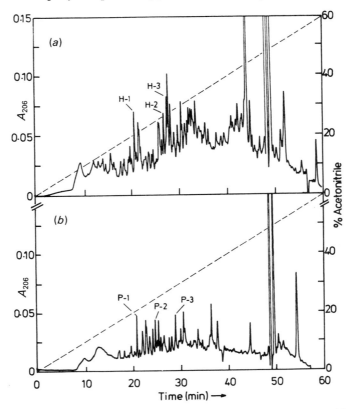

Figure 5.2 Reversed-phase HPLC profiles of tryptic digests of the purified rabbit liver (*a*) and lung (*b*) FMO. Fractions were resolved on a Hibar LiChrospher 100 RP-18 column using an 80 min gradient of acetonitrile (0–80%) in 0.1% trifluoroacetic acid; flow rate was 1 ml/min. Marked peaks indicate peptides selected for NH$_2$-terminal amino acid sequence analysis (Hlavica *et al.*, 1990, with permission).

from the hepatic and pulmonary enzymes display marked differences in the retention times of the individual peptide peaks (Fig. 5.2). Sequence alignment of selected fragments from the two preparations shows significant structural similarity but not identity of the peptides studied (Table 5.5), indicating that the liver and lung protein represent unique entities.

5.4 CONCLUSIONS

All available evidence from kinetic and structural studies suggests that the FMO represents a family of polymorphically expressed isoenzymes

Table 5.5 Sequence alignment of homologous tryptic peptides from rabbit liver and lung FMO

Enzyme source	Peptide peak number	Retention time (min)[a]	NH$_2$-terminal sequence
Liver	H-1	20.5	Phe—⌈Thr—Glu—His—Val—Glu—Gly⌉—Arg
Lung	P-1	20.9	Phe—⌊Lys—Glu—Asn—Val—Glu—Asp—Gly⌋—Arg
Liver	H-2	26.9	Tyr—Leu—Met—Lys—⌈Glu—Pro—Val—Leu⌉—Asn
Lung	P-3	29.3	⌊Glu (Gln)—Val—Leu⌋—Asn
Liver	H-3	27.7	Val—Phe—Ile—Arg—Pro—Ser—⌈Ile—X—Leu—Pro—Ala⌉—Lys
Lung	P-2	24.9	⌊Ile—Leu—Lys—Pro—Leu⌋—Lys

[a] Peptides were eluted from a Hibar LiChrospher 100 RP-18 column at 1 ml/min with an 80 min gradient of acetonitrile (0–80%) in 0.1% trifluoroacetic acid (Hlavica et al., 1990).

characterized by distinct substrate specificity. This is of special interest with respect to the fact that FMO-mediated biotransformation has been recognized to constitute a significant metabolic pathway in the toxication and detoxication of xenobiotics. Therefore, the qualitative as well as quantitative properties of the FMO are important when considering how such compounds will be metabolized in different species and even in different tissues of a given species. The significance of interindividual variations in catalytic capacity of the FMO is best documented by the inherited deficiency of trimethylamine *N*-oxidation in man. The latter process is likely to be brought about mainly by the flavin-containing mono-oxygenase, since conversion of nicotine into its 1'-*N*-oxide, a reaction believed to be exclusively mediated by the FMO (Damani *et al.*, 1988), co-segregates in subjects with deficiency of trimethylamine oxidation (Ayesh *et al.*, 1988). Furthermore, administration of methimazole, a known inhibitor of the FMO, hampers nicotine *N*-oxide formation in human volunteers (Cholerton *et al.*, 1988). In man, the defective enzyme form has been partially characterized, and the apparent K_m value for trimethylamine has been found to be several times higher than that for enzyme derived from the livers of unaffected subjects (Higgins *et al.*, 1972). This suggests a genetically determined change in enzyme structure. Defective enzyme activity is manifested by primary trimethylaminuria (fish-odour syndrome). For a review see Ayesh and Smith (1990).

Attention is also drawn to the fact that FMO participates in the bioactivation of certain nitrogen-containing mutagenic and carcinogenic compounds (Kadlubar *et al.*, 1976; Kimura *et al.*, 1983; Frederick *et al.*, 1985; Ziegler, 1988). The interplay of the various FMO forms and the cytochrome *P*-450-dependent mono-oxygenase system in the organs of animal species is likely to control the mode and/or extent to which activation of these carcinogens occurs, and this may explain the tissue-specific response to the tumorigenic action of these agents. It is obvious that elucidation of the polymorphic expression of the FMO could be a promising area for further studies.

Note added in proof. Recently, full length cDNA clones encoding the FMO from rabbit liver and lung were isolated; the deduced amino acid sequences exhibited only 56% homology. The rabbit liver form was 87% identical to the porcine hepatic enzyme (Lawton *et al.*, 1990).

REFERENCES

Agosin, M. and Ankley, G.T. (1987) Conversion of *N,N*-dimethylaniline to *N,N*-dimethylaniline-*N*-oxide by a cytosolic flavin-containing enzyme from *Trypanosoma cruzi. Drug Metab. Dispos.*, **15**, 200–3.

Arrhenius, E. (1969) Effects of various in vitro conditions on hepatic microsomal *N*- and *C*-oxygenation of aromatic amines. *Chem.-Biol. Interact.*, 1, 361–80.

Ayesh, R. and Smith, R.L. (1990) Genetic polymorphism of trimethylamine *N*-oxidation. *Pharmacol. Ther.*, 45, 387–401.

Ayesh, R., Al-Waiz, M., Crothers, J. *et al.* (1988) Deficient nicotine *N*-oxidation in two sisters with trimethylaminuria. *Br. J. Clin. Pharmacol.*, 25, 664.

Beaty, N.B. and Ballou, D.P. (1981) The oxidative half-reaction of liver microsomal FAD-containing monooxygenase. *J. Biol. Chem.*, 256, 4619–25.

Beckett, A.H., Mitchard, M. and Shihab, A.A. (1971) The influence of methyl substitution on the *N*-demethylation and *N*-oxidation of normethadone in animal species. *J. Pharm. Pharmacol.*, 23, 941–6.

Brodfuehrer, J.I. and Zannoni, V.G. (1987) Flavin-containing monooxygenase and ascorbic acid deficiency. *Biochem. Pharmacol.*, 36, 3161–7.

Cashman, J.R. (1989) Enantioselective *N*-oxygenation of verapamil by the hepatic flavin-containing monooxygenase. *Mol. Pharmacol.*, 36, 497–503.

Cashman, J.R., Proudfoot, J., Pate, D.W. and Högberg, T. (1988) Stereoselective *N*-oxygenation of zimeldine and homozimeldine by the flavin-containing monooxygenase. *Drug. Metab. Dispos.*, 16, 616–22.

Cholerton, S., Ayesh, R., Idle, J.R. and Smith, R.L. (1988) The pre-eminence of nicotine *N*-oxidation and its diminution after carbamizole administration. *Br. J. Clin. Pharmacol.*, 26, 652.

Cummings Ryerson, C., Ballou, D.P. and Walsh, C. (1982) Mechanistic studies on cyclohexanone oxygenase. *Biochemistry*, 21, 2644–55.

Damani, L.A., Pool, W.F., Crooks, P.A. *et al.* (1988) Stereoselectivity in the *N*-oxidation if nicotine isomers by flavin-containing monooxygenase. *Mol. Pharmacol.*, 33, 702–5.

Dannan, G.A. and Guengerich, F.P. (1982) Immunochemical comparison and quantitation of microsomal flavin-containing monooxygenase in various hog, mouse, rat, rabbit, dog and human tissues. *Mol. Pharmacol.*, 22, 787–94.

Devereux, T.R. and Fouts, J.R. (1974) *N*-Oxidation and demethylation of *N,N*-dimethylaniline by rabbit liver and lung microsomes. Effects of age and metals. *Chem.-Biol. Interact.*, 8, 91–105.

Devereux, T.R. and Fouts, J.R. (1975) Effect of pregnancy or treatment with certain steroids on *N,N*-dimethylaniline demethylation and *N*-oxidation by rabbit liver or lung microsomes. *Drug. Metab. Dispos.*, 3, 254–8.

Devereux, T.R., Philpot, R.M. and Fouts, J.R. (1977) The effect of Hg^{2+} on rabbit hepatic and pulmonary solubilized, partially purified *N,N*-dimethylaniline *N*-oxidases. *Chem.-Biol. Interact.*, 18, 277–87.

Dixit, A. and Roche, T.E. (1984) Spectrophotometric assay of the flavin-containing monooxygenase and changes in its activity in female mouse liver with nutritional and diurnal conditions. *Arch. Biochem. Biophys.*, 233, 50–63.

Donoghue, N.A., Norris, D.B. and Trudgill, P.W. (1976) The purification and properties of cyclohexanone oxygenase from *Nocardia globerula* CL1 and *Acinetobacter* NCIB 9871. *Eur. J. Biochem.*, 63, 175–92.

Duffel, M.W., Graham, J.M. and Ziegler, D.M. (1981) Changes in dimethylaniline *N*-oxidase activity of mouse liver and kidney induced by steroid hormones. *Mol. Pharmacol.*, 19, 134–9.

Entsch, B., Ballou, D.P. and Massey, V. (1976) Flavin–oxygen derivatives involved in hydroxylation by *p*-hydroxybenzoate hydroxylase. *J. Biol. Chem.*, 251, 2550–63.

Frederick, C.B., Hammons, G.J., Beland, F.A. *et al.* (1985) *N*-Oxidation of primary aromatic amines in relation to chemical carcinogenesis, in *Biological Oxidation of Nitrogen in Organic Molecules* (eds J.W. Gorrod and L.A. Damani), Ellis Horwood, Chichester, pp. 131–48.

Gigon, P.L. and Bickel, M.H. (1971) *N*-Demethylation and *N*-oxidation of imipramine by rat and pig liver microsomes. *Biochem. Pharmacol.*, **20**, 1921–31.

Gorrod, J.W. and Gooderham, N.J. (1987) The metabolism of *N*-benzyl-4-substituted anilines: factors influencing in vitro *C*- and *N*-oxidation. *Xenobiotica*, **17**, 165–77.

Gorrod, J.W. and Patterson, L.H. (1983) The metabolism of 4-substituted-*N*-ethyl-*N*-methylanilines. II. Some factors influencing α-*C*- and *N*-oxidation. *Xenobiotica*, **13**, 513–20.

Gorrod, J.W., Temple, D.J. and Beckett, A.H. (1975) The differentiation of *N*-oxidation and *N*-dealkylation of *N*-ethyl-*N*-methylaniline by rabbit liver microsomes as distinct metabolic routes. *Xenobiotica*, **5**, 465–74.

Heinze, E., Hlavica, P., Kiese, M. and Lipowsky, G. (1970) *N*-Oxygenation of arylamines in microsomes prepared from corpora lutea of the cycle and other tissues of the pig. *Biochem. Pharmacol.*, **19**, 641–9.

Higgins, T., Chagkin, S., Hammond, K.B. and Humbert, J.R. (1972) Trimethylamine *N*-oxide synthesis: human variant. *Biochem. Med.*, **6**, 392–6.

Hlavica, P. (1970) Studies on the active site of mixed-function oxidases in rabbit liver microsomes. *Biochem. Biophys. Res. Commun.*, **40**, 212–17.

Hlavica, P. (1972) Interaction of oxygen and aromatic amines with hepatic microsomal mixed-function oxidase. *Biochim. Biophys. Acta*, **273**, 318–27.

Hlavica, P. and Hülsmann, S. (1979) Studies on the mechanism of hepatic mircrosomal *N*-oxide formation. *N*-Oxidation of *N,N*-dimethylaniline by a reconstituted rabbit liver microsomal cytochrome *P*-448 enzyme system. *Biochem. J.* **182**, 109–16.

Hlavica, P. and Kehl, M. (1974) Studies on the mechanism of hepatic microsomal *N*-oxide formation. I. Effect of carbon monoxide on the *N*-oxidation of *N,N*-dimethylaniline. *Hoppe-Seyler's Z. Physiol. Chem.*, **355**, 1508–14.

Hlavica, P. and Kehl, M. (1977) Studies on the mechanism of hepatic microsomal *N*-oxide formation. The role of cytochrome *P*-450 and mixed-function amine oxidase in the *N*-oxidation of *N,N*-dimethylaniline. *Biochem. J.*, **164**, 487–96.

Hlavica, P. and Kiese, M. (1969) *N*-oxygenation of *N*-alkyl- and *N,N*-dialkylanilines by rabbit liver microsomes. *Biochem. Pharmacol.*, **18**, 1501–9.

Hlavica, P., Kellermann, J., Henschen, A. *et al.* (1990) Evidence of the existence of structurally distinct hepatic and pulmonary forms of microsomal flavin-containing monooxygenase in the rabbit. *Biol. Chem. Hoppe-Seyler*, **371**, 521–6.

Hofsteenge, J., Weijer, W.J., Jekel, P.A. and Beintema, J.J. (1983) *p*-Hydroxybenzoate hydroxylase from *Pseudomonas fluorescens*. I. Completion of the elucidation of the primary structure. *Eur. J. Biochem.*, **133**, 91–108.

Howell, L.G. and Massey, V. (1970) A non-substrate effector of *p*-hydroxybenzoate hydroxylase. *Biochem. Biophys. Res. Commun.*, **40**, 887–93.

Jenner, P., Gorrod, J.W. and Beckett, A.H. (1973) Species variation in the metabolism of *R*-(+)- and *S*-(−)-nicotine by a α-*C*- and *N*-oxidation in vitro. *Xenobiotica*, **3**, 573–80.

Kadlubar, F.F., Miller, J.A. and Miller, E.C. (1976) Microsomal *N*-oxidation of the hepatocarcinogen *N*-methyl-4-aminoazobenzene and the reactivity of *N*-hydroxy-*N*-methyl-4-aminoazobenzene. *Cancer Res.*, **36**, 1196–206.

Kimura, T., Kodama, M. and Nagata, C. (1983) Purification of mixed-function amine oxidase from rat liver microsomes. *Biochem. Biophys. Res. Commun.*, **110**, 640–5.

Kitchell, B.S., Rauckman, E.J. and Rosen, G.M. (1978) The effect of temperature on mixed-function amine oxidase intrinsic fluorescence and oxidative activity. *Mol. Pharmacol.*, **14**, 1092–8.

Kuwahara, T., White, R.A. and Agosin, M. (1985) A cytosolic FAD-containing enzyme catalyzing cytochrome c reduction in *Trypanosoma cruzi*. I. Purification and some properties. *Arch. Biochem. Biophys.*, **239**, 18–28.

Lawton, M.P., Gasser, R., Tynes, R.E. *et al.* (1990) The flavin-containing mono-oxygenase enzymes expressed in rabbit liver and lung and the products of related but distinctly different genes. *J. Biol. Chem.*, **265**, 5855–61.

McManus, M.E., Stupans, I., Burgess, W. *et al.* (1987) Flavin-containing mono-oxygenase activity in human liver microsomes. *Drug Metab. Dispos.*, **15**, 256–61.

Müller, F., Voordoun, G., van Berkel, W.J.H. *et al.* (1979) A study of p-hydroxy-benzoate hydroxylase from *Pseudomonas fluorescens*. Improved purification, relative molecular mass, and amino acid composition. *Eur. J. Biochem.*, **101**, 235–44.

Nakazawa, K. (1970) Studies on the demethylation, hydroxylation and N-oxidation of imipramine in rat liver. *Biochem. Pharmacol.*, **19**, 1363–9.

Ohmiya, Y. and Mehendale, H.M. (1981) Species differences in pulmonary N-oxidation of chlorpromazine and imipramine. *Fed. Proc.*, **40**, 732.

Ohmiya, Y. and Mehendale, H.M. (1982) Metabolism of chlorpromazine by pulmonary microsomal enzymes in the rat and rabbit. *Biochem. Pharmacol.*, **31**, 157–62.

Ohmiya, Y. and Mehendale, H.M. (1983) N-Oxidation of N,N-dimethylaniline in the rabbit and rat lung. *Biochem. Pharmacol.*, **32**, 1281–5.

Osimitz, T.G. and Kulkarni, A.P. (1982) Oxidative metabolism of xenobiotics during pregnancy: significance of microsomal flavin-containing monooxygenase. *Biochem. Biophys. Res. Commun.*, **109**, 1164–71.

Ozols, J. (1989) Liver microsomes contain two distinct NADPH-monooxygenases with NH_2-terminal segments homologous to the flavin-containing NADPH-monooxygenase of *Pseudomonas fluorescens*. *Biochem. Biophys. Res. Commun.*, **163**, 49–55.

Poulsen, L.L. and Ziegler, D.M. (1979) The liver microsomal FAD-containing monooxygenase. Spectral characterization and kinetic studies. *J. Biol. Chem.*, **254**, 6449–55.

Poulsen, L.L., Hyslop, R. and Ziegler, D.M. (1974) S-Oxidation of thiourylenes catalyzed by a microsomal flavoprotein mixed-function oxidase. *Biochem. Pharmacol.*, **23**, 3431–40.

Poulsen, L.L., Taylor, K., Williams, D.E. *et al.* (1986) Substrate specificity of the rabbit lung flavin-containing monooxygenase for amines: oxidation products of primary alkylamines. *Mol. Pharmacol.*, **30**, 680–5.

Rouer, E., Lemoine, A., Cresteil, T. *et al.* (1987) Effects of genetic or chemically induced diabetes on imipramine metabolism. Respective involvement of flavin monooxygenase and cytochrome P-450-dependent monooxygenases. *Drug Metab. Dispos.*, **15**, 524–8.

Rouer, E., Rouet, P., Delpech, M. and Leroux, J.P. (1988) Purification and comparison of liver microsomal flavin-containing monooxygenase from normal and streptozotocin-diabetic rats. *Biochem. Pharmacol.*, **37**, 3455–9.

Ruenitz, P.C., Bagley, J.R. and Pape, C.W. (1984) Some chemical and biochemical aspects of liver microsomal metabolism of tamoxifen. *Drug. Metab. Dispos.*, **12**, 478–83.

Sabourin, P.J. and Hodgson, E. (1984) Characterization of the purified FAD-containing monooxygenase from mouse and pig liver. *Chem.-Biol. Interact.*, **51**, 125–39.

Sabourin, P.J., Smyser, B.P. and Hodgson, E. (1984) Purification of the flavin-containing monooxygenase from mouse and pig liver microsomes. *Int. J. Biochem.*, **16**, 713–20.

Stefek, M., Benes, L. and Zelnik, V. (1989) *N*-Oxygenation of stobadine, a γ-carboline antiarrhythmic and cardioprotective agent: the role of flavin-containing monooxygenase. *Xenobiotica*, **19**, 143–50.

Tsuji, H., Ogawa, T., Bando, N. and Sasaoka, K. (1986) Purification and properties of 4-aminobenzoate hydroxylase, a new monooxygenase from *Agaricus bisporus*. *J. Biol. Chem.*, **261**, 13203–9.

Tynes, R.E. and Hodgson, E. (1985) Catalytic activity and substrate specificity of the flavin-containing monooxygenase in microsomal systems: characterization of the hepatic, pulmonary and renal enzymes of the mouse, rabbit and rat. *Arch. Biochem. Biophys.*, **240**, 77–93.

Tynes, R.E. and Philpot, R.M. (1987) Tissue- and species-dependent expression of multiple forms of mammalian microsomal flavin-containing monooxygenase. *Mol. Pharmacol.*, **31**, 569–74.

Tynes, R.E., Sabourin, P.J. and Hodgson, E. (1985) Identification of distinct hepatic and pulmonary forms of microsomal flavin-containing monoxygenase in the mouse and rabbit. *Biochem. Biophys. Res. Commun.*, **126**, 1069–75.

Tynes, R.E., Sabourin, P.J., Hodgson, E. and Philpot, R.M. (1986) Formation of hydrogen peroxide and *N*-hydroxylated amines catalyzed by pulmonary flavin-containing monooxygenases in the presence of primary alkylamines. *Arch. Biochem. Biophys.*, **251**, 654–64.

Uehleke, H. (1973) The role of cytochrome P-450 in the *N*-oxidation of individual amines. *Drug. Metab. Dispos.*, **1**, 299–313.

Willi, P. and Bickel, M.H. (1973) Liver metabolic reactions: tertiary amine *N*-dealkylation, tertiary amine *N*-oxidation, *N*-oxide reduction, and *N*-oxide *N*-dealkylation. II. *N,N*-dimethylaniline. *Arch. Biochem. Biophys.*, **156**, 772–9.

Williams, D.E., Hale, S.E., Muerhoff, A.S. and Masters, B.S.S. (1985) Rabbit lung flavin-containing monooxygenase. Purification, characterization, and induction during pregnancy. *Mol. Pharmacol.*, **28**, 381–90.

Williams, D.E., Reed, R.L., Kedzierski, B. *et al.* (1989) The role of flavin-containing monooxygenase in the *N*-oxidation of the pyrrolizidine alkaloid senecionine. *Drug. Metab. Dispos.*, **17**, 380–6.

Williams, D.E., Ziegler, D.M., Nordin, D.J. *et al.* (1984) Rabbit lung flavin-containing monooxygenase is immunochemically and catalytically distinct from the liver enzyme. *Biochem. Biophys. Res. Commun.*, **125**, 116–22.

Ziegler, D.M. (1980) Microsomal flavin-containing monooxygenase: oxygenation of nucleophilic nitrogen and sulfur compounds, in *Enzymatic Basis of Detoxification* (ed. W.B. Jakoby), Vol. I, Academic Press, New York, pp. 201–27.

Ziegler, D.M. (1988) Flavin-containing monooxygenase: catalytic mechanism and substrate specificities. *Drug. Metab. Rev.*, **19**, 1–32.

Ziegler, D.M. and Pettit, F.H. (1964) Formation of an intermediate *N*-oxide in the oxidative demethylation of *N,N*-dimethylaniline catalyzed by liver microsomes. *Biochem. Biophys. Res. Commun.*, **15**, 188–93.

Ziegler, D.M., McKee, E.M. and Poulsen, L.L. (1973) Microsomal flavoprotein-catalyzed *N*-oxidation of arylamines. *Drug Metab. Dispos.*, **1**, 314–21.

Ziegler, D.M., Poulsen, L.L. and McKee, E.M. (1971) Interaction of primary amines with a mixed-function amine oxidase isolated from pig liver microsomes. *Xenobiotica*, **1**, 523–31.

6

Factors regulating the activity of the rabbit lung flavin-containing mono-oxygenase

D.E. Williams

*Department of Food Science and Technology and Toxicology Program,
Oregon State University, Corvallis, OR 97331, USA*

1. Rabbit lung microsomes contain a form of flavin-containing mono-oxygenase (FMO) which is distinct from liver FMO with respect to substrate specificity and immunochemical properties. The 'lung' FMO is also present in rabbit kidney and bladder.
2. As has been determined for the much-studied porcine liver FMO, the rabbit lung enzyme is not inducible by xenobiotics, but does appear to be regulated developmentally.
3. Rabbit lung FMO is higher in adult female rabbits compared to males, as determined by N-oxidation of N,N-dimethylaniline and immuno-quantitation by Western blotting. The level and activity of rabbit lung FMO is increased during late pregnancy. Fetal lung FMO levels are 28–62% of the adult (day 25–30 of gestation) and are increased markedly 1 day after birth. Preliminary evidence indicates that the 'lung' FMO appears to be higher in adult male kidney than in female, but is induced during pregnancy. Fetal FMO was undetectable in kidney. No sex difference or effect of pregnancy was observed with FMO in bladder.
4. Rabbit FMO appears to have a modest diurnal rhythm in lung, exhibiting two-fold higher activity at 12 : 00 noon compared to the low at 8 : 00 pm. No discernible rhythm was observed in rabbit liver. The activities in kidney and bladder were ten-fold lower than lung and liver and were too low to assess accurately a diurnal rhythm.
5. FMO isolated from rabbit lung still contains tightly bound phospholipid, tentatively identified as phosphatidylcholine. Addition of various phospholipids to the purified enzyme had no effect except for phosphatidylinositol-4-monophosphate and -4,5-bisphosphate, which inhibited the enzyme by over 50%. It is unknown if this could be a mechanism of regulation *in vivo*.

6.1 INTRODUCTION

Mammalian microsomal flavin-containing mono-oxygenase (FMO, EC 1.14.13.8) oxygenates a large number of heteroatom-containing xenobiotics (reviewed in Ziegler, 1988). As is the case with the other major mono-oxygenase system, the cytochrome *P*-450-dependent mixed-function oxidase, FMO is found in highest concentration in liver and requires NADPH and O_2 for activity. However, unlike the *P*-450 system, mammalian FMO is not inducible by phenobarbital, 3-methylcholanthrene or any of the other classes of *P*-450 inducers. Mammalian FMO does appear to be regulated developmentally in a number of species and tissues. Examples include sex differences, and variations with age, oestrus cycle or pregnancy (Das and Ziegler, 1970; Heinze *et al.*, 1970; Uehleke *et al.*, 1971; Devereux and Fouts, 1974, 1975; Wirth and Thorgeirsson, 1978; Duffel *et al.*, 1981; Dixit and Roche, 1984; Williams *et al.*, 1985; Dannan *et al.*, 1986; Tynes and Philpot, 1987). Studies employing gonadectomy of young animals indicate a role for the sex steroids in regulation of FMO (Duffel *et al.*, 1981; Dannan *et al.*, 1986).

Until recently it was thought that there was a single form of FMO in the various tissues of a given species. In 1984 and 1985, two laboratories isolated FMO from rabbit lung and found it to be immunochemically and catalytically distinct from the liver enzyme (Williams *et al.*, 1984; Tynes *et al.*, 1985), confirming earlier studies with rabbit lung microsomes which had suggested that the lung form had some unusual properties (Ohmiya and Mehendale, 1983). Previous work had also indicated that rabbit lung FMO was induced during pregnancy (Devereux and Fouts, 1975). These results have been confirmed by assaying both FMO activity and by immunoquantitation on Western blots with guinea-pig antibody to rabbit lung FMO (Williams *et al.*, 1985). Lung FMO is also present in relatively high concentrations in fetal lung microsomes. In addition to lung, this 'extrahepatic' FMO is also present in rabbit kidney and bladder and evidence exists for sex- and developmental-related differences.

The levels and activity of FMO in rabbit lung, liver, kidney and bladder has been examined for possible diurnal rhythms by assaying *N*,*N'*-dimethylaniline *N*-oxidation and by immunoquantitation every 4 hours over a 24 hour period.

Finally, evidence is presented that purified rabbit lung FMO contains tightly associated phospholipid in the form of phosphatidylcholine. Addition of various phospholipids to the purified enzyme has little or no effect on rabbit lung FMO activity with the exception of phosphatidylinositol 4-monophosphate (PIP) and phosphatidylinositol 4,5-bisphosphate (PiP_2), which are both effective inhibitors. It could be speculated that hormonal variations in PiP and PiP_2 levels may be a mechanism for regulating rabbit lung FMO activity.

6.2 PURIFICATION AND DISTINCT PROPERTIES OF RABBIT LUNG FMO

Lung FMO, as purified from pregnant New Zealand White rabbits (day 25–30 of gestation) by a slight modification of a previously published procedure (Williams *et al.*, 1985), is found to elute from ω-amino-octyl Sepharose in the 'wash' buffer if the sodium cholate-solubilized lung microsomes are diluted from a concentration of 0.6% cholate to 0.2% cholate. In this manner rabbit lung FMO is completely resolved from cytochromes *P*-450 on this first column and the FMO is essentially homogeneous (13.4 nmol/mg protein), requiring only concentration, dialysis and detergent removal. This laboratory typically obtains about 1 nmol of purified FMO per gram of lung.

Polyclonal antibodies to rabbit lung FMO, produced both in guinea pig and in goat, do not inhibit catalytic activity of the enzyme. These antibodies can be utilized to perform immunoquantitation by Western blotting of the FMO as described by Burnette (1981) with laser densitometry of the autoradiograms developed after staining with guinea-pig antibody to rabbit lung FMO and ^{125}I-labelled Protein A. FMO activities can most conveniently be measured by *N*,*N*-dimethylaniline (DMA) *N*-oxidation, either colorometrically as described by Gold and Ziegler (1973) or by an HPLC assay on a reverse-phase ACT-1 column using ^{14}C-DMA and radiochemical detection.

Rabbit lung FMO is catalytically and immunochemically distinct from the liver enzyme. Accessibility of the catalytic C4a-peroxyflavin site of rabbit lung FMO appears more restricted than that of the liver enzyme (Nagata *et al.*, 1990) and a number of bulky amines, which are excellent substrates for pig liver FMO, such as imipramine and chlorpromazine, are not oxygenated by the lung enzyme (Williams *et al.*, 1984). In addition, primary alkyamines which are not substrates for liver FMO (although they can act as positive effectors), are *N*-oxygenated by rabbit lung FMO, initially to the *N*-hydroxyamine and then to the oxime (Poulsen *et al.*, 1986).

The amino acid composition of rabbit lung FMO is shown in Table 6.1. Comparison with the previously published amino acid composition of pig liver FMO (Poulsen and Ziegler, 1979) results in a few apparently significant differences. Rabbit lung FMO contains a lower percentage of cysteine and tryptophan and a greater amount of glutamine/glutamate than does the pig liver enzyme. The amino acid sequence of four tryptic peptides of rabbit lung FMO is shown in Fig. 6.1. Computer searches (Protein Identification Resource, National Biomedical Research Fund, Washington, DC) did not find any significant homologies between these sequences and those of other flavoproteins. Preliminary comparison to the amino acid sequence of pig liver FMO does not show high homology (data not shown).

Polyclonal antibodies to rabbit lung FMO do not cross-react with the pig

Table 6.1 Amino acid composition of rabbit lung and pig liver FMO

Amino acid	Rabbit lung[a] (mole %)		Pig liver[b] (mole %)
Ala	6.1	6.3	5.6
Arg	4.8	5.1	3.6
Asx	9.7	8.6	8.8
Cyc/2	0.9	—	2.3
Glx	12.2	11.8	8.8
Gly	6.0	6.3	6.2
His	1.4	1.3	1.7
Ile	2.7	5.7	5.6
Leu	8.7	12.5	9.6
Lys	7.9	5.4	6.8
Met	2.7	2.6	2.3
Phe	6.8	8.7	7.0
Pro	7.0	5.5	6.4
Ser	9.0	6.8	7.0
Thr	4.3	3.5	5.8
Trp	0.8	—	1.7
Tyr	3.0	2.7	3.6
Val	6.1	7.1	7.7

[a] Determined by two commercial laboratories.
[b] Data taken from Poulsen and Ziegler (1979).

or rabbit liver enzymes. Likewise, antibodies to the liver enzyme do not recognize rabbit lung FMO (Williams *et al.*, 1984, 1985).

In addition to differences in substrate specificity, amino acid composition and sequence, and immunochemical properties, the pig liver and rabbit lung FMOs differ markedly in thermostability, pH optima and sensitivity to ionic

Rabbit FMO tryptic peptides

-Trp-Ala-Thr-Gln-Leu-Glu-Gly-Lys-
-Leu-Tyr-Phe-Gly-Pro-Glu-Asn-Ser-Tyr-
-Ala-Ser-Ile-Pro-Gln-Ser-Val-Glu-Thr-Asn-Arg-
-Leu-Phe-Gly-Tyr-Gln-Trp-Ile-Gly-Ala-Thr-Ser-Lys-

Figure 6.1 Amino acid sequences of four tryptic peptides of rabbit lung FMO. Tryptic digestion of rabbit lung FMO and separation of peptides by reverse-phase HPLC (with detection at 205 nm) was essentially as described by Mayes (1984). Four major, well-resolved peaks were submitted for analysis to the Protein/Nucleic Acid Shared Facility of the Cancer Center at the Medical College of Wisconsin.

detergents (Williams *et al.*, 1985). However, even though these forms of FMO display marked physicochemical characteristics, both enzymes operate by the same mechanism and appear to be regulated endogenously during development, perhaps hormonally.

6.3 REGULATION OF MATERNAL AND FETAL RABBIT LUNG FMO DURING PREGNANCY

Rabbit lung FMO has been immunoquantitated by Western blotting in males as well as non-pregnant and pregnant females at 10, 15, 21, 25 and 28 days of gestation and 1, 2 and 3 days post-partum (Figs 6.2*a* and 6.3). In addition, fetal lung FMO was immunoquantitated at 25, 28 and 30 days of gestation and neonatally in 1 day old pups (Fig. 6.2*b*). The levels of FMO in lungs of adult rabbits was 1.4-fold higher in females (0.372 ± 0.040 nmol/mg) than in males (0.259 ± 0.021 nmol/mg). In a previous study it was found that rabbit lung FMO is maximally induced (4-fold) on day 28 of gestation and decreases post-partum (Williams *et al.*, 1985). The results presented in this review suggest that lung FMO is highest at 1 day post-partum and thereafter declines; the level of induction is only two-fold. A specific content of 0.8 nmol FMO per mg protein would represent 5% of the total lung microsomal protein. Fetal (sex undetermined) lung FMO (0.104 ± 0.008 nmol/mg) at 25 days of gestation is 28–40% of the adult level. The levels increase in fetal lung between days 25 and 30 of gestation and then are markedly increased in 1-day-old pups (sex undetermined) to a level (0.259 nmol/mg) equal to that of adult males. This relatively high level of FMO in fetal tissue differs from previous results obtained from fetal rat liver, in which DMA *N*-oxidase levels were very low prior to birth (Uehleke *et al.*, 1971).

Western blots of kidney and bladder microsomes from male, non-pregnant and pregnant (day 28 of gestation) female as well as fetal (day 28) rabbits, immunostained with antibody to rabbit lung FMO, are shown in Fig. 6.4. The levels of FMO appear to be higher in male kidney that in non-pregnant female kidney. It would also appear that the levels in kidney may be induced on day 28 of gestation. Moreover, no 'lung' FMO was detectable in fetal kidney, and no significant sex or pregnancy-related differences were apparent in bladder microsomes.

6.4 DIURNAL REGULATION OF RABBIT FMO

Dixit and Roche (1984) have reported that the activity of FMO in liver microsomes of female mice is about two-fold higher at 4–5 pm as compared

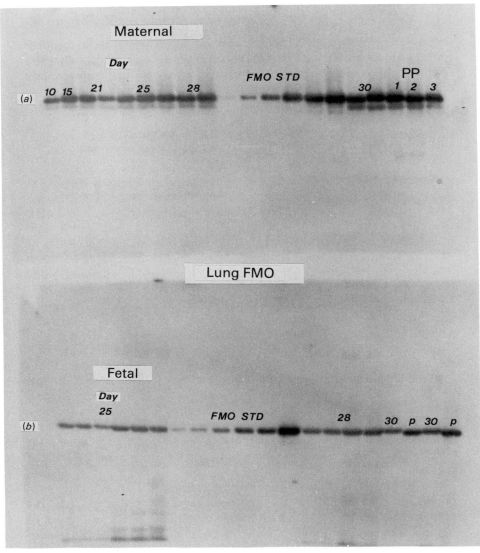

Figure 6.2 Western blotting of maternal and fetal lung microsomes at various times of gestation and post-partum. (*a*) Microsomal proteins (10 μg) from maternal lung of rabbits killed at various times of gestation. Proteins were resolved by SDS-PAGE and electrophoretically transferred to nitrocellulose for subsequent staining with guinea-pig antibody to rabbit lung FMO, followed by ^{125}I-labelled Protein A and autoradiography (Burnette, 1981). The lanes are from left to right: 10, 15 days (single rabbit each), 21 days (2 rabbits), 25 days (3 rabbits), 28 days (2 rabbits), rabbit lung FMO (5 standards, 0.5–10 pmol), 30 days (2 rabbits) and 1, 2 and 3 days post-partum (1 rabbit each). (*b*) Lung microsomes from fetal rabbits at 25 days (pooled from 3 maternal rabbits) at 10 μg and at 50 μg, rabbit lung FMO standards (5 standards, 0.25–5 pmol), 28 days fetal lung (pooled from 2 maternal rabbits) at 10 and 30 μg, and 30 days fetal lung (pooled from a single mother) and 1 day neonatal lung (pooled from 6 kits) at 10 and 50 μg.

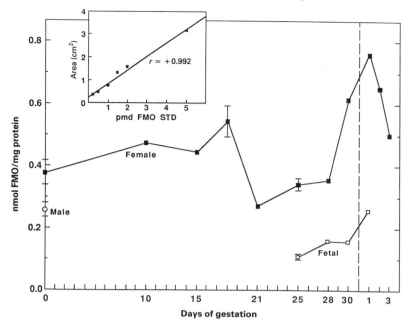

Figure 6.3 Immunoquantitation of the western blots by laser densitometry. The autoradiograms shown in Fig. 6.2 were scanned with a laser densitometer and lung FMO quantitated utilizing a linear regression line (inset) generated from the pure FMO standards.

to 8–9 am. Furthermore, the activity at both time points is about 50% higher in fed than in starved animals.

The diurnal variation of FMO in lung, liver, kidney and bladder of sexually immature New Zealand White rabbits has been determined by assaying the microsomes for activity (^{14}C-DMA *N*-oxidation) and immunoquantitation of FMO by Western blotting. Thus, rabbit lung DMA *N*-oxidation is highest at 12:00 noon (Table 6.2). Activity drops over two-fold between noon and 8:00 pm. Immuno-quantitation of the levels of FMO displays a similar pattern, suggesting that the diurnal variation in rabbit lung FMO is due to differences in microsomal concentrations of the protein. The HPLC assay employed allows also for the determination of the major cytochrome *P*-450 metabolite, *N*-methylaniline. The variation of *P*-450-dependent ^{14}C-DMA *N*-demethylation in rabbit lung resembles the FMO dependent *N*-oxidation with a minimum at 8:00 pm.

No marked variation is observed in liver (Table 6.2) with respect to either *N*-oxidation or *N*-demethylation; maximal activity for both reactions in

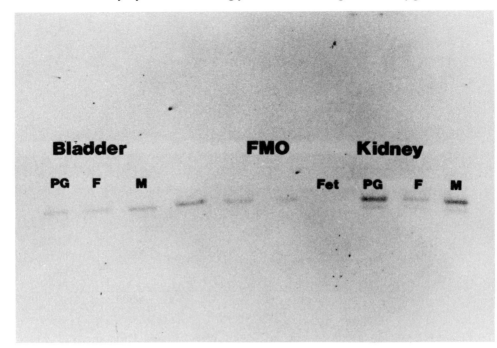

Figure 6.4 Western blot of rabbit kidney and bladder microsomes immunostained with antibody to rabbit lung FMO. Kidney and bladder microsomes (20 μg) were pooled from two or more animals and Western blots performed as described for Fig. 6.2. The microsomes from pregnant (PG) rabbits were pooled from two animals on day 28 of gestation. The three middle lanes were rabbit lung FMO standards at 0.2, 0.5 and 1 nmol.

Table 6.2 Diurnal variation of N,N-dimethylaniline N-oxidation and N-demethylation in rabbit lung and liver[a]

Microsomes	Time	DMA N-demethylation (nmol product/min per mg protein)	DMA N-oxidation	FMO (nmol/mg protein)
Lung	8:00 am	3.49	5.98	0.391
	noon	5.11	7.00	0.479
	8:00 pm	2.52	3.28	0.254
Liver	8:00 am	1.27	3.45	—
	noon	1.36	4.15	—
	8:00 pm	1.20	3.84	—

[a] Williams *et al.* unpublished.

liver is at 4 : 00 pm. The N-oxidation of ^{14}C-DMA in kidney (0.3–0.6 nmol/min/mg) and bladder (0.2–0.5 nmol/min/mg) microsomes is about 10-fold lower than in lung or liver. Activities too low to be accurately determined if a significant variation occurs in these tissues, but the results with bladder suggest a maximum at 4 : 00 pm and a minimum at 8 : 00 am, a pattern opposite to that observed in the lung (Williams *et al.*, unpublished observations).

6.5 PHOSPHOLIPID BINDING TO LUNG FMO AND MODULATION OF ACTIVITY

The relatively high yields and ease of purification of rabbit lung FMO have made possible experiments requiring large amounts of enzyme. ^{31}P-NMR of flavoproteins require concentrations of flavin in excess of 200 μM, the total amount of protein required being approximately 400 nmol. Initial spectra yielded surprising results. Instead of the two upfield pyrophosphate peaks expected from FAD, two peaks at 2.6 and about 0 ppm can be observed (Fig. 6.5*a*). The sharp peak is probably inorganic phosphate. Since the final detergent removal step involves elution of the FMO from hydroxylapatite with 0.3 M-potassium phosphate, the latter presumably represents the source of the inorganic phosphate even though the enzyme is dialysed twice against phosphate-free buffer before analysis. The broad peak at about 0 ppm has a shift similar to that reported previously for phosphatidylcholine (Sotirhos *et al.*, 1986).

Extraction of rabbit lung FMO with organic solvent and TLC of the extracted phospholipid fraction confirmed that the phospholipid present is phosphatidylcholine. This phospholipid must have a relatively tight association with the enzyme to remain bound even after hydrophobic interaction chromatography and a detergent removal step on hydroxyapatite. The absence of the FAD signals could be due to association of the purified enzyme to high molecular weight aggregates. This hypothesis was tested by repeating the analysis in the presence of 1% sodium cholate (Fig. 6.5*b*). Addition of sodium cholate markedly sharpens the peak identified as phosphatidylcholine and gives rise to the appearance of three new peaks, a doublet at −9.97 ppm and −10.41 ppm and one at 3.96 ppm (Fig. 6.5*b*). The upfield doublet has a shift similar to the pyrophosphate peaks of FAD. These shifts are closer to the peaks for free FAD rather than FAD bound to protein as shown in the spectrum of NADPH–cytochrome *P*-450 reductase, another FAD-containing protein (Otvos *et al.*, 1986). It is doubtful that addition of 1% sodium cholate removes FAD from the enzyme, as rabbit lung FMO retains activity under these conditions. The peak at 3.96 ppm has a shift similar to FMN in flavoproteins, however,

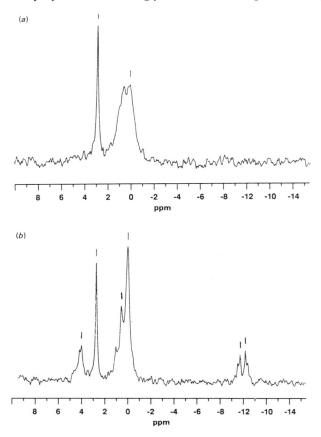

Figure 6.5 ^{31}P-NMR spectrometry of rabbit lung FMO. The spectrum for rabbit lung FMO (250 μM, in 50 mM-Tris-acetate, pH 7.4, 20% glycerol, 0.1 mM EDTA) was obtained at 10°C over approximately 20 h (*a*) At the end of this time, the oxidized spectrum was unchanged and the enzyme retained 85% of its original activity. (*b*) Spectrum obtained on addition of 1% sodium cholate (from a 20% stock of recrystallized cholic acid, pH to 7.5 with NaOH) to the enzyme. Tentative peak identifications are discussed in the text.

previous analysis clearly demonstrates that rabbit lung FMO (as is true for all forms of FMO isolated and analysed to date) contains FAD only (Williams *et al.*, 1985). The origin of this peak must await further analysis.

Previous work has demonstrated that purified pig liver FMO activity can be modulated by phospholipids (Ziegler *et al.*, 1986). Phosphatidylserine is required for activity, whereas the 4-mono- and 4,5-bisphosphate derivatives of phosphatidylinositol are potent inhibitors. A number of different

Table 6.3 Modulation of DMA *N*-oxidation by rabbit lung FMO by phospholipids[a]

Phospholipid added	*^{14}C-DMA N-oxidation* *(% control)*
None	100
PC-bovine liver	93
PC-dilauryl	93
PC-dimyristoyl	81
PC-dipalmitoyl	108
PC-distearoyl	90
PC-diarachoyl	109
PC-lyso	93
PE	99
PS	83
PI	80
PI-4-monophosphate	45

[a] Rabbit lung FMO (0.07 nmol) was preincubated with the above phospholipids (10–15 µg) at room temperature for 10 min prior to the addition of ^{14}C-DMA and the remaining incubation components. Quantitation of ^{14}C-DMA-*N*-oxide was by HPLC. The results shown are averages of duplicate determinations.
Abbreviations: PC, phosphatidylcholine; PE, phosphatidylethanolamine; PS, phosphatidylserine; PI, phosphatidylinositol.

phospholipids have been tested as modulators of the purified rabbit lung FMO (Table 6.3). Phosphatidylcholines with various acyl chain substitutions, as well as phosphatidylcholines extracted from various tissues, are without effect. In addition, phosphatidylserine, phosphatidylethanolamine and phosphatidylinositol have no effect when added to the enzyme at concentrations of 150 µg/nmol FMO. The 4-mono- and 4,5-bisphosphate derivatives of phosphatidylinositol, however, are inhibitory, as previously reported for the pig liver FMO (Table 6.3 and Fig. 6.6). Ziegler *et al.* (1986) have previously reported that the phosphorylated phosphatidylinositols inhibit virtually all activity at a concentration of 21 µg/nmol of enzyme. The degree of inhibition of the rabbit lung FMO by these phospholipids is not as great, concentrations of 500 µg/nmol of FMO only cause slightly more than 50% inhibition (Fig. 6.6). It should be remembered that these results were obtained with rabbit lung FMO which still contains some residual bound phospholipid. More precise determinations of the effect of phospholipids on rabbit lung FMO activity will require enzyme which has been stripped of all residual phospholipid and still retains activity.

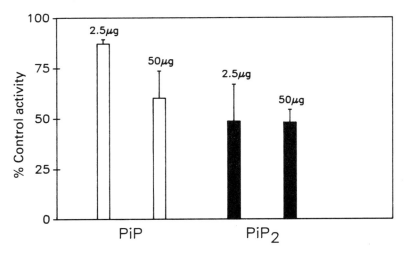

Figure 6.6 Inhibition of rabbit lung FMO ^{14}C-DMA *N*-oxidation by phosphatidylinositol-4-monophosphate   (PiP)   and   Phosphatidylinositol-4,5-bisphosphate (PiP$_2$). Rabbit lung FMO (0.1 nmol) was preincubated with 2.5 or 50 μg of PiP or PiP$_2$ for 10 min at room temperature prior to the addition of the other assay components. The *N*-oxidation of ^{14}C-DMA was then determined by HPLC. The values shown represent mean percentage of control (no phospholipid added) ± the SD of triplicate determinations.

6.6 INHIBITION OF FMO BY TANNIC ACID

Continuing efforts by a number of investigators to find a specific inhibitor for FMO has resulted in the observation by this and Dr Ziegler's laboratory that tannic acid is an effective inhibitor of both pig liver and rabbit lung FMO (Fig. 6.7). However, the usefulness of tannic acid as a specific probe of rabbit lung FMO may be limited, as unpublished evidence from Dr Ziegler indicates that this compound also inhibits cytochrome *P*-450-dependent reactions through inhibition of NADPH-cytochrome *P*-450 reductase, at least in pig liver microsomes.

6.7 CONCLUSIONS

Rabbit lung FMO is a catalytically and immunochemically distinct form of FMO which is also present in a number of other species including the rat, mouse, hamster, guinea pig, rabbit and sheep (Tynes and Philpot, 1987;

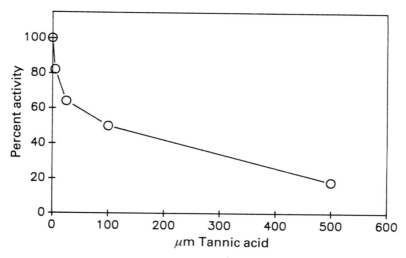

Figure 6.7 Inhibition of rabbit lung FMO ^{14}C-DMA *N*-oxidation by tannic acid. Rabbit lung FMO ^{14}C-DMA *N*-oxidation was assayed in the presence of 0, 5, 25, 100 and 500 μM tannic acid. The values shown are the means of duplicate experiments.

Williams *et al.*, 1985, 1989). This 'extrahepatic' form of FMO is also present in rabbit kidney and bladder. The limited amino acid composition and sequence information available also indicates that this 'extrahepatic' form of FMO differs markedly from the liver enzyme. Rabbit lung FMO is regulated developmentally by age, sex and pregnancy. The relatively high levels of rabbit FMO in fetal and neonatal lung is unusual for mono-oxygenase systems and may play a significant role in metabolism of xenobiotics crossing the placenta or administered through lactation. The molecular mechanism(s) of the developmental regulation of both the liver and the extrahepatic forms of FMO in the rabbit and other species awaits the sequencing of the gene(s) involved. In addition to developmental regulation, rabbit lung FMO may be rapidly modulated, for example diurnally or by hormonal stimulation, by variations in the levels of various phospholipids.

Note added in proof. Recently, Dr Philpot's laboratory has sequenced the cDNAs coding for pig liver, rabbit liver and rabbit lung FMO. The rabbit lung and liver FMO protein sequences are 56% identical; rabbit and pig liver FMO are more closely related (Lawton *et al.*, 1990).

REFERENCES

Burnette, W.N. (1981) 'Western blotting.' Electrophoretic transfer of proteins from sodium dodecyl sulfate-polyacrylamide gels to unmodified nitrocellulose and detection with antibody and radioiodinated Protein A. *Anal. Biochem.*, **112**, 195–203.

Dannan, G.A., Guengerich, F.P. and Waxman, D.J. (1986) Hormonal regulation of rat liver microsomal enzymes. Role of gonadal steroids in programming, maintenance, and suppression of 4-steroid 5-reductase, flavin-containing monooxygenase, and sex-specific cytochromes *P*-450. *J. Biol. Chem.*, **261**, 10728–35.

Das, M.L. and Ziegler, D.M. (1970) Rat liver oxidative *N*-dealkylase and *N*-oxidase activities as a function of animal age. *Arch. Biochem. Biophys.*, **140**, 300–6.

Devereux, T.R. and Fouts, J.R. (1974) *N*-Oxidation and demethylation of *N,N*-dimethylaniline by rabbit liver and lung microsomes. Effect of age and metals. *Chem.-Biol. Interactions*, **8**, 91–105.

Devereux, T.R. and Fouts, J.R. (1975) Effect of pregnancy or treatment with certain steroids on *N,N*-dimethylaniline demethylation and *N*-oxidation by rabbit liver or lung microsomes. *Drug Metab. Dispos.*, **3**, 254–8.

Dixit, A. and Roche, T.E. (1984) Spectrophotometric assay of the flavin-containing monooxygenase and changes in its activity in female mouse liver with nutritional and diurnal conditions. *Arch. Biochem. Biophys.*, **233**, 50–63.

Duffel, M.W., Graham, J.M. and Ziegler, D.M. (1981) Changes in dimethylaniline *N*-oxidase activity of mouse liver and kidney induced by steroid sex hormones. *Mol. Pharmacol.*, **19**, 134–9.

Gold, M.S. and Ziegler, D.M. (1973) Dimethylaniline *N*-oxidase and aminopyrine *N*-demethylase activities of human liver tissue. *Xenobiotica*, **3**, 179–89.

Heinze, E., Hlavica, P., Kiese, M. and Lipowsky, G. (1970) *N*-Oxygenation of arylamines in microsomes prepared from corpora lutea of the cycle and other tissues of the pig. *Biochem. Pharmacol.*, **19**, 641–9.

Lawton, M.P., Gasser, R., Tynes, R.E. *et al.* (1990) The flavin-containing mono-oxygenase enzymes expressed in rabbit liver and lung are products of related but distinctly different genes. *J. Biol. Chem.*, **265**, 5855–61.

Mayes, E.L.V. (1984) Peptide mapping by reverse-phase high pressure liquid chromatography, in *Methods in Molecular Biology*, vol. 1 Proteins (ed. J.M. Walker), Humana Press, USA, pp. 33–9.

Nagata, T., Williams, D.E. and Ziegler, D.M. (1990) Substrate specificities of rabbit lung and porcine liver flavin-containing monooxygenases: differences due to substrate size. *Chem. Res. Toxicol.*, **3**, 372–6.

Ohmiya, Y. and Mehendale, H.M. (1983) *N*-Oxidation of *N,N*-dimethylaniline in the rabbit and rat lung. *Biochem. Pharmacol.*, **32**, 1281–5.

Otvos, J.D., Krum, D.P. and Masters, B.S.S. (1986) Localization of the free radical on the flavin mononucleotide of the air-stable semiquinone state of NADPH-cytochrome *P*-450 reductase using ^{31}P-NMR spectroscopy. *Biochemistry*, **25**, 7220–8.

Poulsen, L.L. and Ziegler, D.M. (1979) The microsomal FAD-containing monooxygenase. Spectral characterization and kinetic studies. *J. Biol. Chem.*, **254**, 6449–55.

Poulsen, L.L., Taylor, K., Williams, D.E. *et al.* (1986) Substrate specificity of the rabbit lung flavin-containing monooxygenase for amines: oxidation products of primary alkylamines. *Mol. Pharmacol.*, **30**, 680–5.

Sotirhos, N., Herslof, B. and Kenne, L. (1986) Quantitative analysis of phospholipids by ^{31}P-NMR. *J. Lipid Res.*, **27**, 386–92.

Tynes, R.E. and Philpot, R.M. (1987) Tissue- and species-dependent expression of multiple forms of mammalian microsomal flavin-containing monooxygenase. *Mol. Pharmacol.*, **31**, 569–74.

Tynes, R.E., Sabourin, P.J. and Hodgson, E. (1985) Identification of distinct hepatic and pulmonary forms of microsomal flavin-containing monooxygenase in the mouse and rabbit. *Biochem. Biophys. Res. Commun.*, **126**, 1069–75.

Uehleke, H., Reiner, O. and Hellmer, K.H. (1971) Perinatal development of tertiary amine *N*-oxidation and NADPH cytochrome *c* reduction in rat liver microsomes. *Res. Commun. Chem. Pathol. Pharmacol.*, **2**, 793–805.

Williams, D.E., Ziegler, D.M., Nordin, D.J. *et al.* (1984) Rabbit lung flavin-containing monooxygenase is immunochemically and catalytically distinct from the liver enzyme. *Biochem. Biophys. Res. Commun.*, **125**, 116–22.

Williams, D.E., Hale, S.E., Muerhoff, A.S. and Masters, B.S.S. (1985) Rabbit lung flavin-containing monooxygenase. Purification, characterization, and induction during pregnancy. *Mol. Pharmacol.*, **28**, 381–90.

Williams, D.E., Meyer, H.H. and Dutchuk, M.S. (1989) Distinct pulmonary and hepatic forms of flavin-containing monooxygenase in sheep. *Comp. Biochem. Physiol.*, **93B**, 465–70.

Wirth, P.J. and Thorgeirsson, S.S. (1978) Amine oxidase in mice – sex differences and developmental aspects. *Biochem. Pharmacol.*, **27**, 601–3.

Ziegler, D.M. (1988) Flavin-containing monooxygenases: catalytic mechanism and substrate specificities. *Drug Metab. Rev.*, **19**, 1–32.

Ziegler, D.M., Taylor, K.L., Nagata, T. and Poulsen, L.L. (1986) Effects of phospholipids on activity of the purified hog liver flavin-containing mono-oxygenase. *Fed. Proc.* (abstract), **45**, 1871.

7

Human pharmacogenetics of nitrogen oxidations

S. Cholerton* and R.L. Smith

Department of Pharmacology and Toxicology, St Mary's Hospital Medical School, Norfolk Place, London W2 1PG, UK

1. Population and family studies of the metabolism of the simple, dietary derived, tertiary aliphatic amine trimethylamine to trimethylamine N-oxide, have demonstrated that this N-oxidation reaction is under genetic control and exhibits polymorphism in a British white population. A single autosomal diallelic gene locus is thought to be responsible with one allele for extensive oxidation and an uncommon variant allele for impaired metabolism.
2. The inheritance of two variant alleles results in primary trimethylaminuria, a condition characterized by an objectionable fish-like odour associated with the breath and sweat of affected individuals. Although these symptoms give rise to a variety of psychological and social problems, treatment of this condition is largely empirical, including the reduction of trimethylamine precursors present in the diet and the use of antibacterial drugs to depress gut floral activity.
3. The Michaelis constant (K_m) of trimethylamine for the hepatic N-oxidizing system of a trimethylaminuric individual was shown to be five times greater than derived from similar systems from the livers of unaffected subjects.
4. Although the implications of this genetically determined defect of N-oxidation in both homozygotes and heterozygotes of primary trimethylaminuria are as yet unknown, this reduced capacity to metabolize trimethylamine to its N-oxide may predispose certain individuals to developing secondary trimethylaminurias.

* Present address: Department of Pharmacological Sciences, The Medical School, Framlington Place, Newcastle-upon-Tyne NE2 4HH, UK.

7.1 INTRODUCTION

Drug and foreign compound biotransformations are mediated by a group of enzymes known collectively as the 'drug-metabolizing enzymes'. These include the microsomal mixed function mono-oxygenases, commonly referred to as the cytochrome *P*-450s but also several non-cytochrome *P*-450 enzymes including the cytosolic alcohol dehydrogenase and the microsomal flavin-containing mono-oxygenase, otherwise known as Ziegler's enzyme. Although such enzymes mediate the oxidation of xenobiotic compounds, many of them also have a known physiological role. Their activity is associated largely with the liver, however their presence has also been demonstrated in other tissues including lung, kidney and intestine. Furthermore, the enzymes of the gut flora are also known to be important mediators of the metabolism of certain xenobiotics.

Thus the human body is well equipped to handle foreign compounds. However, the quantitative and qualitative aspects of metabolism can influence the way in which an individual responds to a specific chemical entity. In this respect it is well established that there is interindividual variation in response to certain compounds and that this may have important implications with respect to drug toxicity and therapeutic outcome. Numerous factors are known to contribute to this variability and these include age, sex, disease state, nutritional status, pharmacodynamic variance, various environmental factors such as smoking and exposure to chemicals in the workplace and lastly the genetic factor. It has long been recognized that some of the drug-metabolizing enzymes of the body exibit polymorphism and that this genetically determined phenomenon can influence an individual's ability to metabolize specific compounds. This has resulted in the emergence of the discipline known as pharmacogenetics.

7.2 HUMAN PHARMACOGENETICS

In 1959 Vogel introduced the term pharmacogenetics for the study of genetically determined variations that are revealed solely by the effect of drug administration. Early examples of therapeutic agents whose metabolism was shown to be governed largely by genetic factors include the muscle relaxant, succinylcholine (Goedde and Altland, 1971) and the antitubercular compound, isoniazid (Drayer and Reidenberg, 1977). As such these two compounds revealed polymorphisms of hydrolysis and acetylation pathways respectively.

In contrast it is only relatively recently that genetic polymorphisms of oxidative metabolism have been demonstrated. Probably the most widely known example of this phenomenon is that which affects the 4-

hydroxylation of debrisoquine (Mahgoub *et al.*, 1977) and the oxidation of sparteine (Eichelbaum *et al.*, 1979). With respect to the metabolism of these compounds a population can be divided into two distinct phenotypes, the extensive metabolizer and the poor metabolizer. Impaired oxidation has been shown to be inherited as an autosomal recessive trait which occurs with a frequency of about 8% in Caucasian subjects (Evans *et al.*, 1983). To date the metabolism of more than 20 other commonly used drugs has been shown to cosegregate with this specific genetic polymorphism. Such observations, in addition to the association of some spontaneous diseases with a specific metabolic phenotype, have resulted in the gene regulating debrisoquine/ sparteine metabolism being perceived as a gene of major consequence and as such there has been considerable interest in the elucidation of the molecular basis of this metabolic defect. The gene has been located on chromosome 22 (Eichelbaum *et al.*, 1987). It has been shown to be multiallelic with at least four mutations all of which can give rise to the poor metabolizer phenotype (Gonzalez *et al.*, 1988).

As a consequence of the wealth of information available regarding the molecular basis and the pharmacological implications of the debrisoquine/ sparteine polymorphism, this tends to be given as the archetypal example of a genetic polymorphism of oxidative metabolism. However, as a phenomenon genetic polymorphism is extremely common and it has been estimated that at least one-third of human gene loci are multiallelic (Harris, 1980). Furthermore, there is growing evidence that polymorphism is more associated with external related functions such as HLA and the immune system and the enzyme processes responsible for the metabolism and detoxification of xenobiotic compounds. In contrast polymorphism, as measured by heterozygosity, is much less common in internal cellular metabolic processes and their respective control mechanisms (Johnson, 1973, 1974). In this respect it seems unlikely that the genes which regulate the enzymes of oxidative metabolism at centres other than carbon are exempt from the phenomenon of genetic polymorphism.

As such recent studies have shown that specific *S*- and *N*-oxidation reactions are also subject to a high degree of genetic control. Population and pedigree studies have shown that the sulphoxidation pathway for the mucolytic agent *S*-carboxymethyl-L-cysteine exhibits polymorphism in a British white population in which impaired sulphoxidation appears as an inherited recessive trait with an incidence of approximately 30% (Mitchell *et al.*, 1984). That a specific polymorphism affects certain *N*-oxidation reactions came to light primarily through studies in which the metabolism of the tertiary amine trimethylamine was investigated. The fact that trimethylamine is not a therapeutic agent distinguishes it from those substrates which revealed the genetic polymorphisms of hydrolysis, acetylation, *C*- and *S*-oxidation. Trimethylamine is a normal dietary constituent, a xenobiotic of no nutritive value at all. Such dietary components in general need to be metabolized for elimination and detoxification purposes, and this appears to be one of the main functions of the 'drug' metabolizing enzymes. Thus to be entirely correct, Vogel's term pharmacogenetics, which he applied to variations in response revealed solely by the effect of drug administration, must be expanded to 'food pharmacogenetics' to encompass the polymorphism associated with the N-oxidation of trimethylamine.

7.3 NITROGEN OXIDATIONS

After carbon and hydrogen, one of the more abundant elements present in drugs and other xenobiotic compounds is nitrogen. In these chemical entities, nitrogen is present in a variety of molecular configurations although it is most common as an amine. These may be either primary, secondary or tertiary alkylamines or arylamines. Several enzymes, including cytochrome P-450s and the flavin-containing mono-oxygenase, have been shown to be involved in the N-oxidation of these chemical groups and there appears to be a relatively high degree of overlap with respect to substrate specificity of these enzymes.

Many tertiary amine-containing drugs and other xenobiotics are metabolically converted into the corresponding N-oxides. These are salt-like compounds which are much less basic than the parent tertiary amine. The simplest N-oxide, trimethylamine N-oxide produced by N-oxidation of the tertiary amine trimethylamine was not only the first N-oxide to be synthesized (Dunstan and Goulding, 1899) but also the first to be discovered as a constituent of living tissues (Suwa, 1909).

It is the enzyme mediated N-oxidation of trimethylamine to trimethylamine N-oxide (Fig. 7.1) with which this chapter is largely concerned, since it is from this metabolic reaction that a polymorphism associated with the oxidation of nitrogen came to light. The enzyme responsible for the production of trimethylamine N-oxide is generally thought to be the non-P-450 flavin-containing mono-oxygenase (Ziegler,

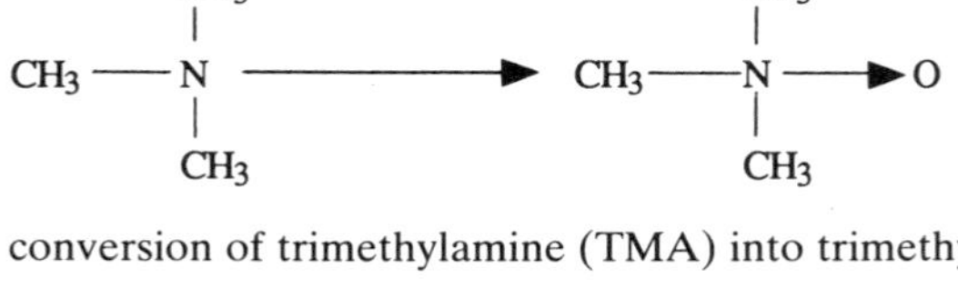

Figure 7.1 The conversion of trimethylamine (TMA) into trimethylamine N-oxide (TMAO).

1980; Sabourin and Hodgson, 1984) which *in vitro* has been shown to mediate the *N*-oxidation of many secondary amines to *N*-hydroxy metabolites and tertiary amines to *N*-oxides (Ziegler, 1985).

7.4 PRIMARY TRIMETHYLAMINURIA: A PHARMACOGENETIC PHENOMENON OF NITROGEN OXIDATION

7.4.1 Primary trimethylaminuria

The first full biochemical and clinical description of primary trimethylaminuria appeared in the literature in 1970 (Humbert *et al.*), although this phenomenon had been recognized in medical circles certainly as early as 1735 (Arbuthnot). To date almost 30 cases of this condition have been observed. Affected individuals excrete significantly increased concentrations of free trimethylamine in their urine. Furthermore, since trimethylamine has a powerful fish-like odour, they are further characterized by an objectional smell of rotting fish associated with their breath and sweat; hence the colloquialism for this condition, 'the fish-odour syndrome'. The presence of trimethylamine in the urine and the sweat of trimethylaminuric individuals has been confirmed by gas chromatographic analysis (Humbert *et al.*, 1970; Al-Waiz *et al.*, 1989).

Although there is no evidence that the elevated levels of free trimethylamine associated with primary trimethylaminuria have any toxicological effects *per se*, the fish-odour nature of this condition confers upon these affected individuals a variety of psychological and social problems. Early in life they have difficulties making friends since they and other children are aware of their peculiar body odour. Such ostracism can result in interrupted schooling and ultimately lead to low academic achievement. As adults they often experience sexual and marital problems associated with this disorder. Throughout their lives trimethylaminurics are continually anxious and embarrassed about their objectionable body odour and, quite understandably, often experience a feeling of rejection. This can lead to a depressive state which may result in a suicide attempt.

7.4.2 Natural occurrence of trimethylamine

Trimethylamine is a strongly basic (pK_a 9.8) simple tertiary aliphatic amine whose natural occurrence in human urine has been recognized for over 100 years (Dessaignes, 1856). An unusually low olfactory threshold, some three orders of magnitude less than that for comparable primary and secondary

Table 7.1 The olfactory threshold values of trimethylamine (and some related aliphatic amines)

Amine	Olfactory threshold concn (ppm)
Trimethylamine	0.00037
N-Methylpropylamine	0.01
Methylethylamine	9.0
Cyclobutylamine	33.0
Propylamine	90.1

Threshold values determined in water buffered to the pK_a of the amine (Willey, 1985).

amines (Table 7.1) and the fact that trimethylamine exists as a vapour at standard atmospheric pressure (boiling point 3°C) result in the readily recognizable odour of this amine. At concentrations of only a few parts per million the smell is fish-like whereas at higher concentrations of between 100–500 ppm the odour becomes acrid and ammoniacal.

Although trimethylamine is a normal constituent of human urine, very little trimethylamine is ingested as such. It is derived mainly from common dietary components containing the trimethylammonium $[(CH_3)_3N^+-]$ function such as choline, carnitine and possibly, to a lesser extent, from lecithin, betaine and ergothioneine (Fig. 7.2). Choline, which is probably the major precursor of trimethylamine, occurs at high levels in eggs, liver and soya beans (Dyer and Wood, 1947) whereas carnitine is mainly associated with meat products (Table 7.2). The extent to which ingested lecithin (phosphatidylcholine) acts as a precursor of trimethylamine is under debate since there is evidence that dietary lecithin is deacylated by pancreatic phospholipase and then absorbed by gut mucosal cells as lysolecithin thus releasing very little free choline.

In the presence of a fish component in the diet, the major source of trimethylamine is trimethylamine N-oxide. Marine fish such as skate and hake can contain up to 0.5% body weight as trimethylamine N-oxide whereas fresh water fish such as salmon and trout have much lower levels (Table 7.2). The actual process whereby these precursors are converted into trimethylamine remains vague, however it is generally thought to be due to enterobacterial degradation since urinary excretion of trimethylamine is greatly reduced by the use of antibiotics such as neomycin (De La Huerga and Popper, 1951; Prentiss *et al.*, 1961). With respect to trimethylamine N-oxide, 50% of a dose of this N-oxide appeared to be well absorbed and excreted unchanged when given to normal volunteers. The remainder was

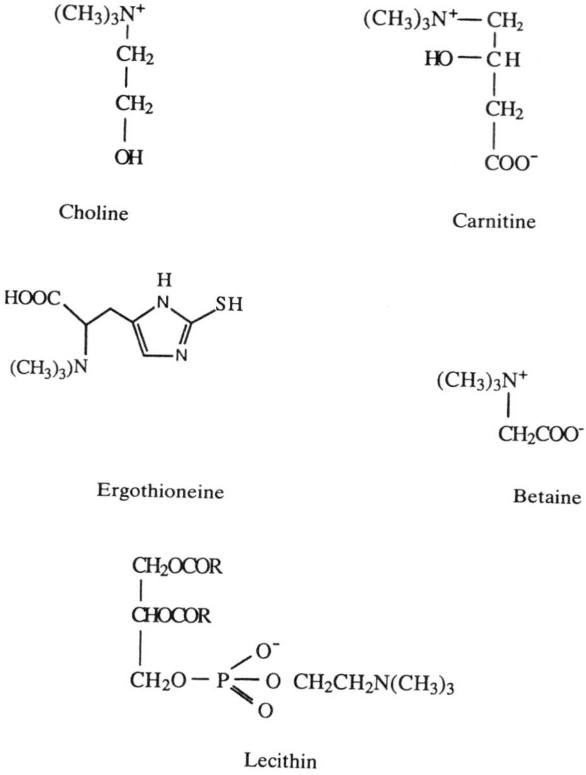

$$(CH_3)_3N^+$$
$$|$$
$$CH_2$$
$$|$$
$$CH_2$$
$$|$$
$$OH$$

Choline

$$(CH_3)_3N^+ - CH_2$$
$$|$$
$$HO - CH$$
$$|$$
$$CH_2$$
$$|$$
$$COO^-$$

Carnitine

HOOC, [imidazole structure] SH

$$(CH_3)_3N$$

Ergothioneine

$$(CH_3)_3N^+$$
$$|$$
$$CH_2COO^-$$

Betaine

$$CH_2OCOR$$
$$|$$
$$CHOCOR$$
$$|$$
$$CH_2O - P \overset{O^-}{\underset{O}{-}} O \ CH_2CH_2N(CH_3)_3$$

Lecithin

Figure 7.2 Precursors of trimethylamine present in the diet.

reduced, probably by gut floral enzymes, to trimethylamine which was absorbed and subsequently metabolized to the *N*-oxide (Al-Waiz *et al*, 1987a).

7.4.3 Metabolism of trimethylamine

It has long been recognized that trimethylamine undergoes *N*-oxidation *in vivo* to form the polar and non-odorous metabolite trimethylamine *N*-oxide and that this reaction is extensive (Lintzel, 1934; Wranne, 1956). A recent study (Al-Waiz *et al.*, 1987b) in healthy volunteers demonstrated that oral doses of ^{14}C-labelled trimethylamine are rapidly absorbed, metabolized to trimethylamine *N*-oxide and excreted in the urine with over 90% of the administered dose recovered in the form of the *N*-oxide. The

Table 7.2 Levels of trimethylamine precursors associated with various foods

Food	Level in food (mg/100 g)			
	Trimethylamine N-oxide	Choline	Lecithin	Carnitine
Dogfish	830	—	—	—
Skate	1103	—	—	—
Plaice	500	—	—	—
Hake	700	—	—	—
Salmon	83	—	—	—
Pickerel	58	—	—	—
Trout	66	—	580	—
Egg yolk	—	1500	—	—
Whole egg	—	—	400	—
Calf liver	—	650	850	3
Ox muscle	—	—	—	60
Lamb muscle	—	—	—	209
Cheese	—	75	—	—
Soya beans	—	273	1480	—
Potato	—	40	<1	—
Cauliflower	—	78	2	<1

Data compiled from Norris and Benoit (1945), Ronald and Jackobson (1947), Shewan (1951), Dyer (1952), Love (1980), Zeisel (1981), Tver and Russell (1981) and Mitchell (1978).

average daily urinary excretion of trimethylamine *N*-oxide for European white Caucasians consuming an average non-fish diet is approximately 50 mg compared to only 1–2 mg of trimethylamine (Al-Waiz *et al.*, 1987d). Thus trimethylamine *N*-oxide appears to be the major form in which trimethylamine is excreted. Although trimethylamine plasma concentration data derived from intravenous and oral dosing have not been used to determine the systemic bioavailability of this amine, its rapid absorption and excretion into urine as the *N*-oxide after oral administration suggests a high degree of first pass metabolism.

7.4.4 Enzymology of trimethylamine metabolism

That the *N*-oxidation of most tertiary amines requires a direct two-electron oxidation of the heteroatom *via* an ionic mechanism and that this reaction is mediated by the flavin-containing mono-oxygenase, since it is the only mammalian enzyme known to be capable of oxygenation by this mechanism, is now well established. However, the exact nature of the enzymology of the *N*-oxidation of the tertiary amine trimethylamine is at present unknown.

Much of our information in this area comes from the results of *in vitro* studies using animal tissues. Ziegler and Mitchell (1972) demonstrated that trimethylamine is a substrate for the flavin-containing mono-oxygenase in porcine liver microsomes. In the same species, the use of $^{18}O_2$ demonstrated that the oxygen incorporated into trimethylamine *N*-oxide is labelled, thus indicating the involvement of mixed function mono-oxygenases in this metabolic pathway (Baker and Chaykin, 1960, 1962). So although it appears that the flavin-containing mono-oxygenase is involved in the metabolic *N*-oxidation of trimethylamine, it looks likely that there is also a *P*-450 contribution. Hlavica and Kehl (1977) have shown this to be the case for this reaction in rabbit liver microsomes.

The nature of the *N*-oxidation of trimethylamine in man is even less well characterized than in animals. In 1972 Higgins *et al.* investigated the metabolic capacity of the trimethylamine *N*-oxidizing system derived from the liver of a trimethylaminuric patient with Noonan's syndrome. The Michaelis constant (K_m) for trimethylamine was found to be five times greater than that characteristic of the enzyme preparation derived from the livers of unaffected subjects. This suggests that a genetically determined qualitative change in the structure of the enzyme had occurred.

With the exception of the above study, much of the information regarding the enzymology of the *N*-oxidation of trimethylamine in man has been assimilated in an indirect manner. There is evidence to suggest that the metabolism of nicotine to its *N*-oxide metabolites is mediated by the flavin-containing mono-oxygenase (Damani *et al.*, 1988). This metabolic pathway has been shown to be defective in individuals with the inherited deficiency of trimethylamine *N*-oxidation (Ayesh *et al.*, 1988). Furthermore, the administration of methimazole, a known inhibitor of the flavin-containing mono-oxygenase, has been shown to reduce the production of the nicotine *N*-oxides from nicotine administered as a chewing gum preparation to healthy volunteers (Cholerton *et al.*, 1988).

7.4.5 Genetic aspects of the *N*-oxidation of trimethylamine

Not all individuals display an extensive capacity for *N*-oxidation. As previously mentioned, a condition known as trimethylaminuria exists in which affected individuals excrete significantly elevated urinary concentrations of free trimethylamine both under normal dietary conditions and following oral challenge with this amine. Ever since the original description of trimethylaminuria by Humbert *et al.* (1970), there have been sporadic reports of other affected individuals which at present amount to almost 30 cases. Several of the early reports suggested that the fish-odour syndrome may be an inherited condition, but this was never established.

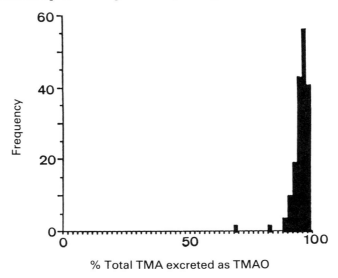

Figure 7.3 Population and frequency distribution in the urinary excretion of TMAO as a proportion of total TMA elimination for 169 normal healthy volunteers.

In fact it was not until the recent population and family studies of Al-Waiz *et al.* (1987c, 1988) that the genetic basis of trimethylaminuria was defined.

It is now well established that the *N*-oxidation of trimethylamine is genetically determined and that it exhibits polymorphism in a white Caucasian population. The ability to *N*-oxidize trimethylamine derived from the diet was found to be skewed in terms of its population distribution when investigated in 169 British white individuals. Using a diagnostic metabolic ratio of trimethylamine/trimethylamine *N*-oxide in 0–24 h urine samples metabolic 'outliers' were revealed (Fig. 7.3). These individuals and their immediate families were subsequently studied. The administration of a 600 mg oral ('load') dose of trimethylamine and the subsequent collection of 0–8 h urine samples made it possible to differentiate between homozygous affected and heterozygous unaffected individuals (carriers). The latter show saturation of *N*-oxidation capacity at this dose whereas a 900 mg dose of trimethylamine was required to precipitate a reduction in the *N*-oxidation of this amine in homozygous unaffected subjects. Thus this 'trimethyamine load test' appeared to be a simple diagnostic method with which to determine the incidence of carriers of this metabolic defect and in the British study this parameter was found to be two out of the 169 subjects studied. The authors suggest, however, that these figures may in fact be an underestimate of those in a normal population.

From the results of these investigations it has been proposed that the *N*-oxidation of trimethylamine is controlled by a single autosomal diallelic locus with one allele for rapid and extensive oxidation and an uncommon variant allele for impaired metabolism. In this respect primary trimethylaminuria is considered to be an inherited metabolic disorder in which the affected individual is homozygous for the allele which determines defective *N*-oxidation.

7.4.6 Case examples of primary trimethylaminuria

Humbert *et al*'s (1970) description of primary trimethylaminuria concerned a 6-year-old girl with the clinical stigmata of Turner's syndrome (Noonan's syndrome) presenting with a chronic chest infection. She also had splenomegaly and various haematological abnormalities. Her mother noticed that the girl intermittently had a peculiar fish-like odour and she was shown to excrete large amounts of free trimethylamine under normal dietary conditions. In addition, after an oral loading dose of the amine, nearly 70% was excreted as unchanged trimethylamine. However, it must be stressed that primary trimethylaminuria can and does appear independently of Noonan's syndrome. Nevertheless, the fact that defective *N*-oxidation of trimethylamine has been shown to be associated with Noonan's syndrome, a condition proposed to be inherited in an autosomal dominant fashion with irregular penetrance, may in fact provide information regarding the location of the gene for the enzyme responsible for the *N*-oxidation of trimethylamine.

The salient features of the published cases of primary trimethylaminuria and also some as yet unpublished cases studied in this laboratory are given in Table 7.3. Information derived from the investigation of these latter individuals draws attention to the occurrence of this condition in early life in both sexes, expression of maternal concern, the embarrassment and anxiety experienced by the sufferers and the possible association of primary trimethylaminuria in females with non-infectious vaginosis. Finally these case reports illustrate the use of the previously mentioned 'trimethylamine load test' to characterize heterozygotes.

Subjects 1 and 2

Subject 1 was a 21-year-old healthy female working as a secretary. Her powerful body odour and the production of urine which had a fish-like smell had been noticed by her mother since the girl's early childhood. Subject 2 was the 18-year-old sister of subject 1. She had also caused her mother considerable concern as a baby because of the persistent graveolant smell

Table 7.3 Clinical details, family history and methods of diagnosis of trimethylaminuria (TMAuria) cases

No. of cases	Sex	Family history of a TMAuria	Family study	Association with other clinical conditions	Method of diagnosis	Treatment	Reference
1	F	None	—	Congenital deformities	Spot urine analysed by GCMS and TMA load	Food restriction	Humbert et al. (1970)
4	2F	None	—	None	Spot urine analysed by GC	Food restriction and neomycin	Danks et al. (1976)
3	1F	2 sibs	—	—	Morning urine analysed by GCMS	Food restriction	Lee et al. (1976)
1	M	None	Choline load test	—	Spot urine analysed by GC after choline load	Food restriction	Marks et al. (1977)
1	M	None	—	—	Urine collected after fish meal analysed by GC	Food restriction	Todd (1979)
1	F	None	—	—	Morning urine analysed by GC after fish meal/ choline load	Food restriction	Spellacy et al. (1979)
1	M	None	—	Copper deficiency preterm baby	Odour	Choline restriction	Blumenthal et al. (1980)

1	3M	None	—	—	Random urine analysed by GC and choline load	Food restriction	Brewster and Schedewie (1983)
1	M	None	—	Congenital deformities	Random urine analysed by GC	Food restriction metronidazole	Shelley and Shelley (1984)
1	M	None	—	—	Random urine analysed by GC	Milk restriction	Rothschild and Hansen (1985)
1	2F	2 sibs	TMA load test	—	24 h urine collection and TMA load test analysed by GC	Food restriction	Al-Waiz *et al.* (unpublished data)
1	M	—	TMA load test	—	24 h urine collection and TMA load test analysed by GC	Food restriction	Al-Waiz *et al.* (unpublished data)
1	F	—	TMA load test	—	24 h urine collection analysed by GC	Food restriction	Al-Waiz *et al.* (unpublished data)

GCMS, gas chromatography – mass spectrometry.

associated with her urine. Both sisters left school at an early age because of social difficulties arising from their bodily odours. Although both sisters were successful in their work, they remained embarrassed and anxious about their condition. Both sisters showed clear biochemical evidence of primary trimethylaminuria as determined by a ratio of urinary tri-methylamine/trimethylamine N-oxide in samples obtained uder normal dietary conditions and following the oral administration of trimethylamine (300 mg). The latter test precipitated a definite 'fish-odour syndrome reaction' which forced the sisters into seclusion throughout the weekend the test was carried out. Their elder sister and both (unrelated) parents were all healthy and did not experience body odour problems. Detailed questionnaires revealed no family history of similar complaints. Routine haematological and biochemical investigations of all family members revealed no abnormalities. Both parents were shown to be heterozygotes (carriers) of the metabolic defect on the basis of the 'trimethylamine load test' (Al-Waiz *et al.*, 1988).

Subject 3

This subject was the healthy 16-year-old daughter of unrelated parents. When the subject was approximately 2 years old her mother first noticed a strong unpleasant smell of fish associated with her daughter which appeared to be unaffected by numerous bathings. Neither parent nor the two younger sisters had any history of the fish-odour syndrome. Normal results were obtained for various biochemical and haematological tests in all family members. The affected daughter was confirmed to be trimethylaminuric on the basis of the urinary trimethylamine/trimethylamine N-oxide ratio under normal dietary conditions. Carrier status in both parents was revealed by the 'trimethylamine load test'.

Subject 4

This subject was an academically high achieving 13-year-old girl. Her mother first noticed her daughter's fish-odour problem when the girl was less than one year old. In later years the mother was able, on the basis of observation, to associate exacerbation of the problem with the consumption of fish and foods rich in choline. Of interest was the observation that following a fish meal the mother complained that her vagina developed a strong fish smell. There is a suggestion that some forms of non-infective vaginosis may occur in carriers of the fish-odour syndrome and that this may be exacerbated by the diet. By contrast neither the father nor the 10-year-old sister experienced body odour problems after dietary challenge. For all family members the routine haematological and biochemical parameters were normal. Analysis of urine produced under normal dietary conditions and also following trimethylamine challenge confirmed that the affected girl was in fact suffering from primary trimethylaminuria. Both parents and the unaffected sister were diagnosed as carriers.

Subject 5

This subject was a 4-year-old hyperactive boy of normal health and mental status for his age. His mother reported that a persistent objectionable body odour resistant to washing had been associated with her son since he was only 2 months old. There was no family history of the fish-odour syndrome and both parents failed to report any symptoms. Biochemical and haematological parameters were normal for the affected boy and his parents. Analysis of urine for trimethylamine/trimethylamine *N*-oxide under normal dietary conditions confirmed that the boy was in fact trimethylaminuric and that his parents were carriers of this defect.

7.4.7 The management of primary trimethylaminuria

Although the greatly elevated concentrations of trimethylamine associated with primary trimethylaminuria do not appear to be toxic *per se*, the wide range of psychosocial conditions which result from this metabolic defect necessitate the need for treatment. Since there have been no controlled studies on the management of the fish-odour syndrome, the treatment of this condition remains essentially empirical. This includes the reduction of trimethylamine precursors present in the diet and the occasional use of antibacterial drugs to depress the activity of the gut flora.

In our hands it appears that dietary restriction, which requires the complete avoidance of foods rich in choline such as eggs, liver, soya beans, kidney and mayonnaise and marine fish which are relatively rich in trimethylamine *N*-oxide, is relatively successful in the management of this metabolic disorder. However, even with rigorous adherence to this regimen, trimethylaminurics experience occasional 'breakthrough' episodes of the fish-odour syndrome for reasons unknown. In contrast, Danks *et al.* (1976) observed that this dietary approach was only successful in one of the four affected individuals investigated. In the same study, a short course of neomycin temporarily abolished the fish-odour in a single patient to whom it was given. Shelley and Shelley (1984) found that a regime of dietary restriction combined with metronidazole was effective in suppressing the body odour. Thus antibacterial therapy coupled with dietary restriction appears to be successful in the treatment of the fish-odour associated with primary trimethylaminuria presumably by suppressing the gut floral degradation of choline and trimethylamine N-oxide.

A quite different approach to the management of the fish-odour syndrome was applied to an individual with trimethylaminuric symptoms associated with suspected viral hepatitis (Ruocco *et al.*, 1989). The administration of either a digestive enzyme complex containing pepsin and papain or the 'acidification of the diet' with yogurt or citrus fruits reduced the intensity of fishy body odour. There was no suggestion as to why this mode of treatment should alleviate the problem of trimethylaminuria.

7.5 SECONDARY TRIMETHYLAMINURIAS: THE CONSEQUENCE OF REDUCED N-OXIDATION CAPACITY

It is now well established that primary trimethylaminuria is a genetically determined condition in which there is a greatly reduced capacity for N-oxidation by a specific metabolic pathway. However, the outwardly obvious signs of the fish-odour syndrome can also occur sporadically with no apparent genetic basis. Relatively little is known about the nature of this condition. However, it appears to be associated with conditions which affect trimethylamine burden, metabolism and elimination. In these cases the condition is known as secondary trimethylaminuria and situations including liver disease, renal failure and precursor overload have all been associated with its occurrence.

In 1978 Marks *et al.* described elevated urinary concentrations of trimethylamine in patients with chronic liver disease, when compared to healthy controls, on a normal diet and also following an oral load dose of choline. A more recent study (Cholerton *et al.* 1989) demonstrated that of 24 patients with liver disease, 12 excreted less trimethylamine N-oxide as a percentage of total trimethylamine than would be expected for a healthy British white population under normal dietary conditions. Ruocco *et al.* (1989) reported the case of an individual who developed the symptoms of trimethylaminuria for the first time in adulthood. The authors suggested that the individual had previously had asymptomatic viral hepatitis, as indicated by the presence of HAV Ab in his serum, and although liver function tests were normal, this had resulted in a defect in trimethylamine metabolism.

It has been proposed that the trimethylaminuria and the reduced N-oxidation capacity associated with liver disease may be due to impaired hepatocellular function or to the presence of a circulatory pathway known as porta-systemic venous shunting through which portal blood may enter the systemic veins and reach the brain without prior metabolism in the liver

(Wranne, 1956; Marks *et al.*, 1978). This may have important implications since trimethylamine is thought to undergo extensive first-pass metabolism; these mechanisms may well increase its systemic bioavailability. In this respect it has been suggested that the resulting increased blood concentrations of trimethylamine and dimethylamine may act synergistically with ammonia, free fatty acids and mercaptans as aetiological factors in the development of the encephalopathy often observed in cirrhotic patients. Furthermore, impaired hepatic trimethylamine metabolism may underlie the development of hepatic fetor since the excessive amounts of the amine associated with hepatic insufficiency will be eliminated not only in the urine of these patients but also in their breath.

Trimethylaminuria and other features of the fish-odour syndrome are also associated with chronic renal failure. Abnormal overgrowth of gut bacteria with concomitant impaired renal function leading to reduced urinary clearance, result in an increased trimethylamine burden in uraemia. As a consequence the amine is cleared by other organs such as the lung, resulting in trimethylamine in the expired air, the presence of which has been confirmed by gas chromatography–mass spectrometry (Simenhoff *et al.*, 1977; Wills and Savory, 1981). With much similarity to its proposed effect in liver disease, it has been suggested that the elevated concentrations of trimethylamine found in uraemic patients may also cross the blood–brain barrier leading to neurological symptoms.

In the case of renal failure two factors, increased trimethylamine burden and reduced urinary clearance, act synergistically to produce secondary trimethylaminuria. However, increased body burden alone has been shown to be sufficient to give rise to fish-odour symptoms. Large doses of choline have been used in the treatment of Alzheimer's disease, tardive dyskinesia, ataxia and Huntington's chorea on the supposition that this could improve cholinergic transmission in these conditions. In such a study, patients with Huntington's chorea were given large doses of choline (8–20 g per day). Although the authors claimed an improvement in the clinical condition of the patients, all complained of an unpleasant 'fish-odour syndrome' (Growden *et al.*, 1977). Furthermore, a patient receiving choline bitartrate for the treatment of Alzheimer's disease developed faecal incontinence and a foul odour possibly due to the excessive production of trimethylamine (Etienne *et al.*, 1978). These observations are probably due to the production of excessive amounts of the amine by enterobacterial degradation of these large doses of choline, a known trimethylamine precursor, which lead to saturation of the *N*-oxidation capacity of the liver and ultimately result in trimethylaminuria.

In cases of bacterial infection such as bacterial vaginosis, the vaginal discharge frequently has a prominent fish-like odour and analysis has confirmed the presence of trimethylamine (Brand and Galask, 1986).

Brewster and Schedewie (1983) observed increased urinary concentrations of trimethylamine in both vaginitis and cervical cancer and this was attributed to local bacterial degradation of trimethylamine *N*-oxide. However, the presence of a fishy-smelling vaginal discharge after a fish meal has been demonstrated in individuals who are known carriers of primary trimethylaminuria (subject 3; section 7.4.6). The association of this clinical condition with the pharmacogenetic phenomenon of defective *N*-oxidation is at present under investigation.

Thus secondary trimethylaminurias appear to arise as a consequence of a specific disorder or treatment regimen. However, it may be the case that certain individuals are more predisposed than others to developing the symptoms of the fish-odour syndrome, when it occurs in this sporadic way, and that this predisposition may have a genetic origin. The ability to saturate the metabolic pathway responsible for the *N*-oxidation of trimethylamine has been observed in those individuals homozygous affected, heterozygous unaffected and homozygous unaffected for primary trimethylaminuria. Decreased *N*-oxidation capacity was apparent after 300 mg, 600 mg and 900 mg single oral doses of trimethylamine respectively. The relatively limited capacity of the affected individuals is apparent even under conditions of normal dietary intake, hence the need for their dietary restriction. In contrast, symptoms of the fish-odour syndrome are not observed in healthy, unaffected heterozygotes under normal dietary conditions. However, these individuals may be more susceptible to secondary trimethylaminurias precipitated by conditions such as liver disease and precursor overload than homozygous unaffected individuals, since their already limited *N*-oxidation capacity would be exacerbated by factors such as chronic hepatic insufficiency and a relatively large body burden of trimethylamine.

Furthermore, it has been proposed that exposure to certain xenobiotics known to be powerful inhibitors of the flavin-containing mono-oxygenase may produce a secondary trimethylaminuria by inhibition of the *N*-oxidation of dietary derived trimethylamine. Such compounds include 1-naphthylthiourea and several goitrogenic substances including thiourea, propylthiouracil, methimazole and the *Brassica* thioglucoside progoitrin, which on ingestion is broken down by the gut bacteria to goitrin [(−)-5-vinyl-2-oxazolidinethione)], a potent goitrogen (Pearson *et al.*, 1980, 1981).

Studies in one breed of chicken (*Gallus domesticus*) with a genetically reduced ability to *N*-oxidize trimethylamine, demonstrated that the metabolic defect was markedly exacerbated if the birds had been fed on a *Brassica* rapeseed meal (Pearson *et al.*, 1979). It was shown that this foodstuff not only contains progoitrin and tannin, both potent inhibitors of trimethylamine *N*-oxidation but it is also rich in sinapine (the choline ester of sinapic acid) and is therefore a potentially rich source of trimethylamine. Under such circumstances of potential inhibition and precursor overload of the enzyme, sufficient unoxidized trimethylamine was generated for the amine to pass into the egg, render it 'tainted' and unfit for human consumption. These observations have led to suggestions that the progoitrin present in *Brassica* species such as cabbage, cauliflower, Brussels sprouts and swede frequently consumed by man may have the potential to inhibit the human *N*-oxidation of trimethylamine (Oginsky *et al.*, 1965; Fenwick *et al.*, 1983a). However, at present it is unknown whether or not the consumption of these *Brassica* vegetables can give rise to secondary trimethylaminurias (Fenwick *et al.*, 1983b). A study involving three healthy individuals consuming up to 100 times the average daily intake of Brussels sprouts and swede revealed no effect on the capacity to *N*-oxidize trimethylamine (Al-Waiz *et al.*, 1988). However, it may be that it is the defective form of the enzyme responsible for the *N*-oxidation of trimethylamine that is susceptible to inhibition by dietary goitrogens and that reduced trimethylamine *N*-oxide production would only be seen in trimethylaminurics and carriers of this genetic defect. Such proposals raise many important questions regarding the nature and nurture of those individuals who develop secondary trimethylaminurias. For example, it seems reasonable to suggest that dietary factors may exacerbate a genetic predisposition to fish-odour symptoms among certain individuals in a way similar to that observed in the specific breed of chicken (Pearson *et al.*, 1979).

The exact physiological role of the flavin-containing mono-oxygenase has not been elucidated, however the decarboxylated metabolite of cysteine, cysteamine, is the only known endogenous substrate (Ziegler, 1980). Thus it has been proposed that this enzyme catalyses the oxidation of cysteamine to cystamine and as such provides oxidizing equivalents necessary for disulphide bond formation during protein synthesis (Ziegler and Poulsen, 1979). Furthermore it has been suggested that this enzyme may influence a number of other complex cellular processes because of its role in maintaining the cellular thiol:disulphide ratio. The implications of a defect in the flavin-containing mono-oxygenase, such as that proposed in primary trimethylaminuria, on these cellular mechanisms are as yet unknown.

7.6 IMPLICATIONS OF REDUCED *N*-OXIDATION CAPACITY ON THE METABOLISM OF OTHER XENOBIOTICS

Studies *in vitro* have shown that flavin-containing mono-oxygenase is responsible not only for the *N*-oxidation of many secondary and tertiary alkylamines, arylamines and hydrazines but also for the *S*-oxidation of the sulphur-containing sulphides, thiols, thioamides and thiocarbamides. Many important groups of drugs contain these specific chemical entities including the antidepressant amines, imipramine and desipramine and the antithyroid thiocarbamides, methimazole and phenylthiourea. Thus it is reasonable to suggest that those individuals who possess a defect in *N*-oxidation, as demonstrated by reduced ability to form the *N*-oxide of trimethylamine, may in fact display compromised metabolism, either *S*- or *N*-oxidation, of other compounds.

In this respect the extent of *N*-oxidation of the tertiary amine nicotine has been shown to be deficient in trimethylaminurics. Unaffected healthy volunteers and two trimethylaminuric sisters (subjects 1 and 2, section 7.4.6) received nicotine in the form of a chewing gum preparation. During an 8-hour period the volunteer group excreted 0.08–0.76 µmol of urinary cotinine, a cytochrome *P*-450-mediated metabolite of nicotine, and 0.27–2.16 µmol of nicotine *N*-oxides, the proposed flavin-containing mono-oxygenase mediated metabolites. In contrast the two sisters excreted 2.0 and 2.4 µmol of cotinine and <0.1 µmol of nicotine *N*-oxides (Ayesh *et al.*, 1988). However, a similar study has revealed the lack of cosegregation between the *N*-oxidation of the antihypertensive agent pinacidil and the trimethylamine *N*-oxidation polymorphism (Ayesh *et al.*, 1989). Both affected and unaffected individuals were seen to excrete approximately 30–50% of an oral dose of the drug in the urine in the form of the *N*-oxidation product, pinacidil *N*-oxide. The lack of cosegregation observed in this study suggests that the *N*-oxidations of trimethylamine and pinacidil are mediated by distinct enzymes. It is suggested that the *N*-oxidation of pinacidil may possibly be mediated by one or more of the *P*-450 isoenzymes.

Under normal dietary conditions trimethylaminuric individuals have a reduced capacity to produce the *N*-oxide metabolite of trimethylamine. Carriers of this defect also have diminished *N*-oxidation capacity, however this is only observed under challenge conditions. It is possible that the *N*-oxidation of other xenobiotic compounds, ingested in relatively large quantities and with high affinity for the flavin-containing mono-oxygenase, may exhibit a certain degree of saturation in heterozygotes for the fish-odour syndrome. This may in part explain both the interindividual variation in response and the toxic side effects of the many xenobiotic compounds whose metabolism is thought to be partially or wholly under the control of the flavin-containing mono-oxygenase.

As a general rule the products of the drug-metabolizing enzymes tend to be compounds which are less toxic than the parent xenobiotic. In this respect, oxidation by the flavin-containing mono-oxygenase of most functional groups containing organic nitrogen results in detoxification, such as the production of the polar hydroxylated products of the phenothiazines. However, the *N*-oxidation of secondary arylamines and 1,1-disubstituted hydrazines results in the production of metabolites biologically more reactive than the parent compound. The porcine liver form of the flavin-containing mono-oxygenase has been shown to catalyse the *N*-oxidation of alkylarylamines to nitrones. This reaction is generally considered an early obligative event in the bioactivation of *N*-methylarylamines to carcinogenic metabolites. In this respect, Kadlubar *et al.* (1976) demonstrated that the *N*-oxidation of *N*-methyl-4-aminoazobenzene, catalysed almost exclusively by the flavin-containing mono-oxygenase, produces the reactive nitrone which, on hydrolysis, yields the chemically reactive primary hydroxylamine.

Like many other enzymes, including those responsible for xenobiotic metabolism the flavin-containing mono-oxygenase appears to exist in multiple forms. An isoenzyme similar in composition and mechanism to the liver form, but immunologically and catalytically distinct, has been isolated from rabbit lung microsomes (Williams *et al.*, 1985; Tynes *et al.*, 1985). Furthermore distinct isoenzymes have been detected in different tissues in mice (Tynes and Hodgson, 1985). The presence of a specific lung form of the flavin-containing mono-oxygenase may provide a potential route for metabolic activation of xenobiotics, since this isoenzyme appears to be solely responsible for the oxidation of primary alkylamines to chemically reactive hydroxylamines (Williams *et al.*, 1985; Tynes *et al.*, 1985). Thus, in such situations the genetically determined defect in the *N*-oxidation pathway present in primary trimethylaminuria may be advantageous, since this deficiency may protect these individuals from the potential toxic effects of specific nitrogen-containing compounds. However, such proposals depend largely on whether or not the properties of the mouse and the rabbit enzymes can be extrapolated to the human form. A study by McManus *et al.* (1987) suggests that for certain flavin-containing mono-oxygenase mediated reactions such similarities exist, although there is no evidence as yet that the human enzyme exists in multiple forms.

REFERENCES

Al-Waiz, M. (1988) 'Genetic, polymorphism of trimethylamine *N*-oxidation in man and its relationship to the fish-odour syndrome'. Ph.D. thesis. London University, pp. 172–3.

Al-Waiz, M., Ayesh, R., Mitchell, S.C. *et al.* (1987a) Disclosure of the metabolic retroversion of trimethylamine *N*-oxide in humans. *Clin. Pharmacol. Ther.*, **42**, 602–12.

Al-Waiz, M., Mitchell, S.C., Idle, J.R. and Smith, R.L. (1987b) The metabolism of ^{14}C-labelled trimethylamine and its *N*-oxidation in man. *Xenobiotica*, **17**, 551–8.

Al-Waiz, M., Ayesh, R., Mitchell, S.C. *et al.* (1987c) A genetic polymorphism of the *N*-oxidation of trimethylamine in humans. *Clin. Pharmacol. Ther.*, **42**, 588–94.

Al-Waiz, M., Ayesh, R., Mitchell, S.C. *et al.* (1987d) The relative importance of N-oxidation and N-demethylation in the metabolism of trimethylamine in man. *Toxicology*, **43**, 117–21.

Al-Waiz, M., Ayesh, R., Mitchell, S.C. *et al.* (1988) Trimethylaminuria ('Fish-odour syndrome'): a study of an affected family. *Clinical Science*, **74**, 231–6.

Al-Waiz, M., Ayesh, R., Mitchell, S.C. *et al.* (1989) Trimethylaminuria: the detection of carriers using a trimethyalmine load test. *J. Inherited Metab. Dis.*, **12**, 80–85.

Arbuthnot, J. (1735) in *An Essay Concerning the Nature of Ailments*, (3rd ed), J. Tonson, London, pp. 82–3.

Ayesh, R., Al-Waiz, M., Crothers, M. J. *et al.* (1988) Deficient nicotine N-oxidation in two sisters with trimethylaminuria. *Br. J. Clin. Pharmacol.*, **25**, 664.

Ayesh, R., Al-Waiz, M., McBurney, A. *et al.* (1989) Variable metabolism of pinacidil: lack of correlation with debrisoquine and trimethylamine C- and N-oxidative polymorphisms. *Br. J. Clin. Pharmacol.*, **27**, 423–8.

Baker, J.R. and Chaykin, S. (1960) The biosynthesis of trimethylamine N-oxide. *Biochim. Biophys. Acta*, **41**, 548–50.

Baker, J.R. and Chaykin, S. (1962) The biosynthesis of trimethylamine N-oxide. *J. Biol. Chem.*, **237**, 1309–13.

Blumenthal, I., Lealman, G.T. and Franklyn, P.P. (1980) Fracture of femur, fish odour and copper deficiency in a preterm infant. *Arch. Dis. Child.*, **55**, 229–31.

Brand, J.M. and Galask, R.P. (1986) Trimethylamine: the substance mainly responsible for the fishy odour often associated with bacterial vaginosis. *Obstet. Gynaecol.*, **68**, 682–5.

Brewster, M.A. and Schedewie, H. (1983) Trimethylaminuria. *Ann. Clin. Lab. Sci.*, **13**, 20–4

Cholerton, S., Ayesh, R., Idle, J.R. and Smith, R.L. (1988) The pre-eminence of nicotine N-oxidation and its diminution after carbimazole administration. *Br. J. Clin. Pharmacol.*, **26**, P652.

Cholerton, S., Ayesh, R., Robinson, H. *et al.* (1989) Secondary trimethylaminurias: effect of liver disease on the N-oxidation of trimethylamine. *Prog. Pharmacol. Clin. Pharmacol.* in press.

Damani, L.A., Pool, W.F., Crooks, P.A. *et al.* (1988) Stereoelectivity in the N-oxidation of nicotine isomers by flavin-containing monooxygenase. *Mol. Pharmacol.*, **33**, 702–5.

Danks, D.M., Hammond, J., Schlesinger, P. *et al.* (1976) Trimethylaminuria: diet does not always control the fishy odour. *N. Engl. J. Med.*, **295**, 962.

De La Huerga, J. and Popper, M. (1951) Urinary excretion of choline metabolites following choline administration in normals and patients with hepatobiliary diseases. *J. Clin. Invest.*, **30**, 364–70.

Dessaignes, M. (1856) Trimethylamin aus menschenharn. *Justis Liebigs Annalen der Chemie*, **100**, 218.

Drayer, D.E. and Reidenberg, M.M. (1977) Clinical consequences of polymorphic acetylation of basic drugs. *Clin. Pharmacol. Ther.* **22**, 251–8.

Dunstan, W.R. and Goulding, E. (1899) The action of alkyl haloids on hydroxylamine. Formation of substituted hydroxylamines and oxamines. *J. Chem. Soc. (Lond.)*, **75**, 792–807.

Dyer, F.E. and Wood, A.J. (1947) Action of enterobacteriaceae on choline and related compounds. *J. Fish. Res. Bd Can.*, **7**, 17–21.

Dyer, D. J. (1952) Amines in fish muscle. VI. Trimethylamine oxide content of fish and marine vertebrates. *J. Fish. Res. Bd Can.*, **8**, 314–24.

Eichelbaum, M., Baur, M.P. and Dengler, H.J. (1987) Chromosomal assignment of human cytochrome *P*-450 (debrisoquine/sparteine type) to chromosome 22. *Br. J. Clin. Pharmacol.,* **23**, 455–8.

Eichelbaum, M., Spannbrucker, N., Steincke, B. and Dengler, H.J. (1979) Defective *N*-oxidation of sparteine in man: a new pharmacogenetic defect. *Eur. J. Clin. Pharmacol.,* **16** , 183–7.

Etienne, P., Gauther, S., Johnson, G. *et al.* (1978) Clinical effects of choline in Alzheimer's disease. *Lancet,* **i**, 508–9.

Evans, D.A.P., Harmer, D., Downtham, D.Y. *et al.* (1983) The genetic control of sparteine and debrisoquine metabolism in man with new methods of analysing bimodal distribution. *J. Med. Genet.* **20**, 321–9.

Fenwick, G.T., Heaney, R.K. and Mullin, J.W. (1983a) Glucosinolates and their breakdown products in food and food plants. *CRC Crit. Rev. Food, Sci. Nutri.* **18**, 123–201.

Fenwick, G.R., Butler, E.J. and Brewster, M.A. (1983b) Are *Brassica* vegetables aggravating factors in trimethylaminuria? *Lancet*, **i**, 916.

Goedde, G.S. and Altland, K. (1971) Suxamethonium sensitivity. *Ann. NY Acad. Sci.,* **179**, 666–70.

Gonzalez, F.G., Skoda, R.C., Kimura, S. *et al.* (1988) Characterization of the common genetic defect in humans deficient in debrisoquine metabolism. *Nature*, **331**, 442–6.

Growden, J.H., Cohen, E.L. and Wurtmann, R.J. (1977) Huntington's disease: clinical and chemical effects of choline administration. *Ann. Neurol.,* **1**, 418–22.

Harris, H. (1980) in *The Principles of Human Biochemical Genetics* (4th edn), North Holland, Amsterdam.

Higgins, T., Chagkin, S., Hammond, K.B. and Humbert, J.R. (1972) Trimethylamine *N*-oxide synthesis; human variant. *Biochem. Med.,* **6**, 392–6.

Hlavica, P. and Kehl, M. (1977) Studies on the mechanism of hepatic microsomal *N*-oxide formation. The role of cytochrome *P*-450 and mixed-function amine oxidase in the *N*-oxidation of *N,N*-dimethylaniline. *Biochem. J.,* **164**, 487–96.

Humbert, J.R., Hammond, K.B., Hathaway, W.E. *et al.* (1970) Trimethylaminuria: the fish-odour syndrome. *Lancet*, **i**, 770–1.

Johnson, G.B. (1973) Importance of substrate variability to enzyme polymorphism. *Nature*, **243**, 151–3.

Johnson, G.B. (1974) Enzyme polymorphism and metabolism. *Science,* **184**, 28–37.

Kadlubar, F.F., Miller, J.A. and Miller, E.C. (1976) Microsomal *N*-oxidation of the hepatocarcinogen *N*-methyl-4-aminoazobenzene and the reactivity of *N*-hydroxy-*N*-methyl-4-aminoazobenzene. *Cancer Res.* **36**, 1196–206.

Lee, C.W.G., Tu, J.S., Turner, B.R. and Murphy, K.E. (1976) Trimethylaminuria: fishy odours in children. *N. Engl. J. Med.,* **295**, 937–8.

Lintzel, W. (1934) Trimethyloxyd. Menschlichen Harn. *Klin. Wochenschr.,* **13**, 304–5.

Love, R.M. (1980) in *The Chemical Biology of Fishes*, Academic Press, London, pp. 458–9.

Mahgoub, A., Idle, J.R., Dring, L.G. *et al.* (1977) Polymorphic hydroxylation of debrisoquine in man. *Lancet*, **ii**, 484–6.

Marks, R., Graves, M.W., Prottey, C. and Hartop, P.J. (1977) Trimethylaminuria: the use of choline as an aid to diagnosis. *Br. J. Dermatol.,* **96**, 399–402.

Marks, R., Dudley, F. and Wan, A. (1978) Trimethylamine metabolism in liver disease. *Lancet*, **i**, 1106–7.

McManus, M.E., Stupans, J., Burgers, W. *et al.* (1987) Flavin-containing monooxygenase activity in human liver microsomes. *Drug Metab. Dispos.,* **15**, 256–61.

Mitchell, M.E. (1978) Carnitine metabolism in human subjects. I. Normal metabolism. *Am. J. Clin. Nutr.* **31**, 293–306.

Mitchell, S.C., Waring, R.H., Haley, C.S. *et al.* (1984) Genetic aspects of the polymodally distributed sulphoxidation of S-carboxymethyl-L-cysteine in man. *Br. J. Clin. Pharmacol.*, **18**, 507–21.

Norris, E.R. and Benoit, G.J. (1945) Studies on trimethylamine oxide. I Occurrence of trimethylamine oxide in marine organisms. *J. Biol. Chem.*, **158**, 433–8.

Oginsky, E.L., Stein, A.E. and Greer, M.A. (1965) Myrosinase activity in bacteria as demonstrated by the conversion of progoitrin to goitrin. *Proc. Soc. Exp. Biol. Med.*, **119**, 360–4.

Pearson, A.W., Butler, E.J., Curtis, R.F. *et al.* (1979) Effects of rapeseed meal on trimethylamine metabolism in the domestic fowl in relation to egg taint. *J. Sci. Food Agric.*, **30**, 799–804.

Pearson, A.E., Fenwick, G.R., Greenwood, N.M. and Butler, E.J. (1980) The effects of goitrogens on the oxidation of trimethylamine in the domestic fowl (*Gallus domestica*). *Comp. Biochem. Biophys.*, **67A**, 397–401.

Pearson, A.W., Greenwood, N.M., Butler, E.J. and Fenwick, G.R. (1981) The inhibition of trimethylamine oxidation in the domestic fowl (*Gallus domesticus*). *Comp. Biochem. Biophys.*, **69C**, 207–12.

Prentiss, P.G., Rosen, H., Brown, N. *et al.* (1961) The metabolism of choline in the germ free rat. *Arch. Biochem. Biophys.*, **94**, 424–9.

Ronald, O.A. and Jakobson, F. (1947) Trimethylamine oxide in marine products. *J. Soc. Chem. Ind.*, **66**, 160–6.

Rothschild, J.C. and Hansen, R.C. (1985) Fish odour syndrome: trimethylaminuria with milk a chief dietary factor. *Pediatr. Dermatol.*, **3**, 38–39c.

Ruocco, V., Florio, M., Grimaldi Filioli, F. *et al.* (1989) An unusual case of trimethylaminuria. *Br. J. Dermatol.*, **120**, 459–61.

Saborin, P.J. and Hodgson, E. (1984) Characterization of purified microsomal FAD-containing monooxygenase from mouse and pig liver. *Chem. Biol. Interact.*, **51**, 125–39.

Shelley, E.D. and Shelley, W.B. (1984) The fish-odour syndrome, trimethyl-aminuria. *J. Am. Med. Assoc.*, **251**, 253–5.

Shewan, J.M. (1951) The chemistry and metabolism of the nitrogen extractives of fish. *Biochem. Soc. Symp.*, **6**, 28–48.

Simenhoff, M.L., Burke, J.F., Sankkonen, J.J. *et al.* (1977) Biochemical profile of uraemic breath. *N. Engl. J. Med.*, **297**, 132–5.

Spellacy, E., Watts, R.W.E. and Goolamali, S.K. (1979) Trimethylaminuria. *J. Inherited Metab. Dis.*, **2**, 85–8.

Suwa, A. (1909) Untersuchungen uber die organextrakte der selachier. I. Die muskelextraktstoffe des dornhaies. *Arch. Ges. Physiol.*, **128**, 421–6.

Todd, A.W. (1979) Psychological problems as the major complication of an adolescent with trimethylaminuria. *J. Pediatr.*, **94**, 936–7.

Tver, D.F. and Russell, P. (1981) in *Nutrition and Health Encyclopaedia*, Von Nostrand Rheinhold, New York, p. 100.

Tynes, R.E. and Hodgson, E. (1985) Catalytic activity and substrate specificity of the flavin-containing monooxygenase in microsomal systems: characterization of the hepatic, pulmonary and renal enzymes of the mouse, rabbit and rat. *Arch. Biochem. Biophys.*, **240**, 77–93.

Tynes, R.E., Sabourin, P.J. and Hodgson, E. (1985) Identification of distinct hepatic and pulmonary forms of microsomal flavin-containing monooxygenase in the mouse and rabbit. *Biochem. Biophys. Res. Commun.*, **126**, 1069–75.

Vogel, F. (1959) Moderne Probleme der Humangenetik. *Ergeb. Inn. Med. Kinderheilk.*, **12**, 52–125.

Willey, G.R. (1985) Trimethylamine – a pungent experience. *Educ. Chem.*, **22**, 178–81.

Williams, D.E., Hale, S.E., Meurhoff, A.S. and Masters, B.B.S. (1985) Rabbit lung flavin-containing monooxygenase. Purification, characterization and induction during pregnancy. *Mol. Pharmacol.*, **28**, 381–90.

Wills, M.R. and Savory, J. (1981) Biochemistry of renal failure. *Annal. Clin. Lab. Sci.*, **11**, 292–9.

Wranne, L. (1956) Urinary excretion of trimethylamine and trimethylamine *N*-oxide following trimethylamine administration to normal and patients with liver disease. *Acta Med. Scand.*, **153**, 433–41.

Zeisel, S.H. (1981) Dietary choline: biochemistry, physiology and pharmacology. *Ann. Rev. Nutr.*, **I**, 95–121.

Ziegler, D.M. (1980) Microsomal flavin-containing monooxygenase: oxygenation of nucleophilic nitrogen and sulphur compounds, in *Enzymatic Basis of Detoxification* (ed. W.B. Jakoby), Academic Press, New York, pp. 201–27.

Ziegler, D.M. (1985) Molecular basis for *N*-oxygenation of *sec*- and *tert*-amines, in *Biological Oxidation of Nitrogen in Organic Molecules – Chemistry, Toxicology and Pharmacology* (eds J.W. Gorrod and L.A. Damani), Ellis Horwood, Chichester, pp. 43–52.

Ziegler, D.M. and Mitchell, C.H. (1972) Microsomal oxidase. IV Properties of a mixed function amine oxidase isolated from pig liver microsomes. *Arch. Biochem. Biophys.*, **150**, 116–25.

Ziegler, D.M. and Poulsen, L.L. (1977) Protein disulphide bond synthesis: a possible intracellular mechanism, *Trends in Biochemical Sciences*, **2**, 79–81.

8

Microbial *N*-hydroxylases

H. Diekmann and H.J. Plattner

Institute of Microbiology, University of Hannover, Schneiderberg 50, 3000 Hannover

1. Genetic and enzymological studies have elucidated the structure and dependence on FAD and NADPH as cofactors of biosynthetic bacterial and fungal *N*-hydroxylases forming N^6-hydroxylysine and N^5-hydroxyornithine from L-lysine or L-ornithine, respectively.
2. The enzyme hydroxylating the peptide-*N* in cyclo-L-leucyl-L-leucyl seems to be different.

8.1 INTRODUCTION

Hydroxylations at the nitrogen atom are common in microorganisms. Ammonia mono-oxygenase (Equation 8.1), intensively studied in *Nitrosomonas europaea*, is thought to be the enzyme also responsible for the oxidation of methane, methanol, ethylene, propylene, benzene, phenol, cyclohexane, bromocarbons and carbon monoxide (Hyman *et al.*, 1988).

$$NH_3 + NADH + H^+ + O_2 \rightarrow NH_2OH + H_2O + NAD \qquad (8.1)$$

Trimethylamine mono-oxygenase (Equation 8.2) was purified from *Pseudomonas aminovorans* and characterized by Boulton *et al.* (1974).

$$(CH_3)_3N + NADPH + H^+ + O_2 \rightarrow (CH_3)_3NO + NADP + H_2O \quad (8.2)$$

Bacterial xanthine oxidase (EC 1.2.3.2) (Equation 8.3) has been purified from *Clostridium, Pseudomonas, Micrococcus* and *Arthrobacter* species (Woolfolk and Downard, 1978; Brons *et al.*, 1987).

$$xanthine + H_2O + O_2 \rightarrow uric\ acid + O_2^{\cdot-} \qquad (8.3)$$

On the other hand, little was known until 1980 about microbial enzymes hydroxylating primary and secondary amines despite the fact that a wide variety of hydroxamates (Winkelmann, 1986) have been found in the culture fluids of micro-organisms grown under iron limitation after the discovery of grisein, terregens factor, coprogen and ferrichrome. Early investigations on the biosynthesis of *N*-hydroxylated compounds (Stevens and Emery, 1966) showed that the source of the hydroxyl amino oxygen in hadacidin, a monohydroxamic antibiotic, is molecular oxygen. The same was later shown by Akers and Neilands (1978) for rhodotorulic acid.

8.2 STUDIES ON *AEROBACTER AEROGENES* 62-1

Starting in 1977 Viswanatha and his group published a series of papers on the synthesis of aerobactin and, as the initial step, on the hydroxylation of L-lysine (Murray *et al.*, 1977). These authors succeeded in establishing a cell-free enzyme system and proposed a hypothetical model (Goh *et al.*, 1989a) accounting for the 1 : 1 stoicheiometry in the consumption of pyruvate and the formation of N^6-hydroxylation product. These studies were hampered by the fact that lysine *N*-hydroxylase is membrane-bound (Goh *et al.*, 1989b) and difficult to solubilize. Moreover, the assay for *N*-hydroxylated products is complicated by the fact that the second enzyme involved in aerobactin biosynthesis, acetyl coenzyme A: alkylhydroxylamine *N*-acetyltransferase, is also present in high amounts (Kusters and Diekmann, 1984).

8.3 STUDIES ON AEROBACTIN BIOSYNTHESIS ENCODED ON pColV PLASMIDS

About 1980 it became evident that aerobactin is the virulence factor in pathogenic strains of *Escherichia coli* (Braun, 1981). Two groups of investigators identified the genes on the aerobactin operon of pColV plasmids (Gross *et al.*, 1984, 1985; Lorenzo *et al.*, 1986) and constructed strains in which single genes were cloned. The lysine N^6-hydroxylase in such strains is overexpressed (Heydel *et al.*, 1987). Minicells carrying the gene for lysine N^6-hydroxylase expressed a 50-kDa protein (Gross *et al.*, 1985).

Lysine N^6-hydroxylase from *E. coli* EN222 (Engelbrecht and Braun, 1986), constituting about 7% of the total amount of soluble protein present after sonication of the cells, was purified to homogeneity (Plattner *et al.*, 1989). Activity of the dialysed enzyme is strictly dependent on the presence of NADPH and FAD, L-lysine being the only substrate to be hydroxylated. L-Ornithine, DL-2,3-diaminopropionic acid, DL-2,6-diaminopimelic acid and N^6-acetyl-L-lysine stimulate oxidation of NADPH ('non-substrate effectors'). The native enzyme is a tetramer of molecular weight 200 000 Da, containing 4 moles of FAD. The K_m-values for L-lysine, FAD and NADPH have been determined to be 1.25 mM, 0.78 μM and 0.2 μM, respectively. SDS-gel electrophoresis indicates the presence of four subunits of apparent molecular mass ~50 kDa, each containing seven titratable sulphydryl groups (Romaguera *et al.*, 1989).

The nucleotide sequence of the gene for lysine N^6-hydroxylase on ColV-K30 plasmid was reported by Herrero *et al.* (1988). Comparison of the amino acid sequence with those of *p*-hydroxybenzoate mono-oxygenase (Wierenga *et al.*, 1983) and *Acinetobacter* cyclohexanone mono-oxygenase (Chen *et al.*, 1988) reveals striking similarities and can be used for identification of the binding sites for NADPH and FAD. From this evidence it can be concluded that bacterial lysine N^6-hydroxylase belongs to the group

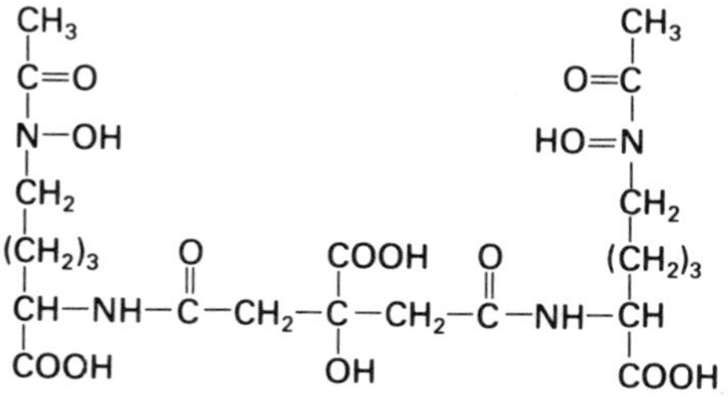

Figure 8.1 Molecular structure of aerobactin.

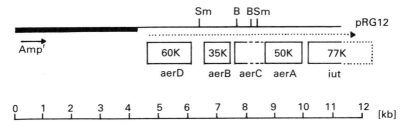

Figure 8.2 Physical map of the aerobactin operon of pRG12 (according to Gross *et al.*, 1985); aerA = lysine *N*-hydroxylase gene.

of 'external' flavoprotein mono-oxygenases (Massey and Hemmerich, 1975) and can be assigned to the EC 1.14.13 enzyme family (Plattner *et al.*, 1989).

Preliminary studies on the reaction mechanism of bacterial lysin *N*-hydroxylase (Waschütza *et al.*, unpublished work) seem to indicate that the bacterial lysine *N*-hydroxylase forms a stable flavin-*C*-(4a)-hydroperoxide (Massey *et al.*, 1982). It is remarkable that enzyme-bound FAD is reduced by NADPH in the absence of substrate and that the apparent K_m value for L-lysine is as high as 1.25 mM. It has as yet not been tested whether lysine *N*-hydroxylase is capable of catalysing *S*-oxidations, as has been reported by Ryerson *et al.* (1982) for cyclohexanone mono-oxygenase.

8.4 STUDIES WITH GRAM-POSITIVE BACTERIA

The biosynthesis of siderophores both bacterial and fungal, mentioned in this review follows the metabolic pattern, namely initial *N*-hydroxylation of the amino acids L-lysine or L-ornithine followed by acetylation at the hydroxylamino group to form the corresponding hydroxamic acid. The same reaction sequence seems to apply to those siderophores from Gram-positive bacteria containing hydroxylated derivatives of diamines, such as cadaverin in arthrobactin (the former terregens factor) or ferrioxamine E (Schafft and Diekmann, 1978). No enzyme activites have so far been detected in these organisms.

8.5 STUDIES ON FUNGI

Using the assay system for bacterial lysine *N*-hydroxylase activity (Plattner *et al.*, 1989), but substituting L-ornithine for L-lysine, *N*-hydroxylase activity can be detected in cell-free extracts of *Aspergillus quadricinctus*, a ferrichrome producer (Diekmann and Krezdorn, 1975), and in *Rhodotorula*

Table 8.1 Specific activities of *N*-hydroxylases in dialysed crude extracts from bacteria and fungi

Source	Protein (mg per assay)	Product formed (nmol)	Specific activity (nmol/mg per min)
E. coli EN222	0.09	46[a]	17
A. quadricinctus	1.4	27[b]	0.22
R. glutinis	2.5	30[b]	0.14

Incubation time was 30 min (a) or 90 min (b).

glutinis, a producer of rhodotorulic acid (Atkin and Neilands, 1968). Activities of crude extracts after dialysis and centrifugation at $100\,000\,g$ are shown in Table 8.1. In the omission test, strict dependence on L-ornithine, FAD and NADPH could be demonstrated. The enzyme from extracts of *R. glutinis* was partially purified by fractional ammonium sulphate precipitation (0.30–0.45 s) and gel filtration on Superose 12HR under FPLC conditions. From the gel filtration experiments molecular weights of about 500 000 Da were calculated for the native *N*-hydroxylases from *A. quadricinctus* and *R. glutinis* (Plattner *et al.*, unpublished data).

From a genomic library of wild-type *Ustilago maydis* DNA on the cosmid transformation vector pCU3, two cosmids were found capable of complementing siderophore auxotrophs of this fungus defective in their ability to hydroxylate L-ornithine (Wang *et al.*, 1989). Although the *N*-hydroxylase gene has not yet been identified there is ample evidence that the gene is cloned on a 2.5 kb fragment.

Pulcherriminic acid is the red pigment of *Candida pulcherrima*. It was shown that, in contrast to the biosynthesis of siderophores, the peptide bonds of the diketopiperazine are formed prior to hydroxylation (MacDonald, 1965; Plattner and Diekmann, 1973). *N*-Hydroxylation of cyclo-L-leucyl-L-leucyl to pulcherriminic acid in washed cells grown under oxygen limitation is induced by vigorous aeration. Induction is repressed by cycloheximide. Hydroxylation *in vivo* is inhibited by potassium cyanide, carbon monoxide and metyrapone, indicating the participation of a cytochrome *P*-450-dependent hydroxylation system (Plattner, unpublished results).

REFERENCES

Akers, H.A. and Neilands, J.B. (1978) Biosynthesis of rhodotorulic acid and other hydroxamate type siderophores, in *Biological Oxidation of Nitrogen* (ed. J.W. Gorrod), Elsevier, Amsterdam, pp. 429–36.

Atkin, C.L. and Neilands, J.B. (1968) Rhodotorulic acid, a diketopiperazine dihydroxamic acid with growth-factor activity. 1. Isolation and characterization *Biochemistry*, **7**, 3734–9.

Boulton, C.A., Crabbee, J.C. and Large, P.J. (1974) Microbial oxidation of amines. Partial purification of a trimethyamine mono-oxygenase from *Pseudomonas aminovorans* and its role in growth on trimethylamine. *Biochem. J.*, **140**, 253–63.

Braun, V. (1981) Eisenversorgung und Virulenz bei *Escherichia coli*. *Forum Mikrobiol.*, **2**, 69–72.

Brons, H.J., Breedveld, M.W., Middelhoven, W.J. *et al.* (1987) Inhibition of bacterial xanthine oxidase from *Arthrobacter* M4 by 5,6-diaminouracil. *Biotechnol. Appl. Biochem.*, **9**, 66–73.

Chen, Y.-C.J., Peoples, O.P. and Walsh, C.T. (1988) *Acinetobacter* cyclohexanone monooxygenase: gene cloning and sequence determination. *J. Bacteriol.*, **170**, 781–9.

Diekmann, H. and Krezdorn, E. (1975) Stoffwechselprodukte von Mikroorganismen. 150. Ferricrocin, Triacetylfusigen und andere Sideramine aus Pilzen der Gattung *Aspergillus*, Gruppe *Fumigatus*. *Arch. Microbiol.*, **106**, 191–4.

Engelbrecht, F. and Braun, V. (1986) Inhibition of microbial growth by interference with siderophore biosynthesis. Oxidation of primary amino groups in aerobactin biosynthesis by *Escherichia coli*. *FEMS Microbiol. Lett.*, **33**, 223–7.

Goh, C.J., Szczepan, E.W., Wright, G. *et al.* (1989a). Investigation of lysine: N-hydroxylation in the biosynthesis of the siderophore aerobactin. *Bioinorg. Chem.*, **17**, 13–27.

Goh, C.J., Szczepan, E.W., Menhart, N. and Viswanatha, T. (1989b). Studies on lysine: N^6-hydroxylation by cell-free systems of *Aerobacter aerogenes* 62–1. *Biochim. Biophys. Acta*, **990**, 240–5.

Gross, R., Engelbrecht, F. and Braun, V. (1984) Genetic and biochemical characterization of the aerobactin synthesis operon on pColV. *Mol. Gen. Genet.*, **196**, 74–80.

Gross, R., Engelbrecht, F. and Braun, V. (1985) Identification of the genes and their polypeptide products responsible for aerobactin synthesis by pColV plasmids. *Mol. Gen. Genet.*, **201**, 204–12.

Herrero, M., Lorenzo, V. de and Neilands, J.B. (1988) Nucleotide sequence of the *iucD* gene of the pColV-K30 aerobactin operon and topology of its product studied with *phoA* and *lacZ* gene fusions. *J. Bacteriol.*, **170**, 56–64.

Heydel, P., Plattner, H. and Diekmann, H. (1987) Lysine N-hydroxylase and N-acetyltransferase of the aerobactin system of pColV plasmids in *Escherichia coli*. *FEMS Microbiol. Lett.*, **40**, 305–9.

Hyman, M.R., Murton, I.B. and Arp, D.J. (1988) Interaction of ammonia monooxygenase from *Nitrososomonas europaea* with alkanes, alkenes, and alkynes. *Appl. Environ. Microbiol.*, 54, 3187–90.

Kusters, J. and Diekmann, H. (1984) Assay for acetyl coenzyme A: alkylhydroxylamine N-acetyltransferase in *Klebsiella pneumoniae* ATCC 25304. *FEMS Microbiol. Lett.*, **23**, 309–11.

Lorenzo, V. de, Bindereif, A. and Neilands, J.B. (1986) Aerobactin biosynthesis and transport genes of plasmid ColV-K30 in *Escherichia coli*. *J. Bacteriol.*, **165**, 570–8.

MacDonald, J.C. (1965) Biosynthesis of pulcherriminic acid. *Biochem. J.*, **96**, 533–8.

Massey, V., Claiborne, A., Detmer, K. and Schopfer, L.M. (1982) Comparative aspects of flavoprotein monooxygenases, *Oxygenases and Oxygen. Metabolism* (eds M. Nozaki, S. Yamamoto, Y. Ishimura, M.J. Coon, L. Ernster and R.W. Eastabrook), Academic Press, New York, pp. 187–205.

Massey, V. and Hemmerich, P. (1975) Flavin and pteridine monooxygenases, in *The Enzymes* (ed. P. Boyer), Academic Press, New York, pp. 191–252.

Murray, G.J., Clark, G.E.D., Parniak, M.A. and Viswanatha, T. (1977) Effect of metabolites on ε-N-hydroxylysine formation in cell-free extracts of *Aerobacter aerogenes* 62-1. *Can. J. Microbiol.*, **55**, 625–9.

Plattner, H. and Diekmann, H. (1973) Stoffwechselprodukte von Mikroorganismen. 124. Einbau von L-Leucin in cyclo-L-Leucyl-L-leucyl, dem Intermediärprodukt der Biosynthese von Pulcherriminsäure, in zellfreien Extrakten von *Candida pulcherrima*. *Arch. Mikrobiol.*, **93**, 363–5.

Plattner, H.-J., Pfefferle, P., Romaguera, A. *et al.* (1989) Isolation and some properties of lysine N^6-hydroxylase from *Escherichia coli* strain EN222. *Biol. Metals*, **2**, 1–5.

Romaguera, A., Waschütza, S., Plattner, H.J. and Diekmann, H. (1989) Reinigung und Eigenschaften der Lysin-N^6-hydroxylase aus *Escherichia coli* EN222. *Forum Mikrobiol.*, **12**, 77, P113.

Ryerson, C.C., Ballou, D.P. and Walsh, C.Z. (1982) Mechanistic studies on cyclohexanone oxygenase. *Biochemistry, * **21**, 2644–55.

Schafft, M. and Diekmann, H. (1978) Cadaverin ist ein Zwischenprodukt der Biosynthese von Arthrobactin und Ferrioxamin E. *Arch. Microbiol.*, **117**, 203–7.

Stevens, R.L. and Emery, T.F. (1966) The biosynthesis of hadacidin. *Biochemistry*, **5**, 74–81.

Wang, J., Budde, A.D. and Leong, S.A. (1989) Analysis of ferrichrome biosynthesis in the phytopathogenic fungus *Ustilago maydis;* cloning of an ornithine-N^5-oxygenase gene. *J. Bacteriol.*, **171**, 2811–18.

Wierenga, R.K., Drenth, J. and Schulz, G.E. (1983) Comparison of the three-dimensional protein and nucleotide structure of the FAD binding domain of *p*-hydroxybenzoate hydroxylase with the FAD- as well as the NADPH-binding domains of glutathion reductase. *J. Mol. Biol.*, **167**, 725–39.

Winkelmann, G. (1986) Iron complex products (siderophores), in *Biotechnology* (eds H. Pape, and H.-J. Rehm), Vol. 4, VCH Weinheim, pp. 215–43.

Woolfolk, C.A. and Downard, J.S. (1978) Bacterial xanthine oxidase from *Arthrobacter* S-2. *J. Bacteriol.*, **135**, 422–8.

9

Roles of aminium radical intermediates in the biotransformation of dihydropyridines, cycloalkylamines, and *N,N*-dimethylanilines by cytochrome *P*-450 enzymes

F. P. Guengerich*, A. Bondon*, R.G. Böcker* and T. L. Macdonald†

* *Department of Biochemistry and Center in Molecular Toxicology, Vanderbilt University School of Medicine, Nashville, Tennessee 37232 and* † *Department of Chemistry, University of Virginia, Charlottesville, Virginia 22901 USA*

1. The oxidation of amines and related compounds by cytochrome *P*-450 enzymes is characterized by reaction mechanisms involving 1-electron pathways and aminium radicals. Further evidence for such a mechanism was provided by the ring expansion of 1-phenylcyclobutylamine to 2-phenyl-1-pyrroline.

2. Cytochrome *P*-450-catalysed N–O bond formation has previously been demonstrated to occur under various conditions, including situations where α-protons are not present (arylamines), Bredt's Rule renders α-protons inaccessible (quinidine), or aminium radical stability is afforded by electronic donation from a neighbouring group (azoprocarbazine). More recent examples include the hydroxylamines of methamphetamine and *N*-(1-phenylcyclobutyl)benzylamine, the pyrrolizidine alkaloid senecionine *N*-oxide, and *N*-(1-phenyl)cyclobutylphenyl nitrone.

3. Many 1,4-dihydropyridines are dehydrogenated by cytochrome *P*-450 enzymes, and the mechanism is believed to involve sequential electron/proton/electron transfer, as indicated by studies showing low kinetic hydrogen isotope effects for several compounds, including nifedipine. The pathway is analogous to that proposed for amine *N*-dealkylation.

4. Rates of N-demethylation of a series of p-substituted N,N-dimethyl-anilines by cytochrome P-450$_{PB-B}$ were measured and analysed using two different approaches: (a) Hammett analysis yields a ϱ value of -0.6 to -0.7, consistent with a positively charged (aminium radical) intermediate; (b) modified Marcus treatment yields estimates of parameters associated with the abstraction of electrons by the putative $[FeO]^{3+}$ complex, namely an effective oxidation potential ($E_{1/2}$) of $+1.7$ to $+2.0$ V and a self-exchange energy (λ) of 22 to 26 kcal/mol.

5. A number of questions remain to be answered concerning the oxidation of amines by cytochrome P-450 enzymes: (a) how much does the oxidation potential vary among individual enzymes and what are the contributing forces? (b) do specific bases in cytochrome P-450 enzymes abstract protons from amines to enhance catalysis? (c) how do the peroxidases differ from the cytochrome P-450 enzymes in their mechanisms of catalysis?

9.1 INTRODUCTION

Cytochrome P-450 (P-450) enzymes catalyse oxidation of a wide variety of amines, many of which are important drugs and carcinogens (Kadlubar and Hammons, 1987; Nelson and Harvison, 1987; Guengerich, 1988; Butler *et al.*, 1989; Shimada *et al.*, 1989). Products of the oxidation of different amines include hydroxylamines, N-oxides, pyridines, dealkylation products and other compounds. A proper understanding of the catalytic mechanism of P-450 is important in rationalizing and predicting what metabolites will be formed and what consequences will result from biotransformation. Although several mechanisms are theoretically possible for the oxidation of amines, several lines of evidence indicate that a pathway involving initial abstraction of a non-bonded electron predominates (Guengerich and Macdonald, 1984). The general mechanism of amine N-dealkylation which is now most widely accepted is shown in Fig. 9.1 where a P-450 high valent iron/oxygen complex is depicted as $[FeO]^{3+}$.

This complex could, at least in principle, exist as any of a number of valence forms, but several lines of evidence suggest that the most likely form is that analogous to peroxidase Compound I, where the iron is in the formal $+4$ state and the residual positive charge is dispersed in the porphyrin ring (McMurry and Groves, 1986). The first step involves abstraction of an electron from the nitrogen. The protons α to such an aminium radical are rather acidic and rearrangement to the methylene radical occurs rapidly, with radical rearrangement ('oxygen rebound') then completing the

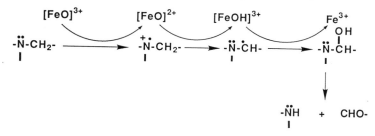

Figure 9.1 Generalized mechanism for amine *N*-dealkylation by *P*-450 enzymes.

formation of the carbinolamine. Some carbinolamines are relatively stable, such as that derived from *N*-methylcarbazole (Gorrod and Temple, 1976; Shea *et al.*, 1982), others decompose rapidly to yield amine and a carbonyl compound. Thus, the first step in the reaction is the same as in the electrochemical oxidation of amines, and the kinetic deuterium isotope effects observed in the enzymatic reactions are similar to those measured in anodic oxidation (Shono *et al.*, 1982; Hall *et al.*, 1989). However, the scheme presented here differs from electrochemical oxidation in that carbinolamine formation is thought to occur via oxygen transfer. Thus, *P*-450 differs from electrochemical oxidation and from the amine oxidations catalysed by many, but not all, peroxidases and flavoproteins.

9.2 FORMATION OF N–O COMPOUNDS

The formation of *N*-oxides and hydroxylamines by *P*-450 enzymes is not thought to occur readily in contrast to the flavoprotein-catalysed *N*-oxidation reactions (Ziegler, 1988). The reason lies in the catalytic mechanism, since the flavin-dependent systems utilize a protonated 4a hydroperoxide instead of a high valent iron/oxygen complex, and dealkylation reactions are rarely seen unless the resulting N–O compounds are unstable. In the case of sparteine (Fig. 9.2, *9.1*) (and isosparteine), the *N*-oxide (*9.2*) was not found to be a product of the *P*-450-dependent oxidation, and the synthetic *N*-oxides did not decompose under physiological conditions to yield the two iminium ions (Δ^6 and Δ^2, *9.3* and *9.4*, respectively) which are isolated from the *P*-450-dependent reaction (Fig. 9.2) (Guengerich, 1984).

Thus, the Polonowski reaction, known to involve Fe^{2+} and acid, does not occur readily under physiological conditions and is not catalysed by *P*-450 enzymes, consistent with the poor ability of iron complexed in porphyrins to act as a Lewis acid (Liebler and Guengerich, 1983). Further, Pandey

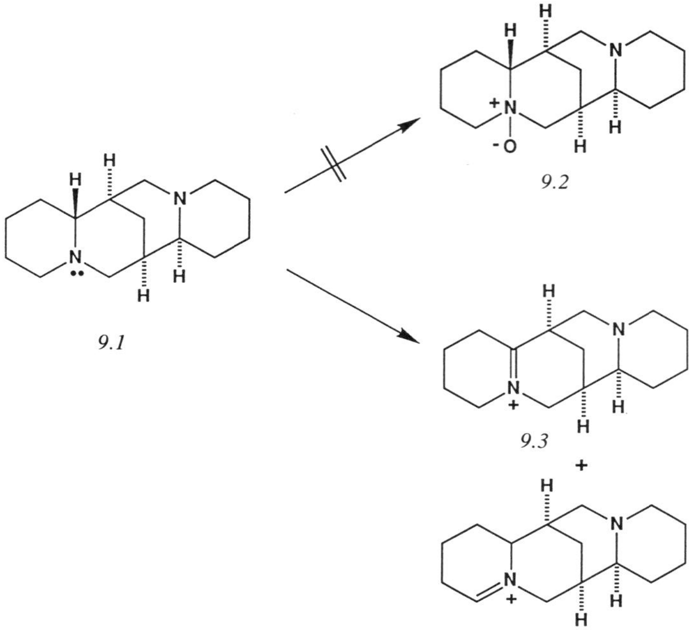

Figure 9.2 Oxidation of sparteine (*9.1*) by *P*-450 enzymes (Guengerich, 1984).

et al. (1989) demonstrated that *P*-450 enzymes do not *N*-demethylate *N,N*-dimethylaniline by a mechanism involving an *N*-oxide. Some metalloporphyrin complexes can cleave amine oxides heterolytically (Shannon and Bruice, 1981), although the rates are very slow and evidence suggests that in the *P*-450-catalysed conversion of *N,N*-dimethylaniline oxide into HCHO a homolytic scission and a formal $[FeO]^{2+}$ complex are more likely to be involved (Burka *et al.*, 1985; Ortiz de Montellano, 1986; Fujimori *et al.*, 1986).

The *P*-450-dependent formation of N–O compounds is known to occur under several conditions. When there are no α-protons available, then the aminium radical recombines with the putative $[FeO]^{2+}$ complex to yield a hydroxylamine, as exemplified in the case of the carcinogen 2-naphthylamine (*9.5*) (Hammons *et al.*, 1985) (Fig. 9.3). The net reaction can also be catalysed by the flavin-containing mono-oxygenase and by horseradish peroxidase, but the mechanisms are at variance (Guengerich, 1990a). In the case of the *N*-oxidation of quinidine (*9.8*) (Fig. 9.4), the abstraction of a proton α to the bridgehead nitrogen would constitute a

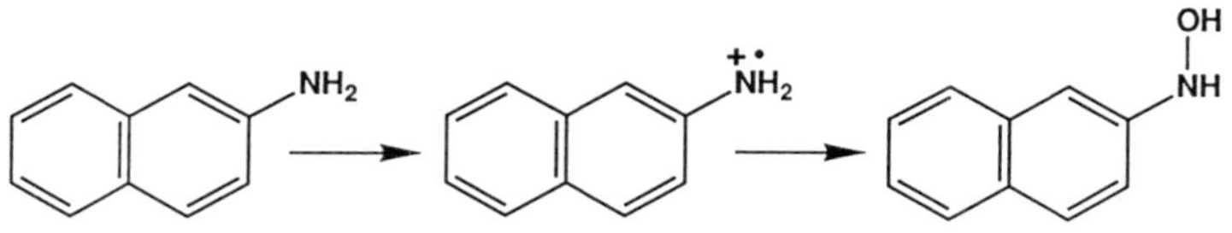

Figure 9.3 *N*-Hydroxylation of 2-naphthylamine (*9.5*) (Hammons *et al.*, 1985).

violation of Bredt's Rule and the reaction proceeds to give the *N*-oxide (*9.9*) as the major product (Guengerich *et al.*, 1986b).

Sometimes, aminium radicals are stabilized by electron donation from a neighbouring atom, as in the case of the formation of the two isomers of azoxyprocarbazine, one of which is shown (*9.12*) (Fig. 9.5) (Prough *et al.*, 1984).

Until recently, situations such as those described above were the only cases in which N–O compound formation could clearly be attributed to *P*-450 enzymes. However, several exceptions to the above guidelines have been documented. Pyrrolizidine alkaloids are oxidized to form both *N*-oxides and pyrroles, the latter being thought to represent the ultimate carcinogens. The conversion of senecionine (*9.13*) to its *N*-oxide (*9.15*) by *P*-450 has been documented (Williams *et al.*, 1989). *P*-450s also catalyse formation of the pyrrole (*9.17*), but *N*-oxide (*9.15*) is not a precursor of the pyrrole, and a pathway showing an aminium ion (*9.14*) as a branch point is presented in Fig. 9.6. Baba *et al.* (1988) have also provided evidence that methamphetamine can be oxidized to the hydroxylamine by *P*-450. The third example, *N*-(1-phenylcyclobutyl)benzylamine (*9.18*) (Fig. 9.7), comes from this laboratory.

One product of the oxidation is the nitrone (*9.22*), which is of the *trans* configuration (Bondon *et al.*, 1989). The formation of this compound can

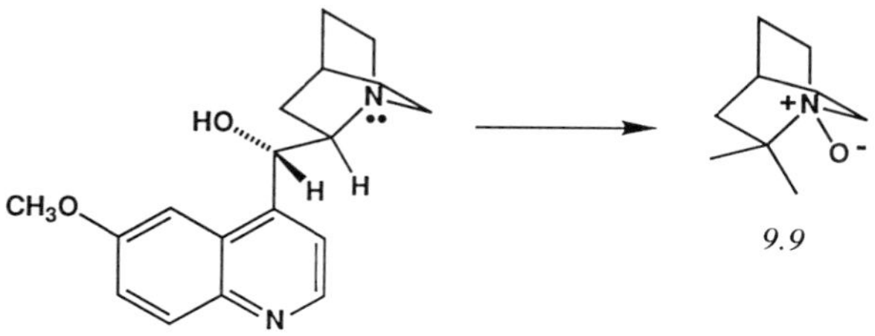

Figure 9.4 Oxidation of quinidine (*9.8*) to the *N*-oxide (*9.9*) (Guengerich *et al.*, 1986b).

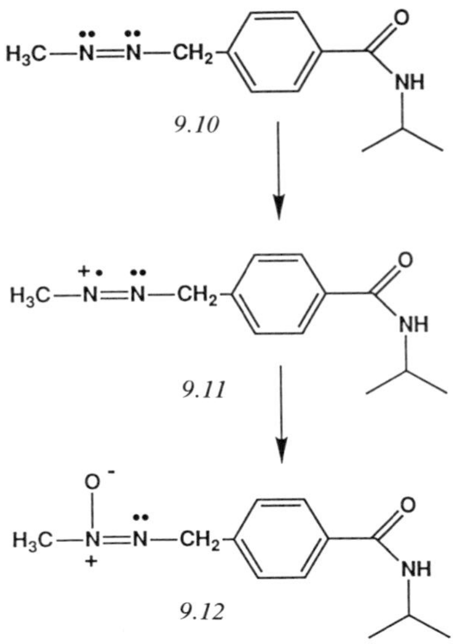

Figure 9.5 Oxidation of azoprocarbazine (*9.10*) to the major product, *N*-isopropyl-α-(2-methyl-*ONN*-azoxy)-*p*-toluamide (*9.12*), by *P*-450 (Prough *et al.*, 1984).

be envisaged to occur either via an imine or a hydroxylamine (*9.20*) intermediate. Both of these compounds were found to accumulate in incubations, the imine being in equilibrium with the carbinolamine (*9.24*) formed en route to the dealkylation products benzaldehyde (*9.25*) and 1-phenylcyclopropylamine (*9.26*). When these two putative nitrone

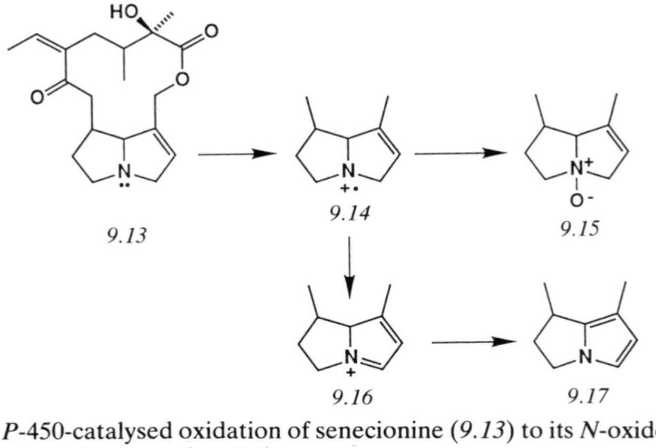

Figure 9.6 *P*-450-catalysed oxidation of senecionine (*9.13*) to its *N*-oxide (*9.15*) and pyrrole (*9.17*) (Williams *et al.*, 1989).

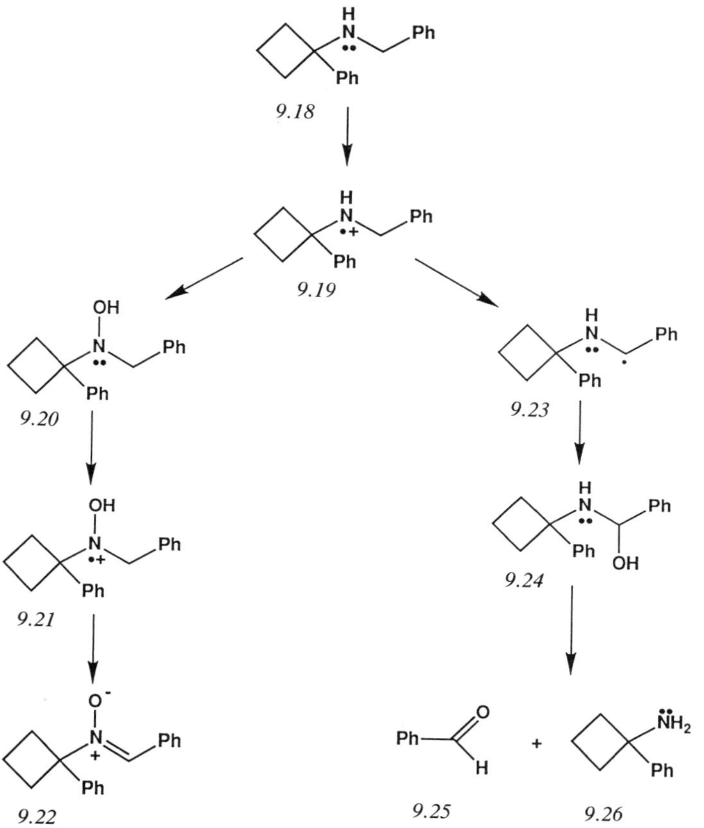

Figure 9.7 Oxidation of *N*-(1-phenylcyclobutyl)benzylamine (*9.18*) to oxygenated and dealkylation products by *P*-450 (Bondon *et al.*, 1989).

precursors were synthesized using radiolabels, only the hydroxylamine (*9.24*) was converted into the nitrone (*9.22*). An aminium radical (*9.21*) is postulated as an intermediate in the oxidation of the hydroxylamine (*9.20*) to the nitrone (*9.22*) owing to lack of a kinetic tritium isotope effect in this reaction (Bondon *et al.*, 1989).

9.3 RING EXPANSION OF A CYCLOBUTYLAMINE

More evidence for the existence of aminium radical pathways was obtained in studies on the further oxidation of *N*-(1-phenylcyclopropyl)benzylamine (*9.18*) by *P*-450 (Bondon *et al.*, 1989). As in the case of flavoprotein

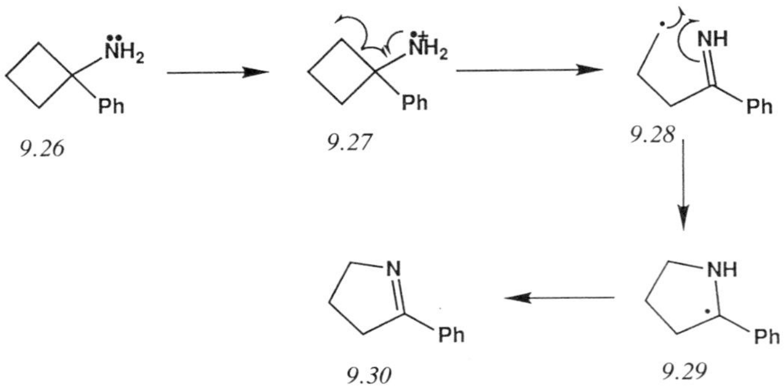

Figure 9.8 Proposed mechanism of ring expansion in the oxidation of 1-phenyl-cyclobutylamine (*9.26*) by *P*-450 (Bondon *et al.*, 1989).

monoamine oxidase (Silverman and Zieske, 1986), oxidation of 1-phenyl-cyclopropylamine (*9.26*) occurs to yield 2-phenyl-1-pyrroline (*9.30*) as identified by its retention time and electron impact fragmentation pattern in combined capillary gas chromatography–mass spectrometry. The observed ring expansion is best rationalized by an aminium radical mechanism for both *P*-450 and monoamine oxidase (Silverman and Zieske, 1986; Bondon *et al.*, 1989) (Fig. 9.8). Mechanism-based *P*-450 inactivation occurs with both cyclopropylamines (Macdonald *et al.*, 1982; Hanzlik and Tullman, 1982; Guengerich *et al.*, 1984) and cyclobutylamines (Bondon *et al.*, 1989), although the products appear to be complex and have not been identified.

9.4 OXIDATION OF DIHYDROPYRIDINES

Another area for consideration is the dehydrogenation of 1,4-dihydropyridines by *P*-450. Although the 2-electron mechanism for the oxidation of dihydropyridines such as NADH is fairly well established with dehydrogenases, including flavoproteins, such a formal hydride transfer reaction would appear difficult for *P*-450s. Several of the Hantzsch dihydropyridines (*9.31*, Fig. 9.9) are used as drugs, and variations in the rates of metabolism of these compounds are of considerable interest, especially in the case of the prototype nifedipine (1,4-dihydro-2,6-dimethyl-4-(2-nitrophenyl)-3,5-pyridinedicarboxylic acid dimethyl ester). In the course of studies on the enzyme involved in the dehydrogenation of nifedipine in human liver (Guengerich *et al.*, 1986a; Beaune *et al.*, 1986), interest was drawn to the chemistry of this reaction. Previous work by

Augusto *et al.* (1982) provided evidence for a 1-electron oxidation mechanism in that alkyl radicals released from the 4-position could be trapped and were also found to react with the pyrrolic nitrogens of the *P*-450 haem. *P*-450 enzymes catalysing nifedipine dehydrogenation are capable of oxidizing a variety of substituted 1,4-dihydropyridines (Böcker and Guengerich, 1986; Guengerich and Böcker, 1988). Little stereoselectivity was seen in the oxidation of a pair of enantiomeric dihydropyridines (+ and − Bayer K8644, 1,4-dihydro-2,6-dimethyl-3-nitro-4(2-[trifluoro-methyl]-phenyl)-5-pyridinecarboxylic acid methyl ester), consistent with the view that a specific base or hydrogen atom acceptor does not have to be in a specific position in the *P*-450 molecule for catalysis to occur. Further, oxidation of the 1,4-dihydro-2,6-dimethyl-4-phenyl-3,5-pyridinedicarb-oxylic acid dimethyl ester model is characterized by very low intrinsic kinetic hydrogen isotope effects, a property of *P*-450-dependent, chemical and electrochemical reactions that proceed via electron abstraction pathways (Shono *et al.*, 1982; Miwa *et al.*, 1983; Miwa and Lu, 1986; Hall *et al.*, 1989). Much larger kinetic hydrogen isotope effects are seen in reactions

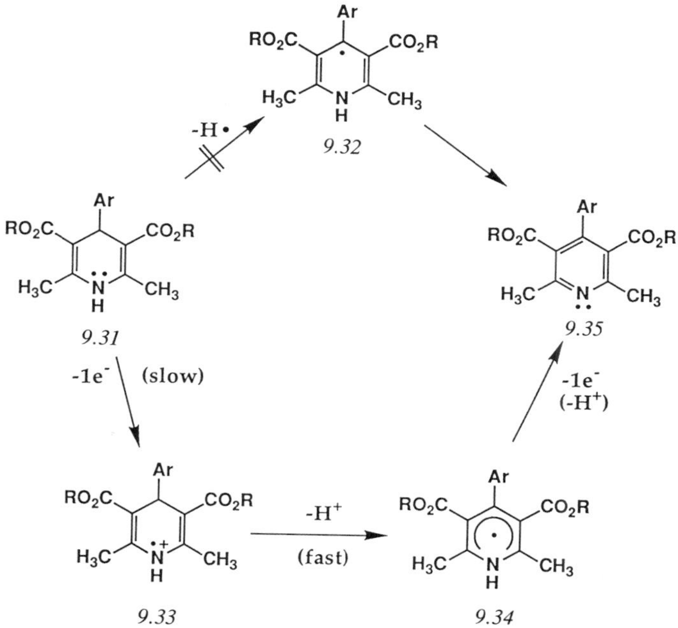

Figure 9.9 Proposed mechanism for dehydrogenation of 1,4-dihydropyridines by *P*-450 and other obligate 1-electron acceptors (Guengerich and Böcker, 1988; Guengerich, 1989a).

that occur via hydrogen atom abstraction (Groves *et al.*, 1978; Ortiz de Montellano, 1986). A scheme consistent with these observations is presented in Fig. 9.9.

More recently, Born and Hadley (1989) have reported that the dehydrogenation of nifedipine itself is characterized by a high kinetic deuterium isotope effect (^DV but not $^D(V/K)$) and have interpreted these results in terms of a hydrogen abstraction mechanism for nifedipine oxidation. However, we have now synthesized [4-^{2}H]- and [4-^{3}H]nifedipine and find that the isotope effects are small (< 2) under a variety of conditions (Guengerich, 1990b). Thus, credence cannot be given to the view that hydrogen abstraction is operative in the case of nifedipine, and the scheme presented in Fig. 9.9 is considered to be general for the dehydrogenation of 1,4-dihydropyridines by *P*-450.

In the proposed stepwise electron/proton/electron transfer mechanism, the deprotonation step is thought to be relatively rapid, otherwise a larger kinetic deuterium isotope effect would have been observed (Carlson *et al.*, 1984; Miller and Valentine, 1988). Sinha and Bruice (1984) have estimated rates of deprotonation in the presence of different general bases, and the values are much higher than those for the overall rate of dehydrogenation observed with *P*-450s. Bäärnhielm and Hannson (1986) found that the dihydropyridine felodipine (1,4-dihydro-2,6-dimethyl-4-(2,3-dichloro-phenyl)-3,5-pyridinedicarboxylic acid dimethyl ester) can be oxidized by peroxidases, and we have found that nifedipine can be oxidized to its pyridine derivative by horseradish peroxidase in the presence of H_2O_2. The apparent kinetic deuterium isotope effect also appears to be low. The question can be raised as to why *N*-dealkylation reactions catalysed by horseradish peroxidase show high kinetic deuterium isotope effects (Miwa *et al.*, 1983). The scheme in Fig. 9.10 compares *N*-dealkylation and dihydropyridine oxidation, which are thought to be similar except for the lack of oxygen insertion in the case of the dihydropyridines (Guengerich and Böcker, 1988; Lee *et al.*, 1988). When the *R* group in the latter class of compounds is alkyl, it tends to leave depending upon the substituent and the particular *P*-450 isoenzyme involved (Lee *et al.*, 1988).

Since horseradish peroxidase shows a high kinetic deuterium isotope effect in *N*-dealkylation but apparently not nifedipine oxidation, the possibility exists that in the peroxidases the removal of the base is rate limiting. Since horseradish peroxidase is thought not to rebound oxygen (Ortiz de Montellano, 1987), the overall mechanism of *N*-dealkylation must differ from that of *P*-450, even though the first step is electron abstraction, which is our current view for peroxidases as well as *P*-450. However, the current dogma is that α-proton loss is rapid (Nelsen and Ippoliti, 1986; Manring and Peters, 1983), although the view may be biased by electrochemical solution studies in which the free base is able to assist in the

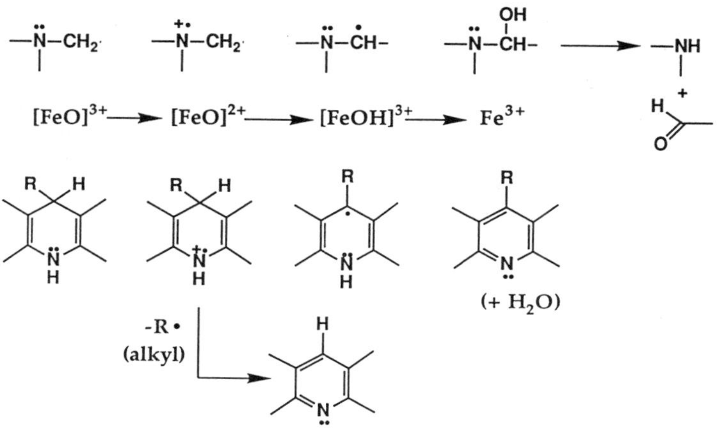

Figure 9.10 Comparison of pathways of *P*-450-catalysed amine *N*-dealkylation and dihydropyridine oxidation (Miwa *et al.*, 1983; Guengerich and Macdonald, 1984; Böcker and Guengerich, 1986; Guengerich and Böcker, 1988; Lee *et al.*, 1988; Guengerich, 1989a).

abstraction process. The work of Hammerich and Parker (1984) and Bordwell *et al.* (1988) suggests that the acidities of aminium radicals themselves may not be much lower than those of their α-protons (Bondon *et al.*, 1989). This observation may point to a role for proteins in assisting through specific base catalysis. Furthermore, it may be possible to rationalize the tendency of some *P*-450s to catalyse the formation of hydroxylamines, *N*-oxides, and nitrones if small differences in the proteins can stabilize aminium radicals of their deprotonated forms through specific acid–base catalysis. Ultimately, more detailed pictures of protein structure and concepts of function will be necessary to address these possibilities.

9.5 ESTIMATION OF THE OXIDATION POTENTIAL OF THE ACTIVE OXYGENATING SPECIES OF *P*-450

Ring substituents have a dramatic influence on rates of *N*-demethylation of *N,N*-dimethylanilines. The results of studies on *N*-demethylation rates measured with P-450$_{PB-B}$ (*P*-450 IIB1) have been utilized in two principal modes. Hammett σ/ϱ analysis has been done, and the ϱ values obtained ranging from -0.6 to -0.7 are indicative of a positively charged intermediate consistent with an aminium radical (Burka *et al.*, 1985). In the other approach, rates of P-450-catalysed N-demethylation (k_{cat}) were plotted *vs.* the 1-electron oxidation potentials of the different substrates (Macdonald *et al.*, 1989). Analysis of the plots by curve fitting to modified Marcus equations yielded an estimated oxidation potential, $E_{1/2}$, of about $+1.8$ V (*vs.* SCE) for electron abstraction by the $[FeO]^{3+}$ complex. $E_{1/2}$ and the λ (self-exchange energy) value were similar regardless of whether the reaction was supported by NADPH–cytochrome P-450 reductase and O_2 or by iodosylbenzene, suggesting a similar mechanism for both reactions as in the case of Hammett analysis. The results provide some insight into the intrinsic ability of P-450 enzymes to abstract electrons. However, it should be considered that the $E_{1/2}$ will be a function of the distance between the reactive centres and the dielectric constant of the protein:

$$E_{1/2 \text{ (apparent)}} = E_{1/2 \text{ (intrinsic)}} + E_{(cf)}$$

and

$$E_{(cf)} = +14.4/r_{1,2} D$$

where $E_{(cf)}$ is the electrostatic contribution (correction factor), as expressed in volts, $r_{1,2}$ is the centre–centre internuclear distance in the transition state between the nuclei of the interacting spheres, and D, as given in Å, is the static dielectric constant in the enzyme active site. $E_{(cf)}$ can be quite appreciable (0.5–1.5 V) when reasonable estimates for $r_{1,2}$ and D are introduced. Electron transfer should be operative over a longer distance than hydrogen atom abstraction, but the distance may vary with different substrates even within the same cytochrome P-450 species. Furthermore, both the inter-radial distance and the dielectric constant may well vary when different cytochrome P-450 enzymes are compared. Thus, the intrinsic $E_{1/2}$ of $[FeO]^{3+}$ protoporphyrin IX may show some alteration in the diverse protein structures (Macdonald *et al.*, 1989).

REFERENCES

Augusto, O., Beilan, H.S. and Ortiz de Montellano, P.R. (1982) The catalytic mechanism of cytochrome P-450. Spin-trapping evidence for one-electron substrate oxidation. *J. Biol. Chem.*, **257**, 11288–95.

Bäärnhielm C. and Hannson, G. (1986) Oxidation of 1,4-dihydropyridines by prostaglandin synthase and the peroxidic function of cytochrome P-450. Demonstration of a free radical intermediate. *Biochem. Pharmacol.*, **35**, 1419–25.

Baba, T., Yamada, H., Oguri, K. and Yoshimura, H. (1988) Participation of cytochrome P-450 isozymes in N-demethylation, N-hydroxylation and aromatic hydroxylation of methamphetamine. *Xenobiotica*, **18**, 475–84.

Beaune, P.H., Umbenhauer, D.R., Bork, R.W. *et al.* (1986) Isolation and sequence determination of a cDNA clone related to human cytochrome P-450 nifedipine oxidase. *Proc. Natl. Acad. Sci. USA*, **83**, 8064–8.

Böcker, R.H. and Guengerich, F.P. (1986) Oxidation of 4-aryl- and 4-alkyl-substituted 2,6-dimethyl-3,5-bis-(alkoxycarbonyl)-1,4-dihydropyridines by human liver microsomes and immunochemical evidence for the involvement of a form of cytochrome *P*-450. *J. Med. Chem.*, **29**, 1596–603.

Bondon, A., Macdonald, T.L., Harris, T.M. and Guengerich, F.P. (1989) Oxidation of cyclobutylamines by cytochrome *P*-450. Mechanism-based inactivation, adduct formation, ring expansion, and nitrone formation. *J. Biol. Chem.*, **264**, 1988–97.

Bordwell, F.G., Chen, J-P. and Bansch, M.J. (1988) Acidities of radical cations derived from remotely substituted and phenyl-substituted fluorenes. *J. Am. Chem. Soc.*, **110**, 2867–72.

Born, J.L. and Hadley, W.M. (1989) Isotopic sensitivity in the microsomal oxidation of the dihydropyridine calcium entry blocker nifedipine. *Chem. Res. Toxicol.*, **2**, 57–9.

Burka, L.T., Guengerich, F.P., Willard, R.J. and Macdonald, T.L. (1985) Mechanism of cytochrome *P*-450 catalysis. Mechanism of *N*-dealkylation and amine oxide deoxygenation. *J. Am. Chem. Soc.*, **107**, 2549–51.

Butler, M.A., Iwasaki, M., Guengerich, F.P. and Kadlubar, F.F. (1989) Cytochrome *P*-450$_{PA}$ (P-450IA2), the phenacetin *O*-deethylase, is primarily responsible for the 3-demethylation of caffeine and the *N*-oxidation of carcinogenic arylamines. *Proc. Natl Acad. Sci. USA*, **86**, 7696–700.

Carlson, B.W., Miller, L.L., Neta, P. and Grodkowski, J. (1984) Oxidation of NADH involving rate-limiting one-electron transfer. *J. Am. Chem. Soc.*, **106**, 7233–9.

Fujimori, K., Takata, T., Fujiwara, S. *et al.* (1986) Intervention of *N*,*N*-dimethylanilinium cation radical in the Polononovski type reaction of *N*,*N*-dimethylaniline *N*-oxide catalyzed by mesotetraphenylporphinato iron/imidazole. *Tet. Lett.*, **27**, 1617–20.

Gorrod, J.W. and Temple, D.J. (1976) The formation of an *N*-hydroxymethyl intermediate in the *N*-demethylation of *N*-methylcarbazole *in vivo* and *in vitro*. *Xenobiotica*, **6**, 265–74.

Groves, J.T., McClusky, G.A., White, R.E. and Coon, M.J. (1978) Aliphatic hydroxylation by highly purified liver microsomal cytochrome *P*-450. Evidence for a carbon radical intermediate. *Biochem. Biophys. Res. Commun.*, **76**, 541–9.

Guengerich, F.P. (1984) Oxidation of sparteines by cytochrome *P*-450: evidence against the formation of *N*-oxides. *J. Med. Chem.*, **27**, 1101–3.

Guengerich, F.P. (1988) Roles of cytochrome *P*-450 enzymes in chemical carcinogenesis and cancer chemotherapy. *Cancer Res.*, **48**, 2946–54.

Guengerich, F.P. (1990a) Enzymatic oxidation of xenobiotic chemicals. *Crit. Rev. Biochem. Mol. Biol.*, **25**, 97–153.

Guengerich, F.P. (1990b) Low kinetic hydrogen isotope effects in the oxidation of 1,4-dihydro-2,6-dimethyl-4-(2-nitrophenyl)-3,5-pyridinedicarboxylic acid dimethyl ester (nifedipine) by cytochrome *P*-450 enzymes are consistent with an electron-proton-electron transfer mechanism. *Chem. Res. Toxicol.*, **3**, 21–6.

Guengerich, F.P. and Böcker, R.H. (1988) Cytochrome *P*-450-catalyzed dehydrogenation of 1,4-dihydropyridines. *J. Biol. Chem.*, **263**, 8168–75.

Guengerich, F.P. and Macdonald, T.L. (1984) Chemical mechanisms of catalysis by cytochromes *P*-450: a unified view. *Acct. Chem. Res.*, **17**, 9–16.

Guengerich, F.P., Martin, M.V., Beaune, P.H. *et al.* (1986a) Characterization of rat and human liver microsomal cytochrome *P*-450 forms involved in nifedipine oxidation, a prototype for genetic polymorphism in oxidative drug metabolism. *J. Biol. Chem.*, **261**, 5051–60.

Guengerich, F.P., Müller-Enoch, D. and Blair, I.A. (1986b) Oxidation of quinidine by human liver cytochrome *P*-450. *Mol. Pharmacol.*, **30**, 287–95.

Guengerich, F.P., Willard, R.J., Shea, J.P. *et al.* (1984) Mechanism-based inactivation of cytochrome *P*-450 by heteroatom-substituted cyclopropanes and formation of ring-opened products. *J. Am. Chem. Soc.*, **106**, 6446–7.

Hall, L.R., Iwamoto, R.T. and Hanzlik, R.P. (1989) Electrochemical models for cytochrome *P*-450. *N*-Demethylation of tertiary amides by anodic oxidation. *J. Org. Chem.*, **54**, 2446–51.

Hammerich, O. and Parker, V.D. (1984) Kinetics and mechanisms of reactions of organic cation radicals in solution. *Adv. Phys. Org. Chem.*, **20**, 55–189.

Hammons, G.J., Guengerich, F.P., Weis, C.C. *et al.* (1985) Metabolic oxidation of carcinogenic arylamines by rat, dog, and human hepatic microsomes and by purified flavin-containing and cytochrome *P*-450 monooxygenases. *Cancer Res.*, **45**, 3578–85.

Hanzlik, R.P. and Tullman, R.H. (1982) Suicidal inactivation of cytochrome *P*-450 by cyclopropylamines. Evidence for cation-radical intermediates. *J. Am. Chem. Soc.*, **104**, 2048–50.

Kadlubar, F.F. and Hammons, G.J. (1987) The role of cytochrome *P*-450 in the metabolism of chemical carcinogens, in *Mammalian Cytochromes P-450* (ed. F.P. Guengerich) Vol. 2, CRC Press, Boca Raton, pp. 81–130.

Lee, J.S., Jacobsen, N.E. and Ortiz de Montellano, P.R. (1988) 4-Alkyl radical extrusion in the cytochrome *P*-450-catalyzed oxidation of 4-alkyl-1,4-dihydropyridines. *Biochemistry*, **27**, 7703–10.

Liebler, D.C. and Guengerich, F.P. (1983) Olefin oxidation by cytochrome *P*-450. Evidence for group migration in catalytic intermediates formed with vinylidene chloride and *trans*-1-phenyl-1-butene. *Biochemistry*, **22**, 5482–9.

Macdonald, T.L., Gutheim, W.G., Martin, R.B. and Guengerich, F.P. (1989) Oxidation of substituted *N,N*-dimethylanilines by cytochrome *P*-450: estimation of the effective oxidation-reduction potential of cytochrome *P*-450. *Biochemistry*, **28**, 2071–7.

Macdonald, T.L., Zirvi, K., Burka, L.T. *et al.* (1982) Mechanism of cytochrome *P*-450 inhibition by cyclopropylamines. *J. Am. Chem. Soc.*, **104**, 2050–2.

Manring, L.E. and Peters, K.S. (1983) Picosecond observation of kinetic vs. thermodynamic hydrogen atom transfer. *J. Am. Chem. Soc.*, **105**, 5708–9.

McMurry, T.J. and Groves, J.T. (1986) Metalloporphyrin models for cytochrome *P*-450, in *Cytochrome P-450* (ed. P.R. Ortiz de Montellano), Plenum Press, New York, pp. 1–28.

Miller, L.L. and Valentine, J.R. (1988) On the electron–proton–electron mechanism for 1-benzyl-1,4-dihydronicotinamide oxidations. *J. Am. Chem. Soc.*, **110**, 3982–9.

Miwa, G.T. and Lu, A.Y.H. (1986) Kinetic isotope effects and 'metabolic switching' in cytochrome *P*-450-catalyzed reactions. *BioEssays*, **7**, 215–19.

Miwa, G.T., Walsh, J.S., Kedderis, G.L. and Hollenberg, P.F. (1983) The use of intramolecular isotope effects to distinguish between deprotonation and hydrogen atom abstraction mechanisms in cytochrome *P*-450 and peroxidase-catalyzed *N*-demethylation reactions. *J. Biol. Chem.*, **258**, 14445–9.

Nelsen, S.F. and Ippoliti, J.T. (1986) On the deprotonation of trialkylamine cation radicals by amines. *J. Am. Chem. Soc.*, **108**, 4879–81.

Nelson, S.D. and Harvison, P.J. (1987) Roles of cytochromes *P*-450 in chemically induced cytotoxicity, in *Mammalian Cytochromes P-450* (ed. F.P. Guengerich), Vol. 2, CRC Press, Boca Raton, pp. 19–79.

Ortiz de Montellano, P.R. (1986) Oxygen activation and transfer, in *Cytochrome P-450* (ed. P.R. Ortiz de Montellano), Plenum Press, New York, pp. 217–71.

Ortiz de Montellano, P.R. (1987) Control of the catalytic activity of prosthetic heme by the structure of hemoproteins. *Acc. Chem. Res.*, **20**, 289–94.

Pandey, R.N., Armstrong, A.P. and Hollenberg, P.F. (1989) Oxidative N-demethylation of N,N-dimethylaniline by purified isozymes of cytochrome P-450. *Biochem. Pharmacol.*, **38**, 2181–5.

Prough, R.A., Brown, M.I., Dannan, G.A. and Guengerich, F.P. (1984) Major isozymes of rat liver microsomal cytochrome P-450 involved in the N-oxidation of N-isopropyl-α-(2-methylazo)-p-toluamide, the azo derivative of procarbazine. *Cancer Res.*, **44**, 543–8.

Shannon, P. and Bruice, T.C. (1981) A novel P-450 model system for the N-dealkylation reaction. *J. Am. Chem. Soc.*, **103**, 4580–2.

Shea, J.P., Valentine, G.L. and Nelson, S.D. (1982) Source of oxygen in cytochrome P-450 catalyzed carbinolamine formation. *Biochem. Biophys. Res. Commun.*, **109**, 231–5.

Shimada, T., Iwasaki, M., Martin, M.V. and Guengerich, F.P. (1989) Human liver microsomal cytochrome P-450 enzymes involved in the bioactivation of pro-carcinogens detected by *umu* gene response in *Salmonella typhimurium* TA1535/pSK1002. *Cancer Res.*, **49**, 3218–28.

Shono, T., Toda, T. and Oshino, N. (1982) Electron transfer from nitrogen in microsomal oxidation of amine and amide. Simulation of microsomal oxidation by anodic oxidation. *J. Am. Chem. Soc.*, **104**, 2639–41.

Silverman, R.B. and Zieske, P.A. (1986) 1-Phenylcyclobutylamine, the first in a new class of monoamine oxidase inactivators: further evidence for a radical intermediate. *Biochemistry*, **25**, 341–6.

Sinha, A. and Bruice, T.C. (1984) Rate-determining general-base catalysis in an obligate 1e-oxidation of a dihydropyridine. *J. Am. Chem. Soc.*, **106**, 7291–2.

Williams, D.E., Reed, R.L., Kedzierski, B. *et al.* (1989) Bioactivation and detoxication of the pyrrolizidine alkaloid senecionine by cytochrome P-450 isozymes in rat liver. *Drug. Metab. Disp.*, **17**, 387–92.

Ziegler, D.M. (1988) Flavin-containing monooxygenases. Catalytic mechanism and substrate specificities. *Drug Metab. Rev.*, **19**, 1–32.

10

The role of cytochrome *P*-450 in the biological nuclear *N*-oxidation of aminoaza-heterocyclic drugs and related compounds

J. W. Gorrod and S. P. Lam*

Chelsea Department of Pharmacy, King's College, University of London, Manresa Road, London SW3 6LX

1. Aminoazaheterocycles are an important group of chemicals which occur in our environment.
2. They exist as natural compounds, pyrolysis products, synthetic medicinal agents and metabolites of complex drug molecules.
3. *N*-Oxidation of aminoazaheterocycles at the aromatic ring nitrogen is a common reaction yielding *N*-oxides. The position of substitution affects the formation of *N*-oxide. In some cases the pyrimidine ring is oxygenated to form both 1 and 3-isomeric *N*-oxides, in others only the 3-isomer is formed.
4. In all cases the evidence strongly implicates the involvement of a phenobarbitone inducible cytochrome *P*-450.
5. The experimental results and the role of physicochemical and enzymic factors in controlling the site of oxidation of aminoazaheterocycles are discussed.

* Present address: The Medicines Control Agency, Department of Health, Market Towers, 1 Nine Elms Lane, Vauxhall, London SW8 5NQ, UK.

10.1 INTRODUCTION

Aminoazaheterocycles are a group of organic molecules having an amino group attached to a carbon adjacent to a ring nitrogen in an aromatic system. The nucleic acid bases adenine, guanine, cytosine and the vitamin thiamine are examples of ortho-aminoazaheterocycles which exist naturally as essential compounds required for the proper physiological and biochemical function of biological systems. A large number of important medicinal agents covering a wide spectrum of pharmacological activities also belongs to this chemical group: e.g. the anti-anaemic, folic acid; the antimalarial, pyrimethamine; the broad spectrum antibacterials, trimethoprim and cefotaxime; the anticancer agent, metoprine; the diuretics, triamterene and amiloride (Fig. 10.1). Many aminoazaheterocycles are also *in vivo* metabolites of medicinal agents (Fig. 10.2). A number of known examples of aminoazaheterocycles, which are among the most mutagenic and carcinogenic compounds, are formed during the cooking process of foods and have been detected in charred meat, broiled fish and in pyrolysates of proteins and amino acids (Fig. 10.3), (Sugimura and Nagao, 1979; Sugimura and Sato, 1983). Sugimura *et al.* (1977) showed that Trp-P-1 (3-amino-1,4-dimethyl-5*H*-pyrido[4,3-b]indole) and Trp-P-2 (3-amino-1-methyl-5*H*-pyrido[4,3-b]indole) are produced during tryptophan pyrolysis and Yamamoto *et al.* (1978) demonstrated that Glu-P-1 (2-amino-6-methyl-dipyrido[1,2-a: 3′,2′-d]imidazole) and Glu-P-2 (2-aminodipyrido[1,2-a: 3′,2′-d]imidazole) are products of glutamic acid pyrolysis. MeIQ (2-amino-3,4-diethylimidazo[4,5-f]quinoline) and IQ (2-amino-3-methylimidazo-[4,5-f]quinoline), having the strongest mutagenic activity (Ishida *et al.*, 1987), were isolated from broiled sardines. Felton *et al.* (1984) and Negishi *et al.* (1984) identified from fried ground beef major mutagens 7,8-DiMeIQx (2-amino-3,7,8-trimethylimidazo[4,5-f]quinoxaline), and MeIQx (2-amino-3,8-dimethylimidazo[4,5-f]quinoxaline) together with at least ten other mutagenic amino azaheterocycles. Felton *et al.* (1986) further isolated from fried ground beef the most abundant mutagen PhIP (2-amino-1-methyl-

Thiamine
Vitamin

Guanine
Nucleic acid base

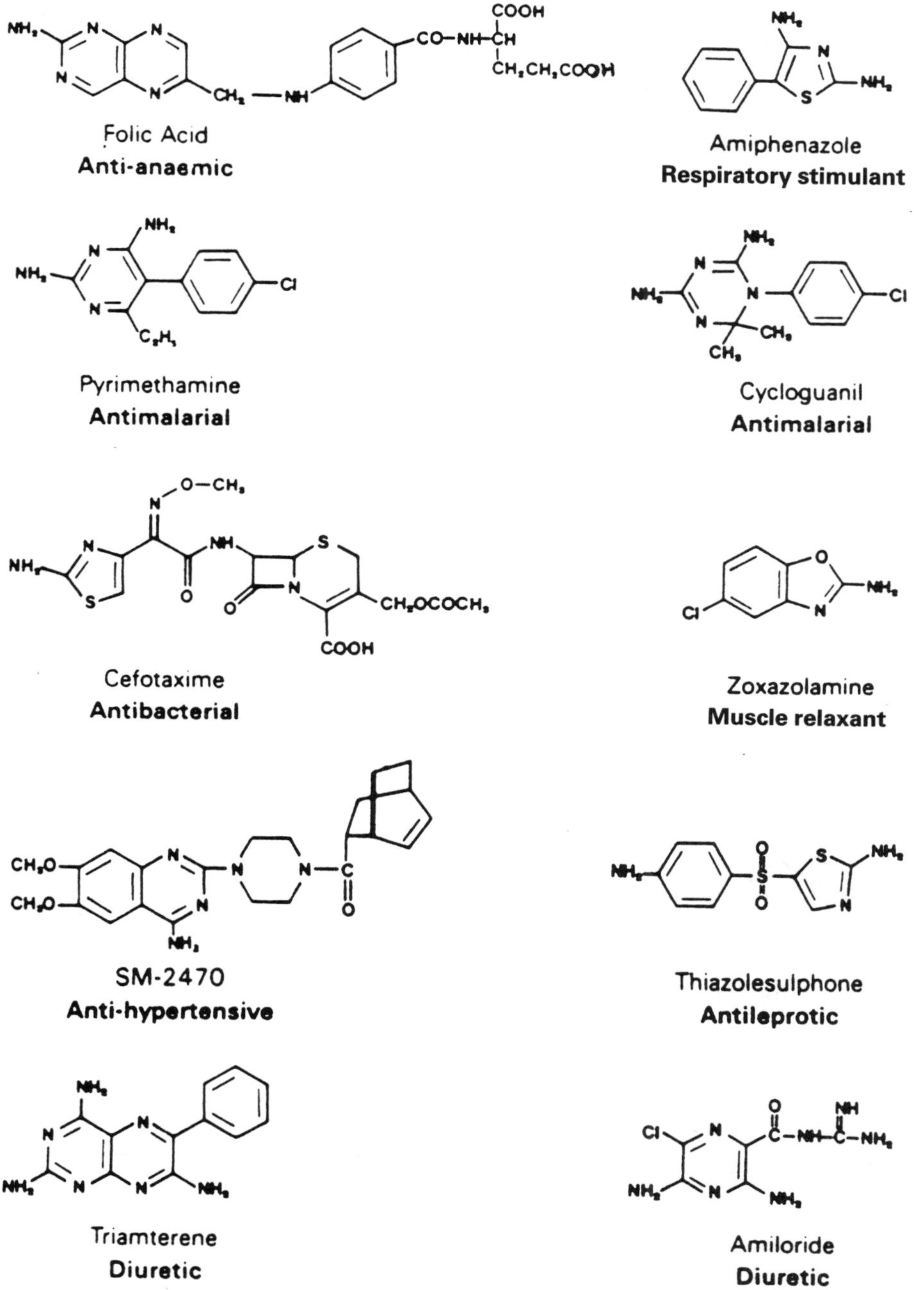

Folic Acid
Anti-anaemic

Amiphenazole
Respiratory stimulant

Pyrimethamine
Antimalarial

Cycloguanil
Antimalarial

Cefotaxime
Antibacterial

Zoxazolamine
Muscle relaxant

SM-2470
Anti-hypertensive

Thiazolesulphone
Antileprotic

Triamterene
Diuretic

Amiloride
Diuretic

Figure 10.1 Some medicinal aminoazaheterocyclic compounds.

6-phenylimidazo[4,5-b]pyridine) which belongs to the amino-imidazo-azaarenes, a new class of mutagenic amino azaheterocycles. The occurrence of aminoazaheterocycles has previously been discussed in greater detail (Gorrod and Lam, 1989, 1990).

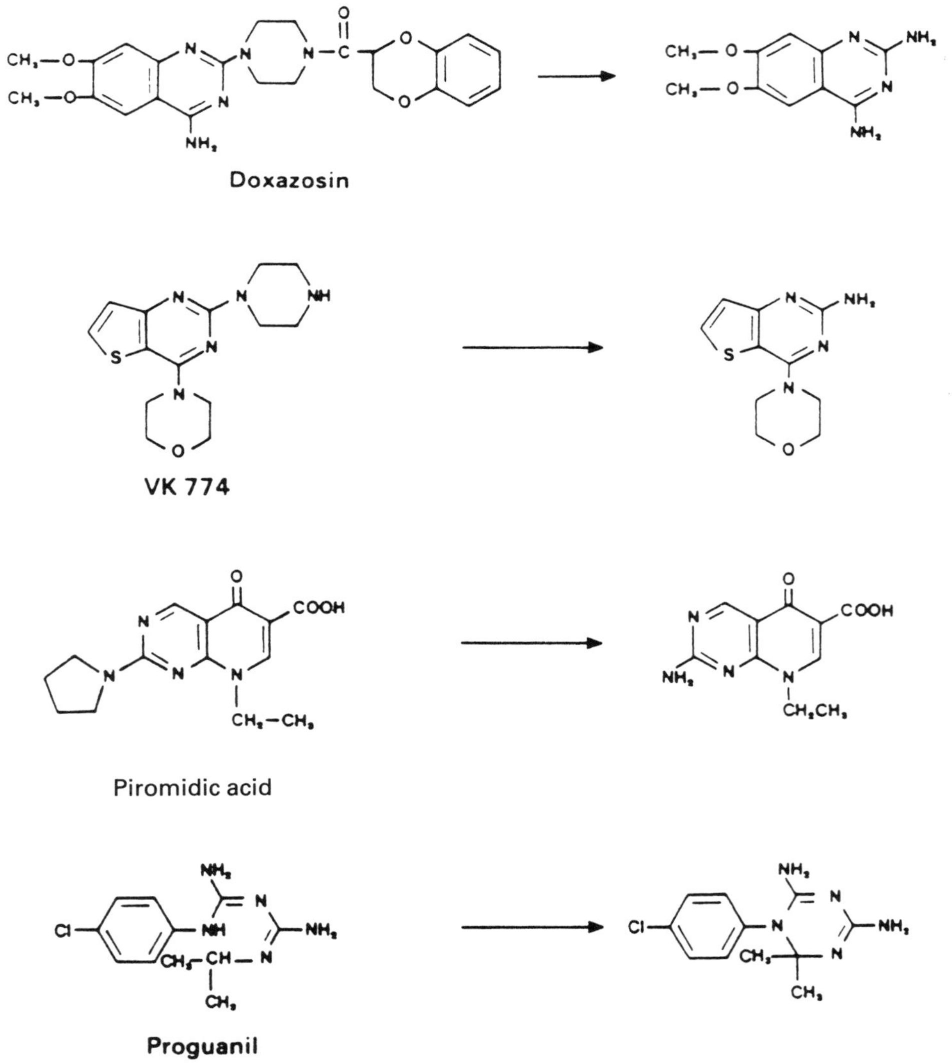

Doxazosin

VK 774

Piromidic acid

Proguanil

Figure 10.2 Examples of aminoazaheterocycles formed from medicinal compounds following metabolic biotransformation.

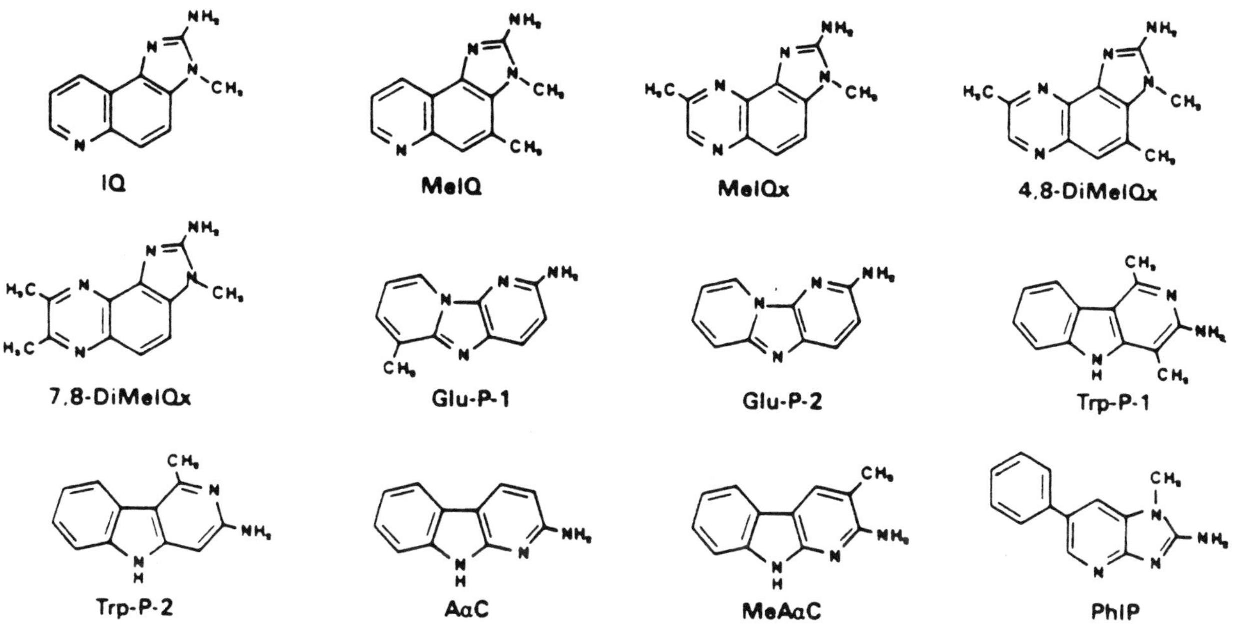

Figure 10.3 Chemical structures of mutagenic aminoazaheterocycles.

10.2 OBSERVATIONS OF METABOLIC N-OXIDATION OF AMINOAZAHETEROCYCLES

Aminoazaheterocycles, when metabolized, may be oxidized either at an endo-ring nitrogen resulting in the formation of an N-oxide or at the exo-amino group to produce a hydroxylamine metabolite. In a number of different studies, the *in vivo* N-oxidation of the widely used medicinal aminoazaheterocycle, trimethoprim (2,4-diamino-5-(3,4,5-trimethoxy-benzyl)-pyrimidine) was investigated in rat, dog and man (Schwartz *et al.*,

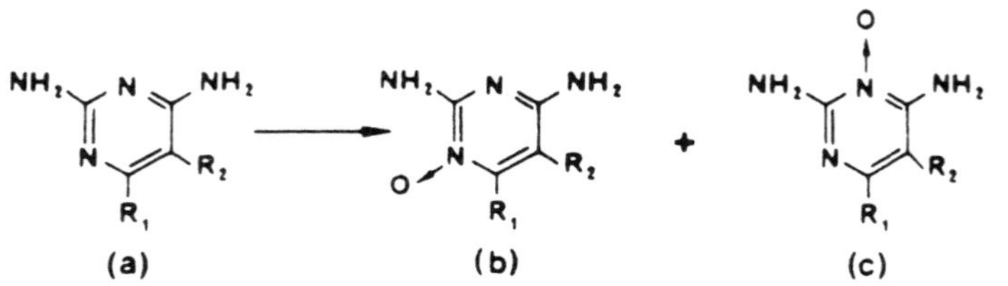

		R_1	R_2
1.	Trimethoprim (a)	H	3,4,5-trimethoxybenzyl
2.	Metoprine (a)	CH_3	3,4-dichlorophenyl
3.	Pyrimethamine (a)	C_2H_5	4-chlorophenyl

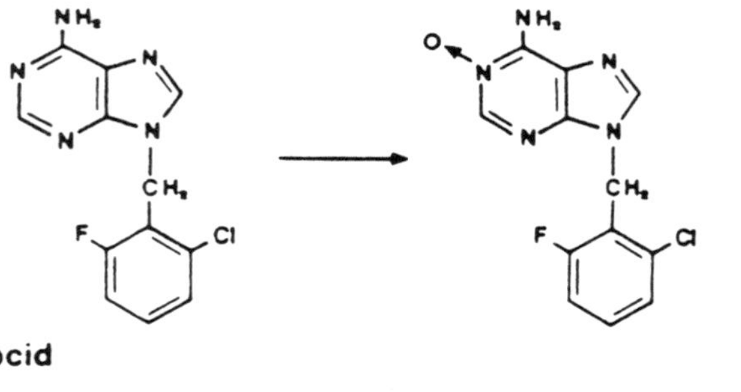

Arprinocid

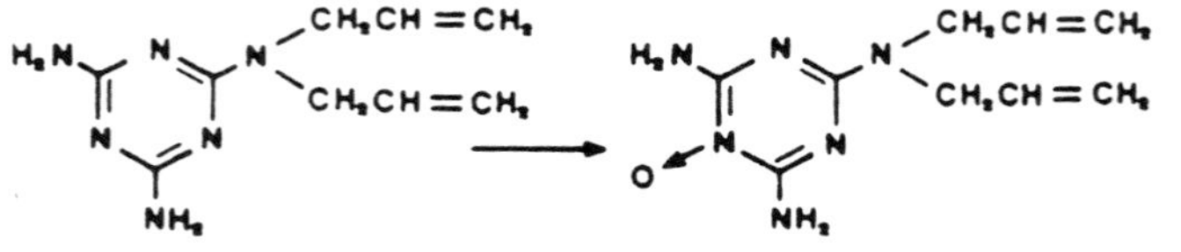

Diallylmelamine

Figure 10.4 Aminoazaheterocycles converted into innocuous amine oxides.

1970; Meshi and Sato, 1972; Sigel, 1973; Brooks *et al.*, 1973; Sigel and Brent, 1973; Hubbel *et al.*, 1978). Results showed that *N*-oxidation occurred at the ring nitrogens leading to the formation of both 1- and 3-*N*-oxides. The formation of 1- and 3-*N*-oxides during the *in vivo* metabolism of the anti-malarial drug pyrimethamine (2,4-diamino-5-(*p*-chlorophenyl)-6-ethyl-pyrimidine) and the anticancer agent metoprine (2,4-diamino-5-(3,4-dichlorophenyl)-6-methylpyrimidine) was also reported (Hubbel *et al.*, 1978, 1980). In another study (Zins, 1965), it was shown that the vasodilator, 2-*N*,*N*-diallylmelamine was converted into a nuclear 5-*N*-oxide. Zins *et al.* (1968) reported that the delayed hypotensive effect was due to the metabolic formation of the active *N*-oxide metabolite. Another example of *N*-oxide formation from an aminoazaheterocycle is the metabolism of an anti-coccidial agent, aprinocid (6-amino-9-(2-chloro-6-fluorobenzyl)purine), a 9-substituted adenine, which forms an active anticoccidial 1-*N*-oxide metabolite (Wolf *et al.*, 1978; Wang and Simashkevich, 1980) (Fig. 10.4). A further example of an aminoazaheterocyclic compound in which the *N*-oxide is an active agent is the hypotensive drug minoxidil (2,4-diamino-6-piperidinopyrimidine-3-*N*-oxide) which is active in rats, dogs, monkeys and humans (Thomas and Hartpoolian, 1975).

Examples of *N*-oxidation occurring at exo-amino groups can be found from metabolic studies of mutagenic aminoazaheterocycles. It has been

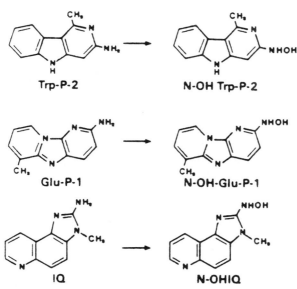

Figure 10.5 Aminoazaheterocycles converted into genotoxic hydroxylamines.

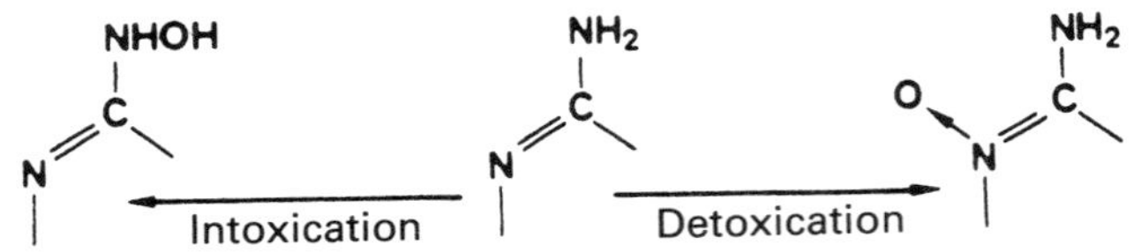

Figure 10.6 Metabolic pathways of aminoazaheterocycles to hydroxylamine or amine-oxide.

shown that these compounds are not themselves mutagens, but require metabolic activation to produce a mutagenic effect (Nemoto *et al.*, 1979). Exo-amino oxidation of Trp-P-1 leading to the formation of the mutagenic 3-*N*-hydroxylamine metabolite was recorded (Yamazoe *et al.*, 1980; Hashimoto *et al.*, 1980a). Similarly, Glu-P-1 following incubation with fortified rat liver microsomes was converted into the 2-hydroxyamino metabolite (Hashimoto *et al.*, 1980b). Another example in which metabolic activation is required at the exo-amino group to form a mutagenic hydroxylamine is the aminoazaheterocyclic IQ (Okamato *et al.*, 1981) (Fig. 10.5).

These observations support the generalization that *N*-oxidation of aminoazaheterocycles at the endo-ring nitrogen results in the formation of an innocuous *N*-oxide whereas *N*-oxidation at the exo-amino group produces potentially genotoxic hydroxylamines (Fig. 10.6).

10.3 ENZYMOLOGY OF NUCLEAR *N*-OXIDATION OF AMINOAZAHETEROCYCLES

It is interesting to observe that these structurally very similar organic molecules, containing the same functional —N=CH—NH$_2$ group when metabolized are capable of being converted into two different kinds of metabolites having such opposite toxicological properties (Fig. 10.6). The enzymology of, and physicochemical factors affecting, these *N*-oxidation processes clearly become important. Our laboratory has been interested, for some years, in the elucidation of the structural, physicochemical and enzymic factors which affect the site of biological *N*-oxidation of aminoazaheterocycles (Gorrod, 1985, 1987; Gorrod and Lam, 1989). Watkins and Gorrod (1987) and Watkins (1988), using hepatic microsomes from rat, rabbit and hamster, showed that following incubations of some 5-substituted, and 5,6-disubstituted 2,4-diaminopyrimidines (trimethoprim, pyrimethamine and metoprine), both 1-, and 3-*N*-oxides were formed. The extent of each isomeric *N*-oxide formed varied in different species. In general, the amount of 3-*N*-oxide produced predominated in any of the

Table 10.1 Effects of enzyme inducers and inhibitors on the *N*-oxide formation from various aminoazaheterocycles using hamster hepatic microsomes

Substrate[a]	N-oxide formed	Treatment						
		Inducers[b]			Inhibitors[c]			
		Pb	βNF	A-1254	SKF-525A	DPEA	OCT	MZ
9BA	1	842	71	1745	13	2	25	101
9BHA	1	383	70	555	21	6	66	85
TMP	1	170	80	—	70	25	18	—
TMP	3	220	100	—	50	25	20	—
PYR	1	195	115	—	45	25	35	—
PYR	3	560	120	—	35	25	50	—
MET	1	390	140	—	45	18	50	—
MET	3	420	90	—	50	55	55	—
DAPPY	3	226	85	—	9	30	82	—
DACPY	3	190	90	—	90	65	60	—
DAMPY	3	240	110	—	61	—	85	—

[a] 9BA, 9-benzyladenine; 9BHA, 9-benzhydryladenine; TMP, trimethoprim; PYR, pyrimethamine; MET, metoprine; DAPPY, 2,4-diamino-6-piperidinopyrimidine; DACPY, 2,4-diamino-6-chloropyrimidine; DAMPY, 2,4-diamino-6-morpholinopyrimidine; N–O, *N*-oxide metabolite; 1 and 3: 1- and 3-*N*-oxides.

[b] Pb, βNF and A-1254, hamsters pretreated with enzyme inducers phenobarbitone, β-napthoflavone and aroclor-1254 respectively.

[c] SKF-525A, DEPA, OCT and MZ; microsomes co-incubated with inhibitors β-diethylaminoethyl-diphenylpropylacetate, 2,4-dichloro-6-phenyl-phenoxyethylamine, *n*-octylamine and methimazole respectively.

Results are expressed as percentage of appropriate control carried out under identical conditions as the test.

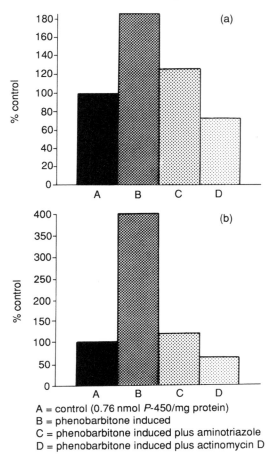

A = control (0.76 nmol *P*-450/mg protein)
B = phenobarbitone induced
C = phenobarbitone induced plus aminotriazole
D = phenobarbitone induced plus actinomycin D

Figure 10.7 The effect of potential protein synthesis inhibitors on cytochrome *P*-450 levels (*a*) and the *in vitro* N-oxidation of 6-piperidino 2,4-diaminopyrimidine (*b*) by hamster liver microsomes (Gorrod and El-Ghomari, unpublished observations).

species investigated. In contrast, a study of a series of 6-substituted 2,4-diaminopyrimidines (El-Ghomari and Gorrod, 1987) only 3-*N*-oxides of some of the substrates were detected; no formation of any 1-*N*-oxide from any of the substrates was observed. A further group of aminoazaheterocycles studied by our laboratory is the 9-substituted 6-aminopurines. Using hepatic microsomes from various species, Lam *et al.* (1987, 1989) showed that only 1-*N*-oxides of some 9-substituted adenines were detected. It is of importance that in all the above studies, no evidence of the biological exo-amino *N*-oxidation producing a hydroxylamine metabolite was obtained.

Investigation of the enzymology of formation of *N*-oxides from these compounds indicated that phenobarbitone inducible form(s) of *P*-450 isoenzymes is largely responsible (Table 10.1). The induction of *P*-450 activity and associated increased *N*-oxidation of 6-substituted 2,4-diamino-pyrimidines and 9-substituted adenines have been shown to be reduced or

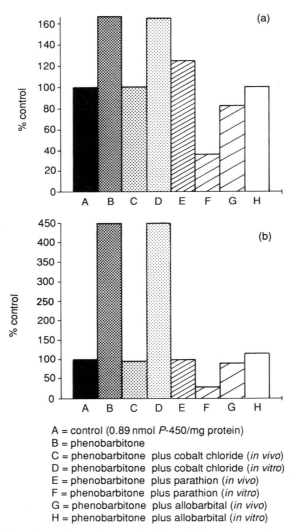

A = control (0.89 nmol *P*-450/mg protein)
B = phenobarbitone
C = phenobarbitone plus cobalt chloride (*in vivo*)
D = phenobarbitone plus cobalt chloride (*in vitro*)
E = phenobarbitone plus parathion (*in vivo*)
F = phenobarbitone plus parathion (*in vitro*)
G = phenobarbitone plus allobarbital (*in vivo*)
H = phenobarbitone plus allobarbital (*in vitro*)

Figure 10.8 The effect of cytochrome *P*-450 modifiers on cytochrome *P*-450 levels (*a*) and the *in vitro* *N*-oxidation of 6-piperidino-2,4-diaminopyrimidine (*b*) by hamster liver microsomes (Gorrod and El-Ghomari, unpublished observations).

inhibited by the co-administration of phenobarbitone together with protein and haem biosynthesis inhibitors. Furthermore, reduction of enhanced N-oxide formation from these aminoazaheterocycles have also been observed when microsomes from phenobarbitone pretreated animals were co-incubated with *P*-450 modifiers. Figures 10.7 and 10.8 illustrate the changes of *P*-450 level and the resultant N-oxide formation from 6-piperidino-2,4-diaminopyrimidine by the utilization of various *in vivo* and *in vitro* agents to investigate the enzymology. Indeed, little evidence was obtained which supported the involvement of *P*-448 isoenzymes. Using specific inhibitors of the flavin-containing mono-oxygenase system, the possibility of N-oxide formation from 9-substituted adenines via this enzyme system was excluded. These results thus strongly suggest that ring N-oxygenation of aminoazaheterocycles is catalysed by phenobarbitone inducible *P*-450 isoenzymes.

10.4   FACTORS INVOLVED IN THE N-OXIDATION OF AMINOAZAHETEROCYCLES

10.4.1   Enzymic factors

Results obtained so far from the studies of N-oxide formation from our laboratory clearly indicate that phenobarbitone-inducible form(s) of *P*-450 isoenzyme(s) is responsible for this particular N-oxidation process. No strong evidence supporting the involvement of a *P*-448 species has been obtained. However, in the case of the mutagenic aminoazaheterocycles, investigations of the enzymology revealed that a different form of cytochrome *P*-450 isoenzyme is involved in the N-oxidation of the exo-amino group to produce hydroxylamines (Yamazoe *et al.*, 1980; Hashimoto *et al.*, 1980a,b; Okamato *et al.*, 1981; Kato *et al.*, 1983; Nagao *et al.*, 1983). Using various specific enzyme inhibitors, inducers, and purified enzymes, Kawajiri *et al.* (1983) demonstrated very clearly that the *P*-448 isoenzymes of cytochrome *P*-450 were mainly responsible for metabolically N-hydroxylating these compounds. In another study Kamataki *et al.* (1983) using purified low and high spin forms of the *P*-448 species demonstrated that the high spin form was capable of N-hydroxylating the aminoaza-heterocycles. Ryan *et al.* (1979) showed that the low-spin form, which was responsible for metabolically activating the carcinogenic benzo[a]pyrenes and other aromatic hydrocarbons, had low activity in the N-hydroxylation of these aminoazaheterocyclic compounds. The involvement of a high spin form of *P*-488 in N-hydroxylation was further substantiated in another study (Yamazoe *et al.*, 1984). Metabolic activation of mutagenic aminoaza-heterocycles has been reviewed (Kato, 1986). These results thus suggest that exo-amino oxidation of aminoazaheterocycles is mediated by the *P*-448 isoenzymes, in particular the high spin form, whereas endo *N*-oxide formation is mainly catalysed via the phenobarbitone-inducible *P*-450 isoenzyme.

10.4.2 Substrate factors

From the above it is clearly important to establish the chemical factors that control the substrate specificity of the enzymes involved in either site of *N*-oxygenation of aminoazaheterocycles. This laboratory has been trying to elucidate the factors controlling nuclear *N*-oxidation; these are considered below.

(a) Amine–imine tautomerism

The concept of tautomerism in affecting *N*-oxidation of aminoazahetero-cycles was suggested by El-Ghomari and Gorrod (1987) in an attempt to explain the metabolic 3-*N*-oxidation of a series of 6-substituted 2,4-diamino-pyrimidines. It was believed that the failure to form any 1-*N*-oxide from these 6-substituted 2,4-diaminopyrimidines could not be due to steric hindrance associated with substituents at the 6-position. This conclusion was supported by results from studies of pyrimethamine and metoprine which had an ethyl and methyl group respectively at the 6-position together with 5-substituents; yet produced both 1- and 3-*N*-oxides (Watkins, 1988). It was suggested that the lack of hydrogen on the tertiary amino group substituted

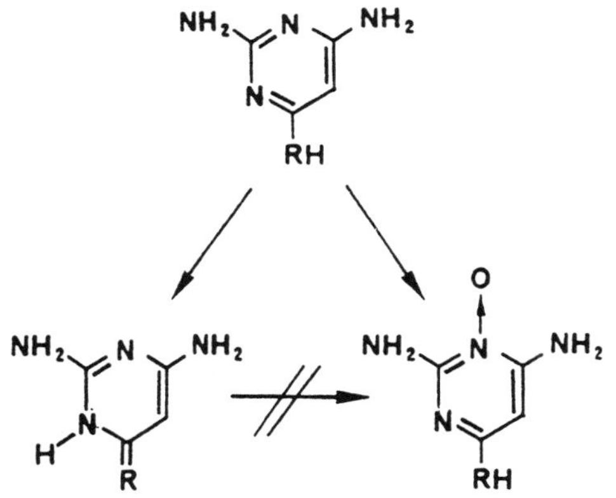

Figure 10.9 Role of tautomerism in the 3-*N* oxidation of 6-substituted 2,4-diaminopyrimidines.

at position 6 of diethylamino-, piperidino-, and morpholino-substituted 2,4-diaminopyrimidines suppresses the formation of any imino tautomeric isomer associated with the 1-N position, thus allowing 3-N-oxide formation. In other analogues, such as the 6-hydroxy and 6-amino substituted compounds, it may be that tautomerism occurs causing a redistribution of electrons at the 1-N position and so suppresses 3-N oxidation (Fig. 10.9).

*(b) Basicity (*pK$_a$*) of aminoazaheterocycles*

The basicity of the nitrogen atom may determine its susceptibility to oxidation. On examination of pK_a values for aminoazaheterocycles (Table 10.2), no direct correlation with metabolic N-oxidation results could be established. The pK_a values for trimethoprim, pyrimethamine and metoprine are 7.2, 7.34 and 7.15 respectively (Nichol *et al.*, 1977). All three compounds form both 1- and 3-N-oxides (Watkins, 1988). In the case of 6-substituted 2,4-diaminopyrimidines (El-Ghomari, 1988), the pK_a values for 6-methyl-, 6-diethylamino-, 6-piperidino-, and 6-morpholino-substituted analogues which form only 3-N-oxides are 7.62, 7.41, 7.05 and 6.63 respectively. However, these values are not dissimilar from those of the 6-amino (pK_a 6.93) and unsubstituted (pK_a 7.33) derivatives which do not form N-oxides. Furthermore, the pK_a of 6-chloro-2,4-diaminopyrimidine,

Table 10.2 Basicity and lipophilicity of aminoazaheterocycles and their metabolic N-oxidation results

Substrate	N-*Oxide*	pK$_a$	*log* P
Adenine	—	4.33	−0.051
9-Methyladenine	—	4.15	−0.027
9-Benzyladenine	1-N	4.02	1.696
9-Benzhydryladenine	1-N	3.74	3.191
9-Trityladenine	—	—	5.079[a]
Metoprine	1-N, 3-N	7.15	2.82
Pyrimethamine	1-N, 3-N	7.34	2.69
Trimethoprim	1-N, 3-N	7.20	0.82
2,4-Diaminopyrimidine	—	7.33	−0.897
2,4-Diamino-6-hydroxypyrimidine	—	3.81	−1.069
2,4,6-Triaminopyrimidine	—	6.93	−1.027
2,4-Diamino-6-methylpyrimidine	3-N	7.62	−0.375
2,4-Diamino-6-morpholinopyrimidine	3-N	6.63	−0.110
2,4-Diamino-6-diethylaminopyrimidine	3-N	7.41	0.639
6-Chloro-2,4-diaminopyrimidine	3-N	3.55	0.686
2,4-Diamino-6-piperidinopyrimidine	3-N	7.05	0.943
Melamine	—	5.21	−0.548

[a] Calculated log *P* (Lam *et al.*, 1989).

which forms a 3-*N*-oxide, is 3.55 and is not much different from the non-*N*-oxidizable 6-hydroxy analogue (pK_a 3.81). In the case of 6-aminopurines (Lam, 1989), the pK_a values for adenine, 9-methyladenine, 9-benzyladenine and 9-benzhydryladenine are 4.33, 4.15, 4.02 and 3.74 respectively; yet only 9-benzyladenine and 9-benzhydryladenine form 1-*N*-oxides. These observations suggest, therefore, that although the pK_a of the nitrogen atom is important in *N*-oxidation generally (Gorrod, 1973, 1978), it does not appear to be a determining factor in the *N*-oxidation of aminoazahetero-cycles. However, in view of the complexity of these molecules having multibase centres, further work on delineating the precise site of protonation and its relevance to the *N*-oxidation processes needs to be carried out.

(c) Lipophilicity (partition coefficient log P)

Lipophilicity of aminoazaheterocycles appears to be an essential requirement for *N*-oxidation. The lipophilicity of 2,4-diaminopyrimidines studied in our laboratory have been determined. Trimethoprim, pyrimethamine and metoprine which form both 1- and 3-*N*-oxides are lipophilic. In the case of 6-substituted 2,4-diaminopyrimidines, most of the substrates which form 3-*N*-oxides are lipophilic, whereas the ones that do not form *N*-oxides are hydrophilic. In studies of 6-aminopurines, the results indicated that the hydrophilic adenine and 9-methyladenine were not *N*-oxidized, whereas the lipophilic 9-benzyladenine and 9-benzhydryl-adenine form the 1-*N*-oxide metabolites. In contrast, 9-trityladenine, a highly lipophilic compound which was not *N*-oxidized, indicates that other factors besides lipophilicity are involved (Table 10.2).

(d) Conformation of aminoazaheterocycles

Using computer graphic and ^{1}H-NMR techniques, the most probable conformations of 9-substituted adenines were established. The stereo-chemical characteristics of these molecules were correlated with their metabolic 1-*N*-oxidation. The results obtained suggested that 9-benzyl- and 9-benzhydryladenines, which form 1-*N*-oxides, have similar favourable conformations (Lam *et al.*, 1989). These conformations may enable favourable binding of the substrates to the enzyme hence allowing *N*-oxidation (Figs 10.10–10.13). In the case of 9-trityladenine even the most favourable conformation showed crowding around the 2-position, not seen with the benzyl and benzhydryl analogues. This suggests that steric hindrance in this molecule may prevent binding to the active site of the enzyme and so prevent *N*-oxidation.

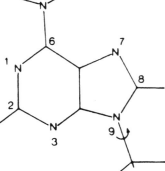

Figure 10.10 Graphic display of 9MA.

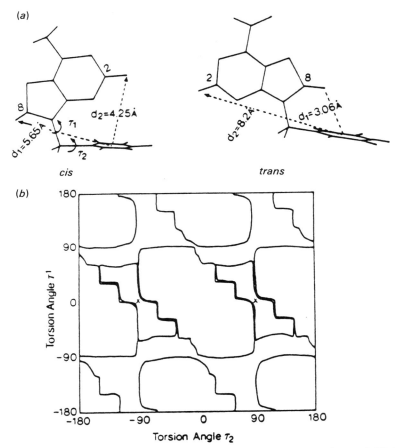

Figure 10.11 (*a*) Graphic displays of *cis* and *trans* conformations of 9BA. (*b*) Variation of the potential energy of 9BA as a function of the torsion angles τ_1 and τ_2. Contours are shown at intervals of 1 kcal/mol with the lowest energy level (-17.6 kcal/mol) indicated by x.

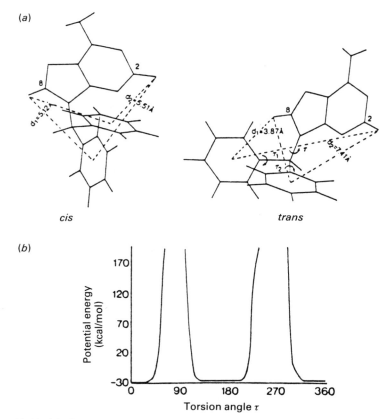

Figure 10.12 (*a*) Graphic displays of *cis* and *trans* conformations of 9BHA. (*b*) Variation of potential energy for 9BHA as a function of the torsion angle τ.

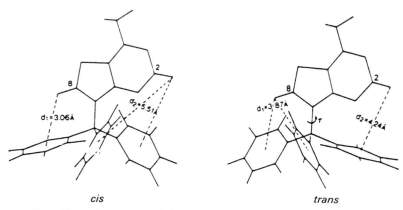

Figure 10.13 Graphic displays of *cis* and *trans* conformations of 9TA.

(e) Electron density on nitrogen

Another factor which may have an influence on the N-oxidation of aminoazaheterocycles is the N-electron density (Gorrod, 1987). The 'lone pair availability' at a particular nitrogen atom could be indicative of its oxidizability. It was calculated that in 2-aminopyridine, the 'lone pair availability' index values were 0.961 and 0.714 for the ring nitrogen and the exo-amino nitrogen respectively. These values, therefore, suggest that the ring nitrogen should oxidize in preference to the exo-amino group. This was found to be the case (Iles, 1988). In the case of 2-aminopyrimidine, values implying a reverse situation to that found for 2-aminopyridine were obtained. The calculated values for the exo-amino group and the ring nitrogen were 0.786 and 0.697 respectively (G.A. Webb, personal communication). The metabolism of isomeric aminopyrimidines is currently under investigation. Until more 'lone pair availability' indexes are available, it is difficult to establish a correlation with metabolic N-oxidation and extrapolate these results to structurally more complicated aminoazaheterocycles. Nevertheless, this method may generate useful information to allow the susceptibility of specific nitrogen atoms in a molecule towards enzymic oxidation to be determined.

An indirect method which could indicate the local electron density of a nitrogen atom in a chemical molecule is by NMR measurement. In a study of 9-substituted adenines, the chemical shift values of the 2-H protons of substituted analogues were compared to that of the adenine molecule. The direction change in chemical shift is an indication of the effect of 9-substitution on the electron density of 1-N nitrogen. It has been shown that 9-substitution generally caused a low field shift of the 2-H proton reflecting an overall electron withdrawing effect. This electron withdrawing effect may cause an increased conjugating overlap of the lone pair of electrons from the 6-amino group with the ring nitrogen thereby allowing 1-N-oxide formation from some of the derivatives (Lam *et al.*, 1989). This method has been used in a study of other 9-substituted adenines. Table 10.3 summarizes the NMR results which support the correlation, suggesting the possible use of chemical shift as a parameter for indication of susceptibility towards N-oxide formation.

(f) Binding spectra of aminoazaheterocycles with P-450

It was thought that the binding interaction of aminoazaheterocycles with microsomal P-450 may correlate with their susceptibility to N-oxidation. Lam (1989) examined the binding characteristics of some 9-substituted adenines with hamster hepatic microsomes. It was found that only weak binding interactions, even at high substrate concentration, were obtained

Table 10.3 ^{1}H NMR characteristics and calculated log P values of various 9-substituted adenines and their *N*-oxidation results

Compound (R)	N-O	Chemical shift (δ) ppm				Calculated log P
		2-H	8-H	Δδ(2H-8H)	Δδ(R-Ad)	
Adenine (Ad)	—	8.28	8.31	−0.03	—	−0.051
Adenosine	—	8.35	8.46	−0.11	0.07	<−0.051
Ara-adenosine	—	8.40	8.59	−0.19	0.12	<−0.051
Adenine monophosphate	—	8.68	8.48	0.20	0.49	<−0.051
Adenine triphosphate	—	8.75	8.55	0.20	0.47	<−0.051
9-Methyladenine	—	8.31	8.17	0.14	0.03	−0.027
9-Ethyladenine	±	8.56	8.46	0.10	0.28	0.053
2-Amino-9-benzyladenine	+	—	7.96	—	—	0.833
9-Phenyladenine	+	8.50	8.42	0.08	0.22	1.157
9-(4-Nitrobenzyl)adenine	+	ND	ND	—	—	1.411
9-Butyladenine	+	8.43	8.30	0.13	0.15	1.563
9-Benzyladenine	+	8.42	8.26	0.16	0.14	1.687
9-(2-Chloro-6-flouro-benzyl)adenine	+	8.42	8.14	0.28	0.14	2.553
9-(2,6-Dichlorobenzyl)adenine	+	8.47	7.98	0.49	0.19	3.076
9-Benzhydryladenine	+	8.34	7.95	0.39	0.06	3.278
9-Trityladenine	—	8.05	8.10	−0.05	−0.23	5.079

N-O, 1-*N*-oxide; —, not detected; +, detected by TLC and HPLC; ±, detected by HPLC but not by TLC; ND, not determined.

from adenine, 9-methyladenine and 9-trityladenine to give typical type II spectra for adenine and 9-methyladenine and a type I spectrum for 9-trityladenine. In the case of 9-benzyladenine, a type I spectral interaction at low substrate concentration and prominent type II spectra at high concentrations was observed. 9-Benzhydryladenine gave a marked and easily recognizable type II binding spectrum at all concentrations. Only the latter two compounds form 1-*N*-oxide metabolites. El-Ghomari and Gorrod (unpublished observations, 1989) also investigated the possible binding interaction of 6-substituted 2,4-diaminopyrimidines with *P*-450. They observed that the *N*-oxide forming 2,4-diamino-6-piperidinopyrimidine produced a type II binding spectrum with *P*-450 whereas the non-*N*-oxidizable 2,4-diamino-6-hydroxypyrimidine did not show any recognizable binding interaction (Fig. 10.14). It would be of interest to investigate the binding interaction of both 5-substituted and 5,6-disubstituted 2,4-diaminopyrimidines, which form both 1-*N* and 3-*N*-oxide metabolites. Results from these observations may allow better understanding of differences in substrate binding with *P*-450 leading to different *N*-oxidation products.

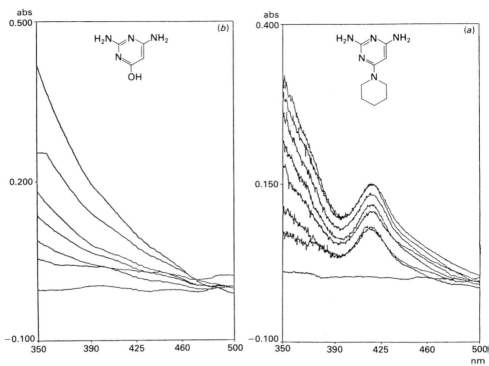

Figure 10.14 The interaction of 6-substituted 2,4-diaminopyrimidines with hamster hepatic microsomal cytochrome *P*-450. (*a*) 6-Piperidino-2,4-diaminopyrimidine; (*b*) 6-hydroxy-2,4-diaminopyrimidine (Gorrod and El-Ghomari, unpublished observations).

10.5 SUBSTRATE–ENZYME COMPLEX FORMATION

The crucial step in a biochemical reaction is the correct formation of a substrate–enzyme complex. The successful formation of the complex and the orientation of the substrate in the enzyme active site are decisive factors that allow substrate specificity and may determine the site of metabolic oxidation in a molecule. The substrate–enzyme complexes of trimethoprim and analogues with dihydrofolate reductase have been extensively investigated. Using ^{1}H and ^{15}N NMR spectroscopy, Bevan *et al.* (1985) showed that trimethoprim is protonated on 1-N when bound to *Lactobacillus casei* dihydrofolate reductase. This specific 1-N binding was further demonstrated when ^{13}C-enriched trimethoprim was used (Cheung *et al.*, 1986). It is believed that 1-N protonated trimethoprim in the complex forms a hydrogen-bonding/charge-charge interaction with the carboxylate

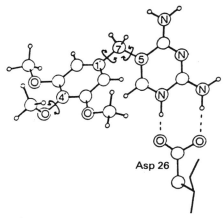

Figure 10.15 The conformation of trimethoprim in its complex with *L. casei* dihydrofolate reductase, together with its relationship to the carboxylate group of Asp-26. The atom numbering for the trimethoprim molecule is also indicated (Searle *et al.*, 1988).

group of Asp-26. Asp-26 also hydrogen bonds to one of the neighbouring protons of the 2-NH$_2$ group to form a stable complex (Fig. 10.15) (Searle *et al.*, 1988). It was also believed that the flexibility of the 5-benzyl group played an important role in the conformation of trimethoprim allowing it to fit into the enzyme active site (Fig. 10.16). A similar situation was observed in the methotrexate–enzyme complex (Hammond *et al.*, 1987).

This binding specificity may also occur in the aminoazaheterocycle–*P*-450–*N*-oxidase system. It may be that in the case of 5-substituted 2,4-diaminopyrimidines, both 1-N and 3-N nitrogens are symmetrical in relation to the 5-substituent. H-Bonding interaction may, therefore, occur either to the 1-N or 3-N site thus allowing *N*-oxygenation at the alternative unbonded nitrogen and hence account for both 1-*N*- or 3-*N*-oxide formation. In the case of susceptible 6-substituted 2,4-diaminopyrimidines, 1-N and 3-N are not symmetrical in relation to the 6-substituent, and it may be that the overall conformation of the molecule in the complex, as influenced by the 6-substituent, only allowed H-bonding to occur at the 1-N site resulting in only 3-*N*-oxide formation. Similarly, in the series of 9-substituted adenines studied, the conformation of the molecule may determine possible binding and complex formation and thereby control *N*-oxide formation. This explanation is in accordance with the conformational analysis determined by computer graphics described by Lam *et al.* (1989).

Although these suggestions may provide an explanation for the different isomeric *N*-oxide formation from similar aminoazaheterocycles, a mechanism to account for the exo-amino oxidation to form hydroxylamines has yet to be established. Amine-imine tautomerism was first suggested by Gorrod (1985) in an attempt to explain how exo and endo N-oxidation of aminoazaheterocycles in molecules of similar structure was possible. It was

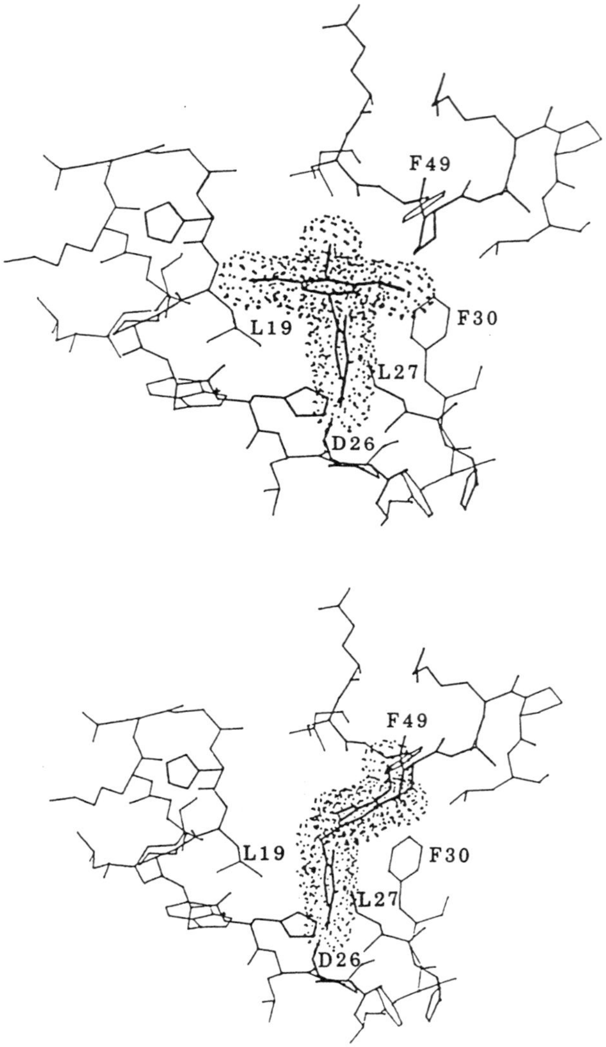

Figure 10.16 The active site region of the *L. casei* dihydrofolate reductase–trimethoprim complex; the van der Waals surface of the trimethoprim molecule is indicated by dots. (Derived by model building and molecular mechanics calculations for altered conformation (Searle *et al.*, 1988).)

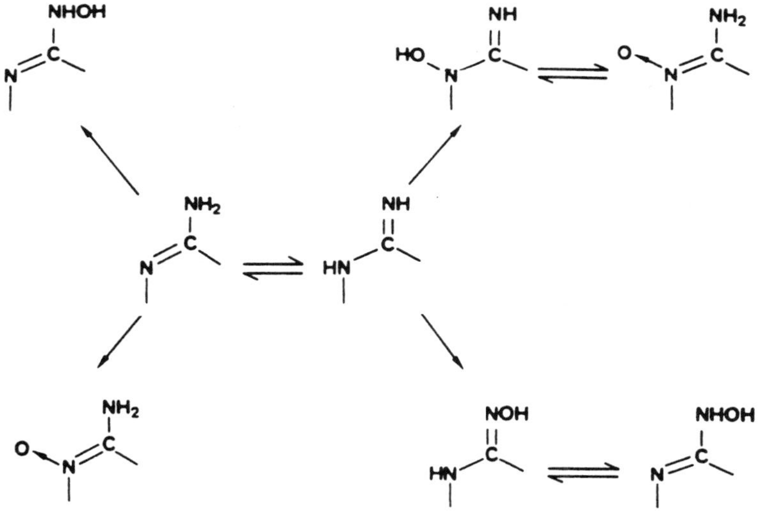

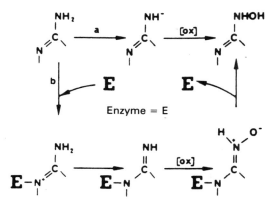

Figure 10.17 Tautomeric forms of amines, amine oxides and hydroxylamines and their possible involvement in the *N*-oxidation of aminoazaheterocycles.

postulated that the substrate binds with a different orientation with different isoenzymes of cytochrome *P*-450 thus producing various products; binding of the substrate with the enzyme at the exo-amino group would lead to *N*-oxidation at the basic ring nitrogen, whereas binding at the ring nitrogen might induce the exo-amino group to tautomerize to the imino form which could then be oxidized to produce the hydroxylamine (Figs 10.17 and 10.18).

Figure 10.18 Possible mechanisms involved in the oxygenation at the *exo*-nitrogen of aminoazaheterocycles.

On the basis of this concept, Lam (1989) reviewed the *N*-oxidation pattern observed with respect to the chemical structures of aminoazaheterocycle substrates. An increasingly apparent feature observed was that ortho aminoazaheterocycles which form *N*-oxides, such as aminopurines, aminopyrimidines and substituted melamines have two or more unsubstituted ring nitrogens in addition to the exo-amino group(s), whereas the aminoazaheterocycles, which are *N*-oxidized at the exo-amino group, have only one unsubstituted ring nitrogen (Figs 10.3–10.5). It is possible that a second unsubstituted ring nitrogen in aminoazaheterocyclic molecules may provide an alternative binding site to the enzyme and direct oxidation to the other ring nitrogen. For compounds having only one available ring nitrogen, it would be expected to bind to an electrophilic part (via H-bonding) of an isoenzyme of *P*-450, allowing the exo-amino group to be *N*-oxidized and vice versa (Fig. 10.19). The preference of a substrate to bind to *P*-450 will depend

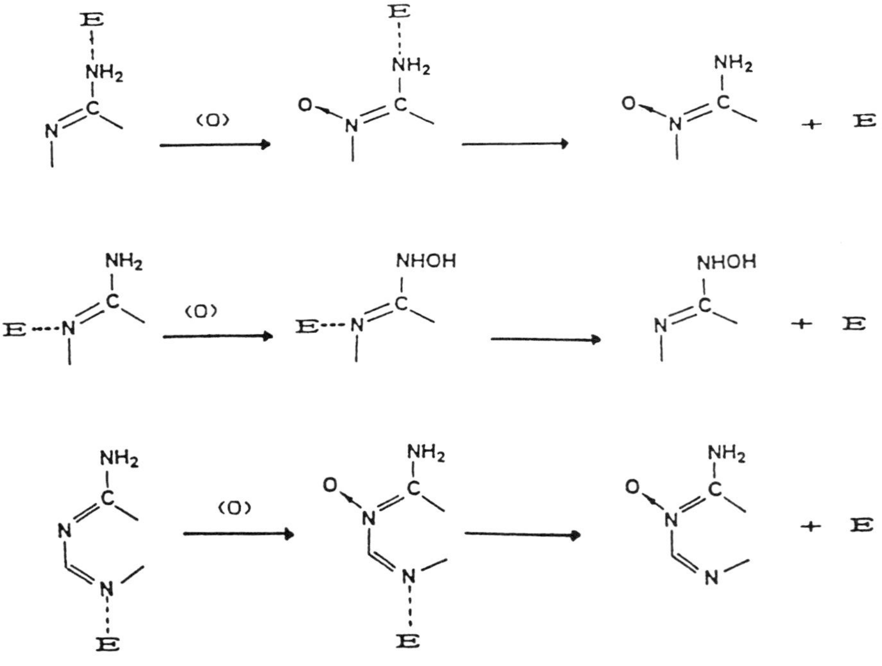

(O) = metabolic oxidation

Part of heterocyclic structures are illustrated

Figure 10.19 Proposed binding of aminoazaheterocycles at different orientation of *N*-oxidase(s) (E) allowing specific formation of *N*-oxidation metabolites.

on the position of the electrophilic site(s) determined by amino acid sequence and the degree of electrophilicity of various *P*-450 isoenzymes. The binding of a substrate to *P*-450 isoenzymes in different orientations may, therefore, account for the specificity of the *N*-oxidation site of amino-azaheterocycles. Metabolic studies using compounds possessing different numbers and spatial orientation of ring nitrogens in relation to their exo amino groups are presently being used to investigate this concept.

10.6 CONCLUSIONS

From this review it is clear that aminoazaheterocycles occur widely in society and can be converted into innocuous *N*-oxides or genotoxic hydroxyl-amines. Whilst the factors which influence the direction and extent of *N*-oxidation are gradually being elucidated, much work remains to unravel the complexities of the control of site of the *N*-oxidation of this interesting group of compounds.

REFERENCES

Bevan, A.W., Roberts, G.C.K., Feeney, J. and Kuyper, L. (1985) ^{1}H NMR and ^{15}N NMR studies of protonation and hydrogen-bonding in the binding of trimethoprim to dihydrofolate reductase. *Eur. Biophys. J.*, **11**, 211–18.

Brooks, M.A., de Silver, J.A.F. and D'Arconte, L. (1973) Determination of trimethoprim and its *N*-oxide metabolites in man, dog, and rat by differential pulse polarography. *J. Pharm. Sci.*, **62**, 1776–9.

Cheung, H.T.A., Searle, M.S., Feeney, J. *et al.* (1986) Trimethoprim binding to *Lactobacillus casei* dihydrofolate reductase: a ^{13}C NMR study using selectively ^{13}C-enriched trimethoprim. *Biochemistry*, **25**, 1925–31.

El-Ghomari, K. (1988) The metabolic *N*-oxidation of aminoazaheterocycles, PhD Thesis, University of London.

El-Ghomari, K. and Gorrod, J.W. (1987) Metabolic *N*-oxygenation of 2,4-diamino-6-substituted pyrimidines. *Eur. J. Drug Metab. Pharmacokinet*, **12**, 253–8.

Felton, J.S., Knize, M.G., Shen, N.H. *et al.* (1986) The isolation and identification of a new mutagen from fried ground beef: 2-amino-1-methyl-6-phenylimidazo-[4,5-b]pyridine (PhIP). *Carcinogenesis*, **7**, 1081–6.

Felton, J.S., Knize, M.G., Wood, C. *et al.* (1984) Isolation and characterization of new mutagens from fried ground beef. *Carcinogenesis*, **5**, 95–102.

Gorrod, J.W. (1973) Differentiation of various types of biological oxidation of nitrogen in organic compounds. *Chem. Biol. Interact.*, **7**, 289–303.

Gorrod, J.W. (1978) The current status of pK_a concepts in the differentiation of enzymic *N*-oxidation, in *Biological Oxidation of Nitrogen* (ed. J.W. Gorrod), Elsevier, Amsterdam, pp. 201–10.

Gorrod, J.W. (1985) Amine-imine tautomerism as a determinant of the site of biological *N*-oxidation, in *Biological Oxidation of Nitrogen in Organic Molecules* (eds J.W. Gorrod and L.A. Damani), Ellis Horwood, Chichester, pp. 219–30.

Gorrod, J.W. (1987) The *in vitro* metabolism of aminoazaheterocycles, in *Drug Metabolism from Molecules to Man* (eds D.J. Benford, J.W. Bridges and G.G. Gibson), Taylor and Francis, London, pp. 456–61.

Gorrod, J.W. and Lam, S.P. (1989) Amino azaheterocycles: friend or foe? in *Molecular Aspects of Human Disease* (eds J.W. Gorrod, O. Albano and S. Papa), Ellis Horwood, Chichester, Vol. 2, pp. 100–12.

Gorrod, J.W. and Lam, S.P. (1990) Microsomal cytochrome *P*-450 mediated *N*-oxygenation of aminoazaheterocycles, in *Biological Oxidation Systems* (eds C.C. Reddy, G.A. Hamilton and K.M. Madyastha), Academic Press, San Diego, 1, 147–62.

Hammond, S.J., Birdsall, B., Feeney, J. *et al.* (1987) Structural comparisons of complexes of methotrexate analogues with *Lactobacillus casei* dihydrofolate reductase by two-dimensional ^{1}H NMR at 500 MHz, *Biochemistry*, 26, 8585–90.

Hashimoto, Y., Shudo, K. and Okamoto, T. (1980a) Activation of a mutagen, 3-amino-1-methyl-5H-pyrido[4,3-b]indole. Identification of 3-hydroxyamino-1-methyl-5H-pyrido[4,3-b]indole and its reaction with DNA. *Biochem. Biophys. Res. Commun.*, 96, 355–62.

Hashimoto, Y., Shudo, K. and Okamoto, T. (1980b) Metabolic activation of a mutagen, 2-amino-6-methyldipyrido[1,2-a: 3',2'-d]imidazole. Identification of 2-hydroxyamino-6-methyldipyrido[1,2-a: 3',2'-d]imidazole and its reaction with DNA. *Biochem. Biophys. Res. Commun.*, 92, 971.

Hubbel, J.P., Henning, M.L., Grace, M.E. *et al.* (1978) *N*-Oxide metabolites of 2,4-diaminopyrimidine inhibitors of dihydrofolate reductase, trimethoprim, pyrimethamine and metoprine, in *Biological Oxidation of Nitrogen* (ed. J.W. Gorrod), Elsevier, Amsterdam, pp. 177–82.

Hubbel, J.P., Kao, C.J., Sigel, C.W. and Nichol, C.A. (1980) Sex dependent disposition of metoprine by mice, in *Current Chemotherapy and Infectious Disease* (eds J.D. Nelson and C. Grassi), American Society for Microbiology, Washington, DC, pp. 1620–1.

Iles, N. (1988) Biological *N*-oxidation of isomeric aminopyridines, PhD Thesis, University of London.

Ishida, Y., Negishi, C., Umemoto, A. *et al.* (1987) Activation of mutagenic and carcinogenic heterocyclic amines by S-9 from the liver of a rhesus monkey. *Toxic. in Vitro*, 1, 45–8.

Kamataki, T., Maeda, K., Yamazoe, Y. *et al.* (1983) A high spin form of cytochrome *P*-450 highly purified from polychlorinated biphenyl-treated rats. *Mol. Pharmacol.*, 24, 146–55.

Kato, R. (1986) Metabolic activation of mutagenic heterocyclic aromatic amines from protein pyrolysates. *CRC Crit. Rev. Toxicol.*, 16, 307–47.

Kato, R., Kamataki, T. and Yamazoe, Y. (1983) *N*-Hydroxylation of carcinogenic and mutagenic aromatic amines. *Environ. Health Perspect.*, 49, 21–5.

Kawajiri, K., Yonehawa, H., Gotoh, O. *et al.* (1983) Contributions of two inducible forms of cytochrome *P*-450 in rat liver microsomes of the metabolic activation of various chemical carcinogens. *Cancer Res.*, 43, 819–23.

Lam, S.P. (1989) Influence of 9-substitution on the microsomal *N*-oxidation of adenine, PhD Thesis, University of London.

Lam, S.P., Barlow, D.J. and Gorrod, J.W. (1989) Conformational analysis of 9-substituted adenines in relation to their microsomal 1-*N*-oxidation. *J. Pharm. Pharmacol.*, 41, 373–8.

Lam, S.P., Devinsky, F. and Gorrod, J.W. (1987) Biological *N*-oxidation of adenine and 9-alkyl derivatives. *Eur. J. Drug Metab. Pharmacokinet.*, **12**, 239–43.

Meshi, T. and Sato, Y. (1972) Studies on sulfamethoxazole/trimethoprim. Absorption, distribution, excretion and metabolism of trimethoprim in rat. *Chem. Pharm. Bull. (Japan)*, **20**, 2079–90.

Nagao, N., Fujita, Y., Wakabayashi, K. and Sugimura, T. (1983) Ultimate forms of mutagenic and carcinogenic heterocyclic amines produced by pyrolysis. *Biochem. Biophys. Res. Commun.*, **114**, 626–31.

Negishi, C., Wakabayashi, K., Tsuda, M. *et al.* (1984) Formation of 2-amino-3,7,8-trimethylimidazo[4,5-f]quinoxaline, a new mutagen, by heating a mixture of creatinine, glucose and glycine. *Mutat. Res.*, **140**, 55–9.

Nemoto, N., Kusumi, S., Takayama, S. *et al.* (1979) Metabolic activation of 3-amino-5H-pyrido[4,3-b]indole, a highly mutagenic principle in tryptophan pyrolysate, by rat liver enzymes. *Chem. Biol. Interact.*, **27**, 191–8.

Nichol, C.A., Cavallito, J.C., Wooley, J.L. and Sigel, C.W. (1977) Lipid-soluble diaminopyrimidine inhibitors of dihydrofolate reductase. *Cancer Treat. Rep.* **61**, 559.

Okamoto, T., Shudo, K., Hashimoto, Y. *et al.* (1981) Identification of a reactive metabolite of the mutagen, 2-amino-3-methylimidazo[4,5-f]quinoline. *Chem. Pharm. Bull.*, **29**, 590.

Ryan, D.E., Thomas, P.E., Karzenlowski, D. and Levin, W. (1979) Separation and characterization of highly purified forms of liver microsomal cytochrome *P*-450 from rats treated with polychlorinated biphenyls, phenobarbital, and 3-methylcholanthrene. *J. Biol. Chem.* **254**, 1365–74.

Schwartz, D.E., Vetter, W. and Englert, G. (1970) Trimethoprim metabolites in rat, dog and man: qualitative and quantitative studies. *Arzneim. Forsch.*, **20**, 1867.

Searle, M.S., Forster, M.J., Birdsall, B. *et al.* (1988) Dynamics of trimethoprim bound to dihydrofolate reductase. *Proc. Natl. Acad. Sci. USA*, **85**, 3787–91.

Sigel, C.W. (1973) Metabolism of trimethoprim in man and measurement of a new metabolite: a new fluorescence assay. *J. Infect. Dis. Suppl.*, **128**, 580–3.

Sigel, C.W. and Brent, D.A. (1973) Identification of trimethoprim 3-*N*-oxide as a new urinary metabolite of trimethoprim in man. *J. Pharm. Sci.*, **62**, 694–5.

Sugimura, T. and Nagao, M. (1979) Mutagenic factors in cooked foods. *CRC Crit. Rev. Toxicol.*, **8**, 189–209.

Sugimura, T. and Sato, S. (1983) Mutagens-carcinogens in foods. *Cancer Res.*, **43**, 2415s–21s.

Sugimura, T., Kawachi, T., Nagao, M. *et al.* (1977) Mutagenic principle(s) in tryptophan and phenylalanine pyrolysis products. *Proc. Jpn. Acad.*, **53**, 58–61.

Thomas, R.C. and Hartpoolian, H. (1975) Metabolism of minoxidil, a new hypotensive agent II: biotransformation following oral administration to rats, dogs and monkeys. *J. Pharm. Sci.*, **64**, 1366–71.

Wang, C.C. and Simashkevich, P.M. (1980) A comparative study of the biological activities of arprinocid and arprinocid-1-*N*-oxide. *Mol. Biochem. Parasitol.*, **1**, 335–45.

Watkins, P.J. (1988) Biological *N*-oxidation of some substituted 2,4-diaminopyrimidines, PhD Thesis, University of London.

Watkins, P.J. and Gorrod, J.W. (1987) Studies on the *in vitro* biological *N*-oxidation of trimethoprim. *Eur. J. Drug Metab. Pharmacokinet.*, **12**, 245–51.

Wolf, F.J., Steffens, J.J., Alvaro, R.F. and Jacob, T.A. (1978) Microsomal conversion of MK-302, arprinocid [6-amino-9-(2-chloro-6-fluorobenzyl)purine] to 6-amino-9-(2-chloro-6-fluorobenzyl)purine-1-*N*-oxide by liver microsomes from the chicken and the dog and to the 2-chloro-6-fluorobenzyl alcohol by liver microsomes from rat and mouse. *Fed. Proc.*, **37**, 814.

Yamamoto, T., Tsuji, K., Kosuge, T. *et al.* (1978) Isolation and structure determination of mutagenic substances in L-glutamic acid pyrolysate. *Proc. Jpn Acad.*, **54**, 248–50.

Yamazoe, Y., Ishii, K., Kamataki, T. *et al.* (1980) Isolation and characterization of active metabolites of tryptophan-pyrolysate mutagen, Trp-P-2, formed by rat liver microsomes. *Chem. Biol. Interact.*, **30**, 125–38.

Yamazoe, Y., Shimada, M., Maeda, K. *et al.* (1984) Specificity of four forms of cytochrome *P*-450 in the metabolic activation of several aromatic amines and benzo[a]pyrene. *Xenobiotica*, **14**, 549–52.

Zins, G.R. (1965) The *in vivo* production of a potent, long acting hypotensive metabolite from diallylmelamine. *J. Pharmacol. Exp. Ther.*, **150**, 109–17.

Zins, G.R., Emmert, D.W. and Walker, R.A. (1968) Hypotensive metabolite production and sequential metabolism of *N,N*-diallylmelamine in rats, dogs and man. *J. Pharmacol. Exp. Ther.*, **159**, 194–205.

11

New aspects of the microsomal N-hydroxylation of benzamidines

B. Clement, M. Immel, H. Pfundner, S. Schmitt and M. Zimmermann

Institut für Pharmazeutische Chemie, Philipps-Universität, D-3550 Marburg, FRG

1. Metabolic investigations with amidines focused attention on chemical, biochemical, pharmacological and toxicological aspects.
2. Cytochrome P-450 is involved in the N-hydroxylation of benzamidines without α-hydrogens to the corresponding amidoximes.
3. Structure–activity relationships and Hammet ϱ constants are in agreement with radical mechanisms proposed for the N-hydroxylation of benzamidines without α-hydrogens and for the N-dealkylation of benzamidines with α-hydrogens. The studies present further evidence for the concept developed on the basis of investigation with benzamidines – N-oxygenation by cytochrome P-450 only in the absence of α-hydrogens.
4. The enzymic reduction of benzamidoxime to benzamidine is demonstrated in experiments *in vitro* and *in vivo*.
5. The N-hydroxylation of benzamidines to benzamidoximes is of pharmacological and toxicological relevance. Thus benzamidoxime exhibits a weak mutagenicity in the Ames test (in the presence of S9), induces DNA-single strand breaks (in rat hepatocytes) and DNA amplification in SV 40-transformed hamster cells.

11.1 INTRODUCTION

11.1.1 Chemical properties of amidines

Amidines are formally derived from carboxylic acids by substitution of the two oxygen atoms for nitrogen atoms. Thus, they possess the same oxidation level as carboxylic acids and their derivatives. Under no circumstances should amidines be confused with imines which are derivatives of ketones and aldehydes or with guanidines which are derived from carbonic acid. This is wrong not only for purely formal reasons but also because of differing chemical properties and differing behaviour in biotransformation reactions. Analogously, the *N*-oxygenated derivatives – oximes, amidoximes, *N*-hydroxyguanidines, respectively – formed by *N*-hydroxylation have very little in common (Clement, 1989a).

This chapter is concerned only with such typically basic amidines which possess high pK_a values as a consequence of forming highly mesomerically stabilized cations; for example, the pK_a value of benzamidine is 11.6 (Albert *et al.*, 1984). Thus, amidines are even more basic than aliphatic amines. This is, of course, only valid for amidines in which none of the nitrogen components form part of an aromatic system or which do not have a strongly electron-accepting substituent on the nitrogen atom (Clement, 1989a).

It is interesting that amidines undergo biotransformations in spite of the fact that, under physiological conditions, they exist almost exclusively in the protonated form, which generally has high aqueous solubility (Clement, 1983).

11.1.2 Pharmacological properties of amidines

Benzamidines, like guanidines and amines, are stable towards hydrolysis and are thus components of several active principles (Grout, 1972; Clement, 1989a). Amidine functional groups are found in drugs of widely different therapeutic classes (Clement, 1989a). In this respect, the aromatic diamidines in particular should be mentioned; these compounds were developed mainly because of their trypanocidal and leishmanicidal activity. The most important representatives of this group are pentamidine and diminazene (Raether *et al.*, 1972; Clement, 1989b) (see Fig. 11.2 for structures). Recently pentamidine has attracted attention because it is one of the drugs of choice for the treatment of pneumonia caused by *Pneumocystis carinii*, an infection which occurs very frequently in AIDS patients (Goa and Campoli-Richard, 1987).

Since several groups of drugs possess amidine functional groups bearing an aromatic moiety bonded to carbon, benzamidines were selected as the model substances for the biotransformation experiments.

11.2 ASPECTS OF METABOLIC INVESTIGATIONS WITH AMIDINES

11.2.1 Chemical aspects

The metabolic investigations with amidines gave rise to a number of challenges. First, the metabolites which could possibly be formed by *N*-oxygenation were synthesized for comparative purposes (Clement, 1983). The structures of the synthetic products were then elucidated by conventional methods. Since the products generally contained several nitrogen atoms, ^{15}N-NMR spectroscopy was also employed. The chemical shifts and coupling constants were used to clarify the question of the constitutions, configurations, positions of tautomeric equilibria, and the sites of protonation (Chapter 2).

Following the preparation of the potential metabolites, preliminary experiments were performed with these synthetic reference substances in order to develop suitable analytical methods and to check the stabilities of the potential metabolites under the conditions prevailing during the metabolic experiments and during the work-up. This procedure with the reference substances is intended to confirm that it is indeed possible to detect the actual formation of an *N*-hydroxylated compound under the experimental conditions used.

11.2.2 Biochemical aspects

The *in vitro* metabolic studies of amidines were performed in the usual manner under aerobic conditions at pH 7.4 with either the $12\,000\,g$ supernatant or the microsomal fraction of liver homogenates in the presence of the necessary cofactors (NADPH, Mg^{2+}). Rabbit liver homogenates were employed for most of the studies. Work-up was performed, as had been tested with the synthetic comparative substances, either by simple shaking with organic solvents or by a freeze-drying process where necessary. This was followed by TLC separation with the help of the comparative substances. Colour visualization of the *N*-hydroxylated metabolites could usually be achieved with Fe^{3+}. Finally, for identification, the appropriate TLC bands were eluted and the mass spectra of the metabolites and reference substances were recorded. Thus, unequivocal structural evidence was provided by a comparison of mass spectral data of the metabolites with those of the synthetic reference compounds. In most of the quantitative investigations, an HPLC analysis was carried out after the work-up (for experimental details see Clement, 1983; Clement and Zimmermann, 1987a, b, 1988).

When new *N*-oxygenated metabolites are detected, quantitative studies with the objective of clarifying the participating enzyme are carried out. Considerations of structure–activity relationships can lead to the formulation of mechanisms which may make it possible to provide a contribution to the problem of the predictability of *N*-oxidative processes.

11.2.3 Pharmacological and toxicological aspects

In co-operation with other research groups, both *in vitro* and *in vivo* studies have been performed and, in so far as pharmacology and toxicology is concerned, the following questions have been examined. Do the *N*-oxygenated compounds still exert a pharmacological activity, or are they responsible for toxicity as a result of metabolic activation? (Clement and Raether, 1985; Clement *et al.*, 1988a). In the course of the toxicological investigations, the genotoxic potentials of the novel *N*-oxygenated compounds were checked (Clement *et al.*, 1988a).

11.3 MICROSOMAL *N*-HYDROXYLATION OF *N,N*-UNSUBSTITUTED BENZAMIDINES

11.3.1 Benzamidine and ring-substituted derivatives

Since Clement (1983) first reported the microsomal *N*-hydroxylation of benzamidines (*11.1*) to benzamidoximes (*11.2*) (Fig. 11.1) this bio-transformation has been the subject of continued intensive investigations. Initially, this *N*-oxygenation was detected in the case of benzamidine itself, as well as ring-substituted but *N,N*-unsubstituted derivatives (*11.1*). This reaction demonstrates that such highly basic and hydrophilic compounds can also be metabolized. Benzamidine represents the most basic centre known to date that takes part in such a reaction.

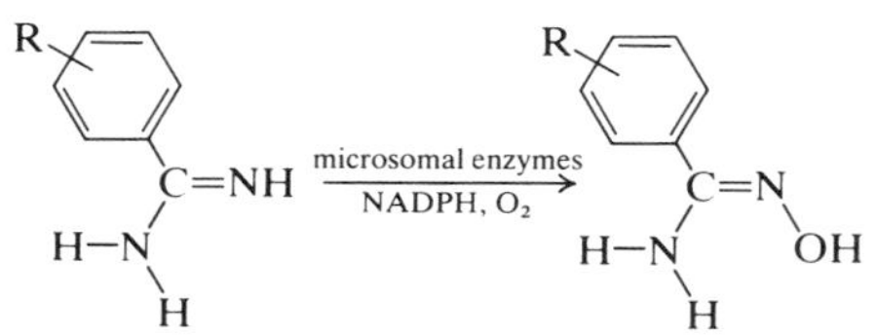

R = H, *p*-CH$_3$, *p*-OCH$_3$, *m*-CH$_3$, *m*-NO$_2$, *o*-CH$_3$, *p*-Cl, *p*-Br, *p*-CN

Figure 11.1 *N*-Hydroxylation of *N,N*-unsubstituted benzamidines (11.1) to benz-amidoximes (11.2).

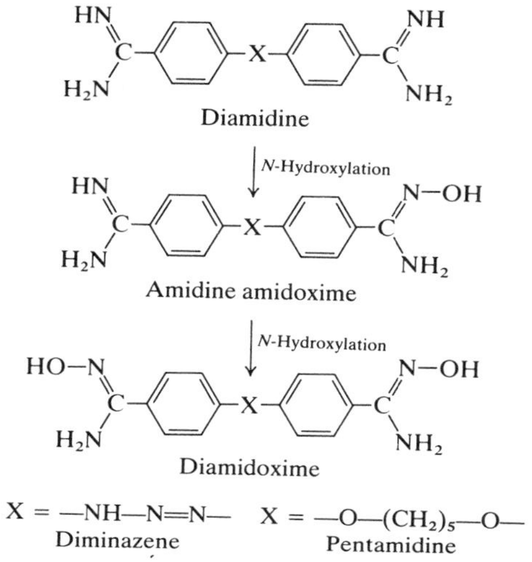

Figure 11.2 *N*-Oxygenation of trypanocidal diamidines *in vitro*.

11.3.2 Pentamidine and diminazene

Drugs possessing *N,N*-unsubstituted amidine functions exhibit *N*-oxygenations in experiments *in vitro*, as can be seen for the examples of the, as yet most intensively investigated, trypanocidic diamidines pentamidine and diminazene (Fig. 11.2).

Once the syntheses (Clement and Raether, 1985) of all potential *N*-hydroxylated metabolites of the monoamidoxime and diamidoxime type had been completed for comparative purposes, the biotransformation investigations showed, after incubation of the diamidines, the presence of both mono- and also diamidoximes as metabolites for both substrates (Clement and Immel, unpublished work). There are no previous reports on the biotransformation of these drugs.

11.3.3 Quantitative studies and enzymology

(a) Characteristics of the microsomal N-hydroxylation of benzamidine to benzamidoxime

The question arose as to which enzyme participated in the *N*-hydroxylation of benzamidines. For quantitative studies the transformation of benzamidine (*11.1*) (R = H) to benzamidoxime (*11.2*) (R = H) (Fig. 11.1) was chosen. On the basis of the observed characteristics – microsomal enzyme, dependence on oxygen and NADPH – it was immediately apparent that only one of the two mono-oxygenases – cytochrome P-450 or the flavin-containing mono-oxygenase – could be taken into consideration. The participation of the FAD-containing mono-oxygenase was excluded by experiments with the purified enzyme (Clement, 1983).

Before studies of the enzymology could be continued, a quantitative analysis of benzamidoxime (11.2) had to be developed. The best results were obtained by reversed phase high-performance liquid chromatography (HPLC) after extraction of benzamidoxime (11.2) with ether (Clement and Zimmermann, 1987b). In the presence of typical inhibitors of the cytochrome P-450 enzyme system – SKF 525-A, metyrapone, CN^- – lower conversion rates were observed. The inhibition by carbon monoxide was studied in detail. The proportion of CO to O_2 was varied and the respective inhibition in comparison to a control determined. It was unambiguously demonstrated that the inhibition of the microsomal N-hydroxylation was dependent on the ratio of CO to O_2 (Clement and Zimmermann, 1987b).

Induction experiments indicated that the isoenzymes involved are not induced by phenobarbital, 3-methylcholanthrene, ethanol or benzamidine itself (Clement and Zimmermann, 1987b). The extent of any role played by hydrogen peroxide or superoxide radicals on the N-oxygenation was investigated by adding catalase or superoxide dismutase to the usual incubation mixtures. It was found that neither catalase nor superoxide dismutase had any significant effect on the reaction.

The formation of benzamidoxime during incubation with the $12\,000\,g$ supernatant of rabbit liver homogenates and NADPH followed Michaelis–Menten kinetics (Clement and Zimmermann, 1987b). The apparent K_m value, calculated from Lineweaver–Burk plots was 1.95 ± 0.30 mM, while the apparent v_{max} value was 0.54 ± 0.17 nmol benzamidoxime produced/min per mg of protein.

Even though it was certain on the basis of the dependence on NADPH that, in fact, only cytochrome P-450 or the FAD-containing mono-oxygenase could be considered as the participating enzyme, other enzymes and active oxygen species have also been tested. Neither H_2O_2 alone, Fenton's reagent, H_2O_2-forming systems (xanthine oxidase + xanthine, glucose oxidase + glucose), nor combinations of H_2O_2 or of the H_2O_2-forming systems with catalase or superoxide dismutase or peroxidase were able to transform benzamidine (11.1) into benzamidoxime (11.2).

(b) Studies with antibodies against NADPH–cytochrome P-450 reductase and with a reconstituted cytochrome P-450 oxidase system

All the characteristics and, above all, the inhibition by CO were in favour of a catalysis by cytochrome P-450. In order to prove this unequivocally, the

NADPH–cytochrome *P*-450 reductase was included in the experimental series. For the experiments with microsomes from rabbit liver homogenates, this reductase has also been purified from this species. The work-up was based on the procedure developed by Jasukochi and Masters (1976) for purification of the enzyme from the rat. The preparation gave only one major band on sodium dodecyl sulphate–polyacrylamide gel electrophoresis. Antibodies have been obtained against this reductase by immunization of a goat. IgG fractions were isolated from serum by precipitation with ammonium sulphate followed by anion exchange chromatography (Clement and Pfundner, unpublished).

The antibodies were added to the usual incubation mixtures with benzamidine. The observation of clear inhibition of the *N*-hydroxylation of benzamidine is considered to be conclusive evidence for a catalysis by the cytochrome *P*-450 enzyme system (Clement and Pfunder, unpublished). Since, as a consequence of the non-inducibility, it was not possible to deduce which isoenzyme or isoenzymes of *P*-450 are responsible for the *N*-hydroxylation of benzamidine, the purification of cytochrome *P*-450 from rabbit liver microsomes has recently been started with the objective of obtaining a preparation which contains all *P*-450 isoenzymes originally present in the active form. Cytochrome *P*-450 has been enriched and separated from other components by means of HIC column chromatography on octyl-Sepharose CL-4B (Kling *et al.*, 1985).

Studies are now being performed with benzamidine and this cytochrome *P*-450 preparation in a system containing added purified rabbit reductase, an NADPH-forming system, and dilauroyl phosphatidylcholine in order to reconstitute the *N*-oxidase activity.

11.4 STRUCTURE–ACTIVITY RELATIONSHIPS AND MECHANISMS OF THE *N*-OXIDATIVE METABOLISM OF AMIDINES

11.4.1 *N*-Hydroxylation of *para*-substituted benzamidines

With the aid of HPLC analyses and simple Michaelis–Menten kinetics, the maximum rates of the microsomal *N*-oxygenations of various *para*-substituted benzamidines (*11.1*) to benzamidoximes (*11.2*) were determined. The presence of electron-donating substituents – methyl group and methoxyl group – increased the rates whereas the presence of electron-accepting substituents – halogenides and the nitrile group – decreased them (Clement and Zimmermann, 1988). In the case of the reaction of *p*-nitrobenzamidine no *p*-nitrobenzamidoxime could be detected (Table 11.1). It was ensured that the selected benzamidines and the formed benzamidoximes did not undergo any additional or further enzymatic or

Table 11.1 Structure-activity relationships in the N-hydroxylation of para-substituted benzamidines (*11.1*) to benzamidoximes (*11.2*)

X (= substituent R)	$v_{max}x$ (nmol benzamidoxime/min/mg protein)
CH_3	1.03 ± 0.14
OCH_3	0.93 ± 0.17
H	0.56 ± 0.19
Cl	0.42 ± 0.08
Br	0.41 ± 0.11
CN	0.15 ± 0.03
NO_2	ND

From Clement and Zimmermann (1988), with permission of Pergamon Press plc.

chemical transformations which could have influenced these quantitative investigations. Only *para*-substituted benzamidines were selected as substrates for the reaction in order to avoid possible steric influences of the substituents (Clement and Zimmermann, 1988).

In accordance with the postulated mechanisms of the N-hydroxylation of benzamidines by cytochrome *P*-450 (see Fig. 11.8) a high electron density on the amidine system is advantageous for the progress of the reaction. The presence of electron donors on the ring renders the removal of an electron from the amidine system more easy (Clement and Zimmermann, 1988).

A significant correlation between the logarithms of the maximum rates with the Hammett σ_p constants was found for a reaction constant of $\varrho = -0.88$ (Fig. 11.3). The experimentally determined ϱ value supports a radical mechanism for the N-oxygenation by the cytochrome *P*-450 enzyme system since reactions with negative or low values of ϱ often proceed by way of radical intermediates (Clement and Zimmermann, 1988).

11.4.2 Studies on the metabolism of N-substituted benzamidines

(a) N-*Substituted benzamidines with* α-*hydrogens*

For the realization of further structure–activity relationships, N-substituted benzamidines have also been investigated as model compounds. In contrast to the N,N-disubstituted derivatives, two different N-oxygenated metabolites are now conceivable from the N-monosubstituted derivatives. Hence, in the case of N-methylbenzamidine (*11.3*), the N-hydroxyamidine (*11.4*) could be formed (Fig. 11.4). On the basis of ^{15}N-NMR investigations, the latter exists predominantly as the α-aminonitrone tautomer (Clement and Kämpchen, 1987). However, a metabolite of this type has not been

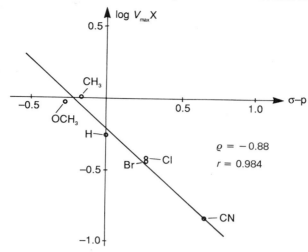

Figure 11.3 Correlations of log $v_{max}x$ with σ_{para} for the *N*-oxygenations of *para*-substituted benzamidines by the $12\,000\,g$ supernatant fractions of rabbit liver homogenates. (From Clement and Zimmermann (1988), with permission of Pergamon Press plc.)

detected from *N*-methylbenzamidine (*11.3*) or from any other *N*-mono-substituted benzamidine (Clement, 1984).

In addition, no benzamidoxime of the type (*11.5*) is formed from *N*-methylbenzamidine. Other *N*-mono- or *N,N*-disubstituted representatives were also not *N*-hydroxylated (Clement, 1984) (Fig. 11.5).

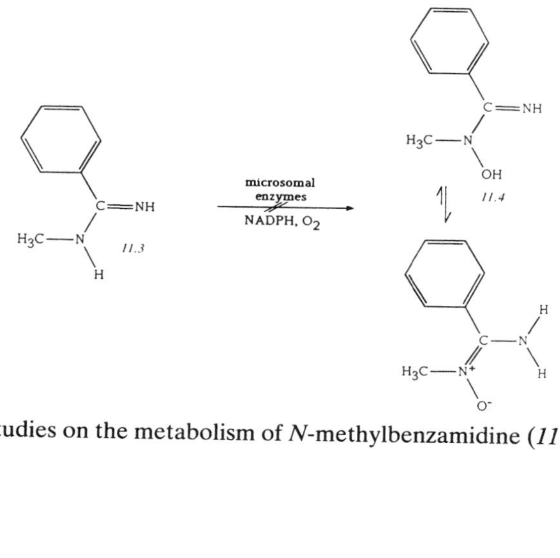

Figure 11.4 Studies on the metabolism of *N*-methylbenzamidine (*11.3*).

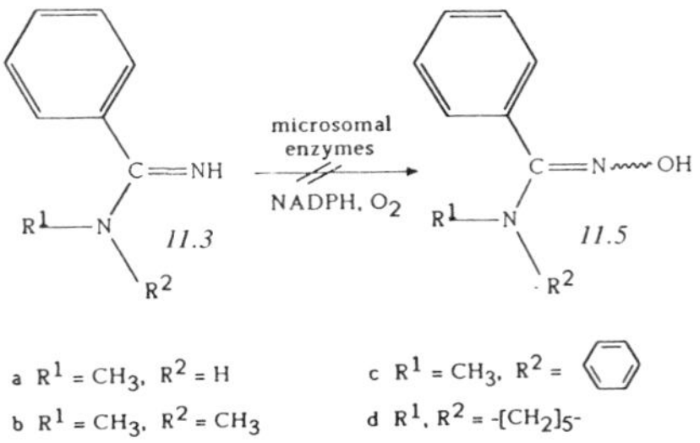

a $R^1 = CH_3$, $R^2 = H$ c $R^1 = CH_3$, $R^2 =$
b $R^1 = CH_3$, $R^2 = CH_3$ d $R^1, R^2 = -[CH_2]_5-$

Figure 11.5 Studies on the N-hydroxylation of benzamidines with α-hydrogens.

It was thus recognized that evidently no carbon atoms bearing hydrogens may be present in the direct neighbourhood of the nitrogen atom, α-hydrogens, when an N-oxygenation process is to be observed. In contrast to the N-unsubstituted examples, CH_3 or CH_2 structural moieties exist in the structural types (*11.3*). The lipophilicity apparently has no influence on the process since some of the examples (*11.3*) are highly lipophilic and all, in contrast to the more hydrophilic benzamidine, were not N-oxygenated.

(b) N-*Substituted benzamidines without* α-*hydrogen atoms*

If the microsomal N-hydroxylations are only dependent on the presence or absence of an α-H atom, it should be possible to N-oxygenate N-mono-substituted derivatives which do not possess α-H atoms. This was confirmed for the example of the reaction of N-*t*-butylbenzamidine (*11.8*) to the corresponding amidoxime (*11.9*) (Fig. 11.6) (Clement and Immel, 1987).

The comparative synthesis of the benzamidine (*11.8*) and the N-oxygenated derivative (*11.9*) are included in Fig. 11.6. Compound (*11.8*) is accessible by the addition of *t*-butylamine to benzonitrile (*11.6*) in the presence of aluminium trichloride and (*11.9*) is best prepared by the reaction of benzohydroxamoyl chloride (*11.7*) with *t*-butylamine (Clement and Immel, 1987).

The microsomal N-hydroxylation of N-*t*-butylbenzamidine (Fig. 11.6) demonstrates that the cytochrome *P*-450 form involved can accept substrates with bulky substituents and suggests that steric aspects may be of minor significance. They, however, cannot be neglected completely since the rate of reaction of the N-*t*-butylbenzamidine is low and moreover the even more bulky *t*-octyl analogue is no longer N-oxygenated (Clement and Immel, 1987).

N-Phenylbenzamidine (*11.10*) represents a further N-monosubstituted amidine lacking α-H atoms and was also N-oxygenated to the corresponding

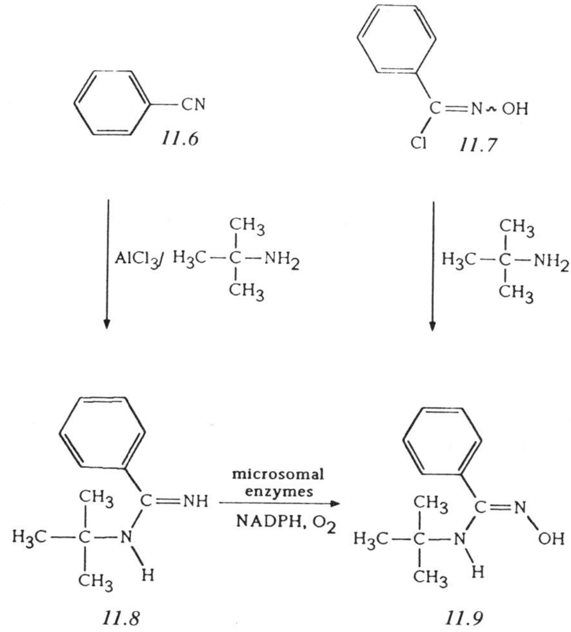

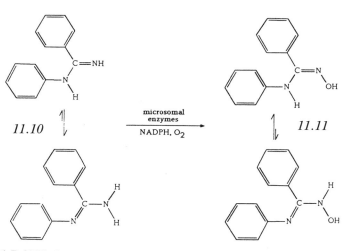

Figure 11.6 Synthesis of *N-t*-butylbenzamidine (*11.8*) and *N-t*-butylbenzamidoxime (*11.9*) and microsomal *N*-hydroxylation of (*11.8*).

Figure 11.7 *N*-Hydroxylation of *N*-phenylbenzamidine (*11.10*).

amidoxime (*11.11*) (Fig. 11.7) (Clement, 1984). Compound (*11.10*) exists predominantly in the tautomeric form in which the double bond is conjugated with the aromatic system (Clement and Kämpchen, 1986), whereas the amidoxime structure in the (*Z*)-configuration is favoured in the *N*-hydroxylated product (*11.11*) (Clement and Kämpchen, 1985).

11.4.3 Mechanism of the *N*-hydroxylation and *N*-dealkylation of benzamidines

All these structure–activity relationships and the Hammett ϱ constants are in full agreement with the radical mechanisms postulated in 1984 (Clement, 1985). These suggested mechanisms for a catalysis by *P*-450 show why an *N*-oxygenation can occur only in the absence of α-H-atoms.

In the case of unsubstituted benzamidines (Fig. 11.8), an electron can be taken up by the active iron–oxygen complex in the last step of the reaction cycle with cytochrome *P*-450 (Ortiz de Montellano, 1986). A radical cation is thus formed, which can be stabilized either by the cleavage of a proton, as formulated in Fig. 11.8, or by the removal of a hydrogen radical. The former

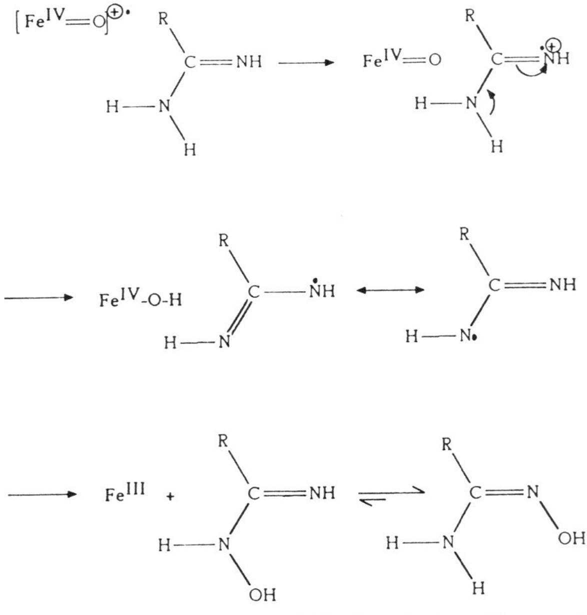

Figure 11.8 Proposed mechanism for the *N*-hydroxylation of benzamidines without α-hydrogens.

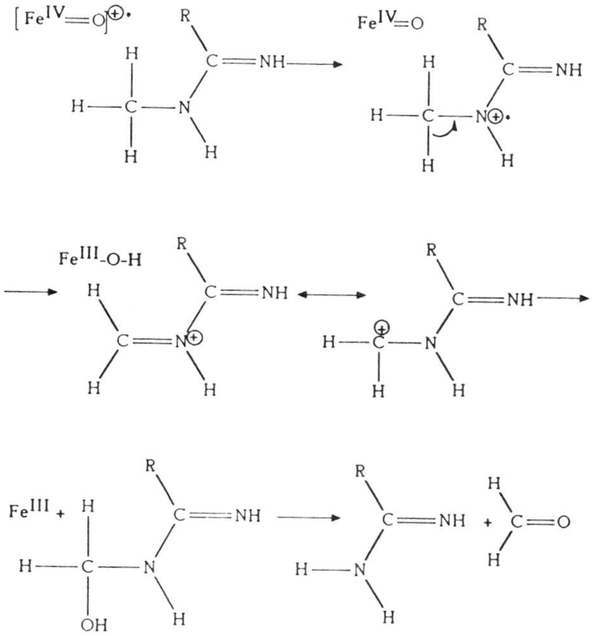

Figure 11.9 Proposed mechanism for the N-dealkylation of benzamidines with α-hydrogens.

process produces a radical which is stabilized by mesomerism, and this can be hydroxylated to give a hydroxyamidine which finally tautomerizes to furnish the amidoxime. The same mechanism could also be in operation for N-phenylbenzamidine and N-t-butylbenzamidine (Fig. 11.8).

In the case of N-alkylbenzamidines with α-hydrogens such as N-methyl-benzamidine (Fig. 11.9) in a carbenium-iminium ion could be formed by the cleavage of a hydrogen radical after removal of an electron. Thus, the hydroxyl group can also be transferred to the carbon atom to form a carbinol-amidine. The latter decomposes immediately to give the N-dealkylated benzamidine and formaldehyde. The result is, the N-alkyl-benzamidines are N-dealkylated instead of being N-oxygenated (Fig. 11.9).

11.4.4 *N*-Demethylation of *N*-methylbenzamidine

Initially, the N-demethylation of N-methylbenzamidine (*11.3*) was merely postulated (Clement, 1985). The postulate was tested by experiment and formaldehyde was detected by the Nash test after incubation of (*11.3*)

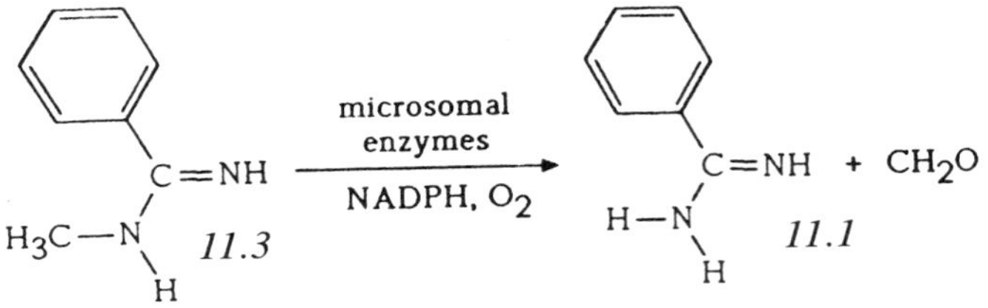

Figure 11.10 Microsomal *N*-demethylation of *N*-methylbenzamidine (*11.3*).

(Fig. 11.10). However, for the characterization of the reaction, a quantitative HPLC analytical method was again developed. Using this method, it was also possible to detect benzamidine (*11.1*) as a metabolite formed from *N*-methylbenzamidine (*11.3*) (Clement and Zimmermann, 1987a). Thus, together with the *N*-oxygenation, the *N*-dealkylation has also been identified as a second, new, metabolic pathway of these strongly basic amidines. These are apparently competing reactions in which the presence or absence of α-hydrogens determines the actual course of the oxidative transformation (Clement 1985, 1989a).

11.5 PREDICTABILITY OF *N*-OXYGENATIONS BY CYTOCHROME *P*-450 DERIVED FROM STUDIES ON AMIDINES

Table 11.2 compares the characteristics of the *N*-hydroxylation of benzamidine (*11.1*) to benzamidoxime (*11.2*) with those of the *N*-demethylation of *N*-methylbenzamidine (*11.3*) to benzamidine (*11.1*) (Clement and Zimmermann, 1987a,b). Neither reaction can be induced. In principle, they are also comparable in other respects although they exhibit very large differences in their K_m and v_{max} values. Furthermore, both reactions also proceed with only lower conversion rates in the presence of all inhibitors studied but the extent and significance of the inhibition differ in some cases.

For example at higher CO/O_2 values a larger sensitivity is determined for the *N*-demethylation than for the *N*-oxygenation (Clement, 1989a). On the other hand kinetic studies, which have been performed so far, demonstrated that the *N*-oxygenation is inhibited by *N*-methylbenzamidine (Clement *et al.*, 1989). Therefore, no final decision as to whether the same isoenzymes are responsible for both reactions can be made on the basis of the available data. Further studies are in progress.

For the concept developed on the basis of investigations with benzamidines – *N*-oxygenation by cytochrome *P*-450 only in the absence of

Table 11.2 Comparison of the characteristics of the *N*-hydroxylation of benzamidine *(11.1)* to benzamidoxime *(11.2)* (reaction A) and of the *N*-demethylation of *N*-methylbenzamidine *(11.3)* to benzamidine *(11.1)* (reaction B)

Reaction	A	B
Effect of inhibition		
SKF 525-A, metyrapone, CN⁻	+++	+++
CO (various CO/O_2 concentrations)	+	+
Effects of inducers		
Phenobarbital	−	−
3-Methylcholanthrene	−	−
Ethanol	−	−
Benzamidines	−	−
Influence of catalase	−	−
and superoxide dismutase	−	−
K_m (mM)	1.95 ± 0.30	23.8 ± 4.2
v_{max} (nmol/min/mg protein)	0.54 ± 0.17	1.98 ± 0.23

Source: Clement and Zimmermann (1987a,b).

α-H atoms – this question is of lesser importance. It is now clear, that the various cytochromes *P*-450 so far examined in detail are discrete gene products, that is, they differ in their amino acid sequences (Black and Coon, 1986). The catalytic course of oxygen activation is, however, identical for all microsomal cytochrome *P*-450 isoenzymes (Ortiz de Montellano, 1986). Hence the mechanism of *N*-oxygenation and the mechanism of *N*-dealkylation with widely different isoenzymes also should not differ. If such a catalytic mechanism is in operation, the product formation (*N*-oxygenated or *N*-dealkylated product) should depend principally on the presence or absence of α-hydrogen atoms.

However, it should be considered here, that the concept *N*-oxygenation or *N*-dealkylation depending on the presence or absence of α-hydrogen atoms cannot make any predictions as to whether or not the respective compound can be accommodated at the substrate binding site of the differing cytochrome *P*-450 isoenzymes (Clement, 1989a). These substrate binding sites can be different in the individual isoenzymes. Thus, it is fairly obvious, that for the manifold isoenzymes and because little is known about the binding positions, it is generally not possible to predict whether a particular substrate can attain access to its binding position and whether a binding will be possible. At present this question can only be answered by experiment. Of course, steric factors, lipophilicity, and other properties (Gorrod, 1985) of the substrate can play a part in this process. However, when the necessary binding does occur and the usual mechanism does come into operation, an

N-oxygenation will occur if no α-H atoms are present on carbon atoms in the direct neighbourhood of the nitrogen. When α-H atoms are present, on the other hand, an N-dealkylation will occur in place of the N-oxygenation.

This concept was postulated in 1984 (Clement, 1985) on the basis of the initial experiments with amidines and has been confirmed by the further investigations presented here. This concept has also been discussed by other authors who have included studies on other functional groups (Guengerich and Macdonald, 1984; Ziegler, 1985). The concept can thus, also be applied to other nitrogen-containing functional groups. There are exceptions in the case of nitrogen radicals, which have increased stabilities and of bicyclic systems (Guengerich and Macdonald, 1984; Clement, 1989a). N-Oxygenations proceeding in the way in which the substrate uncouples the cytochrome P-450, thus generating superoxide and hydrogen peroxide which then transform the substrate (Duncan *et al.*, 1985) can also not be covered by the present concept since other mechanisms are in operation (Clement, 1989a). Of course, the considerations are only valid for cytochrome P-450 and not for the flavin-containing mono-oxygenase since this enzyme exhibits a completely different mechanism (Ziegler, 1985).

11.6 FURTHER TRANSFORMATIONS OF BENZAMIDOXIME AND BENZAMIDINE

11.6.1 Enzyme sources from the rabbit

The results summarized so far were obtained by using the microsomal fractions of rabbit liver homogenates. The conversions of amidines under aerobic conditions in the presence of these enzyme sources and NADPH at pH 7.4 are summarized in Fig. 11.11. Only the N-hydroxylation of benzamidine (*11.1*) and the N-dealkylation of N-methylbenzamidine (*11.3*) were observed. Benzamidine (*11.1*), benzamidoxime (*11.2*) and N-methylbenzamidine (*11.3*) do not undergo any further purely chemical or enzymic transformations under these conditions (Clement *et al.*, 1989). Thus the quantitative results of the N-hydroxylation of benzamidine (*11.1*) to benzamidoxime (*11.2*) and of the N-demethylation of N-methyl-benzamidine (*11.3*) to benzamidine (*11.1*) are not influenced by other transformations. It is possible to study the N-dealkylation and N-hydroxyl-ation reactions as isolated processes and as competing reactions. By means of HPLC analysis and again with use of synthetic material for comparison, both aromatic ring hydroxylation reactions to form phenols and hydrolytic processes can be excluded. Benzamidoxime (*11.2*) is also not further N-hydroxylated to furnish dihydroxybenzamidine (*11.12*), as could be expected from the presence of an NH_2 group and the absence of α-H atoms.

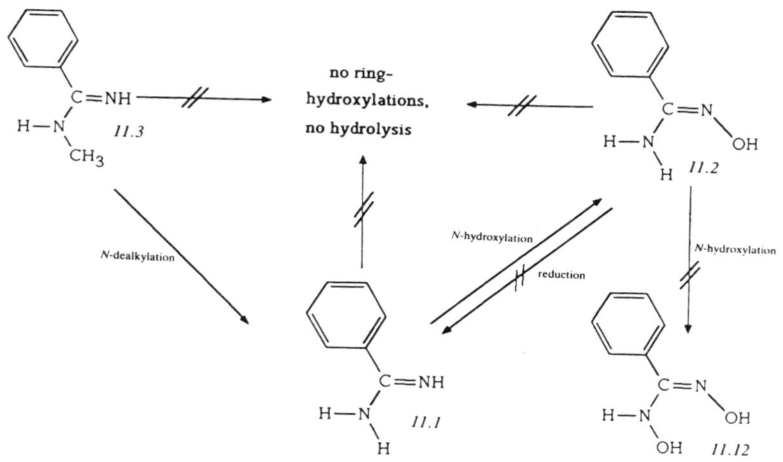

Figure 11.11 Biotransformations of benzamidines (*11.1*) and (*11.3*) and benzamidoxime (*11.2*) by microsomal enzymes of the rabbit (12 000 *g* supernatant or microsomes; NADPH; O$_2$; pH 7.4).

11.6.2 Enzyme sources from the rat

With enzyme sources from the rabbit a reduction of benzamidoxime (*11.2*) back to benzamidine (*11.1*) could also not be detected at pH 7.4. However, when rat liver homogenates were employed under analogous conditions (12 000 *g* supernatant or microsomes, NADPH, aerobic conditions, pH 7.4), this reduction reaction did occur to a considerable extent (Clement *et al.*, 1988b) (Fig. 11.12). An enzymic reduction of another amidoxime was discovered during *in vivo* and *in vitro* investigations on synthetic thrombin inhibitors (Hauptmann *et al.*, 1988).

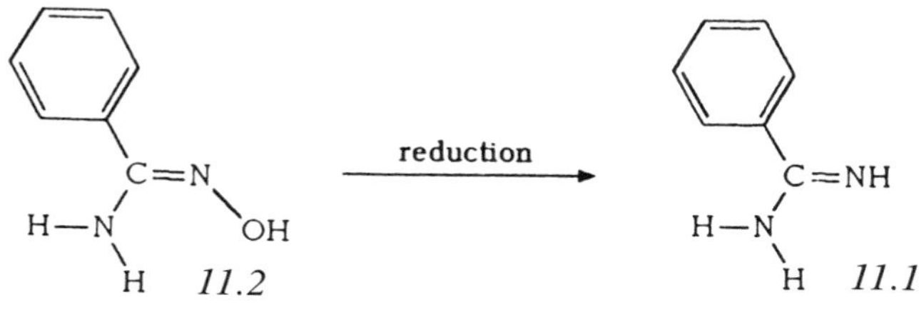

Figure 11.12 Reduction of benzamidoxime (*11.2*) to benzamidine (*11.1*) by microsomal enzymes of the rat (12 000 *g* supernatant or microsomes; NADPH or NADH; O$_2$; pH 7.4).

This novel metabolic transformation exhibits a pH optimum at pH 6.3 and proceeds in the presence of NADPH or NADH. With rabbit liver homogenates, this reduction reaction can first be observed at pH 6.3. With rabbit preparations at pH 7.4, the *N*-hydroxylation predominates whereas with rat preparations the reduction process predominates at this pH value also. Studies for the further characterization of this reduction are in progress.

Thus, there are the first indications that metabolic cycles are in operation during the metabolization of amidines.

11.6.3 *In vivo* studies

Studies have begun in rats using the simple model substances benzamidine (*11.1*) and benzamidoxime (*11.2*) to investigate the newly discovered transformations – *N*-hydroxylation of benzamidines and reduction of benzamidoximes, respectively – *in vivo*. Whereas the reduction of benzamidoxime (*11.2*) was unequivocally demonstrated by the identification of benzamidine (*11.1*) both in urine and in serum after the administration of benzamidoxime (*11.2*) to rats (Clement *et al.*, unpublished), the *N*-hydroxylated compound (*11.2*) could not be detected after the administration of benzamidine (*11.1*) to this species. Of course, this does not mean that the *N*-hydroxylation does not occur *in vivo*. It is equally feasible that the primarily formed benzamidoxime is reduced again very rapidly and therefore cannot be detected in urine or serum.

11.6.4 Phase II reactions

Conjugation reactions of benzamidoximes are feasible *in vivo*, which might also prevent the direct detection of benzamidoximes. Thus, after administration of benzamidine (*11.1*) or benzamidoxime (*11.2*), the urine and serum have also been analysed for the presence of water-soluble conjugates and evidence for the formation of conjugates *in vivo* could be provided, which has to be confirmed by further studies.

In the case of *in vitro* experiments with benzamidoxime and the 9000 *g* supernatant of rabbit liver homogenates, the following results were obtained. When incubations were carried out in the presence of uridine-5'-diphospho-*N*-acetylglucosamine (UDPAG) and uridine-5'-diphospho-glucuronic acid (UDPGA) the amount of extractable free benzamidoxime decreased. However, the isolation of the glucuronide and its comparative synthesis have been unsuccessful to date; these results represent indications for the instability and reactivity of such conjugates (Clement *et al.*, 1988a).

11.7 PHARMACOLOGICAL PROPERTIES OF N-OXYGENATED METABOLITES

The trypanocidal and leishmanicidal activity of the amidoximes of pentamidine (Fig. 11.2) was investigated. These amidoximes were tested against various *Trypanosoma* spp. and *Leishmania donovani* in mice and golden hamsters, respectively. It was found that the amidoximes possessed a very high trypanocidic activity although this was somewhat lower than that of pentamidine (Clement and Raether, 1985). Hence, the further development of these compounds seemed to be of no particular interest.

However, one question is still being investigated further, namely: are the diamidoximes, in contrast to the strongly basic diamidines which exist predominantly in the protonated form under physiological conditions, able to overcome the blood–brain barrier? If this were the case, they could also be active against the so-called second stage of the African sleeping sickness. The central nervous system is affected in this stage of the disease and no suitable therapeutic drugs are as yet available (Clement, 1989b). At present, the activity of the diamidoxime of pentamidine (Fig. 11.2) towards such chronic infections is being investigated.

Hopes for a central activity are coupled with the considerations that, since the diamidoximes, as a consequence of the introduction of oxygen, possess a much lower basicity than the amidines and hence do not exist predominantly in the protonated form there may, in turn, be able to overcome the blood–brain barrier as neutral molecules.

11.8 GENOTOXIC ACTIVITIES OF BENZAMIDOXIME

It is known that several aromatic amines and certain aromatic amides need to be metabolized by N-hydroxylation before they exhibit toxic and carcinogenic properties (Nelson, 1985). This was unknown for strongly basic functional groups such as amidines. Thus the genotoxic potentials of benzamidine (*11.1*) and benzamidoxime (*11.2*) were determined to study the toxicological relevance of the metabolic N-hydroxylation of benzamidines to benzamidoximes (Clement *et al.*, 1988a).

Benzamidoxime exhibits a weak mutagenicity in the Ames test. However, this occurred only in strain TA 98 and only in the presence of the 9000 g supernatant of rabbit liver homogenates. This suggests that a further enzymatic transformation of benzamidoxime (*11.2*) is necessary. It was postulated that unstable conjugates are formed initially and then decompose to furnish electrophilic nitrenium ions. However, it has not yet been possible to prove this hypothesis unequivocally in further experiments. In other tests, benzamidoxime (*11.2*) was found to induce DNA single-strand breaks – performed in rat hepatocytes – and DNA amplification in SV 40-transformed hamster cells. Benzamidine itself, however, was only marginally positive in the DNA single-strand break assay (Clement *et al.*, 1988a).

Thus, as far as the genotoxic potential is concerned, the newly discovered N-hydroxylation of benzamidine (*11.1*) to benzamidoxime (*11.2*) is rather an activating process whereas the reduction may be considered as a detoxification process. Our investigations on benzamidines and their N-hydroxylation reactions will be continued on the basis of these results.

REFERENCES

Albert, A., Goldacre, R. and Phillips, J. (1984) The strength of heterocyclic bases. *J. Chem. Soc.*, 2240–9.

Black, S.D. and Coon, M.J. (1986) Comparative structures of *P*-450 cytochromes, in *Cytochrome P-450* (ed. P.R. Ortiz de Montellano), Plenum Press, New York, London, pp. 161–216.

Clement, B. (1983) The N-oxidation of benzamidines *in vitro*. *Xenobiotica*, **13**, 467–73.

Clement, B. (1984) *In vitro* Untersuchungen zur mikrosomalen N-Oxidation N-substituierter Benzamidine. *Arch. Pharm. (Weinheim)*, **317**, 925–33.

Clement, B. (1985) The biological N-oxidation of amidines and guanidines, in *Biological Oxidation of Nitrogen in Organic Molecules* (eds J.W. Gorrod, and L.A. Damani), VCH and Horwood, Weinheim, Deerfield Beach, Chichester, pp. 253–66.

Clement, B. (1989a) Structural requirements of microsomal N-oxygenations derived from studies on amidines. *Drug Metab. Drug. Interact.*, **7**, 87–108.

Clement, B. (1989b) Verbindungen zur Behandlung von Trypanosomen-infektionen. *Pharm. unserer Zeit*, **18**, 97–111.

Clement, B. and Immel, M. (1987) Untersuchungen zur *in vitro* N-Oxygenierung N-tert.alkylsubstituierter Benzamidine, *Arch. Pharm. (Weinheim)*, **320**, 660–5.

Clement, B. and Kämpchen, T. (1985) ^{15}N-NMR-Studien an 2-Hydroxyguanidinen und Amidoximen. *Chem. Ber.*, **118**, 3481–91.

Clement, B. and Kämpchen, T. (1986) ^{15}N-NMR-Studien der Tautomerie in N-monosubstituierten Amidinen und in N,N′-Diphenylguanidin. *Chem. Ber.*, **119**, 1101–4.

Clement, B. and Kämpchen, T. (1987) ^{15}N-NMR-Studie einer α-Aminonitron/N-Hydroxyamidin-Tautomerie. *Arch. Pharm. (Weinheim)*, **320**, 566–9.

Clement, B. and Raether, W. (1985) Amidoximes of pentamidine: synthesis, trypanocidal and leishmanicidal activity. *Arzneim. Forsch./Drug Res.*, **35**, 1009–14.

Clement, B., Schmezer, P., Weber, H. *et al.*, (1988a) Genotoxic activities of benzamidine and its N-hydroxylated metabolite benzamidoxime in *Salmonella typhimurium* and mammalian cells. *J. Cancer Res. Clin Oncol*, **114**, 363–8.

Clement, B., Schmitt, S. and Zimmermann, M. (1988b) Enzymatic reduction of benzamidoxime to benzamidine. *Arch. Pharm. (Weinheim)*, **321**, 955–6.

Clement, B. and Zimmermann, M. (1987a) Hepatic microsomal N-demethylation of N-methylbenzamidine, N-dealkylation vs N-oxygenation of amidines. *Biochem. Pharmacol.*, **36**, 3127–33.

Clement, B. and Zimmermann, M. (1987b) Characteristic of the microsomal *N*-hydroxylation of benzamidine to benzamidoxime. *Xenobiotica*, **17**, 659–67.

Clement, B. and Zimmerman, M. (1988) Mechanism of the microsomal *N*-hydroxylation of *para*-substituted benzamidines. *Biochem. Pharmacol.*, **37**, 4747–52.

Clement, B., Zimmerman, M. and Schmitt, S. (1989) Biotransformation des Benzamidins und des Benzamidoxims durch mikrosomale Enzyme vom Kaninchen. *Arch. Pharm. (Weinheim)*, **322**, 431–5.

Duncan, J.D., Di Stefano, E.W., Miwa, G.T. and Cho, A.K. (1985) Role of superoxide in the *N*-oxidation of *N*-(2-methyl-1-phenyl-2-propyl)hydroxylamine by the rat liver cytochrome *P*-450 system. *Biochemistry*, **24**, 4155–61.

Goa, K.L. and Campoli-Richard, D.M. (1987) Pentamidine isethionate, a review of its antiprotozoal activity, pharmacokinetic properties and therapeutic use in *Pneumocystis carinii* pneumonia. *Drugs*, **33**, 242–58.

Gorrod, J.W. (1985) Chemical determinants of the enzymology of organic nitrogen oxidation. *Drug Metab. Dispos.*, **13**, 283–6.

Grout, R.J. (1972) Biological reactions and pharmaceutical uses of imidic acid derivatives, in *The Chemistry of Amidines and Imidates* (ed. S. Patai), John Wiley, London, pp. 255–82.

Guengerich, F.P. and Macdonald, T.L. (1984) Chemical mechanisms of catalysis by cytochrome *P*-450: a unified view. *Acc. Chem. Res.*, **17**, 9–16.

Hauptmann, J., Paintz, M., Kaiser, B. and Richter, M. (1988) Reduction of a benzamidoxime derivative to the corresponding benzamidine *in vivo* and *in vitro*. *Pharmazie*, **43**, 559–60.

Jasukochi, J. and Masters, B.S.S. (1976) Some properties of a detergent-solubilized NADPH–cytochrome *c* (cytochrome *P*-450) reductase purified by biospecific affinity chromatography. *J. Biol. Chem.*, **251**, 5337–44.

Kling, L., Legrum, W. and Netter, K.J. (1985) Induction of liver cytochrome *P*-450 in mice by warfarin. **34**, 85–91.

Nelson, S.D. (1985) Arylamines and arylamides: oxidation mechanisms, in *Bioactivation of Foreign Compounds* (ed. M.W. Anders), Academic Press, Orlando, pp. 349–74.

Ortiz de Montellano, P.R. (1986) Oxygen activation and transfer, in *Cytochrome P-450* (ed. P.R. Ortiz de Montellano), Plenum Press, New York, London, pp. 217–71.

Raether, W., Mieth, H. and Loewe, H. (1972) Präparate gegen Protozoen, Mittel gegen Trypanosomen, in *Arzneimittel* (eds G. Ehrhart and H. Ruschig), Verlag Chemie, Weinheim, pp. 138–44.

Ziegler, D.M. (1985) Molecular basis for *N*-oxygenation of sec. and tert.-amines, in *Biological Oxidation of Nitrogen in Organic Molecules* (eds J.W. Gorrod and L.A. Damani), VCH and Horwood, Weinheim, Deerfield Beach, Chichester, pp. 43–52.

12

Studies on the *N*-oxidation of phentermine: evidence for an indirect pathway of *N*-oxidation mediated by cytochrome *P*-450

A.K. Cho, J.D. Duncan and J.M. Fukuto

Department of Pharmacology and Jonsson Comprehensive Cancer,
UCLA Center for Health Sciences, Los Angeles, CA, USA

1. An overview of the metabolism of arylalkylamines related to amphetamine is presented.
2. *N*-Hydroxyamphetamine and other alpha substituted phenylhydroxylamines are oxidized to the nitroso state, which forms a stable complex with cytochrome *P*-450, inhibiting its function as a mono-oxygenase.
3. *N*-Hydroxyphentermine on the other hand is further metabolized to the nitro state, the latter being a major urinary metabolite of phentermine in man.
4. Mechanistic details of the above are discussed, with summaries of data using microsomes and reconstituted systems.

12.1 INTRODUCTION

This chapter summarizes results of our investigations on the oxidation of aliphatic nitrogen functions. For some time we have been studying the metabolism of arylalkylamines related to amphetamine. Our focus has been on amphetamine and phentermine and their *N*-hydroxy derivatives.

The structures of the two compounds are shown in Fig. 12.1 and differ by the presence of an additional alpha methyl group on phentermine. The added alpha carbon substitution has very little effect on base strength but appears to have a significant steric effect that is discernible in terms of the interaction of the nitroso states of these structures with cytochrome *P*-450.

N-Hydroxyamphetamine and other alpha substituted phenethylhydroxyl amines are oxidized to the nitroso state which forms a stable complex with cytochrome *P*-450, inhibiting its function as a mono-oxygenase (Lindeke and Paulsen-Sorman, 1988). Additionally, in rat liver microsomes, *N*-hydroxyamphetamine is oxidized to its oxime by a non *P*-450 pathway (Matsumoto and Cho, 1982). Phentermine, with its two alpha methyl groups appears to be too hindered to form the complex and as a result undergoes additional metabolic transformations. Phentermine is readily *N*-oxidized by cytochrome *P*-450 (Duncan and Cho, 1982). Although amphetamine undergoes *N*-hydroxylation as well, the conversion appears to be much slower (Duncan *et al.*, 1983). *N*-Hydroxyphentermine is further oxidized *in vitro* by cytochrome *P*-450 to the nitro compound. Caldwell *et al.* (1975) have reported that significant amounts (30% of the dose) of the nitro metabolite of *p*-chlorophentermine were excreted by human subjects so that this transformation appears to occur *in vivo* as well. Excretion of nitro compounds is somewhat unusual in that most aromatic nitro compounds are reduced before excretion (Rickert *et al.*, 1984). Since the *N*-oxidation pathway appeared to be an important pathway in the overall metabolism of this type of compound, its biochemical and chemical details were scrutinized.

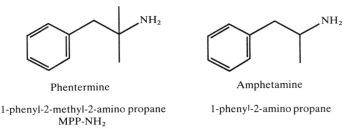

Phentermine

1-phenyl-2-methyl-2-amino propane
MPP-NH$_2$

Amphetamine

1-phenyl-2-amino propane

Figure 12.1 Structures of phentermine and amphetamine.

12.2. BIOCHEMICAL STUDIES WITH MICROSOMES

One of the early observations was the sensitivity of the oxidation of *N*-hydroxyphentermine (MPP-NHOH) to enzyme(s) present in the cytosol. Thus, Sum and Cho (1979) found that washed microsomes were capable of oxidizing MPP-NHOH to the corresponding nitro compound (MPP-NO_2) but when the microsomes were contaminated with cytosol, the *N*-hydroxy compound was stable. As will be shown later, the stability was probably due to the presence of superoxide dismutase present in the cytosol.

Subsequent experiments demonstrated that the oxidation of *N*-hydroxyphentermine by liver microsomes was cytochrome *P*-450 dependent. The reaction was inhibited by carbon monoxide and the *P*-450 inhibitor, DPEA, but differed from other oxidations by this mono-oxygenase in its sensitivity to superoxide dismutase (SOD) (Maynard and Cho, 1981). Thus, superoxide dismutase inhibited the oxidation of MPP-NOH to MPP-NO_2 but not the conversion of MPP-NH_2 into MPP-NHOH. This study also demonstrated that the conversion of MPP-NHOH into MPP-NO_2 could be effected by superoxide generated by xanthine oxidase. The peroxide-generating system, glucose oxidase, was not capable of the oxidation reaction. These observations were consistent with the generation of superoxide by interaction of MPP-NHOH with cytochrome *P*-450 and its subsequent oxidation by superoxide.

Cytochrome *P*-450 is capable of generating superoxide by an uncoupling process in which the catalytic cycle is terminated after a one electron reduction of the haem-dioxygen complex (Estabrook and Werringloer, 1977). This aborted action reduces oxygen to superoxide. For example, benzphetamine, a substrate for *N*-demethylation by cytochrome *P*-450 also uncouples cytochrome *P*-450 and generates hydrogen peroxide. Evidence for MPP-NHOH-mediated hydrogen peroxide formation by microsomes and by purified and reconstituted cytochrome *P*-450 (Cho *et al.*, 1982) was obtained in experiments with rat liver microsomes and cytochrome *P*-450 (Table 12.1).

In separate experiments, MPP-NHOH was shown to increase superoxide levels in washed microsomes, evidenced by the formation of adrenochrome from norepinephrine (Maynard, 1982). The hydrogen peroxide generated in these systems presumably results from the proton-induced disproportionation of superoxide. In contrast, increasing concentrations of *N*-hydroxyamphetamine does not result in increasing levels of hydrogen peroxide in microsomal systems. Instead this hydroxylamine blocks the cytochrome *P*-450-dependent increase in hydrogen peroxide. There is, however, a constant basal level of hydrogen peroxide generated at moderate *N*-hydroxyamphetamine levels. *N*-Hydroxyphentermine was regarded as a

Table 12.1 Hydrogen peroxide generation by *N*-hydroxyphentermine in reconstituted systems

Reductase (units/ml)	Cytochrome P-450 (nmol/ml)	Rate of H_2O_2 formation (nmol/min)		
		Control	NOHA	NOHP
0	0	0.063	0.132	0.76
1000	0	10.04	8.91	10.80
1000	0.05	13.90	6.30	10.40
1000	0.10	15.50	6.36	13.10
1000	0.20	17.90	6.17	18.30
1000	0.30	18.50	6.77	21.50
1000	0.40	17.30	6.64	19.70

Hydrogen peroxide levels were measured in the mixture of proteins in the presence and absence of hydroxylamine. Concentrations of *N*-hydroxyphentermine (NOHP) and *N*-hydroxyamphetamine (NOHA) were 1mM. (From Cho *et al.*, 1982.)

direct uncoupler of cytochrome *P*-450 whereas *N*-hydroxyamphetamine was not. Thus, there appear to be fairly specific structural requirements for the uncoupling of cytochrome *P*-450 by these types of substrates.

12.3 BIOCHEMICAL STUDIES WITH RECONSTITUTED SYSTEMS

The biochemical details of the oxidation reaction were examined further with reconstituted systems. The oxidation of MPP-NHOH to MPP-NO$_2$ is a 4-electron change (Fig. 12.2) so that the nitroso state is a likely intermediate. Initial experiments demonstrated that the reaction sequence shown in Fig. 12.2 occurred in reconstituted cytochrome *P*-450 systems using both 3-methylcholanthrene- and phenobarbital-induced haem. A systematic evaluation of the components of the cytochrome *P*-450 system indicated that the flavoprotein, cytochrome *P*-450 reductase had limited activity in oxidizing MPP-NHOH to the nitro compound but in the presence of a haem the activity increased substantially (Table 12.2).

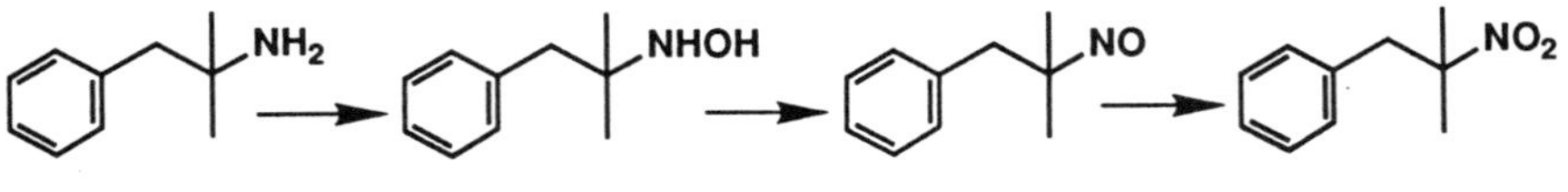

Figure 12.2 Metabolic reaction sequence for phentermine in reconstituted cytochrome *P*-450 systems.

Table 12.2 Oxidation of *N*-hydroxyphentermine by reconstituted cytochrome *P*-450 systems

Components	Nitro compound (nmol/10 min)
P-450, reductase, phospholipid	0
P-450, phospholipid, NADPH	0
Reductase, phospholipid, NADPH	14 ± 10
P-450, reductase, phospholipid, NADPH	70 ± 7

The cytochrome *P*-450 was purified from livers obtained from phenobarbital pretreated animals. (From Duncan *et al.*, 1985.)

The involvement of cytochrome *P*-450 in the microsome-mediated oxidation was also demonstrated by the concentration-dependent inhibition of the reaction by anti-*P*-450 IgG (Duncan *et al.*, 1985). The formation of the nitroso compound, MPP-NO, was analysed by HPLC and shown to exhibit the same sensitivity to SOD and catalase as that of MPP-NO$_2$. These experiments demonstrated that the reaction sequence shown in Fig. 12.2 occurred. Phentermine-type arylalkyl hydroxylamines appear to uncouple the catalytic cycle of cytochrome *P*-450 and undergo indirect oxidation by the oxygen species generated by reduction of molecular oxygen with NADPH. These compounds, with their quaternary alpha carbon, appear to differ from the amphetamine-type compound as the latter structure inhibits cytochrome *P*-450 function. However, in experiments with rabbit liver microsomes, Florence (1985) demonstrated that the nitro compound resulting from amphetamine oxidation was also generated by a superoxide dismutase sensitive pathway so that once superoxide is formed, the analogous reaction can occur. The experimental evidence indicates that MPP-NH$_2$ is oxidized by cytochrome *P*-450 to MPP-NHOH which then uncouples the cytochrome *P*-450 cycle to generate superoxide. Subsequent oxidation of MPP-NHOH results from its reaction with superoxide. Therefore, in the cytochrome *P*-450-dependent oxidation of MPP-NH$_2$ to MPP-NO or MPP-NO$_2$, the initial oxidation, MPP-NH$_2$ to MPP-NHOH, is performed by a typical cytochrome *P*-450 oxidative process while all subsequent oxidations of MPP-NHOH and MPP-NO are the result of oxidation by the active oxygen species generated from cytochrome *P*-450 uncoupling. The intimate chemical details of this unusual oxidative process were elucidated by examining the individual chemical reactions between the active oxygen species which would result from the uncoupling and the compounds.

12.4 SUPEROXIDE-MEDIATED REACTIONS: POSSIBLE MECHANISMS

Since superoxide was clearly involved, the nature of superoxide-mediated reactions was examined. This one-electron reduction product of dioxygen contains an unshared electron and a negative charge and its chemistry reflects these features. Superoxide is a relatively poor oxidant (Sawyer and Valentine, 1981) and its ability to oxidize organic compounds by hydrogen abstraction mechanisms appears to be limited to those functions containing easily extractable hydrogen atoms such as phenols and hydroquinones. Epinephrine, for example is oxidized by superoxide to adrenochrome (Udenfriend, 1962). Superoxide is also a weak base and exists in equilibrium with its conjugate acid with a pK_a of about 4.7. As such it can also abstract a proton from a dissociable function. Although its low pK_a makes such an acid–base reaction highly unlikely in aqueous media, it may be possible in the non-polar nature of the lipid matrix in which cytochrome P-450 exists. Thus, there are two possible mechanisms for the superoxide-mediated oxidation of these arylalkylhydroxylamines. The reactions are summarized in Fig. 12.3.

Reactions 1, 2, 3 and 4 represent the proton abstraction pathway in which superoxide anion is acting as a base to remove the hydroxyl proton

Anionic pathway

$$MPPNHOH + O_2^{-} \longrightarrow MPPNHO^{-} + HOO\cdot \quad (1)$$

$$2HOO\cdot \longrightarrow O_2 + HOOH \quad (2)$$

$$MPPNHO^{-} + HOOH \longrightarrow MPPNO + HO^{-} + H_2O \quad (3)$$

$$MPPNO + O_2 \xrightarrow{\text{SLOW}} MPPNO_2 \quad (4)$$

Radical pathway

$$MPPNHOH + O_2^{-}\cdot \longrightarrow MPPNHO\cdot + HOO^{-} \quad (5)$$

$$MPPNHO\cdot \longrightarrow 1/2 MPPNO + 1/2 MPPNHOH \quad (6)$$

$$MPPNHOH + HOO^{-} \longrightarrow MPPNO + HO^{-} + H_2O \quad (7)$$

$$MPPNO + HOO^{-} \longrightarrow MPPNO_2 + HO^{-} \quad (8)$$

Figure 12.3 Possible mechanisms for the superoxide-mediated oxidation of arylalkylhydroxylamine.

generating the hydroxylamine anion and hydroperoxide radical. The hydroperoxide radical, in turn, disproportionates rapidly to oxygen and hydrogen peroxide. As the glucose oxidase experiments showed, hydrogen peroxide is not capable of oxidizing the hydroxylamines but peroxide anion can oxidize the hydroxylamine to a mixture of nitroso and nitro compound. The alternate pathway (reactions 5, 6, 7 and 8) reflects the radical character of superoxide and in reaction 5, a hydrogen atom is removed from the hydroxylamine. The generated nitroxyl radical disproportionates to the hydroxylamine and the nitroso compound (reaction 6). The nitroso compound can then be oxidized by oxygen or by hydrogen peroxide anion (reactions 7 and 8).

With these two possibilities in mind, the oxidation of hydroxylamine and nitroso compound by the different oxygen species was then examined in polar and non-polar media (Fukuto *et al.*, 1985). One of the complicating factors in the study was the ability of nitroso compounds to dimerize (Cho and Fukuto *et al.*, 1988). Nitrosoalkanes dimerize in polar media so that their reactivity depends on the dissociation to a monomer. However, in the lipid microenvironment of the cytochrome *P*-450 active site, behaviour more like that in non-polar solvents might be expected. Therefore, the various oxidants were examined in polar and non-polar media. Water or methanol were used as the polar medium and diethyl ether was used as the non-polar medium. Superoxide was not capable of oxidizing the nitroso compound but could oxidize the hydroxylamine to the nitroso state in both polar and non-polar media. Hydroperoxide anion but not hydrogen peroxide was capable of oxidizing the hydroxylamine to the nitroso state and the nitroso compound, to the nitro state. Dioxygen could oxidize both functions but was more effective in oxidizing the nitroso compound in non-polar media. Thus, the oxidation of the hydroxylamine appears to be a sequence of reactions in which the first step is an oxidation to the nitroso state with subsequent oxidation to the nitro state.

The two proposed mechanisms for the oxidation of MPP-NHOH by superoxide, the anionic and radical pathways differ in two ways. First, in aprotic media, the anionic mechanism will generate molecular oxygen (reaction 2), whereas the radical pathway will not. Second, the overall stoichiometry of the processes differ; the anionic mechanism would yield maximally 0.5 equivalent of oxidized products per equivalent of superoxide since only 0.5 equivalent of hydrogen peroxide is generated (reaction 2). The other product, molecular oxygen was shown to be a poor oxidizing agent for both MPP-NHOH and MPP-NO. The radical pathway, on the other hand, would yield between 0.5 and 1.0 equivalent of oxidized products per equivalent of superoxide depending on the relative rates of reactions 7 and 8. These differences were utilized to distinguish between the two mechanisms. In ether, it was demonstrated that oxidation of MPP-NHOH

Table 12.3 Stoicheiometry of N-hydroxyphentermine oxidation by superoxide

Equivalents of O_2^-	% remaining (a)	% total product (b)
0.5	48	52
1	25	75

The indicated equivalents of superoxide were allowed to react with MPP-NHOH in ether and the mixture analysed by GC for oxidized product (MPP-NO and MPP-NO$_2$). (a) Refers to the substrate remaining. (b) Refers to the percentage of the substrate that was oxidized either to the nitroso or nitro state. (From Fukuto *et al.*, 1985.)

by superoxide is accomplished without the evolution of oxygen. Also, analysis of the reaction products gave stoicheiometries more consistent with the radical pathway; one equivalent of superoxide gave 0.74 equivalents of total oxidized products (MPP-NHOH and MPP-NO) (Table 12.3). These experiments indicate that the oxidation of MPP-NHOH occurs by the radical mediated pathway outlined in reactions 5, 6, 7, and 8. The feasibility of an oxidative radical process was also demonstrated by reacting MPP-NHOH with TEMPO (2,2,6,6-tetramethylpiperidinyloxy) a nitroxyl radical species. TEMPO oxidized the hydroxylamine in both polar and non-polar media by abstracting a hydrogen atom to form the MPP-nitroxide which disproportionates to MPP-NHOH and MPP-NO with the stoicheiometry predicted in reaction 6.

12.5 CONCLUSIONS

These studies have identified a unique pathway of nitrogen oxidation in which an amine is first N-hydroxylated by cytochrome P-450 then the formed hydroxylamine uncoupled the mono-oxygenase system to generate superoxide. Aliphatic hydroxylamines and hydroxylamine itself react with superoxide (Nanni and Sawyer, 1980) through a radical pathway to form the nitroxide. The nitroxide then disproportionates to the hydroxylamine and nitroso compound, which is oxidized further by either the peroxide side product or atmospheric oxygen. The reaction sequence appears to be limited to aliphatic nitrogen functions. Although aromatic hydroxylamines are N-hydroxylated by cytochrome P-450 (Lotlikar and Zaleski, 1975), their reactivity precludes these reactions. Aromatic hydroxylamines are readily oxidized to the nitroso state, and microsomal incubations, contain a mixture of both (Fukuto *et al.*, 1986). They do not appear to uncouple the cytochrome P-450 system to superoxide or to hydrogen peroxide. However, it is possible that other substrates with the appropriate chemical reactivities can participate in this indirect pathway of oxidation.

REFERENCES

Caldwell, J., Koster, U., Smith, R.L. and Williams, R.T. (1975) Species variations in the N-oxidation of chlorphentermine. *Biochem. Pharmacol.*, **24**, 2225–32.

Cho, A.K. and Fukoto, J.M. (1988) Chemistry of organic nitrogen compounds, in *Biotransformation of Organic Nitrogen Compounds* (eds A.K. Cho and B. Lindeke), S. Karger AG, Basel.

Cho, A.K., Maynard, M.S., Matsumoto, R.M. *et al.*, (1982) The opposing effects of N-hydroxyamphetamine and N-hydroxyphentermine on the H_2O_2 generated by hepatic cytochrome P-450. *Mol. Pharmacol.*, **22**, 465–70.

Duncan, J.D. and Cho, A.K. (1982) N-Oxidation of phentermine to N-hydroxyphentermine by a reconstituted cytochrome P-450 oxidase system from rabbit liver. *Mol. Pharmacol.*, **22**, 235–8.

Duncan, J.D., Di Stefano, E.W., Miwa, G.T. and Cho, A.K. (1985) Role of superoxide in the N-oxidation of N-(2-methyl-1-phenyl-2-propyl)hydroxylamine by the rat liver cytochrome P-450 system. *Biochemistry*, **24**, 4155–61.

Duncan, J.D., Hallstrom, G., Florence, V.M. *et al.* (1983) N- and 'C-oxidation of' - alkyl substituted phenethylamines by a reconstituted cytochrome P-450 oxidase system from rabbit liver. *Acta Pharm. Suec.*, **20**, 331–40.

Estabrook, R.W. and Werringloer, J. (1977) Cytochrome P450 – its role in oxygen activation for drug metabolism, in *Drug Metabolism Concepts* (ed. D.M. Jerina), ACS symposium series No. 44, ACS, Washington, DC, pp. 1–26.

Florence, V.M. (1985) PhD Thesis, UCLA.

Fukoto, J.M., Brady, J.F., Burstyn, J.N. *et al.* (1985) Direct formation of complexes between cytochrome P-450 and nitrosoarenes. *Biochemistry*, **25**, 2714–19.

Lindeke, B. and Paulsen-Sörman, U. (1988) Nitrogenous compounds as ligands to hemoporphyrins – the concept of metabolic–intermediary complexes, in *Progress in Basic and Clinical Pharmacology* (eds P. Lomax and E.S. Vesell), Karger, Basel, **1**, 63–102.

Lotlikar, P.D. and Zaleski, K. (1975) Ring- and N-hydroxylation of 2-acetamindifluorene by rat liver reconstituted cytochrome P-450 enzyme system. *Biochem. J.*, **150**, 561–4.

Matsumoto, R.M. and Cho, A.K. (1982) Conversion of N-hydroxyamphetamine to phentylacetone oxime by rat liver microsomes. *Biochem. Pharmacol.*, **31**, 105–8.

Maynard, M.S. (1982) PhD Thesis, UCLA.

Maynard, M.S. and Cho, A.K. (1981) Oxidation of N-hydroxyphentermine to 2-methyl-2-nitro-1-phenylpropane by liver microsomes. *Biochem. Pharmacol.*, **30**, 1115–19.

Nanni, J.E. and Sawyer, D.T. (1980) Superoxide-ion catalyzed oxidation of hydroxyphenazines, reduced flavins, hydroxylamine and related substrates via hydrogen atom transfer. *J. Am. Chem. Soc.*, **102**, 7591–3.

Rickert, D.E., Butterworth, B.P. and Popp, J.A. (1984) Dinitrotoluene: acute toxicity, oncogenicity, genotoxicity and metabolism. *CRC Crit. Rev. Toxicol.*, **13**, 217–34.

Sawyer, D.T. and Valentine, J.S. (1981) How super is superoxide? *Accts. Chem. Res.*, **14**, 393–400.

Sum, C.Y. and Cho, A.K. (1979) The metabolism of N-hydroxyphentermine by rat liver microsomes. *Drug Metab. Dispos.*, **7**, 65–9.

Udenfriend, S. (1962) *Fluorescence assay in Biology and Medicine*. Academic Press, New York, pp. 40–151.

Wajer, Th.A.J.W., Mackor, A. and DeBoer, Th.J. (1969) C-nitroso compounds. VII, an ESR study of alkylnitroxides. *Tetrahedron*, **25**, 175.

13

Molecular activation mechanisms involved in arylamine cytotoxicity: peroxidase products

P. J. O'Brien, S. Jatoe, L. G. McGirr and S. Khan

Faculty of Pharmacy, University of Toronto, Toronto, Ontario, Canada

1. Three enzyme systems have been recognized as being involved in N-oxidative bioactivation of acetaminophen. These include cytochrome P-450, peroxidases and prostaglandin H synthase.
2. Peroxygenase activity of cytochrome P-450 produces small amounts of N-acetyl-p-benzoquinoneimine, acetaminophen polymers being the main reaction products.
3. The horseradish peroxidase–H_2O_2 system catalyses oxidation of acetaminophen to dimers, trimers and polymers with only small amounts of N-acetyl-p-benzoquinoneimine.
4. Prostaglandin H synthase mediates arachidonate-dependent formation of low concentrations of an acetaminophen–glutathione adduct.
5. Hypohalites or myeloperoxidase–H_2O_2 in the presence of halides catalyse the oxidation of acetaminophen largely to N-acetyl-p-benzoquinoneimine.
6. Peroxidative activation of acetaminophen and aminophenol in isolated hepatocytes in the presence of halides represents a metabolic pathway that is considerably more toxic than the other routes of bioactivation.

13.1 INTRODUCTION

The induction of cancer by xenobiotics is believed to be initiated by metabolism of the xenobiotic to electrophilic metabolites which covalently link to nucleophilic groups on informational cellular macromolecules such as DNA (Miller, 1970). Much attention has been focused on the *N*-oxidation of arylamines and arylamides by cytochrome *P*-450-dependent hepatic mixed function oxidases. However, peroxidases (donor-H_2O_2 oxidoreductases, EC 1.11.1.7) also catalyse the *N*-oxidation of arylamines. The latter oxidation usually differs in being a one-electron oxidation to free radicals which often form dimeric or polymeric products. Peroxidases are also effective at catalysing the irreversible binding of xenobiotics (particularly arylamines) to DNA and protein, and are therefore considered to be potentially genotoxic or cytotoxic. Comprehensive reviews have appeared on the activation of xenobiotics by peroxidases (O'Brien, 1984, 1985, 1988a,b; Meunier, 1987) or the peroxidase activity of prostaglandin H synthase (Boyd and Eling, 1985; Josephy, 1988). Furthermore, suggestive evidence for *in vivo* peroxidase activation has recently appeared. Thus about 20% of the DNA adducts formed in the urothelium of dogs treated with the bladder carcinogen 2-naphthylamine are estimated to be of peroxidative origin (Yamazoe *et al.*, 1985). This may also be true for the bladder carcinogen benzidine as the peroxidase product benzidine diimine readily form DNA adducts similar to that formed *in vivo* (O'Brien *et al.*, 1985; Yamazoe *et al.*, 1989).

In the following, acetaminophen has been selected as the nitrogen-containing drug for detailed discussion. The available research literature for the enzyme catalysed *N*-oxidation of acetaminophen by cytochrome *P*-450 peroxidases or prostaglandin synthetase is reviewed. The cytotoxicity of acetaminophen oxidized by peroxidases will be compared with a new halide-dependent peroxidase activation pathway that is considerably more toxic than other routes of activation.

13.2 ENZYMIC MECHANISMS OF ACETAMINOPHEN BIOACTIVATION AND CYTOTOXICITY

Acetaminophen is a good antipyretic drug that effectively relieves mild and moderate pain and has a minor anti-inflammatory action. In most countries it is available without prescription because it lacks the occasional side effects of aspirin, e.g. gastrointestinal ulceration with haemorrhage and Reyes syndrome. It is probably now the most widely sold and safest drug. However, fatal hepatic necrosis can occur after a deliberate or accidental overdose. This is believed to result from the oxidation of acetaminophen which escapes conjugation to sulphates and glucuronides as a result of saturation of the phase II conjugating enzymes. The N-acetyl-p-benzo-quinoneimine formed by oxidation is detoxified by endogenous GSH to form a GSH conjugate. The catalytic action of GSH-S-transferase can lead to the complete depletion of hepatic GSH when the rate of acetaminophen oxidation exceeds the rate of GSH synthesis. Once GSH is more than 70% depleted, essential protein thiols are alkylated by the N-acetyl-p-benzo-quinone imine and cytotoxicity ensues later. Without specific therapy on the third day after an acetaminophen overdose, fewer than 10% of patients develop severe liver damage, 1% suffer acute renal failure and 1–2% die in hepatic failure (Prescott, 1983). N-Acetylcysteine, cysteamine or methionine given within 10 hours can prevent impending liver toxicity and N-acetylcysteine is now the preferred therapy (Smilkstein *et al.*, 1988). These antidotes are believed to act primarily by stimulating glutathione synthesis and hence facilitate the glutathione conjugation of the N-oxidation of acetaminophen.

The three enzymic systems previously described that catalyse the N-oxidation of acetaminophen include cytochrome P-450, peroxidase and prostaglandin synthetase. Thus a small amount of acetaminophen N-oxidation to N-acetyl-p-benzoquinoneimine has been demonstrated for cumene hydroperoxide and microsomes (from phenobarbital-pretreated rats) as a result of the peroxygenase activity of cytochrome P-450 (Dahlin *et al.*, 1984). However, others have shown that acetaminophen polymers were the main products formed by the hydroperoxide system and that microsomes and NADPH could catalyse the oxidation of acetaminophen to 3-(glutathion-S-yl)-acetaminophen to a lesser degree but without polymer formation (Potter and Hinson, 1987a). Furthermore, reconstituted mixed function oxidase systems containing P-450$_{UT-A}$, the major isoenzyme from untreated male rats, or P-450$_{BNF-B}$ from β-naphthoflavone pretreated rats catalysed a small amount of GSH conjugate formation much more readily than P-450$_{PB-B}$ from phenobarbital pretreated male rats (Harvison *et al.*, 1988). It is, therefore, not clear why microsomes from phenobarbital treated rats were more effective than those from control rats. In all of these cytochrome P-450 systems, however, the yield of N-oxidation product detected was 0.5–2%.

The horseradish peroxidase and the H_2O_2 system catalyses the oxidation of acetaminophen to dimers, trimers and polymers. In the presence of GSH, 3-(glutathion-S-yl)-acetaminophen was also formed in a 2% yield. The low yield of GSH conjugate presumably reflected futile reduction of the phenoxyl radical by GSH to form thiyl radicals and GSSG (Ross *et al.*, 1985). In the absence of GSH, N-acetyl-p-benzoquinoneimine was isolated by HPLC in a 2% yield. ESR fast-flow systems were also used to detect an intense, three-line (ESR) spectrum with a characteristic g value of 2.0043,

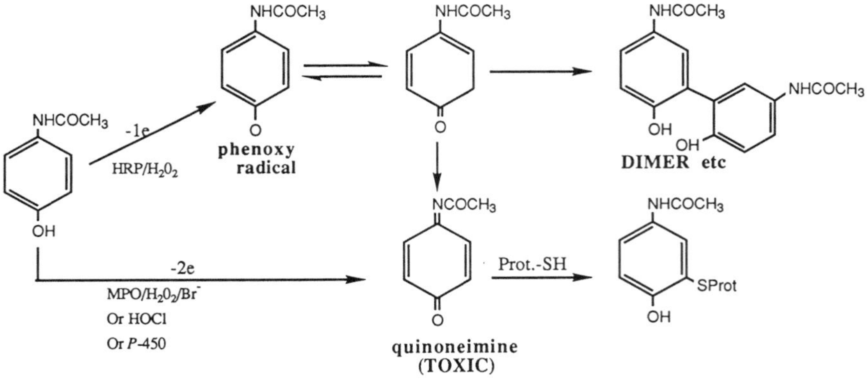

Figure 13.1 Acetaminophen activation mechanism: one electron versus two electron oxidation.

which was assigned to an oxygen-centred phenoxyl free radical. The observed three-line pattern is characteristic of parasubstituted phenoxyl radicals with a dominant 5-G hyperfine coupling to the two equivalent ortho-hydrogens which results in this $1:2:1$ triplet (Mason and Fischer, 1986). As shown in Fig. 13.1, after radical formation, a radical coupling reaction occurs primarily at the 3–5 position which have the highest unpaired electron density. This dimerization is followed by an enolization to regenerate a stable acetaminophen dimer. As shown in Fig. 13.1 some of the radicals are also disproportionate to equimolar amounts of N-acetyl-p-benzoquinoneimine and acetaminophen (Potter and Hinson, 1987a). Introduction of methyl groups in the 3 and 5 positions of acetaminophen does not change hepatotoxicity significantly (Fernando *et al.*, 1980), but markedly increases stability of the phenoxyl free radical by decreasing dimer formation owing to steric effects. The N-acetyl-3,5-dimethyl-p-benzo-quinoneimine formed by disproportionation is also much more stable than N-acetyl-p-benzoquinoneimine (Mason and Fischer, 1986).

Prostaglandin-synthetase and arachidonate were also shown to catalyse the formation of an acetaminophen–glutathione conjugate in a low yield (Moldeus and Rahimtula, 1980; Moldeus *et al.*, 1982). The arachidonate or H_2O_2 and prostaglandin-synthetase pathway for catalysing the N-oxidation of acetaminophen has been proposed for the kidney medulla. This could explain clinical chronic analgesic nephropathy, lesions of the renal papilla, induced by the chronic use of analgesic mixtures that contain acetaminophen. It could also explain the renal tubular necrosis in conjunction with hepatotoxicity that has been observed with rodents. The very high content of prostaglandin-synthetase of the renal inner medullary interstitial cells could explain the high binding of acetaminophen in the renal inner medulla compared to the renal cortex particularly because cytochrome *P*-450 is located in the renal cortex and cannot be detected in the renal inner medulla (Mohandas *et al.*, 1981; Boyd and Eling, 1981). Furthermore the anti-inflammatory action of acetaminophen and other non-steroidal anti-inflammatory drugs can be attributed to their effect at inhibiting prostaglandin-synthetase when operating under low peroxide tone (Hamel and Lands, 1982). Acetaminophen presumably inhibits by acting as an effective substrate of prostaglandin-synthetase and thereby reduces peroxide initiators (Markey *et al.*, 1987). Potter and Hinson (1987b) have shown that the principal products formed are acetaminophen polymers and that polymer formation was decreased 80% and GSH conjugate was formed in a $<0.4\%$ yield when GSH was present. The metabolism of acetaminophen by microsomal prostaglandin H synthase and arachidonic acid was inhibited by indomethacin but the activity with H_2O_2 was not inhibited by indomethacin.

In all the above systems, the yields of *N*-acetylbenzoquinoneimine and GSH conjugate are low. Furthermore, different oxidation mechanisms seem to be involved with these peroxidase systems. Thus GSH conjugate formation catalysed by prostaglandin-synthetase is only slightly decreased by increasing concentrations of GSH, ascorbic acid or NADPH, suggesting that prostaglandin H synthetase catalyses both the one- and two-electron oxidation of acetaminophen (Potter and Hinson, 1987b). Since, on the other hand, GSH conjugate formation by horseradish peroxidase and H_2O_2 is inhibited by ascorbic acid or NADPH, it was concluded that peroxidase–H_2O_2 catalyses only one-electron oxidation of acetaminophen. The low yield of GSH conjugate formed by both systems presumably reflects the futile reduction of the phenoxy radical intermediates by GSH resulting in extensive GSH oxidation (Ross *et al.*, 1985). Polymer formation is similar for both systems, but does not occur during acetaminophen oxidation by microsomal mixed function oxidase, suggesting that cytochrome *P*-450 catalyses a two electron oxidation (Potter and Hinson, 1987a). The low yield of GSH conjugate, and thus presumably *N*-acetylbenzoquinoneimine, formed in the latter system is $<2\%$. This presumably reflects the competitive rapid and futile reduction of *N*-acetylbenzoquinoneimine by NADPH and NADPH cytochrome *P*-450 reductase. By contrast, the high polymer formation catalysed by cytochrome *P*-450 and cumene hydroperoxide presumably reflects a one-electron oxidation by alkoxy radicals formed by the homolytic decomposition of the hydroperoxide. The ready alkylation of microsomal proteins by *N*-acetylbenzoquinoneimine could also partly explain the low levels of quinoneimine detected.

Peroxidases in the target organ or peroxidases in activated circulating polymorphonuclear leukocytes, monocytes, or eosinophils also metabolize drugs or carcinogens to cytotoxic or genotoxic products. Furthermore, polymorphonuclear leukocytes, monocytes and eosinophils readily infiltrate the tissues of patients with a variety of infectious disorders, allergies, autoimmune disorders and tumours. They also infiltrate tissues associated with inflammatory disease where they become activated: e.g. rheumatoid arthritis, myocardial reperfusion injury, respiratory distress syndrome, ulcerative colitis, alcoholic liver cirrhosis. After infiltration, these cells become activated by perturbation of the plasma membrane by bacteria bodies (e.g. the chemotactic *N*-formylmethionyl-leucyl-phenyl-alanine), chemotactic complement fragments, leukotrienes, platelet-activating factors, immobilized immune complexes, or contact with opsonized bacteria. A cyanide-resistant respiratory burst ensues, in which oxygen is converted into hydrogen peroxide and degranulation resulting in peroxidase release occurs (O'Brien, 1988b). Eosinophils are not a striking feature of the acute inflammatory response but are involved in allergic and immune responses and accumulate in areas of helminthic infestations. Extensive degranulation of eosinophils and deposition of eosinophil peroxidase also occurs in certain pathological conditions such as Hodgkin's disease, chronic myelogenous leukaemia, atopic dermatitis and endomyo-cardial fibrosis (Samoszuk *et al.*, 1988). Anti-inflammatory drugs may act by being metabolized by the peroxidase in these cells to products which prevent the tissue destructive events in inflammatory disease. However, occasionally these products may cause adverse reactions such as bone marrow toxicity (leukopenia, agranulocytosis, neutropenia, etc.), lupus, hypothyroidism, or thyroid cancer (Uetrecht, 1988).

Chronic inflammation has been associated with the subsequent development of cancer. Thus tumour formation has also often been attributed to chronic irritation and malignant tumours occur at a variety of inflammatory sites (Gordon and Weitzman, 1988). Many tumour promoters are also powerful activators of leukocyte oxygen metabolism and thereby activate chemical carcinogens to DNA-reactive or genotoxic intermediates which contribute to the carcinogenetic process. Thus, it was shown that leukocytes, activated by the tumour promoter phorbol myristate acetate, oxidize the carcinogens [^{14}C]methylaminoazobenzene, benzidine, aminofluorene or phenol and cause the irreversible binding of these carcinogens to nuclear DNA (O'Brien *et al.*, 1985; Tsuruta *et al.*, 1985); binding is dependent on oxygen and prevented by azide and cyanide, indicating that myeloperoxidase catalyses oxidative activation of these xenobiotics. Other investigators have shown that [^{14}C]chloraniline and 4-methylaniline bind to RNA to a much greater extent than to DNA in activated leukocytes (Corbett and Corbett, 1988). Activated leukocytes also catalysed the binding of benzo(a)pyrene-7,8-dihydrodiol to exogenous

DNA (Twerdok and Trush, 1988), and [^{14}C]phenol and 1-naphthol are extensively bound to endogenous protein in activated leukocytes (Eastmond *et al.*, 1987). Phenol binding was attributed to a myeloperoxidase mechanism, but naphthol binding may also involve a superoxide-anion-radical-catalysed oxidation to 1,4-naphthoquinone.

The physiological role of mammalian peroxidases is likely to (*a*) remove toxic peroxides, (*b*) defend against microbes in saliva, milk, plasma or after phagocytosis by leukocytes, (*c*) defend against tumour cells, parasites, worms, etc. (*d*) haemolyse old erythrocytes in the spleen, and (*e*) inactivate thyroxine, oestradiol, chemotactic factors, and leukotrienes. For all these functions, halides increase the effectiveness of the peroxidase or are essential in biosynthesis (e.g. thyroxine). The halide effect can be attributed partly to the peroxidase catalysed formation of hypohalites. However, for thyroid peroxidase function or extrathyroidal peroxidase activity, iodide is essential for the iodination and coupling of the tyrosine residues involved in thyroxine formation. The blood contains approximately 0.1 M-chloride, 50–100 μM-bromide and 0.02 μM-iodide (Weiss *et al.*, 1982). Some peroxidases can utilize the halides as substrates and oxidize them to the corresponding hypohalous acids.

$$H_2O_2 + X^- + H^+ \rightarrow HOX + H_2O$$
$$(X^- = Cl^-, Br^-, I^-, \text{ and } SCN^-)$$

Their order of effectiveness is iodide > bromide > chloride. However, only myeloperoxidase and chloroperoxidase are effective at utilizing chloride. Eosinophils and lactoperoxidase probably utilize bromide *in vivo* whereas most peroxidases can utilize iodide. It has also been estimated that 0.16 mM-hypochlorite is generated in one hour by activated neutrophils (4×10^6 cells/ml) thereby making hypochlorite the primary oxygen metabolite (Grisham *et al.*, 1984). Recent evidence suggests that much of the antibacterial, antiparasitic, antiprotozoal, antitumour cell functions and proinflammatory activity of neutrophils *in vivo* is largely due to the oxidant properties of hypochlorite (Weiss, 1986). Much of this activity with eosinophils has also been attributed to the formation of hypobromite (Weiss *et al.*, 1986).

Recently studies on the metabolism of xenobiotics by activated polymorphonuclear leukocytes have also shown that halides play an important role. A wide variety of leukocyte nitrogen-containing compounds are *N*-oxidized by hypochlorite to form *N*-chloroamines, e.g. taurine, ammonium salts, NAD(P)H, cytosine, histidine (reviewed by O'Brien, 1988b). The anti-inflammatory drugs, dapsone or sulphonamides, and the antiarrhythmic drug, procainamide have also been shown to undergo a peroxidase-catalysed halide-dependent and halide-independent

N-oxidation by leukocytes to products which may be responsible for the major serious side effects of these drugs in certain individuals (Uetrecht, 1989).

A new method involving the use of isolated hepatocytes as model target cells has led to the discovery of a highly toxic and novel N-oxidation pathway for acetaminophen involving H_2O_2, mammalian peroxidases, and physiological concentrations of halides. A similar halide requirement was also recently found for the cytotoxicity of phenacetin metabolites (O'Brien *et al.*, 1990). The activation mechanism involves N-oxidation by hypohalites generated by the peroxidases.

13.3 BIOACTIVATION OF ACETAMINOPHEN AND AMINOPHENOL IN HEPATOCYTES IN THE PRESENCE OF HALIDES

As shown in Table 13.1, acetaminophen is not toxic to isolated rat hepatocytes even at a concentration of 20 mM. Furthermore,

Table 13.1 Peroxidative activation of acetaminophen in isolated rat hepatocytes

Additions	Cytotoxicity			
	60 min	*120 min*	*180 min*	*240 min*
None	15 ± 2	18 ± 2	19 ± 2	20 ± 2
Acetaminophen (20 mM)	17 ± 3	18 ± 2	19 ± 3	19 ± 3
Acetaminophen (1 mM) + HRP + H_2O_2 (1 mM)	18 ± 3	21 ± 2	22 ± 3	22 ± 4
Acetaminophen (0.5 mM) + HOCl (1 mM)	32 ± 3	50 ± 5	86 ± 7	96 ± 4
Acetaminophen (0.3 mM) + HOCl (0.3 mM) + KBr (0.3 mM)	38 ± 3	45 ± 4	66 ± 5	92 ± 4
Acetaminophen (0.3 mM) + MPO + H_2O_2 (0.3 mM) + KCl (0.1 M) + KBr (0.2 mM)	33 ± 4	44 ± 5	57 ± 7	72 ± 6
HOCl (1 mM)	15 ± 3	18 ± 3	19 ± 3	20 ± 3
HOCl (0.6 mM) + KBr (0.6 mM)	15 ± 3	18 ± 3	17 ± 3	20 ± 4
N-Acetylbenzoquinoneimine (0.3 mM)	37 ± 3	47 ± 4	63 ± 6	96 ± 4
Radical coupling products (5 mM)	15 ± 4	18 ± 4	19 ± 3	20 ± 4

Reaction conditions: isolated hepatocytes (10^6 cells/ml) in Krebs–Henseleit buffer.

acetaminophen is not cytotoxic even when oxidized with H_2O_2 and horseradish peroxidase to phenoxy radicals, dimers and polymers. Preincubation of acetaminophen with H_2O_2 and horseradish peroxidase before addition to hepatocytes is also not cytotoxic, indicating that the dimeric and polymeric products are not cytotoxic even at high concentrations. Acetaminophen is also not cytotoxic when oxidized by H_2O_2 and myeloperoxidase. However, the acetaminophen oxidation products are highly cytotoxic if formed in the presence of physiological concentrations of chloride and bromide. Acetaminophen becomes toxic when oxidized by hypochlorite particularly in the presence of physiological concentrations of bromide. It can also be seen that hypochlorite with or without bromide is not cytotoxic in the absence of acetaminophen even at higher concentrations than those used above.

To understand the cytotoxic mechanisms involved, knowledge of the reactive products is required. As shown by other investigators, peroxidase catalyses the oxidation of acetaminophen to a complex range of dimers and polymer products (Potter and Hinson, 1987a) and little quinoneimine formation is observed spectrally. Because of this, the peroxidase products of the 3,5-dimethyl-substituted analogue were investigated so as to diminish the complications of the dimerization and polymerization reactions. Spectral studies have shown that 3,5-$(CH_3)_2$-acetaminophen is oxidized by horseradish peroxidase and H_2O_2 to 2,6-$(CH_3)_2$-N-acetylbenzoquinoneimine (Mason and Fischer, 1986). As shown in Table 13.1, 3,5-$(CH_3)_2$-acetaminophen, which also induces hepatotoxicity (Fernando *et al.* 1980) becomes highly cytotoxic when oxidized by this peroxidase system or by hypochlorite. Figure 13.2(*a*) shows that one equivalent of hypochlorite oxidizes 3,5-$(CH_3)_2$-acetaminophen to 3,5-$(CH_3)_2$-N-acetylbenzoquinoneimine. However, with five equivalents of hypochlorite the 3,5-$(CH_3)_2$-N-acetylbenzoquinoneimine formed is converted into 2,6-$(CH_3)_2$-p-benzoquinone (Fig. 13.2b). However, two equivalents of hypochlorite are required to oxidize acetaminophen. The products formed include N-acetylbenzoquinoneimine, p-benzoquinone as well as ring-chlorinated products. In the absence of hypochlorite, N-acetylbenzoquinoneimine slowly hydrolyses stoichiometrically to benzoquinone and acetamide. A mechanism for this reaction involving an ipso attack by water to form an intermediate carbinolamide has been suggested (Novak *et al.*, 1986).

The presence of physiological concentrations of bromide in an acetaminophen-myeloperoxidase-chloride mixture dramatically increases the yield and stability of the N-acetylbenzoquinoneimine formed. As shown in Fig. 13.3(*a*) the presence of bromide also dramatically increases the yield and stability of N-acetylbenzoquinoneimine formed with one equivalent of hypochlorite. A nearly stoichiometric formation of N-acetylbenzoquinoneimine is found (O'Brien *et al.*, 1991). The subsequent addition of

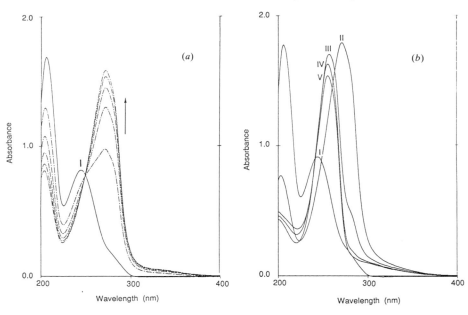

Figure 13.2 3,5-$(CH_3)_2$-Acetaminophen (I) oxidation by hypochlorite. Reaction conditions: 3,5-$(CH_3)_2$-acetaminophen (100 μM) and (a) hypochlorite (100 μM) or (b) hypochlorite (500 μM) in 3.0 ml of 50 mM-phosphate buffer, pH 7.4 at 20°C. The spectra were recorded immediately and every 2 min (a, – · –) or 5 min (b, III, IV, V) after the addition of hypochlorite.

one equivalent of hypochlorite rapidly oxidizes *N*-acetylbenzoquinoneimine to benzoquinone (Fig. 13.3*b*). Furthermore, as shown in Table 13.1, bromide dramatically increases the cytotoxicity of acetaminophen oxidized by hypochlorite or a myeloperoxidase, H_2O_2, chloride system to a level of cytotoxicity observed with *N*-acetylbenzoquinoneimine. As the cytotoxicity correlates with the extent of quinoneimine formation, it is likely that quinoneimine is responsible for the cytotoxicity. The mechanism for this remarkable effect of bromide can be attributed to the formation of hypobromite by the following reaction:

$$HOCl + Br^- \rightarrow HOBr + Cl^-$$

The hypobromite formed would be a weaker oxidizing agent because of its lower oxidation potential, but would be a stronger halogenating or bleaching agent than hypochlorite (Rice and Gomez-Taylor, 1986). The stoichiometric yield of *N*-acetylbenzoquinoneimine formed with hypobromite could be explained because hypobromite, unlike hypochlorite,

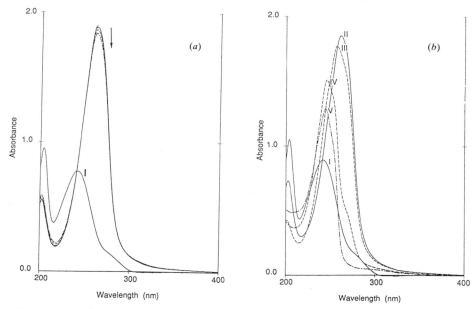

Figure 13.3 Oxidation of acetaminophen (I) by hypochlorite in the presence of bromide. (*a*) Acetaminophen (100 μM), potassium bromide (100 μM) and hypochlorite (100 μM) in 3.0 ml of 50 mM-phosphate buffer, pH 7.4, at 20°C. The spectra were recorded immediately and every 2 min after the addition of hypochlorite. (*b*) As in (*a*), except subsequent addition of one equivalent of hypochlorite after the formation of *N*-acetylbenzoquinoneimine (II) and spectra recorded every 5 min (III, IV, V).

does not oxidize *N*-acetylbenzoquinoneimine. Little quinoneimine is formed in the absence of halides.

In the kidney acetaminophen undergoes *N*-deacetylation to the nephrotoxin β-aminophenol (Newton *et al.*, 1983). As the oxidative activation of *p*-aminophenol has been implicated in acetaminophen-induced interstitial nephritis and papillary necrosis in the kidney, the effect of a peroxidase–H_2O_2 system on the cytotoxic effectiveness of *p*-aminophenol was therefore tested. As shown in Table 13.2, *p*-aminophenol, at the concentrations used, is not cytotoxic to isolated hepatocytes unless oxidized by peroxidase–H_2O_2 or hypochlorite. As shown in Fig. 13.4(*a*), the products formed with the peroxidase–H_2O_2 reaction system are identical to *p*-benzoquinoneimine and were formed in stoichiometric amounts. However, the products formed with one equivalent of hypochlorite were identified as a mixture of benzoquinoneimine and *N*-chloro-benzoquinoneimine (O'Brien *et al.*, 1991). With two equivalents of hypochlorite, however,

Table 13.2 Peroxidative activation of aminophenol in isolated hepatocytes

Additions	Cytotoxicity			
	60 min	120 min	180 min	240 min
None	15 ± 2	18 ± 2	19 ± 2	20 ± 2
4-Aminophenol (0.6 mM)	15 ± 3	18 ± 3	20 ± 3	19 ± 3
4-Aminophenol (0.3 mM) +HOCl (1 mM)	34 ± 4	55 ± 5	82 ± 7	96 ± 4
4-Aminophenol (0.2 mM) + HRP + H₂O₂	37 ± 6	64 ± 5	76 ± 8	96 ± 4
Benzoquinoneimine (0.15 mM)	31 ± 3	48 ± 3	75 ± 6	96 ± 4
Chlorobenzoquinoneimine (0.25 mM)	35 ± 4	51 ± 4	73 ± 6	96 ± 4

Reaction conditions: isolated hepatocytes (10^6 cells/ml) in Krebs–Henseleit buffer.

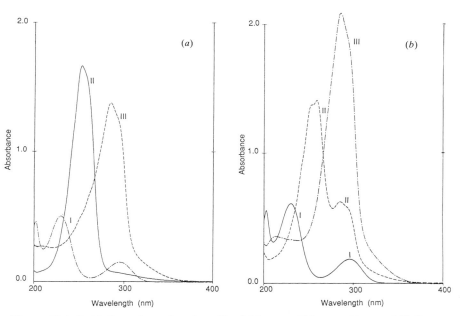

Figure 13.4 Oxidation of *p*-aminophenol by (*a*) horseradish peroxidase and H₂O₂ or (*b*) hypochlorite. Reaction conditions: (*a*) *p*-aminophenol (100 μM, I) in 3.0 ml of 50 mM-phosphate buffer, pH 7.4 at 20 °C was oxidized by horseradish peroxidase (3 μg/ml) and H₂O₂ (100 μM) to benzoquinoneimine (II) and formation of *N*-chlorobenzoquinoneimine (III) after the addition of hypochlorite (100 μM). (*b*) *p*-Aminophenol (100 μM, I) in 3.0 ml of 50 mM-phosphate buffer, pH 7.4 at 20°C with hypochlorite at 100 μM (II) or 200 μM (III).

N-chloro-benzoquinoneimine is the major product formed (Fig. 13.4(*b*)). One equivalent of hypochlorite also rapidly oxidizes benzoquinoneimine to *N*-chloro-benzoquinoneimine (O'Brien *et al.*, 1991). However, *N*-chloro-benzoquinoneimine was found to be less cytotoxic than benzoquinoneimine (Table 13.2).

13.4 CONCLUSIONS

Hepatotoxicity (centrilobular necrosis) and nephrotoxicity (renal tubular necrosis) that can occur following an acetaminophen overdose in humans and in laboratory animals has been attributed to oxidation of some of the acetaminophen through the cytochrome *P*-450-dependent mixed-function oxidase system to *N*-acetyl-*p*-benzoquinoneimine which depletes cellular glutathione stores and binds covalently to proteins. However, the lack of cytotoxicity when rat hepatocytes are incubated with high concentrations of acetaminophen indicate that cytochrome *P*-450-dependent oxidase activity is not sufficient to cause acetaminophen toxicity over the five-hour incubation period studied. Other investigators have shown that cultured or isolated hepatocytes can be susceptible to acetaminophen if prepared from rats in which hepatic *P*-450 isoenzymes have been induced by pretreatment of the animals with 3-methylcholanthrene or phenobarbital (Porubek *et al.*, 1987; Kyle *et al.*, 1988).

The marked cytotoxicity that ensues when acetaminophen is oxidized by H_2O_2 and myeloperoxidase in the presence of chloride or bromide appears to be due to an increased formation of *N*-acetylbenzoquinoneimine. Furthermore, H_2O_2 and myeloperoxidase or horseradish peroxidase in the absence of chloride or bromide does not oxidize acetaminophen to cytotoxic products, indicating that the phenoxy radical (Mason and Fischer, 1986) or radical coupling products (Potter and Hinson, 1987b) are not cytotoxic. Thus no cytotoxicity is observed if the hepatocytes are present during the peroxidase catalysed oxidation of acetaminophen even though most of the acetaminophen is oxidized. The stable radical-coupling product dimers or 'melanin' polymers are also not cytotoxic. The oxidation of acetaminophen to *N*-acetylbenzoquinoneimine when the peroxidase oxidizing system contains chloride and bromide can be attributed to a change from a mechanism involving a one electron oxidation of acetaminophen to a peroxidase catalysed two electron oxidation of acetaminophen by hypobromite formed by a peroxidase catalysed oxidation of chloride/ bromide. Evidence for this is that methionine, a hypohalite trap, prevents acetaminophen oxidation by the peroxidase–H_2O_2–halide system.

Harvison *et al.* (1988) proposed that cytochrome *P*-450 oxidizes the amide nitrogen of acetaminophen by a one-electron abstraction followed by rapid recombination of the radical pair to form a cytochrome *P*-450 haem site-bound hydroxamic acid, which dehydrates to generate *N*-acetyl-*p*-benzoquinoneimine. Peroxidases, however, probably catalyse a one-electron oxidation of the phenolic oxygen to form a phenoxy radical which disproportionates to *N*-acetyl-*p*-benzoquinoneimine (Potter and Hinson, 1987a). In Fig. 13.5(*a*), a mechanism for the oxidation of acetaminophen by hypobromite is outlined. The first step involves an attack on the acetaminophen amide nitrogen by hypobromite to form an unstable intermediate, *N*-bromoacetaminophen. This then eliminates HBr to give *N*-acetyl-*p*-benzoquinoneimine as the major product. Benzoquinone may then be formed by further oxidation of this product with elimination of acetamide. In Fig. 13.5(*b*) a similar mechanism for the oxidation of *p*-aminophenol by hypochlorite to *p*-benzoquinoneimine is outlined. The absence of an acetyl group on this quinoneimine renders it susceptible to *N*-chlorination by excess hypochlorite to give a stable *N*-chlorobenzoquinoneimine.

Peroxidases in the liver seem to be located in the endoplasmic reticulum and nuclear envelope of the phagocytic reticuloendothelial cells (Kupffer cells) which line the bile ducts in the liver (Fahimi *et al.*, 1976) and form as

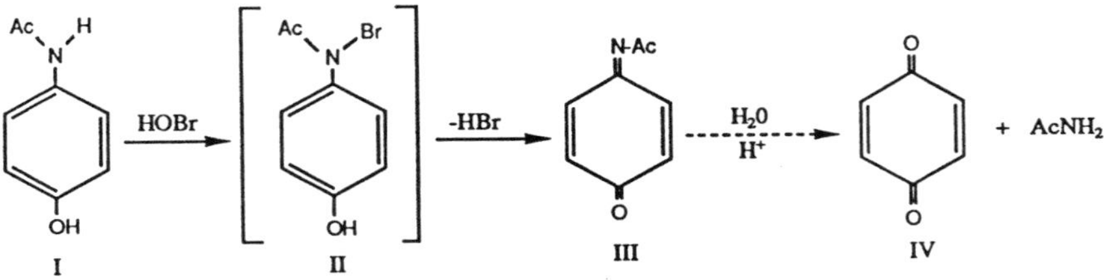

(*a*)

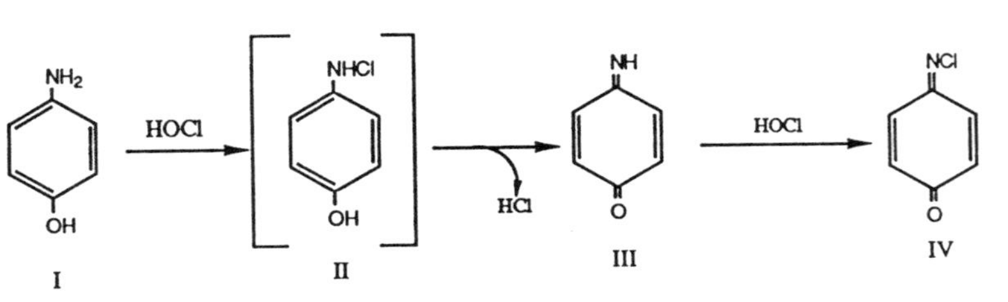

(*b*)

Figure 13.5 (*a*) Mechanism of oxidation of acetaminophen (I) by HOBr. Products formed were *N*-bromoacetaminophen (II); *N*-acetylbenzoquinoneimine (III) and 1,4-benzoquinone (IV). (*b*) Mechanism of oxidation of *p*-aminophenol (I) by HOCl. Products formed were *N*-chloro-*p*-aminophenol (II); benzoquinoneimine (III) and *N*-chlorobenzoquinoneimine (IV).

much as 14% of the liver. These non-parenchymal cells play an essential role as a selective filter to rapidly remove blood-borne abnormal particulate matter and gut derived antigens (e.g. endotoxin from the portal circulation), primarily through the process of phagocytosis. Kupffer cells contain 30 times higher peroxidase specific activity than the hepatocytes (Van Berkel, 1974) and could, therefore, play a role in the peroxidase catalysed activation of acetaminophen. The reduced spectra of this peroxidase is also similar to myeloperoxidase (Fahimi *et al.*, 1976) which could also indicate that the peroxidase can form hypochlorite, although it is not known whether the peroxidase is released from Kupffer cells.

However, mononuclear phagocytes with the ability to release peroxidases can be recruited into the liver and activated by bacterial, viral or parasitic infections, lipopolysaccharide endotoxins, or by autoimmune responses to liver antigens (Ferluga and Allison, 1978). This may arise because damaged cells in the liver release chemotactic and activating factors towards polymorphonuclear leukocytes, Kupffer cells and monocytes (Laskin *et al.*, 1986), particularly if sensitized by previous infections. In response to these inflammatory stimuli, the peroxidase-containing mononuclear phagocytes migrate into the liver from the blood and bone marrow and become activated. These cells have more peroxidase than resident Kupffer cells. They are also 10–15 times more phagocytic when activated, release 30% more hydrogen peroxide and other activated oxygen species than resident Kupffer cells (Pilaro and Laskin, 1986; Laskin *et al.*, 1988), and may promote hepatic damage, e.g. hepatitis, through the release of toxic secretory products (Ferluga and Allison, 1978). Laskin *et al.* (1986) have proposed that the release of activated oxygen species contributes to hepatic injury, tumour promotion and carcinogenesis.

Treatments of rats with acetaminophen results in the accumulation of activated mononuclear phagocytes in the centrilobular regions of the liver within 24 hours in the absence of necrosis. Levels of migration of mononuclear phagocytes were 4–7 times greater than the migration of resident Kupffer cells. Furthermore, cultured hepatocytes treated with acetaminophen released chemotactic and activating factors towards Kupffer cells and monocytes (Laskin *et al.*, 1986). In mice, Jaeschke and Mitchell (1988) have found a rapid 15-fold increase in neutrophil accumulation between 1.5 and 3 hours post-dose, i.e. after glutathione depletion and macromolecule alkylation. They suggested this increased the ensuing necrosis as a result of plugging the hepatic microvasculature and causing ischaemic infarction. A variety of other agents including ethanol, phenobarbital, carbon tetrachloride, halothane, phenytoin, and para-aminosalicylate also induce the accumulation of mononuclear phagocytes in the periportal region but, unlike acetaminophen, this is accompanied by the induction of zone-specific injury or hepatitis. Phenobarbital or endotoxin also induce accumulation of activated mononuclear phagocytes in the liver (Laskin *et al.*, 1988; Pilaro and Laskin, 1986). The increased susceptibility of alcoholics to acetaminophen toxicity (McClain *et al.*, 1980; Licht *et al.*, 1980) has been attributed to decreased hepatic glutathione levels (Lauterburg and Veley, 1988) or to the induction of a *P*-450 isoenzyme that activates acetaminophen (Sato *et al.*, 1981). However, distinctive cytological features of alcoholic hepatitis is the infiltration of polymorphonuclear leukocytes associated with necrotic and Mallory-body-containing liver cells (MacSween, 1981). Infiltrated or recruited mononuclear phagocytes could be highly effective in causing liver necrosis as a result of oxidation of acetaminophen by hypohalites.

Idiosyncratic or adverse drug reactions such as drug-induced hepatic damage, lupus syndrome, agranulocytosis, fever, lymphadenopathy, or skin rash, are believed to be hypersensitivity reactions involving the immune system because prior exposure to the drug markedly enhances the symptoms when the person is re-exposed to the drug. Alternatively, a lag period of more than a week between starting the drug and toxicity development may exist (Uetrecht, 1989). Another characteristic is the presence of eosinophilia. Reactive metabolites released by hepatocytes would be expected to bind to the cell membrane of neighbouring cells and induce antibody formation. It is possible that the antibodies bind to hepatocyte cell membranes and lead to T-cell-mediated hepatocyte destruction. However, activated eosinophils and mononuclear phagocytes in the liver at the time of drug re-exposure could lead to the potentially highly toxic activation of drug by hypohalites.

Acetaminophen or phenacetin abuse in humans is also associated with interstitial nephritis and papillary necrosis in the kidney. One toxic mechanism proposed is the *N*-deacetylation of acetaminophen to *p*-amino-phenol (Newton *et al.*, 1983), as *p*-aminophenol was previously found to cause extensive renal tubular necrosis and interstitial nephritis (Calder *et al.*, 1975). Data presented in this review demonstrate that the cytotoxicity of *p*-aminophenol is markedly enhanced by oxidation to *p*-benzoquinoneimine by peroxidase–H_2O_2 or hypochlorite. Furthermore, prostaglandin hydroperoxidase has been located in the kidney medulla (Zenser *et al.*, 1979) and prostaglandin hydroperoxidase has been shown readily to oxidize *p*-aminophenol (Josephy *et al.*, 1983).

Recently, Mayeno *et al.* (1989) have shown that the halogenating ability of human eosinophils is markedly hampered in the absence of bromide. The unique features of eosinophil peroxidase enables eosinophils preferentially to utilize bromide to generate a brominating agent in the presence of at least a 1000-fold excess of chloride. This has been suggested as an explanation for their effectiveness in destroying parasitic worms and damaging host tissues in inflammatory disease (Weiss *et al.*, 1986). It was also suggested that bromide deficiency, as may occur in patients receiving long-term total parenteral nutrition, may decrease eosinophil effectiveness (Mayeno *et al.*, 1989). Myeloperoxidase has different properties so that polymorphonuclear leukocytes do not preferentially utilize bromide. However, the results presented here suggest that polymorphonuclear leukocytes, by generating hypochlorite in the presence of physiological concentrations of bromide, could oxidize acetaminophen stoichiometrically to the highly cytotoxic *N*-acetylbenzoquinoneimine.

In conclusion, acetaminophen is far more cytotoxic if *N*-oxidized by hypohalites generated by mammalian peroxidases from halides. Although the target organs, liver and kidney, contain cells with endogenous peroxidases, the source of oxidizing hypohalite is more likely to be invading or migrating eosinophils, polymorphonuclear leukocytes, monocytes or macrophages.

REFERENCES

Boyd, J.A. and Eling, T.E. (1981) Prostaglandin endoperoxide synthetase dependent cooxidation of acetaminophen to intermediates which covalently bind to rabbit renal medullary microsomes. *J. Pharmacol. Exp. Ther.*, **219**, 659–64.

Boyd, J.A. and Eling, T.E. (1985) Metabolism of aromatic amines by prostaglandin H synthase. *Environ. Health Perspect.*, **64**, 45–51.

Calder, I.C., Williams, P.J., Woods, R.A. *et al.* (1975) Nephrotoxicity and molecular structure. *Xenobiotica*, **5**, 303–7.

Corbett, M.D. and Corbett, B.R. (1988) Nucleic and binding of arylamines during the respiratory burst of human gramlocytes. *Chem. Res. Toxicol.*, **1**, 356–63.

Dahlin, D.C., Miwa, G.T., Lu, A.Y.H. and Nelson, S.D. (1984) *N*-Acetyl-*p*-benzoquinoneimine: a cytochrome *P*-450 mediated oxidation product of acetaminophen. *Proc. Natl. Acad. Sci.*, **81**, 1327–31.

Eastmond, D.A., French, R.C., Ross, D. and Smith, M.T. (1987) Metabolic activation of 1-naphthol and phenol by human leukocytes. *Chem.-Biol. Interact.*, **63**, 47–62.

Fahimi, H.D., Gray, B.A. and Herzog, V.K. (1976) Cytochemical localization of catalase and peroxidase in sinusoidal cells of rat liver. *Lab. Invest.*, **34**, 192–201.

Ferluga, J. and Allison, A. (1978) Role of mononuclear infiltrating cells in pathogenetics of hepatitis. *Lancet*, **ii**, 610–11.

Fernando, C.R., Calder, I.C. and Ham, K.N. (1980) Studies on the mechanism of toxicity of acetaminophen. *J. Med. Chem.*, **23**, 1153–8.

Gordon, L.I. and Weitzman, S.A. (1988) *The Respiratory Burst and Carcinogenesis and its Physiological Significance* (eds A.J. Sbarra and R.R. Strauss), Plenum Press, New York, pp. 277–98.

Grisham, M.B., Jefferson, M.M., Melton, D.F. and Thomas, E.L. (1984) Chlorination of endogenous amines by isolated neutrophils. *J. Biol. Chem.*, **259**, 10404–12.

Hamel, A.M. and Lands, W.E.M. (1982) Modification of anti-inflammatory drug effectiveness by ambient lipid peroxides. *Biochem. Pharmacol.*, **31**, 3307–11.

Harvison, P.J., Guengerich, F.P., Rashed, M.S. and Nelson, S.D. (1988) Cytochrome *P*-450 isozyme selectivity in the oxidation of acetaminophen. *Chem. Res. Toxicol.*, **1**, 47–52.

Jaeschke, H. and Mitchell, J.R. (1988) Acetaminophen toxicity. *N. Engl. J. Med.*, **319**, 1601–2.

Josephy, D. (1988) Activating aromatic anines by prostaglandin synthetase. *Free Rad. Biol. Med.*, **6**, 533–42.

Josephy, D., Eling, T. and Mason, R.P. (1983) Oxidation of *p*-aminophenol by prostaglandin synthetase. *Mol. Pharmacol.*, **23**, 461–6.

Kyle, M.E., Nakae, D., Serroni, A. and Farber, J.L. (1988) Chloroethyl nitrosourea potentiates the toxicity of acetaminophen in cultured hepatocytes. *Mol. Pharmacol.*, **34**, 584–9.

Laskin, D.L., Robertson, F.M., Pilaro, A.M. and Laskin, J.D. (1988) Activation of liver macrophages following phenobarbital treatment of rats. *Hepatology*, **8**, 1051–5.

Laskin, D.L., Pilaro, A.M. and Sungchul, J. (1986) Potential role of activated macrophages in acetaminophen hepatotoxicity. *Toxicol. Appl. Pharmacol.*, **86**, 216–26.

Lauterburg, B.H. and Veley, M.E. (1988) Glutathione deficiency in alcoholics: risk factor for paracetamol hepatotoxicity. *Gut*, **29**, 1153–7.

Licht, H., Seeff, I.B. and Zimmermann, H.J. (1980) Apparent potentiation of acetaminophen hepatotoxicity by alcohol. *Ann. Intern. Med.*, **92**, 511.

MacSween, R.N.M. (1981) Alcoholic liver disease: morphological manifestations: review by an international group. *Lancet*, 707.

Markey, C.M., Alward, A., Weller, P.E. and Marnett, L.J. (1987) Quantitative studies of hydroperoxide reduction by prostaglandin H synthetase. *J. Biol. Chem.*, **262**, 6266–79.

Mason, R.P. and Fischer, V. (1986) Free radicals of acetaminophen: their subsequent reactions and toxicological significance. *Fed. Proc.*, **45**, 2493–9.

Mayeno, A.N., Curran, A.J., Roberts, R.L. and Foote, C.S. (1989) Eosinophils preferentially use bromide to generate halogenating agents. *J. Biol. Chem.*, **264**, 5660.

McClain, C.J., Kromhout, J.P., Peterson, F.J. and Holtzman, J.L. (1980) Potentiation of paracetamol hepatotoxicity by alcohol. *J. Am. Med. Assoc.*, **244**, 251–3.

Meunier, B. (1987) Horseradish peroxidase: a useful tool for modeling the extra-hepatic biooxidation of oxogens. *Biochimie*, **69**, 3–9.

Miller, J.A. (1970) Carcinogenesis by chemicals: an overview. *Cancer Res.*, **30**, 559–70.

Mohandas, J., Duggin, G.G., Horvath, J.S. and Tiller, D.J. (1981) Metabolic oxidation of acetaminophen mediated by cytochrome *P*-450 mixed function oxidase and prostaglandin endoperoxide synthetase in rabbit kidney. *Toxicol. Appl. Pharmacol.*, **61**, 252–9.

Moldeus, P. and Rahimtula, A. (1980) Metabolism of paracetamol to a glutathione conjugate catalysed by prostaglandin synthetase. *Biochem. Biophys. Res. Commun.*, **96**, 469–75.

Moldeus, P., Andersson, B., Rahimtula, A. and Berggren, M. (1982) Prostaglandin synthetase catalysed activation of paracetamol. *Biochem. Pharmacol.*, **31**, 1363–70.

Newton, J.F., Juo, C.H., Yoshimoto, M. and Hook, J.B. (1983) Acetaminophen nephrotoxicity in the rat. Renal metabolic *in vitro*. *Toxicol. Appl. Pharmacol.*, **70**, 433–44.

Novak, M., Pelecanou, M. and Pollack, L. (1986) Hydrolysis of the model carcinogen N-pivaloyloxy. 4:methooxyacetanilide: involvement of N-acetyl p-benzoquinoneimine. *Am. Chem. Soc.*, **108**, 112–20.

O'Brien, P.J. (1984) Multiple mechanisms for the metabolic activation of carcinogenic arylamines in *Free Radicals in Biology* VI (ed. W.A. Pryor), Academic Press, London, New York, pp. 289–322.

O'Brien, P.J. (1985) Free-radical mediated DNA binding. *Env. Health Perspect.*, **64**, 219–32.

O'Brien, P.J. (1988a) Radical formation during the peroxidase catalyzed metabolism of carcinogens and xenobiotics. *Free Radical Biol. Med.*, **4**, 169–83.

O'Brien, P.J. (1988b) Oxidants formed by the respiratory burst: their physiological role and their involvement in the oxidative metabolism and activation of drugs, in *The Respiratory Burst and Its Physiological Significance* (eds A.J. Sbarra and R.R. Strauss), Plenum Press, New York, pp. 203–32.

O'Brien, P.J., Khan, S., Jatoe, S. and McGirr, L.G. (1990) Activation of xenobiotics by hypohalites or peroxidases, H_2O_2 and halide, in *Biological Oxidation Systems* (eds C.C. Reddy, G.A. Hamilton and K.M. Madyastha), Academic Press, New York.

O'Brien, P.J., Gregory, B., Fanney, R. et al. (1985) *Microsomes and Drug Oxidations* (eds A.R. Boobis, J. Caldwell, F. de Matteis and C.R. Elcombe), Taylor and Francis, London, pp. 100–12.

O'Brien, P.J., Khan, S. and Jatoe, S.D. (1991) Formation of biological reactive intermediates by peroxidases: halides mediated acetaminophen oxidation and cytotoxicity, in *Biological Reactive Intermediates* IV (eds C.M. Witmer, R. Snyder, D. Jollow, G.F. Kalf, J.J. Koksis and I.G. Snipes), Plenum Publication Corporation, New York, pp. 51–64.

Pilaro, A.M. and Laskin, D.L. (1986) Accumulation of activated mononuclear phagocytes in the liver following lipopolysaccharide treatment of rats. *J. Leukocyte Biol.*, **40**, 29–41.

Porubek, D.J., Rundgren, M., Harrison, P.J. et al. (1987) Investigation of mechanisms of acetaminophen toxicity in isolated rat hepatocytes with acetaminophen analogues. *Mol. Pharmacol.*, **31**, 647–53.

Potter, D.W. and Hinson, J.A. (1987a) Mechanisms of acetaminophen oxidation to N-acetyl-p-benzoquinoneimine by horseradish peroxidase and cytochrome P-450. *J. Biol. Chem.*, **262**, 966–73.

Potter, D.W. and Hinson, J.A. (1987b) The one and two-electron oxidation of acetaminophen catalysed by prostaglandin H synthase. *J. Biol. Chem.*, **262**, 974–80.

Prescott, L.G. (1983) Paracetamol overdosage. *Drugs*, **25**, 290–314.

Rice, R.G. and Gomez-Taylor, M. (1986) Occurrence of by-products of strong oxidants reacting with drinking water contaminants. *Environ. Health Perspect.*, **69**, 31–44.

Ross, D., Norbeck, K. and Moldeus, P. (1985) The generation and subsequent fate of glutathionyl radicals in biological systems. *J. Biol. Chem.*, **260**, 15028–32.

Samoszuk, M.K., Gidanian, F. and Rietveld, C. (1988) Cytotoxic properties of human eosinophil peroxidase plus major basic protein. *Am. J. Pathol.*, **132**, 455–60.

Sato, C., Matsuda, Y. and Lieber, C.S. (1981) Increased hepatotoxicity of acetaminophen after chronic ethanol consumption in the rat. *Gastroenterology*, **80**, 140–8.

Smilkstein, M.J., Knapp, G.L., Kulig, K.W. and Rumack, B.H. (1988) Efficacy of oral N-acetylcysteine in the treatment of acetaminophen overdose. *N. Engl. J. Med.*, **319**, 1557–62.

Tsuruta, Y., Subrahmanyam, V.V., Marshall, W. and O'Brien, P.J. (1985) Peroxidase mediated irreversible binding of arylamine carcinogens to DNA intact polymorphonuclear leukocytes activated by a tumour promoter. *Chem.-Biol. Interact.*, **53**, 25–35.

Twerdok, L.E. and Thrush, M.A. (1988) Neutrophil-derived oxidants as mediators of chemical activation in bone marrow. *Chem. Biol. Interact.*, **65**, 261–73.

Uetrecht, J.P. (1988) Drug-induced agranulocytosis and other effects mediated by peroxidases during the respiratory burst. *The Respiratory Burst and Its Physiological Significance* (eds A.J. Sbarra and R.R. Strauss), Plenum Press, New York, pp. 233–44.

Uetrecht, J.P. (1989) Idiosyncratic drug reactions: possible role of reactive metabolites generated by leukocytes. *Pharm. Res.*, **6**, 265–73.

Van Berkel, T.J. (1974) Difference spectra, catalase and peroxidase activities of isolated parenchymal and non-parenchymal cells from rat liver. *Biochem. Biophys. Res. Commun.*, **61**, 204–9.

Weiss, S.J. (1986) Chlorinated oxidants generated by leukocytes. *Adv. Free Radical Biol. Med.*, **2**, 91–106.

Weiss, S.J., Klein, R., Slivka, A. and Wei, M. (1982) Chlorination of taurine by human neutrophils. *J. Clin. Invest.*, **70**, 598–607.

Weiss, S.J., Test, S.T., Eckmann, C.M. *et al.* (1986) Brominating oxidants generated by human eosinophils. *Science*, **234**, 200–3.

Yamazoe, Y., Miller, D.W., Weiss, C.C. *et al.* (1985) DNA adducts formed by ring-oxidation of the carcinogen 2-naphthylamine with prostaglandin H synthase *in vitro* and in dog urothelium *in vivo*. *Carcinogenesis*, **6**, 1379–87.

Yamazoe, Y., Zenser, T.V., Miller, D.W. and Kadlubar, F.F. (1989) The benzidine: DNA adduct formed by peroxidase and H_2O_2. *Carcinogenesis*, **9**, 1635–41.

Zenser, T.V., Mattamal, M.B. and Davis, B.B. (1979) Cooxidation of benzidine by renal medullary prostaglandin cyclooxygenase. *J. Pharmacol. Exp. Ther.*, **211**, 460–5.

PART 3

Reductions and Conjugations of *N*-Oxygenated Compounds

14

Reduction and conjugation reactions of N-oxides

L.A. Damani

Chelsea Department of Pharmacy, King's College London, Manresa Road, London SW3 6LX, UK

1. This chapter is an overview on the further metabolism of N-oxygenated metabolites of tertiary amines, e.g. N-oxides.
2. There are three options open to N-oxygenated intermediates – further oxidation, reductions and conjugations, all of which may either be enzymic or non-enzymic. N-Oxides undergo enzymic reductions, and some are conjugated. Additionally they undergo non-enzymic rearrangements, even under physiological conditions.
3. Reductive metabolism of N-oxides *in vivo* may lead to futile recycling of the parent compound. Nitrogenous compounds that undergo this reversible oxidation/reduction may, therefore, have long half-lives. Some drugs may be designed as N-oxygenated pro-drugs (e.g. N-oxides, nitro compounds), that undergo selective bioreductions (activation) under anaerobic conditions in appropriate target tissues (e.g. tumours).
4. Conjugation of certain types of N-oxygenated compounds (e.g. hydroxamic acids) is now well established, and such reactions are known to confer unusual reactivity on the products formed towards cellular macromolecules. More recently, novel N-oxide conjugations (O-sulphations) have also been described for some heteroaromatic N-oxides; the implications of such reactions to drug pharmacology and toxicology are discussed, although it is not as yet clear whether this is a general route of metabolism for all types of N-oxides.

14.1 INTRODUCTION

In the biological *oxidation* of nitrogen functionalities in xenobiotics, two processes are discernible – those where there is removal of electrons and protons, and those where there is addition of oxygen. This chapter is restricted to the further metabolism of products of the latter process, i.e. *N*-oxygenated compounds, in particular the *N*-oxides. Such compounds may be produced as oxygenated metabolites of nitrogenous compounds, e.g. *N*-oxides, hydroxylamines, hydroxamic acids etc. (Damani, 1982; Hlavica, 1982; Gorrod and Damani, 1985; Cho and Lindeke, 1988). Alternatively, *N*-oxo functionalities may be present in synthetic or biosynthetic drug and other xenobiotics molecules (Jenner, 1978). The presence of an *N*-oxo group in a drug molecule may be a deliberate design feature, with bioreduction in appropriate target tissues as an activation process. Whereas a considerable amount of data has now accumulated on nitrogen oxidation (see Bridges *et al.*, 1972; Gorrod, 1978; Gorrod and Damani, 1985 and references cited therein), the further metabolism of the *N*-oxygenated intermediates has not been studied as extensively or systematically. In particular, the reversible nature of certain *N*-oxygenations under physiological conditions *in vivo* has not been fully established. However, a clear understanding of the role of *N*-oxidation of nitrogenous compounds *in vivo* is only possible if the contribution of metabolically reversible reactions is fully appreciated.

The metabolic options available to any *N*-oxygenated compound are determined by the chemical nature of the functional group and the oxidation state of the constituent nitrogen. In general there are three options; (1) further oxidation to an *N*-oxygenated metabolite with the nitrogen in a higher oxidation state, e.g. oxidation of an arylhydroxylamine (Ar-NHOH, oxidation state -1) to an arylnitroso metabolite (Ar-NO, oxidation state $+1$); (2) reduction of the *N*-oxygenated functionality, e.g. reduction of *N*-oxides (oxidation state -1) to the parent tertiary amines (oxidation state -3); (3) conjugation at the *N*-oxygenated groups, e.g. *N-O*-glucuronidation, or direct *N*-glucuronidation of arylhydroxylamines. This review only addresses the latter two options, i.e. reductions and conjugations of *N*-oxides. These further transformations may in some instances be non-enzymic, occurring *in vivo* or *in vitro* or during sample storage/analysis due to the instability of the *N*-oxygenated intermediate, e.g. non-enzymic aliphatic *N*-oxide reductions during sample analysis (Chapter 3). Even when these reactions are *biotransformations*, they need not always be mediated by mammalian enzyme systems. In many instances the microbial flora in the gastrointestinal tract plays a significant role in the bioreduction of *N*-oxygenated compounds. Reductions and conjugations, and subsequent hydrolysis of the conjugates by microflora, afford the opportunity for metabolic recycling. These reversible processes undoubtedly limit the elimination of certain drugs from the body and contribute to their long half-lives.

This review examines the reduction and conjugation reactions of tertiary aliphatic and heteroaromatic N-oxides. The enzymic basis for these reactions are discussed with respect to published *in vitro* data, and evidence for the occurrence of such reactions *in vivo* is presented. The implications of such reactions to drug pharmacology and toxicology are discussed. The further metabolism of hydroxylamines, hydroxamic acids, nitroso and nitro compounds, oximes and nitrones is not addressed here, but is reviewed by Hlavica (1982) and Cho and Lindeke (1988).

14.2 FURTHER TRANSFORMATIONS OF N-OXIDES

14.2.1 Metabolic options open to N-oxides

N-Oxides are N-oxygenated products of tertiary amines, where the nitrogen lone pair of electrons is in the bonding orbital linking the nitrogen and oxygen atoms. There are essentially four types of N-oxides (Fig. 14.1): (1) tertiary aliphatic amine N-oxides, e.g. trimethylamine N-oxide; (2) N,N-dialkylarylamine N-oxides, e.g. N,N-dimethylaniline N-oxide; (3) heteroaromatic N-oxides, e.g. pyridine N-oxide; and finally (4) imino-N-oxides (or nitrones), e.g. methaquolone N-oxide. Further transformations of N-oxides are often dependent on which of these classes they belong to (see Fig. 14.2). For example, tertiary aliphatic and N,N-dialkylarylamine N-oxides are usually much less stable, chemically and metabolically, than the heteroaromatic N-oxides. In view of their lability, such N-oxides are readily reduced back to the parent amine, often non-enzymically during sample storage, workup and analysis (Damani, 1985; Chapter 3). N-Oxide reduction can also be enzymic, both *in vitro* and *in vivo*, and can be mediated by microbial and mammalian enzyme systems. N-Oxides have been reported to undergo chemical transformation reactions

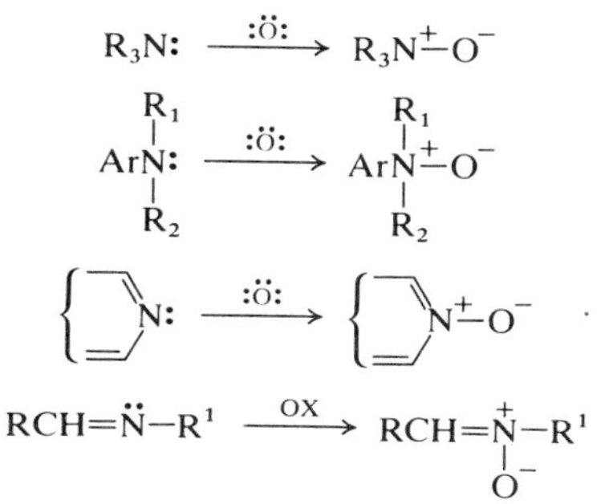

$$R_3N: \xrightarrow{\;:\ddot{O}:\;} R_3\overset{+}{N}{-}O^-$$

$$\overset{\displaystyle R_1}{\underset{\displaystyle R_2}{ArN:}} \xrightarrow{\;:\ddot{O}:\;} \overset{\displaystyle R_1}{\underset{\displaystyle R_2}{Ar\overset{+}{N}{-}O^-}}$$

$$RCH{=}\ddot{N}{-}R^1 \xrightarrow{\;OX\;} RCH{=}\overset{+}{N}{-}R^1 \atop O^-$$

Figure 14.1 N-Oxygenation of different types of tertiary amines to amine N-oxides.

— Reductions

$$R_3N^{\pm}O^- \xrightarrow[2H^+]{2e} R_3N\colon + H_2O$$

Chemical (non-enzymic)

Enzymic (microbial, mammalian)

— *N*-oxide conjugations

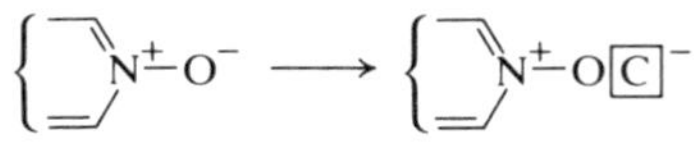

$$\left\{ \underset{=}{\overbrace{}}N^{\pm}O^- \right\} \longrightarrow \left\{ \underset{=}{\overbrace{}}N^{\pm}O\boxed{C}\, \right\}^-$$

— Rearrangements

Cope eliminations; Meisenheimers

arrangement; Polonovski reaction

Figure 14.2 Transformation options open to *N*-oxides.

(Lindeke, 1982). Cope elimination reactions may occur *in vivo* or *in vitro* under physiological conditions as further transformations of metabolically formed tertiary amine *N*-oxides (Cashman *et al.*, 1988). Certain *N*-oxides can also be converted into reactive intermediates non-enzymically in parts of the body exposed to sunlight. This type of further transformation of imino-*N*-oxides (e.g. chlordiazepoxide) is discussed by Bei Jersbergen Van Henegouwen (Chapter 26). Novel *N*-oxide sulphate conjugates have also recently been reported (Johnson *et al.*, 1982), but such *N*-*O*-sulphates appear to occur only in certain heteroaromatic *N*-oxides.

14.2.2 Reduction of *N*-oxides

(a) In vivo *reduction of* N-*oxides*

All the four classes of *N*-oxides (Fig. 14.1) can be reduced to the corresponding tertiary amines (Bickel, 1969; Hewick, 1982; Ziegler, 1988) in a variety of *in vitro* and *in vivo* test systems. *N*-Oxide reduction *in vivo* undoubtedly occurs, but the data can be difficult to interpret. The reduction may be carried out non-enzymically or enzymically by mammalian tissues, or it may occur in the gastrointestinal tract, since the gut microflora have a large capacity for reductive metabolism. Whether *N*-oxide reduction actually occurs often depends on the *N*-oxide type, the specific animal species being studied and the route of drug administration. As a general rule, tertiary aliphatic and *N*,*N*-dialkylarylamine *N*-oxides are more easily reduced, chemically and metabolically, than the heteroaromatic *N*-oxides. Another general rule is that *N*-oxide reduction is likely to be more extensive

14.1   *14.2*

if the compound is dosed orally, or if it undergoes extensive biliary excretion after parenteral dosing.

Beckett and co-workers carried out some elegant studies on the metabolism of nicotine-1'-*N*-oxide (*14.1*), which illustrate some of these general comments. Administration of this *N*-oxide intravenously to man leads to complete recovery of the intact *N*-oxide in the urine within 12 hours. However, after oral administration of nicotine-1'-*N*-oxide, reduction clearly occurs, with nicotine and cotinine detectable in urine under conditions of acidic urinary pH (Beckett *et al.*, 1970; Jenner *et al.*, 1973). These data appear to suggest that reduction of nicotine-1'-*N*-oxide in man is mediated almost entirely by the gut microflora. However, the same group have carried out studies on reduction *in vivo* of nicotine-1'-*N*-oxide in germfree and conventional rats, which indicate that *N*-oxide reduction in this species occurs both in the mammalian tissues and in the gut microflora (Dajani *et al.*, 1975a, b).

The early studies on the simplest of the tertiary aliphatic *N*-oxides, trimethylamine *N*-oxide, also clearly illustrate the general comments (section 14.2.1). Administration of this simple *N*-oxide, or the parent trimethylamine, orally or parenterally to rats, leads to a nearly quantitative recovery of the *N*-oxide in urine (Norris and Benoit, 1945). The parent amine is rapidly absorbed and extensively *N*-oxidized *in vivo*, the resultant metabolite being eliminated in the urine. The dosed *N*-oxide probably suffers some reduction in the gut (after oral dosing) or in the tissues (after oral or parenteral dosing), but the trimethylamine so produced is again extensively re-*N*-oxidized *in vivo*. This type of 'metabolic cycling' probably occurs with a large number of amine drugs (Ziegler, 1988). There is evidence of repeat oxidations/reductions, or reductions/oxidations, with a variety of tertiary amines and amine *N*-oxides, respectively.

N-Oxide reduction *in vivo* is not always a major route of metabolism, even in the case of tertiary aliphatic amine *N*-oxides. This may depend on the presence of other more metabolically vulnerable sites on the drug molecule. For example, intravenous dosing of cocaine *N*-oxide (*14.2*) to the rat (50 mg/kg) yielded ecgonine *N*-oxide methyl ester as its major urine metabolite (Misra *et al.*, 1979). This metabolite is produced by ester hydrolysis, the *N*-oxide function remaining intact during this metabolism and subsequent elimination of the ecgonine N-oxide. Other minor metabolites were cocaine (0.5%), norcocaine (1%), benzoylecgonine, ecgonine, ecgonine N-oxide, and minor amounts of unmetabolized compound.

Schmidt and Oelschläger (1989) suggested that as the basicity is reduced by N-oxidation, the distribution coefficient of N-oxides at pH 7.4 is likely to be greater than that for the amines. They suggest that N-oxides of various types should readily serve as substrates for further biotransformations. They have examined the *in vitro* metabolism of N-oxides of pyridine, a pyrrolidine derivative and some morpholine derivatives. With N-(4-fluorobenzyl)-pyrrolidine as a model the transformation of amine, amine N-oxide and lactam into one another has been investigated. Their preliminary data support the intramolecular, enzyme-catalysed transfer of oxygen (^{18}O) from nitrogen to the adjacent carbon-atom of N-(4-fluorobenzyl)-pyrrolidine-N-oxide, to afford the corresponding lactam. This would appear to be a novel reaction leading to lactam formation, and its generality in cyclic amine to cyclic lactam formation needs clarification.

Although heteroaromatic N-oxides are more stable metabolically than the tertiary aliphatic amine N-oxides, their reduction does occur to some extent. For example, oral dosing of trimethoprin 1-N-oxide (*14.3*) to rats resulted in a pattern of urinary metabolites similar to that for trimethoprin (Hubbell *et al.*, 1978). This suggests that the N-oxide can be reduced to the parent drug and recycled *in vivo*. After intravenous dosing of [^{14}C]pyridine N-oxide to rats, almost 95% of the dose was recovered unchanged in 24 hour urine. However, after oral dosing, although 90% of radioactivity was recovered in 24 hour urine, only 50% of the urinary radioactivity was the N-oxide, the remaining radioactivity being various other metabolites of pyridine (Damani, 1984 unpublished data). This suggests that hetero-aromatic N-oxides may be subject to gut microbial reduction in this species.

Overall, the *in vivo* metabolism of N-oxides has not been studied extensively or systematically. Hawes *et al.* (Chapter 15), give a list of the N-oxides that have been studied to date, with emphasis on their studies on amitriptyline N-oxide and chlorpromaine N-oxide. Ziegler (1988) has

14.3   14.4   14.5

proposed a speculative, but rather plausible, model for this metabolic cycling of amines/amine N-oxides. From data on the flavin mono-oxygenases (FMO) studies and the limited data on the N-oxide reductase(s), it would seem that tertiary amine drugs may undergo rapid N-oxidation in oxygenated tissues. The reverse reaction, i.e. N-oxide reduction, is considerably slower in these tissues. Since the free amine is usually the pharmacologically active agent, rapid N-oxidation and limited reduction of the N-oxide metabolite may be a 'metabolic buffer' that increases the therapeutic index of such tertiary amine drugs. The extent to which this type of metabolic cycling actually occurs *in vivo* needs urgent clarification, and represents a challenging pharmacokinetic problem.

(b) In vitro *reduction of* N-*oxides*

(i) REDUCTION OF HETEROAROMATIC N-OXIDES. There is ample evidence for heteroaromatic N-oxide reduction from *in vitro* data. Pyridine N-oxide is reduced by *P. lindneri* (*Thermobacterium mobile*) (Neuberg, 1954), in fermenting sucrose solutions using baker's yeast (May, 1957), and by anaerobic incubation with rat faeces (Damani, 1984 unpublished observations). Resting cells of *Escherichia coli* reduce nicotinic acid and N-oxide (*14.4*) to nicotinic acid (Tatsumi and Kanamitsu, 1961). Nicotinic acid N-oxide is utilized by certain micro-organisms which normally require nicotinic acid. Tatsumi and Kanamitsu (1961) suggest that these micro-organisms can adaptively form enzymes which reduce the N-oxide, so that the nicotinic acid formed can be utilized. Another vitamin derivative, pyridoxine N-oxide has also been shown to be reduced by yeast, and also after oral dosing to rats (Sakuragi and Kummerow, 1959, 1960). Replacement of dietary pyridoxine with the N-oxide supported the growth of rats almost efficiently as did an equivalent amount of the parent compound. This clearly suggests than an efficient reduction of the N-oxides occurs, mediated either by the gut microflora or by hepatic/extra-hepatic enzymes *in vivo* after absorption of the N-oxide. Pyridylalanine N-oxides inhibit the growth of *E. coli* (Sullivan *et al.*, 1968). 4-Pyridylalanine N-oxide was demonstrated to be the most potent isomer. This isomer was also the one most rapidly reduced to 4-pyridylalanine, the most likely *E. coli* growth inhibitor. *L. arabinosus* does not reduce any of the pyridylalanine N-oxide isomers, and interestingly the growth of these microbes is unaffected at comparable concentrations. Microbial reductions have been described for many other heteroaromatic N-oxides.

Reductions of heteroaromatic N-oxides is also mediated *in vitro* by mammalian tissues. Chaykin and Bloch (1959) were the first to demonstrate the *in vitro* reduction of nicotinamide N-oxide (*14.5*) by hog liver homogenates. The same group later purified an enzyme from hog liver that reduced nicotinamide N-oxide to nicotinamide (Murray and Chaykin, 1966a). The reductase is a metalloflavoprotein and shows dependence on NADH and some other cytosolic cofactors. This enzyme had properties similar to those of xanthine oxidase. Subsequently, xanthine oxidase prepared from liver and milk was demonstrated to catalyse the reduction of nicotinamide N-oxide (Murray and Chaykin, 1966b; Kitamura and Tatsumi, 1984), the reaction being sensitive to oxygen and inhibited by cyanide. Murray *et al.* (1966) have carried out some detailed mechanistic studies on the reduction of nicotinamide-[1-^{18}O]oxide. They have demonstrated that one mechanism of xanthine oxidase reduction was by direct oxygen transfer from the N-oxide. These authors speculate on the role of heteroaromatic N-oxides as biological oxygenating agents, in view of their ability to act as oxygen donors. The reduction of many purine N-oxides in the presence of xanthine oxidase was demonstrated by Stohrer and Brown (1969). This N-oxide reduction represents a detoxification mechanism for this class of compounds, many of which have oncogenic actions (Brown, 1971). Xanthine oxidase also mediates reduction of some aliphatic amine-N-oxides (see below). Whereas detailed studies have been carried out on the cytochrome *P*-450-mediated reduction of aliphatic N-oxides (Sugiura *et al.*, 1976), it is not clear whether heteroaromatic N-oxides are reduced by this system.

(ii) REDUCTION OF ALIPHATIC AMINE N-OXIDES. There is indirect evidence, from *in vivo* intravenous/oral N-oxide dosing studies, that aliphatic amine N-oxides are subject to microbial N-oxide reduction, e.g. nicotine-1'-N-oxide reduction in rats occurs both in the gut microflora and in the mammalian tissues (Dajani *et al.*, 1975a,b). There are numerous reports of such microbial reductions of a variety of aliphatic amine N-oxides. The reduction of this class of N-oxides *in vitro* by mammalian systems was first described as early as 1927 by Ackermann *et al.*, who demonstrated reduction of trimethylamine N-oxide with liver preparations. The interpretation of data on aliphatic amine N-oxide reduction requires care, since reduction of this class of N-oxides may be enzymic or non-enzymic and/or be carried out by gut microflora depending on the substrate and the specific animal species. For example, trimethylamine N-oxide can be reduced by Fe^{2+}, cysteine, or reduced glutathione (Ackermann *et al.*, 1927). Similarly, Fe^{2+} catalyses the reduction of chlorpromazine N-oxide (Coccia and Westerfield, 1967). Fe^{2+} ions complexed to haem can also effect N-oxide reduction; the N-oxides of N,N-dimethylaniline, N,N-dimethylaminoazobenzene and imipramine are readily reduced in whole red blood cells, or in oxygenated solutions of haemoglobin (Uehleke and Stahn, 1966; Bickel *et al.*, 1968; Kiese *et al.*, 1971). *In vivo* this erythrocyte-mediated N-oxide reduction may make an important overall contribution to N-oxide reductase activity.

The early systematic work on *N*-oxide reduction came from Bickel, (1971a; Bickel *et al.*, 1968) and a quote from one of the early reviews (Bickel, 1971b) aptly summarizes the complexities of this metabolic reaction, '*N*-oxide reduction has been reported to be enzymatic, non-enzymatic, or both; aerobic or anaerobic; heat-labile or heat-stable; dependent on, or not connected with xanthine oxidase; or dependent on various co-factors'. The studies by Dajani *et al.* (1975b) on the hepatic and extrahepatic reduction of nicotine-1'-*N*-oxide (*14.1*) in various tissues of the rat also indicated the complex nature of this metabolic route. Whereas most tissues studied had some *N*-oxide reductase activity, the activity was highest in the liver. *N*-Oxide reduction was observed in both the soluble and the microsomal fraction of rat liver and small intestine, the reductase(s) being non-specific with regard to their requirement for NADH or NADPH. From various induction, inhibition and heat inactivation studies, Dajani *et al.* concluded that nicotine-1'-*N*-oxide reduction was mediated not only by cytochrome *P*-450, but was also linked partially to NADPH-dependent flavoprotein enzymes, such as NADPH-cytochrome *P*-450 (c) reductase, or xanthine oxidase.

The work from Kato's group (Sugiura *et al.*, 1976, 1977; Kato *et al.*, 1978) was the first that resulted in the clarification of the enzymology of aliphatic amine *N*-oxide reduction. They examined the reduction of imipramine *N*-oxide (*14.6*), tiaramide *N*-oxide (*14.7*), and *N,N*-dimethylaniline *N*-oxide (*14.8*) with rat liver subcellular preparations in the presence of NADPH. Most of the reductase activity was located in the microsomal fraction, although reduction of all three substrates was clearly demonstrated in the mitochondrial, lysosomal and soluble fractions. The microsomal NADPH-dependent *N*-oxide reductase activity was inducible by phenobarbitone pretreatment, was oxygen sensitive, and inhibited by carbon monoxide, *n*-octylamine, DPEA and various other *P*-450 inhibitors, suggesting involvement of cytochrome *P*-450 in *N*-oxide reduction. A combination of NADPH and NADH was found to stimulate *N*-oxide reduction to a greater extent, prompting a closer investigation of the NADH-dependent *N*-oxide reductase activity (Sugiura *et al.*, 1977). Removal of NADH-cytochrome b_5

14.6

14.7

14.8

14.9

reductase or cytochrome b_5 from microsomes with trypsin or subtilisin reduced the NADH-dependent N-oxide reductase activity by 77%. The reaction was found to be sensitive to carbon monoxide and was inhibited 85% by antibodies against NADH-cytochrome b_5 reductase. These results suggest that reducing equivalents for N-oxide reduction are transferred from NADH to cytochrome P-450 mainly via NADH–cytochrome b_5 reductase and cytochrome b_5 in the microsomal membranes. The maximum rates of NADH-dependent reduction of imipramine N-oxide, tiaramide N-oxide and N,N-dimethylaniline N-oxide are about 50% of those seen with NADPH (Sugiura *et al.*, 1976, 1977).

Iwasaki *et al.* (1977) have provided proof of the ability of cytochrome P-450 to mediate N-oxide reduction, from studies with reconstituted systems containing highly purified rabbit liver microsomal cytochrome P-450 and NADPH–cytochrome P-450 (*c*) reductase. Cytochromes P-450 and P-448 reduced tiaramide N-oxide (*14.7*) at comparable rates, and P-450R isolated from rat, rabbit and *Pseudomonas putida* had comparable reducing activity despite widely different substrate specificities in oxidation reactions. Iwasaki *et al.* suggest that the N-oxide did not bind to the normal substrate-binding site, but interacts directly with the cytochrome haem itself. With purified enzyme systems, reduced methyl viologen was shown to be capable of replacing both NADPH and the NADPH-cytochrome P-450 reductase. Indeed, with reduced methyl viologen as electron donor, tiaramide N-oxide reduction proceeds at a 7000-fold higher rate in the absence of NADPH–cytochrome P-450 reductase. This clearly means that the reduced methyl viologen can directly reduce the P-450 haem, a step which is critical for catalytic N-oxide reduction. Kato and his colleagues have proposed a reaction mechanism based upon the sequential two-electron reduction of cytochrome P-450, in a manner analogous to the reduction of cytochrome P-450 when it is acting as a mixed-function oxidase (Fig. 14.3) (Sugiura *et al.*, 1976). The rate-limiting step in the reduction of tertiary amine N-oxides has been suggested to be the reduction of cyto-chrome P-450 (Kato *et al.*, 1978). The overall reaction mechanism for

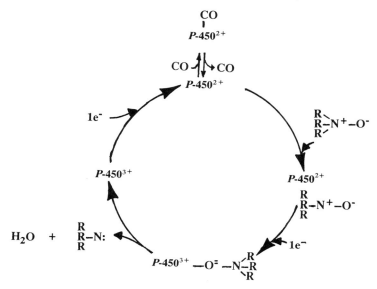

Figure 14.3 Proposed mechanism of tertiary amine *N*-oxide reduction catalysed by cytochrome *P*-450 (Sugiura *et al.*, 1976).

NADPH-dependent *N*-oxide reduction by cytochrome *P*-450 is as follows: (*a*) reduction of cytochrome *P*-450 by NADPH, via NADPH–cytochrome *P*-450 reductase; (*b*) direct co-ordination of the *N*-oxide substrate to the reduced cytochrome haem region; and (*c*) a further one-electron reduction of the reduced cytochrome *P*-450-*N*-oxide complex, followed by two-electron reduction of the tertiary amine *N*-oxide to its corresponding tertiary amine and water (Fig. 14.3).

Cytochrome *P*-450 involvement has also been demonstrated in the case of indicine *N*-oxide (*14.9*) reduction (Powis and Wincentsen, 1980) with rat liver microsomes. Indicine *N*-oxide is a pyrrolizidine alkaloid *N*-oxide with anticancer activity. As with tiaramide *N*-oxide reduction (see above), indicine *N*-oxide reduction by rat hepatic microsomal fractions occurs maximally under anaerobic conditions and is completely inhibited by carbon monoxide. However, in contrast to tiaramide *N*-oxide reduction, NADH is almost as effective as NADPH in supporting indicine *N*-oxide reduction. The NADH-dependent reduction is inhibited 48% by 0.5 mm-KCN and 45% by 0.8 m-acetone, whereas the NADPH-dependent reduction is inhibited only 3% by KCN and stimulated 28% by acetone. Phenobarbitone pretreatment produces a selective increase in the maximal rate of the NADPH-dependent *N*-oxide reduction. Powis and Wincentsen (1980) rationalized these observations in terms of a different isoenzymic form of cytochrome *P*-450 catalysing indicine *N*-oxide reduction, this isoenzyme being able to accept electrons from NADH to effect *N*-oxide reduction, but which does not contribute to oxidative microsomal drug metabolism.

In addition to the microsomal *N*-oxide reductase(s), the cytosol also appears to be a site for reduction of certain tertiary aliphatic amine *N*-oxides. Kataoka and Naito (1979) provided evidence that a purified rat liver xanthine oxidase preparation, like milk xanthine oxidase, was responsible for the xanthine-dependent reduction of benzydamine *N*-oxide. In crude rat liver preparations, the enzymic reduction seems to involve xanthine oxidase (xanthine-dependent) and/or cytochrome *P*-450 (NADPH- or NADH-dependent), because allopurinol, an inhibitor of xanthine oxidase, and *n*-octylamine, an inhibitor of cytochrome *P*-450 *N*-oxide reductase block the reduction of benzydamine *N*-oxide. Johnson and Ziegler (1986) have described a cytosolic *N*-oxide reductase for N,N-dimethylaminoazobenzene; this purified protein was shown to be a cytosolic, NADPH-dependent *N*-oxide reductase.

The total number of enzymes involved in *N*-oxide reduction is therefore probably large, but their substrate specificities, mechanisms and their relative contributions to *N*-oxide reductions *in vivo* are still not fully known. This would seem a fruitful area for future research.

14.2.3 Conjugation of *N*-oxides

(a) In vivo *conjugation of* N-*oxides*

A quantitatively important pathway of metabolism common to most aromatic azaheterocycles is *N*-oxygenation, affording water-soluble *N*-oxide metabolites (Damani and Case, 1984; Damani, 1985). Until recently, such *N*-oxides were regarded as metabolically stable, and were thought to be excreted without further modification of the *N-O*-function. However, data have accumulated over the last decade suggesting that conjugation may occur directly at the *N*-oxide oxygen in some heterocyclic *N*-oxides. *In vivo* data implicating *N-O*-glucuronidation and *N-O*-sulphation as potential metabolic routes for *N*-oxides has come from studies with amino heteroaromatic *N*-oxides, e.g. minoxidil (Fig. 14.5), and from work on the 1- and 3-*N*-oxides of pyrimethamine and metoprine (Fig. 14.4).

Hubbel *et al.* (1978) have described the urinary metabolic profiles of pyrimethamine and metoprine in rats. In the case of pyrimethamine, a major metabolite (34%) was the 3-*N*-oxide of a compound which had been hydroxylated in the α-carbon of the β-ethyl group (R^1, Fig. 14.4). The 1- and 3-*N*-oxides were also present in urine as free compounds, representing 4% and 1–4% of dose. Interestingly, β-glucuronidase treatment of the urine

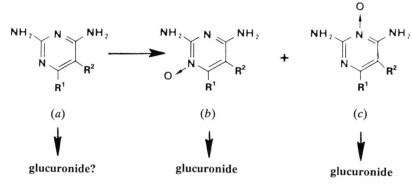

(a)   (b)   (c)

glucuronide?   glucuronide   glucuronide

Figure 14.4 Metabolic conversion of metoprine (*a*, R^1 = CH$_3$, R^2 = 3,4-dichlorophenyl) and pyrimethamine (*b*, R^1 = C$_2$H$_5$, R^2 = 4-chlorophenyl) to 1- and 3-*N*-oxides (*b* and *c*), and subsequently to glucuronide conjugates. The exact structure of the glucuronide conjugates has not been established.

resulted in increase of these two *N*-oxides, by 1.4% and 1% respectively, indicating that the *N*-oxides are at least partially further metabolized to glucuronic acid conjugates. Metoprine was also metabolized in the rat by *N*-oxidation to form the 1-*N*-oxide; the free form of this accounted for 60% of the dose, and a further 7% was present as a glucuronic acid conjugate.

The earlier studies on minoxidil (2,4-diamino-6-piperidinopyrimidine 3-*N*-oxide, Fig. 14.5) by Gottlieb *et al.* (1972) and Thomas and Harpootlian (1975) had also implicated glucuronidation as a metabolic route for this *N*-oxide drug. These authors studied the biotransformation of minoxidil in the rat, dog, monkey and human. Whereas qualitatively the urinary metabolic profiles were similar in all these species, there were major quantitative differences. In monkeys and humans, the major metabolite was 'minoxidil glucuronide', which afforded minoxidil on β-glucuronidase, treatment of urine.

The structures of the glucuronic acid conjugates of the aminopyrimidine *N*-oxides have not been fully established. The three possibilities (Fig. 14.6) are: (*a*) reaction of glucuronic acid with the *exo*-amino groups to afford *N*-glucuronides, (*b*) tautomerization of the *N*-oxide and *exo*-amino group *in vivo* to an *N*-hydroxyimine structure, followed by reaction of the glucuronic acid with the *N*-hydroxy function to afford the *N-O*-glucuronide, and (*c*) direct reaction of glucuronic acid with the *N*-oxide oxygen, to afford an *N-O* glucuronide. Although Thomas and Harpootlian (1975) favoured the amine-imine tautomerism (*b* above), this view may have to be modified in view of recent findings that minoxidil can undergo sulphation directly at the *N*-oxide function. [^{14}C]Minoxidil *N-O*-sulphate has been isolated from

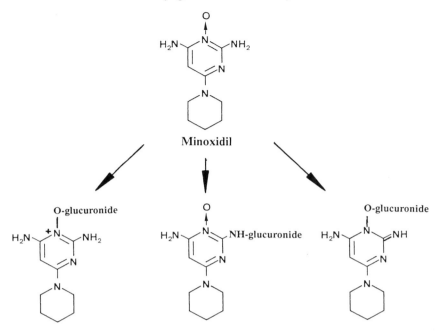

Figure 14.5 Metabolic conversion of minoxidil to a glucuronic acid conjugate, a major metabolite in monkey and man. The exact structure of the glucuronide conjugate has not been established.

the bile of rats following i.v. administration of [^{14}C]minoxidil providing another example of N-oxide conjugation *in vivo* (more details are given in section 14.2.3(*b*).

(b) In vitro *conjugation of* N-*oxides*

The only conclusive *in vitro* data on N-oxide conjugation are those for N-oxide sulphation; it is not clear whether N-oxide functions can be sites for other types of conjugations (e.g. glucuronidations, methylations) in *in vitro* test systems. The evidence for N-oxide sulphation comes from studies on minoxidil (Johnson *et al.*, 1982; McCall *et al.*, 1983). Minoxidil is a clinically effective hypotensive agent that lowers blood pressure by dilating peripheral arteriolar blood vessels. Several lines of evidence indicated that minoxidil produces a pharmacologically active metabolite, including the following facts: (*a*) the pharmacological half-life of minoxidil exceeds the serum half-life; (*b*) the onset of hypotensive effect is delayed 60–90 minutes following

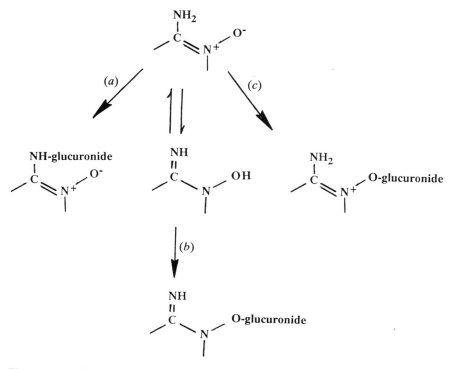

Figure 14.6 Types of glucuronide conjugates that are possible for amino azaheterocycle *N*-oxides.

drug administration; (*c*) minoxidil does not exhibit a direct relaxant effect on vascular smooth muscle. None of the minoxidil metabolites identified by Thomas and Harpootlian (1975) exhibited hypotensive activity. An observation that pretreatment of rats with paracetamol (4-acetamido-phenol), a sulphate scavenger, markedly decreased the hypotensive activity of minoxidil, suggested the formation of an active sulphate metabolite. This was conclusively demonstrated by Johnson *et al.* (1982) using *in vitro* test systems, containing enzyme source, [35S]PAPS and minoxidil. The 100 000 *g* supernatant fraction of rat liver homogenate contains a sulphotransferase activity which catalyses the *N-O*-sulphation of minoxidil. Synthetic minoxidil *N-O*-sulphate has been prepared and fully characterized, the structure being unambiguously assigned by X-ray crystallography (McCall *et al.*, 1983). The enzyme synthesized product had identical chromato-graphic (HPLC) and mass spectral characteristics and was converted into minoxidil when treated with sulphatase (Johnson *et al.*, 1982).

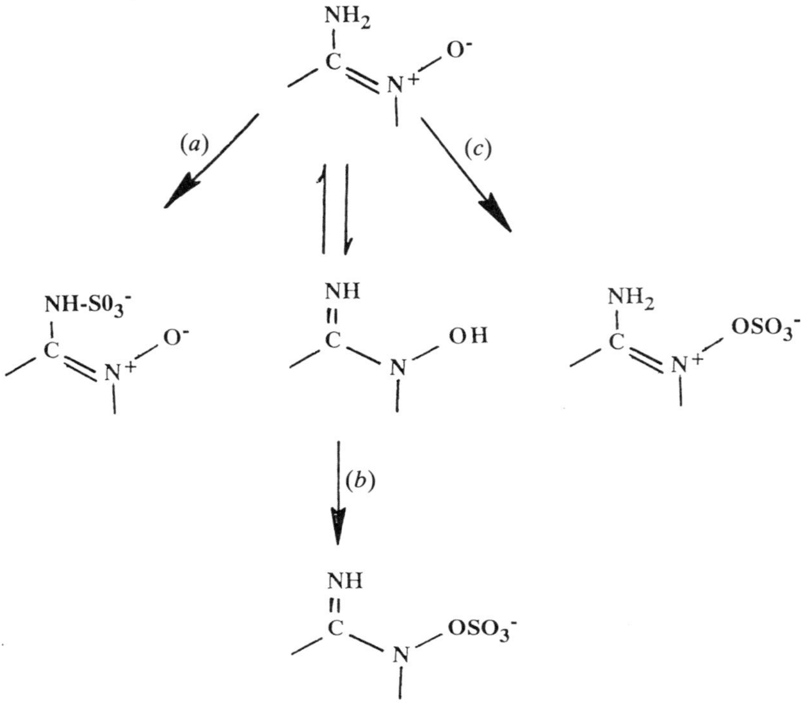

Figure 14.7 Types of sulphate conjugates that are possible for amino azaheterocycle *N*-oxides.

As in the case of glucuronidation of aminopyrimidines, there are several types of sulphation reactions possible (Fig. 14.7); (*a*) the *N*-sulphation of the *exo*-amino groups to afford sulphamic acids; (*b*) amine-imine tautomerization, followed by *N-O*-sulphation of the *N*-hydroxyimine function; (*c*) direct *N-O*-sulphation of the *N*-oxide oxygen; and (*d*) sulphation at any available electron-rich carbons, e.g. C-5 in minoxidil.

Chemically and enzymically only possibility (*c*) is realized. There are clear pharmacological implications of this type of *N*-oxide-sulphate conjugate. First, these novel conjugates are inner salts and therefore more lipophilic than their *N*-oxide precursors; drug elimination is therefore likely to be affected, with the possibility of appreciable circulating levels of these *N-O*-sulphates. These novel chemical entities may have interesting pharmacology; the *N-O*-sulphate of minoxidil appears to be a direct acting vasodilator, and may well be the active principle of minoxidil. McCall *et al.* (1983) have synthesized heterocyclic *O*-sulphates from several di- and triaminopyrimidine *N*-oxides and from 2,4-diamino-6-(diallylamino) triazine 3-oxides; these *O*-sulphates are hypotensives and apparently act by direct vasodilation.

Johnson *et al.* (1982) examined the sulphotransferase activity in rat liver 100 000 *g* supernatant in some detail, using *p*-nitrophenol and minoxidil as substrates. Whereas the pH optima for the *O*-sulphation of *p*-nitrophenol was as expected pH 6.4, the pH optima for minoxidil *N-O*-sulphation was between pH 8.0 and pH 8.5. The enzyme activity in the crude 100 000 *g* fraction was maintained for several months during storage at −76°C. In contrast, enzyme activity in partially purified preparations (0–50% ammonium sulphate precipitated fraction) was lost during storage under similar conditions. The *N*-oxide sulphotransferase activity is not limited to minoxidil. The ability of other pyrimidine, as well as pyridine-, triazine- and imidazole-*N*-oxides to serve as substrates was investigated using the soluble liver preparation and PAPS [^{35}S] (Table 14.1). The variety of structures that served as substrates indicate that the heterocyclic *N*-oxides are generally substrates for this enzyme activity. Johnson *et al.* (1982) draw attention to the fact that the extent of sulphation noted may be only an estimate of the activity, due to the possible instability of the resulting *N-O*-sulphates. To date, detailed structure–metabolism relationship studies on *O*-sulphation of azaheterocyclic *N*-oxides have not been carried out, but clearly such studies are needed before one can propose predictive rules for this unique route of metabolism.

14.3 CONCLUSIONS

This chapter overviews data which suggest that aliphatic and heteroaromatic *N*-oxides are not necessarily always metabolic end-products, but can themselves serve as substrates for further biotransformations. There are limited data at present on *N*-oxide conjugation (heteroaromatic *N*-oxide sulphation), but considerable data on *in vivo* and *in vitro* *N*-oxide reduction. Whereas *N*-oxide reduction is a major route of metabolism, particularly for aliphatic *N*-oxides, it is prudent to examine biological samples for metabolites where the *N*-oxide function has survived intact. In some cases metabolism may occur at other more metabolically vulnerable sites (e.g. cocaine *N*-oxide → ecgonine *N*-oxide methyl ester). In the case of heteroaromatic amines, *N*-oxidation clearly changes the chemical and presumably biochemical reactivity of the ring carbons. Azaheteroaromatic *N*-oxides may, therefore, serve as substrates for either cytochrome *P*-450 mediated electrophilic enzymic oxidations at electron-rich carbons, or molybdenum hydroxylase(s) mediated nucleophilic enzymic oxidations at

Table 14.1 *N*-Oxide sulphotransferase activity with heterocyclic *N*-oxides

Structure	Z	R^1	R^2	W	X	mM	% Converted
Pyrimidines (Y = N, W = CH)	A						
1 Minoxidil	0	$-NH_2$	$-NH_2$	$-CH$	Piperidino	30	7.9
2		$-NH_2$	$-NH_2$	$-CH$	Piperidino	30	0
3	0	Acetylamino	Acetylamino	$-CH$	Piperidino	10	0.4
4	0	$-NH_2$	$-NH_2$	$-CH$	4'-Hydroxypiperidino	15	1.4
						30	5.3

No.								
5		0	-NH$_2$	-CH$_3$	-CH	Morpholino	15	3.3
6		0	-NH$_2$	-NH$_2$	-C-NH$_2$	Piperidino	15	0
Pyridines (Y = W = CH)	A							
7		0	-H	-H	-CH	-OCH$_3$	10	2.8
8		0	-CH$_3$	-H	-CH	-Cl	5	4
9		0	-H	-H	-CH	-NO$_2$	10	0
10		0	-CH$_3$	-H	-CH	-NO$_2$	15	1
Triazines (Y = W = N)	A							
11*		0	-NH$_2$	-NH$_2$	N	-N-(CH$_2$CH=CH$_2$)$_2$	7.5	0.8
12*		0	-NH$_2$	p-Chlorobenzyl-	-N	-N-(CH$_2$CH=CH$_2$)$_2$	30	0.2
13*		0	-NH$_2$	-CH$_3$	N	Morpholino	30	0.8
Imidazoles	B					Morpholino		
14			CH$_3$CH$_2$-			-OH	7.5	5.6
15			-CH$_3$			-OH	1.5	15.8
16			-CH$_3$			-H	15	50
Others								
17	C						30	0
18	D						15	8.4

From Johnson et al. (1982), with permission.
* The pH of the incubation mixture was 7.5; all other compounds were tested at pH 8.0.

electron-deficient carbons. These possibilities have not as yet been examined.

In addition to the enzymic reductions and conjugates described in this review, N-oxides can also undergo non-enzymic metal-catalysed N-dealkylation, Cope elemination, Meisenheimer rearrangement and the Polonowski reaction. These post-enzymatic reactions have been fully described by Lindeke (1982). Whereas these chemical reactions should not uncritically be applied to a physiological situation, there is evidence for their occurrence under such conditions (see below), but they occur mostly during sample workup and analysis. There is evidence that amine N-oxides undergo Cope elimination reactions to produce secondary amines, hydroxylamines and olefins under physiological conditions. For example, Patterson and Gorrod (1978) proposed a Cope elimination to explain the preponderance of N-de-ethylation over N-demethylation in the *in vitro* microsomal metabolism of p-chloro N-ethyl, N-methylaniline. However, these authors only measured the secondary aniline; the secondary hydroxylamine and ethene were not assayed. Cashman *et al.* (1988) have also examined the possibility that the Cope elimination of N-oxides contributes to the overall N-dealkylation in the case of homozimeldine and verapamil N-oxides. Their work points to this alternate mechanism for N-dealkylation; flavin-containing mono-oxygenase catalysed N-oxygenation leading subsequently in some cases to non-enzymic Cope elimination products. They propose that N-dealkylation metabolites of homozimeldine, verapamil and possibly other homoallylic or homobenzylic tertiary amines arise via N-oxygenation.

The recent work on further transformation of imino–N-oxides (Chapter 26) in parts of the body exposed to sunlight opens up yet other possible routes for disposition of N-oxides. In the case of these imino-N-oxides, production of reactive oxaziridine intermediates leads to covalent binding to tissue macromolecules. It is therefore clear that non-enzymic reactions of tertiary amine N-oxides under 'physiological conditions' can often lead to novel and unanticipated products. It is hoped that in the future systematic studies on the fate of N-oxides will emerge, which more clearly define the role of both the enzymic pathways (reductions, conjugations) and non-enzymic pathways (reductions, rearrangements, Cope eliminations etc.) in the overall *in vitro* and *in vivo* transformation of different types of N-oxides.

REFERENCES

Ackermann, D., Poller, K. and Linnewen, W. (1927) Über das Verhalten des Trimethylaminoxyds im intermediären stoffwechsel als biolugischer Wasser-stoffg – Acceptor, besonders Sulfhydrylgruppen gegenuber, *Z. Biol.*, **85**, 435.
Beckett, A.H., Gorrod, J.W. and Jenner, P. (1970) Absorption of (−) nicotine-1′-N-oxide in man and its reduction in the gastrointestinal tract. *J. Pharm. Pharmacol.*, **22**, 722–3.

Bickel, M.H. (1969) The pharmacology and biochemistry of *N*-oxides. *Pharmacol. Rev.*, **21**, 325–55.

Bickel, M.H. (1971a) Liver metabolic reactions: tertiary amine *N*-dealkylation, tertiary amine *N*-oxidation, *N*-oxide reduction, and *N*-oxide *N*-dealkylation. *Arch. Biochem. Biophys.*, **148**, 54–62.

Bickel, M.H. (1971b) *N*-Oxide formation and related reactions in drug metabolism. *Xenobiotica*, **1**, 313–19.

Bickel, M.H., Weder, H.J. and Aebi, H. (1968) Metabolic interconversions between imipramine, its *N*-oxide, and its desmethyl derivative in rat tissues *in vitro*. *Biochem. Biophys. Res. Commun.*, **33**, 1012.

Bridges, J.W., Gorrod, J.W. and Parke, D.V. (eds) (1972) *Biological Oxidation of Nitrogen in Organic Molecules*. Taylor and Francis, London.

Brown, G.B. (1971) Metabolic aspects of purine *N*-oxide derivatives. *Xenobiotica*, **1**, 361.

Cashman, J.R., Proudfoot, J., Pate, D.W. and Högberg, T. (1988) Stereoselective *N*-oxygenation of zimeldine and homozimeldine by the flavin-containing monooxygenase. *Drug Metab. Dispos.*, **16**, 616–22.

Chaykin, S. and Bloch, K. (1959) The metabolism of nicotinamide *N*-oxide. *Biochim. Biophys. Acta.*, **31**, 213–16.

Cho, A.K. and Lindeke, B. (eds) (1988) *Biotransformation of Organic Nitrogen Compounds*, S. Karger, Basel.

Coccia, P.F. and Westerfeld, W.W. (1967) The metabolism of chlorpromazine by liver microsomal enzyme systems. *J. Pharmacol. Exp. Ther.*, **157**, 446.

Dajani, R.M., Gorrod, J.W. and Beckett, A.H. (1975a) Reduction *in vivo* of (−)-nicotine-1′-*N*-oxide by germ-free and conventional rats. *Biochem. Pharmacol.*, **24**, 648–50.

Dajani, R.M., Gorrod, J.W. and Beckett, A.H. (1975b) *In vitro* hepatic and extrahepatic reduction of (−)-nicotine-1′-*N*-oxide in rats. *Biochem. Pharmacol.*, **24**, 109–17.

Damani, L.A. (1982) Oxidation at nitrogen centres, in *Metabolic Basis of Detoxication* (eds W.B. Jakoby, J.R. Bend and J. Caldwell), Academic Press, New York, London, pp. 127–49.

Damani, L.A. (1985) Oxidation of tertiary heteroaromatic amines, in *Biological Oxidation of Nitrogen in Organic Molecules* (eds J.W. Gorrod and L.A. Damani), Ellis Horwood, Chichester, UK, pp. 205–18.

Damani, L.A. and Case, D.E. (1984) Metabolism of heterocycles, in *Comprehensive Heterocyclic Chemistry*, Chapter 1. 09, Vol. 1 (ed. O. Meth-Cohn), Pergamon Press, Oxford, pp. 223–46.

Gorrod, J.W. (ed.) (1978) *Biological Oxidation of Nitrogen*, Elsevier/North Holland, Amsterdam.

Gorrod, J.W. and Damani, L.A. (eds) (1985) *Biological Oxidation of Nitrogen in Organic Molecules*, Ellis Horwood, Chichester, UK.

Gottlieb, T.B., Thomas, R.C. and Chidsey, C.A. (1972) Pharmacokinetic studies of minoxidil, *Clin. Pharmacol. Ther.*, **13**, 436–40.

Hewick, D.S. (1982) Reductive metabolism of nitrogen-containing functional groups, in *Metabolic Basis of Detoxication* (eds W.B. Jakoby, J.R. Bend and J. Caldwell), Academic Press, New York, London, pp. 151–70.

Hlavica, P. (1982) Biological oxidation of nitrogen in organic compounds and disposition of *N*-oxidized products. *CRC Crit. Rev. Biochem.*, **12**, 39–101.

Hubbell, J.P., Henning, M.L., Grace, M.E. *et al.* (1978) *N*-Oxide metabolites of the 2,4-diaminopyrimidine inhibitors of dihydrofolate reductase, trimethoprim, pyrimethamine and metoprine, in *Biological Oxidation of Nitrogen* (ed. J.W. Gorrod), Elsevier/North Holland, Amsterdam, pp. 177–82.

Iwasaki, K., Noguchi, H., Kato, R. *et al.* (1977) Reduction of tertiary amine N-oxide by purified cytochrome *P-450*, *Biochem. Biophys. Res. Commun.*, **77**, 1143.

Jenner, P. (1978) Synthetic and metabolic *N*-oxidation products in centrally active pharmacological agents, in *Biological Oxidation of Nitrogen* (ed. J.W. Gorrod), Elsevier/North Holland, Amsterdam, pp. 383–98.

Jenner, P., Gorrod, J.W. and Beckett, A.H. (1973) The absorption of nicotine-1′-N-oxide and its reduction in the gastrointestinal tract in man. *Xenobiotica*, **6**, 341–9.

Johnson, G.A., Barsuhn, K.J. and McCall, J.M. (1982) Sulfation of minoxidil by liver sulfotransferase. *Biochem. Pharmacol.*, **31**, 2949–54.

Johnson, P.R.L. and Ziegler, D.M. (1986) Properties of a *N,N*-dimethyl-*p*-amino-azobenzene oxide reductase purified from rat liver cytosol. *J. Biochem. Toxicol.*, **1**, 15–27.

Kataoka, S. and Naito, T. (1979) Reduction of benzydamine *N*-oxide by rat liver xanthine oxidase. *Chem. Pharm. Bull (Tokyo)*, **27**, 2913–20.

Kato, R., Iwasaki, K. and Noguchi, H. (1978) Reduction of tertiary amine *N*-oxides by cytochrome *P-450*. Mechanism of the stimulatory effect of flavins and methyl viologen. *Mol. Pharmacol.*, **14**, 654.

Kiese, M., Renner, G. and Schlaeger, R. (1971) Mechanism of the autocatalytic formation of ferrihaemoglobin by *N,N*-dimethylaniline *N*-oxide. *Naunyn Schmiedebergs Arch. Exp. Pathol. Pharmakol.*, **268**, 247.

Kitamura, S. and Tatsumi, K. (1984) Involvement of liver aldehyde oxidase in the reduction of nicotinamide *N*-oxide. *Biochem. Biophys. Res. Commun.*, **120**, 602.

Lindeke, B. (1982) The non- and postenzymatic chemistry of *N*-oxygenated compounds. *Drug Metab. Rev.*, **13**, 71–121.

May, A. (1957) Bioreduction of pyridine *N*-oxide. *Enzymology*, **18**, 142–5.

McCall, J.M., Aiken, J.W., Chidester, C.G. *et al.* (1983) Pyrimidine and Triazine 3-oxide sulfates: a new family of vasodilators. *J. Med. Chem.*, **26**, 1791–3.

Misra, A.L., Pontani, R.B. and Vadlamani, N.L. (1979) Metabolism of norcocaine, *N*-hydroxy norcocaine and cocaine *N*-oxide in the rat, *Xenobiotica*, **9**, 189–99.

Murray, K.N. and Chaykin, S. (1966a) The enzymatic reduction of nicotinamide *N*-oxide. *J. Biol. Chem.*, **241**, 2029.

Murray, K.N. and Chaykin, S. (1966b) The reduction of nicotinamide *N*-oxide by xanthine oxidase. *J. Biol. Chem.*, **241**, 3468.

Murray, K.N., Watson, G.J. and Chaykin, S. (1966) Catalysis of the direct transfer of oxygen from nicotinamide *N*-oxide to xanthine by xanthine. *J. Biol. Chem.*, **241**, 4798.

Neuberg, C. (1954) Bioreduction of trimethylamine oxide. *Bull. Res. Coun. Isr.*, **4**, 12–20.

Norris, E.R. and Benoit, G.J. (1945) Studies on trimethylamine oxide. I. Occurrence of trimethylamine oxide in marine organisms. *J. Biol. Chem.*, **158**, 433–8.

Powis, G. and Wincentsen, L. (1980) Pyridine nucleotide cofactor requirements of indicine *N*-oxide reduction by hepatic microsomal cytochrome *P-450*. *Biochem. Pharmacol.*, **29**, 347–51.

Sakuragi, T. and Kummerow, I.A. (1959) Pyridoxine *N*-oxide. *J. Org. Chem.*, **24**, 1032.

Sakuragi, T. and Kummerow, F.A. (1960) Utilization of pyridoxine *N*-oxide by rats. *Proc. Soc. Exp. Biol. Med.*, **103**, 185.

Schmidt, N. and Oelschlager, H. (1989) *N*-Oxides as substrates for further biotransformations, Abstract No. 018, *4th Int. Sym. Biol. Oxid. Nitrogen in Org. Mol.*, Munich, Germany.

Stohrer, G. and Brown, G.B. (1969) Purine *N*-oxides XXVIII. The reduction of purine N-oxides by xanthine oxidase. *J. Biol. Chem.*, **244**, 2498–502.

Sugiura, M., Iwasaki, K. and Kato, R. (1976) Reduction of tertiary amine N-oxides by liver microsomal cytochrome P-450. *Mol. Pharmacol.*, **12**, 322–34.

Sugiura, M., Iwasaki, K. and Kato, R. (1977) Reduced nicotinamide adenine dinucleotide-dependent reduction of tertiary amine N-oxide by liver microsomal cytochrome P-450. *Biochem. Pharmacol.*, **26**, 489.

Sullivan, P.T., Kester, M. and Norton, S.J. (1968) Synthesis and study of pyridylalanine *N*-oxides. *J. Med. Chem.*, **11**, 1172.

Tatsumi, H. and Kanamitsu, O. (1961) Biological activity of nicotinic acid N-oxide. IV. Reduction of nicotinic acid *N*-oxide by the resting cells of *E. coli* K.12, *Yakugaku Zasshi*, **81**, 1767.

Thomas, R.C. and Harpootlian, H. (1975) Metabolism of minoxidil, a new hypotensive agent. II: Biotransformation following oral administration to rats, dogs and monkeys. *J. Pharm. Sci.*, **64**, 1366–71.

Uehleke, H. and Stahn, V. (1966) Zur biologischen Bildung von N,N-Dimethylanilin-N-Oxyd und N,N-Dimethyl-aminoazobenzol-N₄-Oxyd aus den Aminen sowie ihre Reaktionen mit Hämoglobin und mit Lebermikrosomen. *Naunyn Schmiedebergs Arch. Pharmakol. Exp. Pathol.*, **255**, 287.

Ziegler, D.M. (1988) Flavin-containing monooxygenase: catalytic mechanism and substrate specificities. *Drug Metab. Rev.*, **19**, 1–32.

15

In vivo metabolism of N-oxides

**E.M. Hawes, T.J. Jaworski, K.K. Midha, G. McKay,
J.W. Hubbard and *E.D. Korchinski**

*Colleges of Pharmacy and *Medicine, University of Saskatchewan, Saskatoon,
Saskatchewan, Canada S7N 0WO*

1. The *in vivo* metabolism of N-oxides is discussed, with emphasis mainly on aliphatic tertiary amine N-oxides. Illustration is with psychotropic drugs; specifically, a review of the literature pertaining to amitriptyline N-oxide and work from our laboratories on chlorpromazine N-oxide is presented.
2. These two compounds were extensively metabolized in all the species investigated. With both these compounds, in the rat there was absence of N-oxides in urine and faeces, whereas such compounds were found in the excreta of dog and man.
3. The fact that there is concurrent occurrence of N-oxidation and N-oxide reduction makes determination of the contribution of each process *in vivo* difficult.
4. The presence of other metabolizable sites within the aliphatic tertiary amine N-oxide compound should be considered. In fact, generally for such compounds the major metabolic routes involve a combination of metabolism at the tertiary amine N-oxide group and metabolism at other metabolizable sites.

15.1 INTRODUCTION

The *in vivo* metabolism of compounds which contain N-oxide groups, either aliphatic or heteroaromatic tertiary amines, has not been extensively studied. Just as there are differences in the enzymes involved in the oxidation of the two different types of tertiary amines to their N-oxides, it might be expected that there will be differences in the effect such a group has on the overall metabolism of substrates. Therefore, in order to simplify discussion, the focus in this chapter will be on only one type, namely the N-oxides of aliphatic tertiary amines. However, it should be realized that compounds containing a heteroaromatic tertiary amine group are important as is illustrated by two drugs which possess such a functionality, namely chlordiazepoxide and minoxidil. In both these cases some of the major metabolites in man either retain the N-oxide group as such or form conjugates thereof (Thomas and Harpootlian, 1975; Johnson *et al.*, 1983; Boxenbaum *et al.*, 1977; also see Chapter 4).

Regarding compounds which contain an aliphatic tertiary amine group, many of them are converted into the N-oxide metabolite under *in vivo* conditions (Bickel, 1969; Jenner, 1971). However, just as easily as the N-oxide metabolites are formed, they are reduced to their parent compounds (Ziegler, 1988). Thus, this concurrent occurrence of N-oxidation and N-oxide reduction will mask the extent of contribution from each individual process. The presence of other functional groups will also influence the significance that these simultaneously occurring oxidative and reductive processes have on the overall disposition of the parent compound.

Few studies have been conducted which thoroughly investigate the effect of all these factors on the overall metabolism of the parent aliphatic tertiary amine compound. The few N-oxide compounds which have been extensively investigated *in vivo* most commonly belong to the psychotropic drug class. These compounds include the N-oxide of the antidepressant drugs amitriptyline (Breyer-Pfaff *et al.*, 1978; Brodie *et al.*, 1978; Midgley *et al.*, 1978; Melzacka and Danek, 1983; Kuss *et al.*, 1985) and imipramine

15.1

15.2

(Nagy, 1978; Nagy and Hansen, 1978), and the antipsychotic agent chlorpromazine (Alfredsson *et al.*, 1977; Jaworski *et al.*, 1988; 1990). This chapter will focus on the studies of amitriptyline *N*-oxide (*15.1*) and chlorpromazine *N*-oxide (*15.2*); a review of the published work with the former, and the recent published and unpublished work of our research group in the latter case. The ultimate objectives of such studies are to understand better the disposition of both *N*-oxides and the tertiary amines; this may aid us in understanding and improving the therapeutic benefit of the tertiary amine drugs.

Examples of some aliphatic tertiary amine *N*-oxide compounds which have been administered *in vivo* are given in Table 15.1. Over the years a number of such compounds have been administered to various species and Table 15.1 represents a fairly comprehensive list.

15.2 *IN VIVO* METABOLISM OF TRICYCLIC ANTIDEPRESSANT *N*-OXIDES

Investigations regarding the *in vivo* metabolism of the tricyclic anti-depressant *N*-oxides have involved primarily the agents amitriptyline *N*-oxide and imipramine *N*-oxide. Since the observations regarding the metabolism of these two compounds have been found to be generally similar, discussion is here limited to the more extensively studied agent, namely amitriptyline *N*-oxide. Comparison of these studies to the work involving chlorpromazine *N*-oxide is discussed in section 15.4. The metabolism of amitriptyline *N*-oxide has been investigated in rat, dog and man. In work conducted by Brodie *et al.* (1978), the excretion, plasma concentrations of radioactivity and nature of the radioactive components in excreta were compared in rats and dogs after administration of an oral dose of [^{14}C]amitriptyline *N*-oxide. Some of the results of this study are summarized in Table 15.2. In rats there was a far greater amount of radioactivity excreted in faeces than urine. This difference can most likely be explained by biliary excretion of the administered drug since [^{14}C]amitriptyline *N*-oxide when administered orally to bile-duct cannulated rats, is excreted mainly (79.2%, $n = 2$) via the bile during a 2-day period. It is evident from this study that the majority of the dose isolated from rat faeces originated via bile that was delivered into the intestine rather than from unabsorbed drug. The difference observed between the quantity of radioactive material excreted via the bile and that observed in the faeces is most likely due to enterohepatic recycling.

Amitriptyline *N*-oxide was extensively metabolized in rat as no significant amount of the administered compound could be identified in faeces, plasma or urine. It should be noted that amitriptyline *N*-oxide has been quantitated

Table 15.1 *In vivo* metabolism of some aliphatic tertiary amine *N*-oxide compounds

Compound	Species	Metabolites[a]	References
Amitriptyline *N*-oxide	Rat	*N*-Desmethylnortriptyline 10-Hydroxyamitriptyline Nortriptyline	Brodie *et al.* (1978)
	Dog	*N*-Desmethylnortriptyline 10-Hydroxyamitriptyline	Melzacka and Danek (1983) Brodie *et al.* (1978)
	Man	10-Hydroxyamitriptyline 10-Hydroxynortriptyline 10-Hydroxyamitriptyline 10-Hydroxyamitriptyline *N*-oxide 10-Hydroxynortriptyline Nortriptyline	Breyer-Pfaff *et al.* (1978) Midgley *et al.* (1978)
Arecoline *N*-oxide	Rat	Nortriptyline Arecaidine Arecaidine *N*-oxide *N*-Acetyl-S-(3-carboxy-1-methyl-piperid-4-yl)-L-cysteine	Kuss *et al.* (1985) Nery (1971)
Chlorcyclizine *N*-oxide	Rat	Norchlorcyclizine	Kuntzman *et al.* (1967)
Chlorpromazine *N*-oxide	Rat	7-Hydroxychlorpromazine *N*-Desmethylchlorpromazine Chlorpromazine sulphoxide	Alfredsson *et al.* (1977)
		7-Hydroxychlorpromazine *N*-Desmethylchlorpromazine *N*-Desmethylchlorpromazine sulphoxide	Jaworski *et al.* (1988)
	Dog	Chlorpromazine *N*,*S*-dioxide Chlorpromazine sulphoxide 7-Hydroxychlorpromazine *N*-Desmethylchlorpromazine *N*-Desmethylchlorpromazine sulphoxide	Jaworski *et al.* (1990)

Compound	Species	Metabolites	Reference
	Man	Chlorpromazine N,S-dioxide Chlorpromazine sulphoxide 7-Hydroxychlorpromazine 7-Hydroxy-N-desmethyl-chlorpromazine N-Desmethylchlorpromazine N-Desmethylchlorpromazine sulphoxide	Jaworski et al. (1990)
N,N-Dimethylamphetamine N-oxide	Man		Jenner (1971)
N,N-Dimethyldodecylamine N-oxide	Rat	Various products of N-oxide reduction, ω,β-oxidation and C-hydroxylation	Turan and Gibson (1981)
	Rabbit	Various products of N-oxide reduction and ω,β-oxidation	Turan and Gibson (1981)
N,N-Dimethyloctylamine N-oxide	Man		
Heliotrine N-oxide	Rat		Dehner et al. (1968)
	Mouse		Powis et al. (1979a)
Imipramine N-oxide	Rat	N-Desmethylimipramine	Nagy (1978)
	Man	N-Desmethylimipramine	Nagy and Hansen (1978)
	Man		Ames and Powis (1978)
Indicine N-oxide			Powis et al. (1979b)
	Rabbit		Ames and Powis (1978)
			Powis et al. (1979b)
	Mouse		Powis et al. (1979a)
Levopropoxyphene N-oxide	Dog	N-Desmethyl levopropoxyphene	McMahon and Sullivan (1977)
	Rat	N-Desmethyl levopropoxyphene	McMahon and Sullivan (1964)
Morphine N-oxide	Rat	Morphine-3-O-glucuronide Normorphine	Misra and Mitchell (1971)
Nicotine-1'-N-oxide	Rat	Cotinine	Heimans et al., 1971
	Man	Cotinine	Dajani et al. (1975) Jenner et al. (1973)
Trimethylamine N-oxide	Rat		Norris and Benoit (1945)
	Man		Al-Waiz et al. (1987a, b)

<hr>

[a] Metabolites column does not include the administered compound or its reduced tertiary amine derivative.

Table 15.2 Mean excretion and plasma kinetics of radioactivity after a single oral dose of [^{14}C]amitriptyline N-oxide to rat[a], dog[b] and man[c]

	Rat	Dog	Man
n (sex)	6 (3M,3F)[d]	4 (2M,2F)	4 (M)
Dose (mg/kg)	2	2	0.36[e]
% Dose excreted in (days)			
Urine	38.0 ± 5.1(5)	64.2 ± 6.4(5)	88.3 ± 2.9(9)
Faeces	60.7 ± 6.0(5)	26.6 ± 3.6(5)	4.9 ± 1.8(3–6)
Total in urine	99.6 ± 2.9[f]	90.8 ± 6.4	93.2 ± 4.2
and faeces			
% Dose excreted in urine	—[g]	9(1)	34(2)
as amitriptyline N-oxide			
(days)			
Measured plasma radio-			
activity as total activity			
t_{max} (h)	4	0.75	1.25
$t_{1/2}$ (h)	16	2	10
Measured plasma radio-			
activity as amitriptyline			
N-oxide			
t_{max} (h)	—[g]	0.5	1.25
% of total at t_{max}	<10	30	75

[a] A Sprague–Dawley derived strain.
[b] Beagle.
[c] Data taken from Brodie *et al.* (1978) and Midgley *et al.* (1978).
[d] Different groups of rats were used to obtain the excretion and plasma data.
[e] 25 mg dose to 67.0–70.5 kg weight range healthy volunteers.
[f] Includes also that in the carcass (0.9 ± 0.2%) and expressed air (<0.1%).
[g] Levels too low to be quantitated.

in plasma after its oral administration to rats (Melzacka and Danek, 1983), however the dose reported was ten times higher than in the study under discussion. Examination of rat urine showed the occurrence of a very complex metabolic pattern involving at least 20 metabolites, with 10-hydroxyamitriptyline, nortriptyline and N-desmethylnortriptyline being identified. It should be pointed out that the amount of N-desmethyl-nortriptyline increased after incubating the urine with pH 5.0 buffer. This is indicative of an acid-labile conjugate such as an N-glucuronide being present. The faecal extracts from the rat contained many components with the two major being 10-hydroxyamitriptyline and N-desmethyl-nortriptyline. Both of these compounds were also the major metabolites found in bile where they were largely present as conjugates.

In many respects dog differed from rat in the disposition of amitriptyline N-oxide. In dog a major proportion of the radioactivity was excreted in urine rather than faeces. In addition, amitriptyline N-oxide was also a major component of urine and plasma after oral administration to dogs. One similarity between both rat and dog was the observation of N-desmethylnortriptyline and 10-hydroxyamitriptyline as major metabolites; in dog both were isolated from urine and the latter from faeces. In addition, 10-hydroxyamitriptyline was largely present in urine as conjugate(s). Thus, as well as N-oxide reduction, both N-dealkylation and alicyclic hydroxylation are major routes of metabolism of amitriptyline N-oxide in dog and rat.

Similar work conducted with man (Midgley *et al.*, 1978) is also summarized in Table 15.2. In this study, most of the radioactivity from the tracer oral dose of amitriptyline N-oxide was excreted in urine rather than faeces. However, almost half the dose was eliminated in urine during the first 6 h and the major component in this early phase of elimination was unchanged drug (approx. 30% of dose). These results were reflected in plasma analysis in that at early sampling times, such as the peak concentration of radioactivity, amitriptyline N-oxide accounted for most of the total radioactivity.

Six major components were identified in urine, namely, amitriptyline N-oxide, 10-hydroxyamitriptyline, 10-hydroxynortriptyline, two compounds that were likely the O-glucuronide or O-sulphate conjugates of these 10-hydroxy compounds and 10-hydroxyamitriptyline N-oxide. Of these compounds the administered agent was present in the greatest amount (34%), with the other hydroxylated components present in quantities that were at most one-third of this. The excretion of amitriptyline N-oxide largely occurred early after administration as illustrated by a comparison of its excretion in the 0–3 and 12–24 h urine collection periods: namely as 18.7 and 1.1%, respectively, of the administered dose. However, excretion was prolonged and radioactivity could still be detected in the urine on day 9. Measurement of the 10-hydroxy metabolites indicated that a far greater proportion of the slow elimination phase was due to these hydroxylated metabolites rather than the parent drug.

Man was similar to both rat and dog in that [^{14}C]amitriptyline N-oxide appeared to be completely absorbed by the oral route. In other regards man more closely resembled dog than rat. Thus over at least 5 days the mean percentage of the dose excreted in urine was 38%, 64% and 88% for rat, dog and man, respectively, whereas these figures for faeces were 61%, 27% and 5%. The latter figures suggest that biliary excretion is most important in rat, less so in dog and least important in man. The most important interspecies difference in the nature of the excreted compounds was that whereas N-oxides were at most minor components of rat excreta, they were major components in dog and man. Thus amitriptyline N-oxide was a major component of the urine and plasma of dog and man, while

10-hydroxyamitriptyline N-oxide was a major metabolite in human urine (ca. 10%). Further evidence for the presence of large amounts of the parent compound in man is also reflected in the fact that the peak plasma concentration of amitriptyline N-oxide after oral administration of 25 mg of amitriptyline N-oxide to healthy subjects is much higher than (at least 25 times) the peak concentration of amitriptyline or nortriptyline after administration of 25 mg of amitriptyline to the same subjects (unpublished results discussed by Midgley *et al.*, 1978). Despite these differences in N-oxidation and/or N-oxide reduction and differences in the sequence of metabolite events, for the examined species there was resemblance in that N-dealkylation and alicyclic hydroxylation were major metabolic routes in all three species.

Other single dose comparisons between amitriptyline N-oxide and amitriptyline have been carried out in healthy volunteers. In the first such study two separate doses of each, equivalent to 100 mg of amitriptyline were administered to one volunteer (Breyer-Pfaff *et al.*, 1978), whereas in the second 50 mg doses of each were given to 11 volunteers (Kuss *et al.*, 1985). The results generally substantiated the previous work (Midgley *et al.*, 1978) and only data from the latter study which involved plasma analysis of parent drug and metabolites are given here. Thus the more rapid absorption of amitriptyline N-oxide is reflected in a half-life of absorption for this agent $(0.31 \pm 0.10\,h)$ that was more than four times less than that of amitriptyline $(1.36 \pm 0.49\,h)$. The data also indicated that although extremely high levels of amitriptyline N-oxide are attained after administration of this agent, the levels of amitriptyline (and nortriptyline) are significant but somewhat less than after administration of amitriptyline. For example, after administration of the parent compounds the mean C_{max} of amitriptyline N-oxide $(483.8 \pm 105.2\,ng/ml)$ was about 25 times greater than that of amitriptyline $(19.0 \pm 5.5\,ng/ml)$. In addition, the levels of amitriptyline were some 2–3-fold higher after amitriptyline than amitriptyline N-oxide administration as reflected in the mean C_{max} $(19.0 \pm 5.5$ c.f. $8.4 \pm 2.6\,ng/ml)$ and AUC_0^{24} $(271 \pm 61$ c.f. 98.2 ± 28.7 in units not given).

15.3 *IN VIVO* METABOLISM OF CHLORPROMAZINE N-OXIDE

15.3.1 Background

Chlorpromazine, the prototype phenothiazine antipsychotic agent, is of interest since not only is the aliphatic tertiary amine functional group capable of undergoing metabolic conversion but other areas of the molecule such as the phenothiazine ring are capable of being metabolically altered.

For example, the occurrence of hydroxylation and sulphoxidation of the phenothiazine ring will affect the extent of metabolism which takes place at the side-chain tertiary amine group.

With chlorpromazine, the metabolic interconversion between chlorpromazine and chlorpromazine *N*-oxide has been established. However, only one previous report (Alfredsson *et al.*, 1977) examined the *in vivo* metabolism of chlorpromazine *N*-oxide. This report documents the investigation of metabolite levels in the brain and serum of rat after the administration of chlorpromazine *N*-oxide. It is not known whether all the metabolites present in either tissue were identified. Only three compounds were verified as being present in brain and serum, these being chlorpromazine, 7-hydroxychlorpromazine and *N*-desmethylchlorpromazine, with no mention being made of any additional compounds. Therefore, a systematic investigation was carried out to characterize the metabolites in urine and faeces after administration of single oral doses of chlorpromazine *N*-oxide to rat, dog and man (Jaworski *et al.*, 1988; 1990). In order to address some of the questions raised in this work, further metabolic studies were also carried out in the rat.

15.3.2 Study design

Experiments were designed for the identification of the metabolites in rat, dog and man after oral administration of chlorpromazine *N*-oxide. The procedures utilized for these studies have been reported previously (Jaworski *et al.*, 1988; 1990), but a summary of these is outlined below. Since no quantitation of metabolites was carried out a quantitative comparison of metabolites between excreta would only be approximate.

(a) Studies in rats

Female Lewis rats ($n = 8$) which had been fasted for 12 h were administered chlorpromazine *N*-oxide orally (20 mg/kg as a solution of the maleate salt) and placed in individual metabolic cages which allowed for the separate collection of urine and faeces. Pooled (0–24 h) urine and faeces were separately collected from each rat and immediately frozen after collection. Both urine and faeces were initially freeze dried and the residues were then shaken with methanol in order to separate the drug-related material from the majority of endogenous material present in the samples.

Subsequent solvent/solvent extraction procedures that included use of a relatively polar mixture of pentane–dichloromethane–2-propanol (46 : 49 : 5, by vol.) were employed for the isolation of metabolites from the methanolic extracts of urine and faeces. The isolated metabolite mixture from each excreta was then separated into individual components with the aid of HPLC such that the fractions of eluent from the HPLC which corresponded to the individual peaks were collected and evaported to dryness. This purification by HPLC was repeated where necessary and each evaporated residue was then analysed by mass spectrometry, in most cases as direct probe samples in the electron impact and/or chemical ionization (ammonia) mode. In addition to carrying out these procedures with unknown samples, blank samples and blank samples to which had been added authentic drug and metabolite standards (synthesized or donated) were used as controls.

(b) Studies in dogs

Female mixed breed dogs ($n = 5$) were dosed orally with chlorpromazine N-oxide (2 mg/kg as an extemporaneously prepared capsule of the maleate salt). The isolation and identification procedures used in these experiments were very similar to those used for rats.

(c) Studies in humans

Under an approved IND (Health Protection Branch, Ottawa, Ontario: IND no. 58P875200) and protocol healthy adult males ($n = 3$) were administered chlorpromazine N-oxide (equivalent to 50 mg chlorpromazine as an extemporaneously prepared capsule of the maleate salt) orally. The procedures employed for the examination of human excreta (urine 0–24 h; faeces 0–60 h) were different from those used for both rat and dog in that there was direct examination of the methanolic extracts by HPLC–MS with the aid of a plasmaspray interface as opposed to the tedious manual HPLC isolation followed by separate MS identification employed for the other two species. The most important benefit of using HPLC–MS is a major reduction in the time required for separation and identification of metabolites.

(d) Further metabolic experiments in the rat

Using rats of the same strain, sex, age and weight range and using chlorpromazine N-oxide at the same dose (20 mg/kg) and dosage form as before, additional experiments were carried out:

1. To determine the relative proportion of drug and metabolites eliminated in the urine and faeces (Jaworski *et al.*, 1988)
2. To determine the effect that the avoidance of the gastrointestinal tract in the administration of chlorpromazine N-oxide would have on its disposition
3. To determine what metabolites, if any, were eliminated in bile (Jaworski *et al.*, unpublished work)

In the first of these additional experiments, [³H]chlorpromazine N-oxide was administered to rats ($n = 5$) and the urine and faeces were collected at 24 h intervals for a total of 7 days. The collected samples were individually freeze dried and subsequent sample preparation prior to radioactive determination was based on the procedure of Mahin and Lofberg (1966). Briefly, the procedure used for the present work involved initially heating the samples with either nitric acid or a mixture of nitric acid and hydrogen peroxide prior to quantitative radioactive determination by liquid scintillation counting with the aid of a radioactive standard to correct for sample quenching.

In the second of these additional experiments, rats ($n = 3$) were given chlorpromazine N-oxide intraperitoneally. Urine and faeces were then both collected separately and subsequently subjected to similar extraction procedures and HPLC separation as described previously for rat. However, mass spectral identification was obviated since it was performed in the analogous experiment involving oral administration. Therefore, in the present experiment, metabolites were separated by HPLC and identified by comparison of their retention times with those of authentic standards.

In the third additional experiment, bile was collected from each rat after administration of chlorpromazine N-oxide by either intraperitoneal ($n = 5$), intravenous ($n = 1$) or oral ($n = 1$) routes. Collection of bile was for up to 8 h via an inserted cannula and then samples were directly injected into a gradient HPLC system. The components of the injected samples were separated and the appropriate fractions of mobile phase were collected. The identity of the individual metabolites was obtained by direct insertion probe mass spectrometry using chemical ionization or fast atom bombardment ionization. Aliquots of bile were also extracted with pentane–dichloromethane–2-propanol and the extracts subjected to isolation, separation and identification procedures as described previously for metabolite identification in urine and faeces.

15.3.3 Biotransformation of chlorpromazine *N*-oxide

Examination of HPLC chromatograms from urine and faecal (Fig. 15.1) extracts of the rat revealed that the same five metabolites were present in both excreta after oral administration of chlorpromazine N-oxide. The five metabolites chlorpromazine, 7-hydroxychlorpromazine, chlorpromazine sulphoxide, N-desmethylchlorpromazine and N-desmethylchlorpromazine sulphoxide were identified by comparison of their HPLC retention times and mass spectral data with those of authentic reference samples. An additional experiment was performed that involved hydrolysing urine samples with β-glucuronidase and extracting the sample with organic

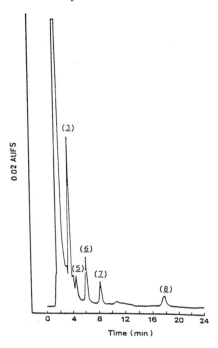

Figure 15.1 HPLC chromatogram of a faecal extract from a rat dosed orally with chlorpromazine N-oxide. Structures and names of metabolites are given in Fig. 15.4.

solvent; however, the presence of any increase in previously identified metabolites (including 7-hydroxychlorpromazine) or the presence of a previously undetected metabolite was not observed.

The results observed with the dog were somewhat different from those obtained with the rat. In the case of dog seven compounds were identified in urine and six in faeces. The compound present in dog urine but not faeces was chlorpromazine N,S-dioxide. The six metabolites identified in both excreta were chlorpromazine, 7-hydroxychlorpromazine, chlorpromazine sulphoxide, N-desmethylchlorpromazine, chlorpromazine N-oxide and N-desmethylchlorpromazine sulphoxide. In addition, treatment of urine samples with β-glucuronidase revealed a relative increase in the HPLC peak height for the peak corresponding to 7-hydroxychlorpromazine.

The results from the human data demonstrated that there were eight compounds present in the urine extracts (Fig. 15.2) as compared to five metabolites in the faecal extracts. The three compounds present in only urine were 7-hydroxy-N-desmethylchlorpromazine, chlorpromazine

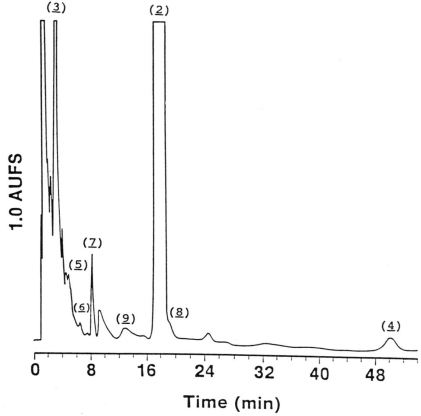

Figure 15.2 HPLC chromatogram of a urine extract from a man dosed orally with chlorpromazine *N*-oxide. The structures and names of metabolites are given in Fig. 15.4.

N-oxide and chlorpromazine *N,S*-dioxide, whereas the five identified in both excreta were chlorpromazine, 7-hydroxychlorpromazine, chlorpromazine sulphoxide, *N*-desmethylchlorpromazine and *N*-desmethylchlorpromazine sulphoxide. It was also shown that treatment of human urine with β-glucuronidase revealed a relative increase in the HPLC peak height for 7-hydroxychlorpromazine.

Regarding the first of the three additional experiments carried out using rats, administration of [³H]chlorpromazine *N*-oxide indicated that virtually all the quantifiable radioactivity was excreted in 0–72 h for urine and 0–96 h for faeces and totalled approximately 80% (79.0 ± 11.7%) of the orally administered dose. Approximately twice the amount of radioactivity was

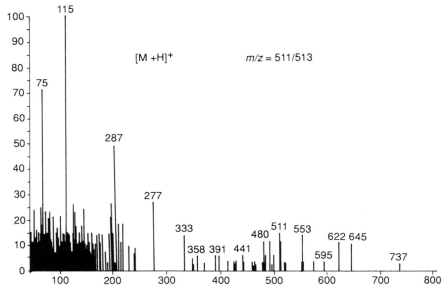

Figure 15.3 Mass spectrum of 7-hydroxychlorpromazine glucuronide using fast atom bombardment (FAB) mode of ionization.

eliminated in the faeces (52.1 ± 9.7%) as compared to the urine (26.9 ±·7.2%).

Intraperitoneal administration of chlorpromazine N-oxide gave the same metabolic profile as before; i.e. the same five metabolites were present in urine and faeces of rats administered chlorpromazine N-oxide either orally or intraperitoneally.

Finally when bile, or extracts of bile from rats administered chlorpromazine N-oxide (by oral, intraperitoneal and intravenous routes) were examined, six compounds were isolated in each case. These included chlorpromazine, 7-hydroxychlorpromazine, chlorpromazine sulphoxide, chlorpromazine N-oxide, chlorpromazine N,S-dioxide and 7-hydroxy-chlorpromazine glucuronide (Fig. 15.3).

(a) Metabolism of chlorpromazine N-oxide in the rat

The work conducted in the rat is the first systematic investigation of the metabolism of chlorpromazine N-oxide reported for any species. Since five metabolites, but not the administered compound, could be identified in urine and faeces indicates that chlorpromazine N-oxide is extensively metabolized in this species. Furthermore, the inability to find any compound which retained the *N*-oxide functional group suggests that the secondary and tertiary metabolites identified in urine and faeces are most likely formed by initial *N*-oxide reduction and subsequent metabolism.

It is important to note that no *N*-oxidized compounds were identified even though recovery studies using both excreta resulted in efficient extraction of polar (including the *N*-oxide and *N,S*-dioxide) and non-polar metabolites while using the appropriate solvent mixture. The fact that chlorpromazine *N*-oxide has been shown to be unstable during preparation of biological samples for analysis (Krieglstein *et al.*, 1979; Hubbard *et al.*, 1985; McKay *et al.*, 1985; Hawes *et al.*, 1986) was also not overlooked in this work. Studies were initially conducted to establish that samples of chlorpromazine *N*-oxide and chlorpromazine *N,S*-dioxide when added to blank urine and faeces were stable when subjected to the isolation and HPLC procedures used in these studies.

(b) Metabolism of chlorpromazine in the dog

Chlorpromazine *N*-oxide was also extensively metabolized in dog since all five of the metabolites identified in rat were characterized in the urine and faeces of this species. However, unlike in rats, chlorpromazine *N*-oxide was found to be present in the extracts of dog urine and faeces even though the proportion present in faeces was estimated to be smaller than that found in urine. In addition, chlorpromazine *N,S*-dioxide appeared to be present in large quantity in urine but none could be detected in faeces. These observations suggest that *N*-oxides may be undergoing reduction as they travel down the intestinal tract to the colon since this has been observed previously in various species (Scheline, 1973). Prior work in these laboratories has demonstrated that the dog metabolizes chlorpromazine extensively to chlorpromazine *N*-oxide (P.K.F. Yeung, J.W. Hubbard and K.K. Midha, unpublished results). It is, therefore, most likely that elimination of significant amounts of *N*-oxide compounds in urine is due to a combination of any unmetabolized administered dose and that formed metabolically.

The observation of *N*-oxide compounds in the excreta of dog but not rat is not the only interspecies difference observed from the results of this work, since an apparent relative decrease in the amount of sulphoxide metabolites in extracts of faeces when compared to extracts of urine of rat was not seen in dog. However, in the case of the dog, the difficulties encountered in acquiring a suitable mass spectrum for each of the two sulphoxide metabolites, especially chlorpromazine sulphoxide, in faeces, indicated that in the excreta these metabolites were present only in trace quantities. Nevertheless, as previously pointed out similarity did exist to the observation regarding sulphoxide metabolites in the urine and faeces of rat in that lesser amounts of *N*-oxidized compounds appeared to be present in faeces than in urine of dog.

(c) Metabolism of chlorpromazine in man

Man, as with the other two species examined, exhibited extensive metabolism of chlorpromazine *N*-oxide. In terms of the metabolic profile, however, there was more similarity to that of dog than rat (Fig. 15.4). The only difference observed in the urine of man and dog was the presence of 7-hydroxy-*N*-desmethylchlorpromazine in human compared to its absence in dog. Whereas, the only difference seen with the faeces of the two species was the presence of unchanged chlorpromazine *N*-oxide in dog compared to its absence in human, although the amount in dog faeces was much smaller

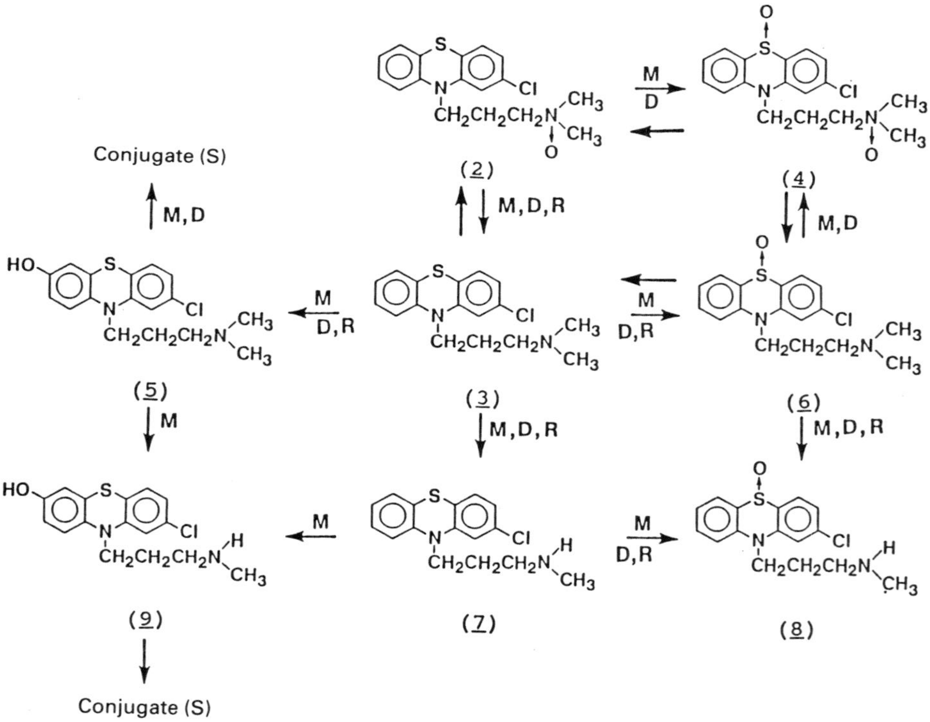

Figure 15.4 Metabolic pathways of chlorpromazine *N*-oxide in rat (R), dog (D) and man (M). *2*, chlorpromazine *N*-oxide; *3*, chlorpromazine; *4*, chlorpromazine *N,S*-dioxide; *5*, 7-hydroxychlorpromazine; *6*, chlorpromazine sulphoxide; *7*, *N*-desmethylchlorpromazine; *8*, *N*-desmethylchlorpromazine sulphoxide; *9*, 7-hydroxy-*N*-desmethylchlorpromazine.

than that present in dog urine. Indeed in the urine of both species there appeared to be large amounts of the N-oxide and N,S-dioxide compounds. This observation may not be due to a lack of metabolism of the N-oxide group in these species but rather due to greater capacity of these species to form N-oxides from the tertiary amines produced. It is known that like the dog, man is also an extensive producer of chlorpromazine N-oxide upon administration of chlorpromazine (Gruenke *et al.*, 1985; Yeung *et al.*, 1987).

In man, the observation of greater amounts of N-oxide compounds in urine as compared to faeces is also similar to that observed in dog. This can likewise be explained by the reduction of the N-oxides on transit down the intestinal tract to the colon. The same comparison cannot be done with rat studies due to the lack of detection of any N-oxide-containing compounds in either excreta of this species. Urine from man was also treated with β-glucuronidase which as in dog led to a relative increase in the amount of 7-hydroxychlorpromazine in extracts of urine. This suggests that a significant proportion of this phenolic metabolite conjugates prior to its excretion. One last similarity between man and dog was the difficulty encountered in deducing from the HPLC chromatograms the relative proportion of sulphoxide metabolites, namely chlorpromazine sulphoxide and N-desmethylchlorpromazine sulphoxide, between urine and faeces.

(d) Further metabolic data for chlorpromazine N-oxide in the rat

The first of the additional experiments carried out in the rat involved determination of the mass balance of [^{3}H]chlorpromazine N-oxide. The results indicate that most of the orally administered dose was eliminated via the urine and faeces in that ca. 80% was accounted for in these excreta over the first 3–4 days. It is speculated that the remainder of the radioactivity was still present in the tissues and would undergo elimination over a longer period of time. It is important to note that twice as much radioactivity was eliminated via the faeces than the urine.

The second experiment in the rat which involved intraperitoneal administration of chlorpromazine N-oxide was designed to investigate the question of whether or not reduction of chlorpromazine N-oxide was occurring solely in the gastrointestinal tract. Intraperitoneal administration enabled the reductive effects of the intestine to be initially by-passed. The result was that the same five metabolites identified after oral administration of chlorpromazine N-oxide were also produced after intraperitoneal administration. This indicates that a significant amount of chlorpromazine N-oxide is reduced even when intraperitoneal administration is used and that the N-oxide and its metabolites could be absorbed from this site of administration. More importantly, these results imply that with oral administration of chlorpromazine N-oxide, the reductive process may not be a simple matter of chlorpromazine N-oxide undergoing complete reduction within the gastrointestinal tract followed by absorption of the reduced species. It is still not possible to identify all the sites and tissues where reduction of the compound occurs and the importance of each site or tissue. Therefore, the situation is probably more complicated than might be speculated if only the results of the qualitative experiment involving oral administration of chlorpromazine N-oxide are considered.

Administration of [^{3}H]chlorpromazine N-oxide allowed determination that the greatest proportion of eliminated drug and metabolites was found in faeces and, consequently, it was postulated that the metabolites isolated from faeces after oral administration of chlorpromazine N-oxide could have been excreted into the intestine via the bile. Therefore, to test this hypothesis an experiment was undertaken to identify the metabolites of chlorpromazine N-oxide, if any, present in rat bile. In fact, six compounds were identified in bile, each time, whether chlorpromazine N-oxide was administered intraperitoneally, intravenously or orally. Thus between these routes of administration there appeared to be no qualitative differences in the metabolic profile. When making a comparison of the metabolites eliminated in urine and faeces with those eliminated in bile, it is important to remember that urine and faeces were collected for 0–24 h after drug administration, whereas bile was only collected at most up to 8 h. The metabolites which were found in urine and faeces but not in bile were N-desmethylchlorpromazine and N-desmethylchlorpromazine sulphoxide. It may be that these particular metabolites are not excreted in bile; however, it is possible that the collection period for bile was insufficient to enable identification of such compounds and as noted from the previous experiment both appeared to be present in lesser amounts in faeces as compared to urine.

There were also three compounds present in rat bile which were not detected in rat urine or faeces. These were chlorpromazine N-oxide, chlorpromazine N,S-dioxide and 7-hydroxychlorpromazine glucuronide. To the best of the author's knowledge, this is the first unequivocal evidence for the identification of intact 7-hydroxychlorpromazine glucuronide in any species. It was previously noted that when the urine from a rat administered chlorpromazine N-oxide was treated with β-glucuronidase, the HPLC chromatogram of an extract of such treated urine showed no apparent increase in peak height for 7-hydroxychlorpromazine. Therefore, further studies are required in order to clarify this apparent conflict. In addition, it is noteworthy that the sulphate conjugate of 7-hydroxychlorpromazine could not be identified in bile.

An important observation is the fact that, unlike in faeces or urine, chlorpromazine N-oxide and chlorpromazine N,S-dioxide were detected in the bile of the rat. This is the only direct evidence which indicated that, in fact, compounds containing an *N*-oxide group were present in the excretion fluids of the rat. It also gives the best and most direct evidence that to a significant extent chlorpromazine *N*-oxide was probably absorbed as such into the body. The fact that neither chlorpromazine *N*-oxide nor chlorpromazine *N*,*S*-dioxide were detected in the faeces of the non-cannulated rat can be most simply explained in terms of their excretion in bile, followed by subsequent reduction of the *N*-oxide group of these compounds on passage down the intestinal tract to the colon.

15.4 COMPARISON OF STUDIES INVOLVING AMITRIPTYLINE *N*-OXIDE AND CHLORPROMAZINE *N*-OXIDE

Some interesting similarities are observed when comparison is made in the metabolic profiles of rat, dog and man between amitriptyline *N*-oxide (Breyer-Pfaff *et al.*, 1978; Brodie *et al.*, 1978; Midgley *et al.*, 1978; Melzacka and Danek, 1983; Kuss *et al.*, 1985) and chlorpromazine *N*-oxide (Jaworski *et al.*, 1988; 1990; unpublished results). First, in the only case where comparison of quantitative data is possible, for the rat there was similarity in the distribution of the radioactive dose between urine and faeces (1 : 1.6 and 1 : 1.9 for amitriptyline *N*-oxide and chlorpromazine *N*-oxide, respectively). Second, for both these *N*-oxides, compounds with an *N*-oxide functional group could be identified in the excreta of man and dog but not rat. In addition for man and dog, these *N*-oxide compounds were especially eliminated in urine rather than faeces. Finally, for both chlorpromazine *N*-oxide and amitriptyline *N*-oxide, biliary recycling appears to be very important in the rat.

That these similarities in metabolism exist between chlorpromazine *N*-oxide and amitriptyline *N*-oxide is not surprising in that structurally they are also similar (*15.1*, *15.2*). They are both compounds where a $(CH_3)_2NO$ moiety is attached to a tricyclic heterocyclic nucleus via a three carbon chain. However, a structural difference which affects important metabolic difference is that the cyclic 10,11-ethyl bridge of amitriptyline *N*-oxide is altered in the case of chlorpromazine *N*-oxide to that of the 5-position sulphur atom of the phenothiazine ring. In fact, alicyclic 10-hydroxylation and *S*-oxidation at these sites are respective major metabolic routes of these compounds. That this does result in metabolic difference is well illustrated in the case of amitriptyline *N*-oxide in that the importance of 10-hydroxylation is firmly established such that in man it is proposed (Midgley *et al.*, 1978) that the major metabolic route involves the following conversions:

amitriptyline *N*-oxide → 10-hydroxyamitriptyline *N*-oxide

10-hydroxynortriptyline ← 10-hydroxyamitriptyline

↓   ↓

O-glucuronide   *O*-glucuronide

That indeed this latter route is the major metabolic route implies that to a great extent in man amitriptyline *N*-oxide is not a prodrug of amitriptyline. This is also implied from the single dose studies which showed that the levels of amitriptyline and its active metabolite nortriptyline were far lower after administration of amitriptyline *N*-oxide than a comparable dose of amitriptyline (Breyer-Pfaff *et al.*, 1978; Midgley *et al.*, 1978; Kuss *et al.*, 1985). Furthermore, on chronic oral dosing with amitriptyline *N*-oxide to depressed patients the mean steady-state levels of amitriptyline and nortriptyline were some 2–3-fold less than after the same dose of amitriptyline to comparable patients (Kuss *et al.*, 1984). In this and other clinical studies in which both these drugs have been compared and effectively used at the same or very similar doses, generally for amitriptyline *N*-oxide, the antidepressant response was at least as good and somewhat earlier, and side effects were less marked (especially anticholinergic and sedative effects) (Rapp, 1978; Borromei, 1982; Cassano *et al.*, 1983; Kuss *et al.*, 1984; Platz, 1987). The question arises as to which compound(s) is responsible for those comparative favourable effects of amitriptyline *N*-oxide? Notwithstanding its conversion *in vivo* into active *N*-deoxygenated compounds, the *N*-oxide is itself far less potent than amitriptyline in *in vitro* screens used to test antidepressant drugs (Hyttel *et al.*, 1980). The possibility that a major contribution to these favourable effects is from 10-hydroxy metabolites is likely in that in general such metabolites are known to be active (Bertilsson *et al.*, 1979; Hyttel *et al.*, 1980; Robinson *et al.*, 1985) and the available data indicate that as compared to amitriptyline the presence of the *N*-oxide functional group effectively diverts the metabolism from the dimethylaminopropylidene side chain to the tricyclic ring system (Midgley *et al.*, 1978; Breyer-Pfaff *et al.*, 1978).

Only a qualitative study was carried out of the metabolism of chlorpromazine *N*-oxide in man (Jaworski *et al.*, 1990), but the limited evidence would indicate that since *N*-oxygenated compounds appear to be important metabolites in the excreta, it is not a simple prodrug of chlorpromazine in man. It is also likely that *N*-oxide metabolites of chlorpromazine such as those found, namely the *N*-oxide and *N,S*-dioxide, are inactive as antipsychotic agents (Lewis *et al.*, 1983). Further study would be required before making a decision as to whether a comparative clinical evaluation of chlorpromazine and chlorpromazine *N*-oxide would be worthwhile.

15.5 CONCLUSIONS

It was often believed that the processes involved in *in vivo* metabolism of aliphatic tertiary amine *N*-oxides were simplistic in nature, namely reduction occurring solely within the gastrointestinal tract followed by oxidation of the reduced species. Question was even raised as to whether tertiary amine *N*-oxides could be absorbed from the gastrointestinal tract in their intact form.

Generally, tertiary amine *N*-oxides are well absorbed as such from the intestine. Furthermore, notwithstanding vast interspecies differences, the *in vivo* metabolism of *N*-oxides cannot be generally presented by the simplistic scenario just described. This was amply illustrated by the previous discussions of the metabolism of amitriptyline *N*-oxide and chlorpromazine *N*-oxide. Thus in the first place the concurrent occurrence of *N*-oxidation and *N*-oxide reduction hampers our understanding as to the extent of contribution of each individual process. In addition, other than unique cases such as trimethylamine *N*-oxide (Al-Waiz *et al.*, 1987a,b), the presence of other metabolizable sites within the molecule has great influence as to the significance of *N*-oxide reduction in the overall disposition of the parent compound. Such is the situation with the alicyclic hydroxylation of the tricyclic ring of amitriptyline *N*-oxide. Thus the disposition and quantitative metabolism of an *N*-oxide can be markedly different in many regards from that of its corresponding aliphatic tertiary amine compound. The former is not generally simply the prodrug of the latter, as was described with the examples detailed in this report. Hence, each interconvertible pair of compounds are best regarded as distinctly different compounds which merit separate investigation as drugs. The current situation is that generally the aliphatic tertiary amine of such compounds is marketed, whereas the corresponding *N*-oxide is not. Study of the *N*-oxide can produce a potentially useful therapeutic entity with a different side-effect profile. Such a study can also give clinically useful information about the disposition of a marketed or investigational tertiary amine compound.

REFERENCES

Alfredsson, G., Wiesel, F.-A. and Skett, P. (1977) Levels of chlorpromazine and its active metabolites in rat brain and the relationship to central monoamine metabolism and prolactin secretion. *Psychopharmacology*, **53**, 13–18.

Al-Waiz, M., Ayesh, R., Mitchell, S.C. *et al.* (1987a) Disclosure of the metabolic retroversion of trimethylamine *N*-oxide in humans: a pharmacogenetic approach. *Clin. Pharmacol. Ther.*, **42**, 608–12.

Al-Waiz, M., Mitchell, S.C., Idle, J.R. and Smith, R.L. (1987b) The metabolism of ^{14}C-labelled trimethylamine and its *N*-oxide in man. *Xenobiotica*, **17**, 551–8.

Ames, M.M. and Powis, G. (1978) Determination of indicine *N*-oxide and indicine in plasma and urine by electron-capture gas-liquid chromatography. *J. Chromatogr.*, **166**, 519–26.

Bertilsson, L., Mellström, B. and Sjöqvist, F. (1979) Pronounced inhibition of noradrenaline uptake by 10-hydroxy metabolites of nortriptyline. *Life Sci.*, **25**, 1285–92.

Bickel, M.H. (1969) The pharmacology and biochemistry of *N*-oxides. *Pharmacol. Rev.*, **21**, 325–55.

Borromei, A. (1982) A new antidepressant agent: amitriptyline *N*-oxide. *Adv. Biochem. Psychopharmacol.*, **32**, 43–7.

Boxenbaum, H.G., Geitner, K.A., Jack, M.L. *et al.* (1977) Pharmacokinetic and biopharmaceutic profile of chlordiazepoxide HCl in healthy subjects: multiple-dose oral administration. *J. Pharmacokinet. Biopharm.*, **5**, 25–39.

Breyer-Pfaff, U., Ewert, M. and Wiatr, R. (1978) Comparative single-dose kinetics of amitriptyline and its *N*-oxide in a volunteer. *Arzneimittelforschung*, **28**, 1916–20.

Brodie, R.R., Chasseaud, L.F., Hawkins, D.R. and Midgley, I. (1978) The pharmacokinetics and metabolism of ^{14}C-amitriptylinoxide in rat and dog. *Arzneimittelforschung*, **28**, 1907–10.

Cassano, G.B., Conti, L., Massimetti, G. *et al.* (1983) A controlled clinical trial between amitriptyline and its *N*-oxide metabolite. *Psychopharmacol. Bull.*, **19**, 98–103.

Dajani, R.M., Gorrod, J.W. and Beckett, A.H. (1975) Reduction *in vivo* of (−)-nicotine-1′-*N*-oxide by germ-free and conventional rats. *Biochem. Pharmacol.*, **24**, 648–50.

Dehner, E.W., Machinist, J.M. and Ziegler, D.M. (1968) Effect of SKF 525-A on the excretion of amine oxides by the rat. *Life Sci.*, **7**, 1135–46.

Gruenke, L.D., Craig, J.C., Klein, F.D. *et al.* (1985) Determination of chlorpromazine and its major metabolites by gas chromatography/mass spectrometry: application to biological fluids. *Biomed. Mass Spectromet.*, **12**, 707–13.

Hawes, E.M., Hubbard, J.W., Martin, M. *et al.* (1986) Therapeutic monitoring of chlorpromazine III: minimal interconversion between chlorpromazine and metabolites in human blood. *Ther. Drug. Monit.*, **8**, 37–41.

Heimans, R.L.H., Fennessy, M.R. and Gaff, G.A. (1971) Some aspects of the metabolism of morphine-*N*-oxide. *J. Pharm. Pharmacol.*, **23**, 831–6.

Hubbard, J.W., Cooper, J.K., Hawes, E.M., *et al.* (1985) Therapeutic monitoring of chlorpromazine I: pitfalls in plasma analysis. *Ther. Drug Monit.*, **7**, 222–8.

Hyttel, J., Christensen, A.V. and Fjalland, B. (1980) Neuropharmacological properties of amitriptyline, nortriptyline and their metabolites. *Acta Pharmacol. Toxicol.*, **47**, 53–7.

Jaworski, T.J., Hawes, E.M., McKay, G. and Midha, K.K. (1988) The metabolism of chlorpromazine *N*-oxide in the rat. *Xenobiotica*, **18**, 1439–47.

Jaworski, T.J., Hawes, E.M., McKay, G. and Midha, K.K. (1990). The metabolism of chlorpromazine *N*-oxide in man and dog. *Xenobiotica*, **20**, 107–15.

Jenner, P. (1971) The role of nitrogen oxidation in the excretion of drugs and foreign compounds. *Xenobiotica*, **1**, 399–418.

Jenner, P., Gorrod, J.W. and Beckett, A.H. (1973) The absorption of nicotine-1′-*N*-oxide and its reduction in the gastro-intestinal tract in man. *Xenobiotica*, **3**, 341–9.

Johnson, G.A., Barsuhn, K.J. and McCall, J.M. (1983) Minoxidil sulfate, a metabolite of minoxidil. *Drug Metab. Dispos.*, **11**, 507–8.

Krieglstein, J., Rieger, H. and Schütz, H. (1979) Effects of chlorpromazine and some of its metabolites on the EEG and on dopamine metabolism of the isolated perfused rat brain. *Eur. J. Pharmacol.*, **56**, 363–70.

Kuntzman, R., Phillips, A., Tsai, I. *et al.* (1967) N-Oxide formation: a new route for inactivation of the antihistaminic chlorcyclizine. *J. Pharmacol. Exp. Ther.*, **155**, 337–44.

Kuss, H.J., Jungkunz, G. and Holsboer, F. (1984) Amitriptyline: looking through the therapeutic window. *Lancet*, 464–5.

Kuss, H.-J., Jungkunz, G. and Johannes, K.-J. (1985) Single oral dose pharmacokinetics of amitriptylinoxide and amitriptyline in humans. *Pharmacopsychiatry*, **18**, 259–62.

Lewis, M.H., Widerlov, E., Knight, D.L. *et al.* (1983) N-Oxides of phenothiazine antipsychotics: effects on *in vivo* and *in vitro* estimates of dopaminergic function. *J. Pharmacol. Exp. Ther.*, **225**, 539–45.

Mahin, D.T. and Lofberg, R.T. (1966) A simplified method of sample preparation for determination of tritium, carbon-14 or sulphur-35 in blood or tissue by liquid scintillation counting. *Anal. Biochem.*, **16**, 500–9.

McKay, G., Cooper, J.K., Hawes, E.M. *et al.* (1985) Therapeutic monitoring of chlorpromazine II: pitfalls in whole blood analysis. *Ther. Drug Monit.*, **7**, 472–7.

McMahon, R.E. and Sullivan, H.R. (1964) The oxidative demethylation of 1-propoxyphene and 1-propoxyphene N-oxide by rat liver microsomes. *Life Sci.*, **3**, 1167–74.

McMahon, R.E. and Sullivan, H.R. (1977) Reduction of levopropoxyphene N-oxide to propoxyphene by dogs *in vivo* and rat liver microsomal fraction *in vitro*. *Xenobiotica*, **7**, 377–82.

Melzacka, M. and Danek, L. (1983) Pharmacokinetics of amitriptyline N-oxide in rats after single and prolonged oral administration. *Pharmacopsychiatry*, **16**, 30–4.

Midgley, I., Hawkins, D.R. and Chasseaud, L.F. (1978) The metabolic fate of the antidepressant agent amitriptylinoxide in man. *Arzneimittelforschung*, **28**, 1911–16.

Misra, A.L. and Mitchell, C.L. (1971) Determination of morphine-N-methyl-^{14}C oxide in biological materials, its excretion and metabolites in the rat. *Biochem. Med.*, **5**, 379–83.

Nagy, A. (1978) The kinetics of imipramine-N-oxide in rats. *Acta Pharmacol. Toxicol.*, **42**, 68–72.

Nagy, A. and Hansen, T. (1978) The kinetics of imipramine-N-oxide in man. *Acta Pharmacol. Toxicol.*, **42**, 58–67.

Nery, R. (1971) The metabolic interconversion of arecoline and arecoline 1-oxide in the rat. *Biochem. J.*, **122**, 503–8.

Norris, E.R. and Benoit, G.J. (1945) Studies on trimethylamine N-oxide. III. Trimethylamine oxide excretion by the rat. *J. Biol. Chem.*, **158**, 443–8.

Platz, W.E. (1987) Amitriptylinoxide versus amitriptyline. *Int. J. Neurosci.*, **32**, 700.

Powis, G., Ames, M.M. and Kovach, J.S. (1979a) Relationship of the reductive metabolism of indicine N-oxide to its antitumor activity. *Res. Commun. Chem. Pathol. Pharmacol.*, **24**, 559–69.

Powis, G., Ames, M.M. and Kovach, J.S. (1979b) Metabolic conversion of indicine N-oxide to indicine in rabbits and humans. *Cancer Res.*, **39**, 3564–70.

Rapp, W. (1978) Comparative trial of amitriptyline N-oxide and amitriptyline in the treatment of out-patients with depressive syndromes. *Acta Psychiatr. Scand.*, **58**, 245–55.

Robinson, D.S., Cooper, T.B., Howard, D. *et al.* (1985) Amitriptyline and hydroxylated metabolite plasma levels in depressed outpatients. *J. Clin. Psychopharmacol.*, **5**, 83–8.

Scheline, R.R. (1973) Metabolism of foreign compounds by gastrointestinal microorganisms. *Pharmacol. Rev.*, **25**, 451–523.

Thomas, R.C. and Harpootlian, H. (1975) Metabolism of minoxidil, a new hypotensive agent II: biotransformation following oral administration to rats, dogs and monkeys. *J. Pharm. Sci.*, **64**, 1366–71.

Turan, T.S. and Gibson, W.B. (1981) A comparison of the elimination and biotransformation of dodecyldimethylamine oxide (DDAO) by rats, rabbits and man. *Xenobiotica*, **11**, 447–58.

Yeung, P.K.F., Hubbard, J.W., Korchinski, E.D. and Midha, K.K. (1987) Radioimmunoassay for the *N*-oxide metabolite of chlorpromazine in human plasma and its application to a pharmacokinetic study in healthy humans. *J. Pharm. Sci.*, **76**, 803–8.

Ziegler, D.M. (1988) Flavin-containing monooxygenases: catalytic mechanism and substrate specificities. *Drug Metab. Rev.*, **19**, 1–32.

16

Purification and characterization of rat hepatic acetyltransferase

C.M. King and S.J. Land

Department of Chemical Carcinogenesis, Michigan Cancer Foundation, 110 East Warren Avenue, Detroit, Michigan 48201, USA

1. Rat hepatic acetyltransferase is an acidic, sulphydryl-dependent enzyme that is believed to be involved in mediating some of the toxic effects of aromatic amines. This potential derives from the enzyme's ability to O-acetylate arylhydroxylamines to yield products that are capable of reacting spontaneously with nucleic acid and protein.

2. The enzyme has now been purified in studies that have employed three distinct assays: N,O-acetyltransfer of the acetyl group of N-hydroxy-2-acetylaminofluorene (AHAT) and acetyl-CoA-dependent O- or N-acetylation with N-hydroxy-3,2'-dimethyl-4-aminobiphenyl (OAT) and 2-aminofluorene (NAT), respectively.

3. Purification from cytosols by sequential salt precipitation, ion-exchange chromatography and gel filtration gives preparations with ratios of AHAT : OAT : NAT of $\sim 1:3:20$, i.e. approximately the same as previously observed for human cytosols.

4. Immunoaffinity chromatography on supports prepared from mouse monoclonal antibody (mAb)-1A3 retards elution of the enzyme, which allows it to be purified without loss of activity or a change in the ratio of activities. SDS-PAGE of the immunoaffinity-purified enzyme shows a single, silver-stained band at ~ 32 kDa. Western blots with five different monoclonals, separately or as a mixture, reveals only a single peptide at each stage of purification. An immunoaffinity column made with mouse mAb-1F2 has a higher binding affinity that precludes elution from this column without loss of enzyme activity. The eluted protein shows a single, immunoreactive, silver-stained band at ~ 32 kDa.

5. These data support the conclusion that a single peptide of rat liver is capable of N- and O-acetylation as well as N,O-acetyltransfer and are consistent with previous observations suggesting a similar relationship in the analogous rabbit and mouse enzymes.

16.1 INTRODUCTION

A wide variety of N-substituted aromatic compounds have been shown to be carcinogenic, mutagenic and/or toxic. Exposure to these agents is likely to come from occupational, environmental, dietary and pharmacological sources. Structurally these compounds are similar in that they all have, or can be metabolized to arylhydroxylamine derivatives. Biological experiments revealed that these metabolites were likely to be more closely related to the adverse genotoxic properties of aromatic amines or nitroarenes than the parent compounds themselves. Although it was demonstrated that some of these potential metabolites could react with nucleic acid in acidic media (e.g. pH 5) (Kriek, 1965), many had only marginal reactivity at pH values that were likely to be encountered in physiological systems.

Insight into the enzymatic mechanisms by which arylhydroxylamines might be transformed to reactive derivatives came from experiments in which N-hydroxy-2-acetylaminofluorene was shown to be converted by rat cytosol into nucleic acid bound adducts with concomitant loss of the N-acetyl moiety (King and Phillips, 1968). Subsequent studies demonstrated that this activation was likely to involve N,O-acetyltransfer with the formation of reactive N-acetoxyarylamines (Bartsch *et al.*, 1972, 1973; King, 1974). Importantly, these and additional studies revealed that this enzyme activity was widely distributed in other tissues of the rat and those of many other species (King and Allaben, 1980).

Some of these tissues are known to produce tumours on exposure to potential precursors of these metabolites (Garner *et al.*, 1984). Because of the susceptibility of many organs of the rat to a broad range of aromatic amines, this species has been used extensively in mechanistic studies in attempts to relate metabolic activation to tumour induction. The relationship of the formation of reactive N-acetoxyarylamines to tumour induction was approached by use of hydroxamic acids that differed in their ability to be activated by rat mammary N,O-acyltransferase because they possessed different N-acyl groups (Weeks *et al.*, 1980; Shirai *et al.*, 1981; Allaben *et al.*, 1982). The results of these studies showed that rat mammary tumour induction by these derivatives was related to their ability to be activated by cytosolic N,O-acyltransferase. The major DNA adduct formed in the mammary gland and liver of rats exposed to aromatic amines is consistent with formation via an acetyltransferase-mediated reaction (Allaben *et al.*, 1983). Collectively, these data support the conclusion that the enzymatic formation of reactive N-acetoxyarylamines in the rat is responsible for tumour formation in the mammary gland and possibly other tissues.

16.2 ENZYMES RESPONSIBLE FOR FORMATION OF REACTIVE *N*-ACETOXYARYLAMINE METABOLITES

Studies in the 1950s demonstrated that carcinogenic aromatic amines could undergo *N*-acetylation (Garner *et al.*, 1984). Thus, *N*-acetylated metabolites were explored for their potentials to elicit tumour formation and serve as substrates in the metabolic disposition of these agents. Booth (1966) demonstrated that rat liver cytosol could catalyse transfer of the *N*-acetyl group of *N*-hydroxy-*N*-acetylarylamines, i.e. arylhydroxamic acids, to other arylamines. Following the demonstration that rat cytosol also catalysed the formation of arylhydroxamic acid adducts with nucleic acid with loss of the *N*-acetyl group (King and Phillips, 1968), evidence was obtained suggesting that this process involved an *N,O*-acetyltransfer (Bartsch *et al.*, 1972, 1973; King, 1974) and was therefore a reaction analogous to that described by Booth (1966). Recognition that an enzyme(s) capable of this activation was present in many tissues of the rat and other species led to attempts to characterize the protein responsible for this activity.

Three basic types of assay have been used to characterize the enzymes that are capable of acetylating aromatic amine derivatives (Fig. 16.1) (Land *et al.*, 1989). The acetyl CoA-dependent *N*-acetylation of primary aromatic amines yields stable acetamides that can conveniently be measured as the

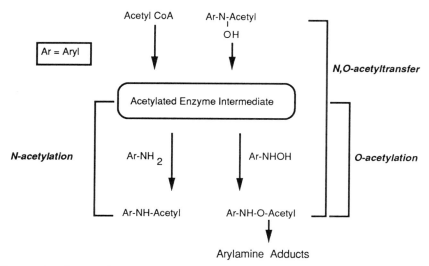

Figure 16.1 Enzymatic acetyltransfer reactions.

production of organic soluble radioactivity on incubation with labelled acetyl-CoA. N,O-Arylhydroxamic acid acetyltransferase activity is measured by incubation of ring-labelled arylhydroxamic acid in the presence of nucleic acid that serves as a trapping agent for the reactive O-acetoxy-arylamine product. Direct O-acetylation is best demonstrated by determining the incorporation of radioisotope into nucleic acid in assays that contain labelled N-hydroxy-3,2'-dimethyl-4-aminobiphenyl and unlabelled acetyl-CoA. The N-acetylated hydroxamic acid that may be formed in this reaction cannot serve as an acetyl donor as in the N,O-arylhydroxamic acid acetyltransfer assay, due to the presence of the ortho methyl group.

Initial efforts to characterize the rat enzyme were hampered by the extreme lability of the enzyme (King, 1974). The enzyme was considered to require a sulphydryl group for activity, since reagents that reacted with these groups resulted in loss of activity. Loss of activity could be reduced if pyrophosphate buffers and sulphydryl protective agents (i.e. dithiothreitol) were used (King, 1974). Greater retention of activity was also achieved by carrying out all manipulations under an argon atmosphere. These techniques permitted extensive purification by sequential salt precipitation, chromatography on ion-exchange and gel filtration columns and preparative electrophoresis for enzymes from both rabbit (Glowinski et al., 1980) and rat liver (Allaben and King, 1984).

The results of these studies demonstrated that the polymorphic N-acetylation of aromatic amines by rabbit liver cytosol was paralleled by a similar polymorphism of the N,O-acetyltransfer activity (Glowinski et al., 1980). Furthermore, these two activities were inseparable even when purified to homogeneity, as demonstrated by migration as a 33 kDa peptide on sodium dodecyl sulphate–polyacrylamide gel electrophoresis (SDS–PAGE). Subsequent studies with mouse liver acetyltransferase provided additional evidence that N-, O- and N,O-acetyltransfer were carried out on the same peptide (Mattano et al., 1989), although the production of reactive N-acetoxyarylamine products by the mouse enzyme is relatively low as compared to the analogous rabbit (Glowinski et al., 1980), rat or human enzymes (Land et al., 1989).

Conclusive evidence that all three acetylation activities can be carried out by a single peptide from rat liver has been obtained by use of immunochemical techniques. Purification of the acetylase by the above mentioned techniques discloses that these activities are inseparable. Development of a panel of monoclonal antibodies permitted purification by the use of immunoaffinity chromatography to yield a single, silver-stained peptide on SDS–PAGE of 32 kDa. The use of a monoclonal antibody with a lower binding affinity, 1A3, permitted recovery of all three enzyme activities as a peak that was resolved from the earlier-eluting, unbound protein contaminants. Affinity columns prepared from antibody with a higher binding affinity, 1F2, precluded elution of the three enzyme activities from the column. Elution of the 1F2 column with diethylamine at pH 11.2 yielded a single peptide of 32 kDa on both silver-stained SDS-PAGE and western blots of replicate gels.

Cytosolic material partially purified by fractional precipitation with ammonium sulphate yielded a major immunoreactive component that had an isoelectric point of approximately 4.4 with a trace of material that was slightly more basic. Direct enzyme assay of eluates of the gel segments demonstrated that the area with the antigenic components was the only fraction with activity. The other more highly purified preparations showed the presence of increasingly greater proportions of immunoreactive peptides at higher isoelectric points.

Although the enzyme is dependent on the presence of sulphydryl groups for these activities, and these groups may be responsible in part for the extreme lability of the enzyme, these data suggest that there is a progressive loss of acidic groups on purification that may be a more important factor in the loss of catalytic activity. Since the loss in activity occurs without perceptible change in behaviour on SDS-PAGE, proteolysis is unlikely to play an important role in the process.

16.3 CONCLUSION

The data presented support the conclusion that, in rat liver, a single peptide can account for all of the *N*-acetylation, *O*-acetylation and *N,O*-acetyl-transfer activities of this organ. Although the possibility cannot be excluded, on the basis of the characterizations given here, that microheterogeneity is responsible for these results, the methodology described should aid in further molecular biological and biochemical approaches that can resolve this issue.

REFERENCES

Allaben, W.T. and King, C.M. (1984) Purification of arylhydroxamic acid *N,O*-acyltransferase. *J. Biol. Chem.*, **259**, 12128–34.

Allaben, W.T., Weeks, C.E., Weis, C.C. *et al.* (1982) Rat mammary gland carcinogenesis after local injection of *N*-hydroxy-*N*-acyl-2-aminofluorenes: relationship to metabolic activation. *Carcinogenesis*, **3**, 233–40.

Allaben, W.T., Weis, C.C., Fullerton, N.F. and Beland, F.A. (1983) Formation and persistance of DNA adducts from the carcinogen *N*-hydroxy-2-acetylamino-fluorene in rat mammary gland *in vivo*. *Carcinogenesis*, **4**, 1067–70.

Bartsch, H., Dworkin, M., Miller, J.A. and Miller, E.C. (1972) Electrophilic *N*-acetoxyaminoarenes derived from carcinogenic *N*-hydroxy-*N*-acetylamino-arenes by enzymatic deacetylation and transacetylation in liver. *Biochim. Biophys. Acta*, **286**, 272–98.

Bartsch, H., Dworkin, C., Miller, E.C. and Miller, J.A. (1973) Formation of electrophilic *N*-acetoxyarylamines in cytosols from rat mammary gland and other tissues by transacetylation from the carcinogen *N*-hydroxy-2-acetylamino-biphenyl. *Biochim. Biophys. Acta*, **304**, 42–55.

Booth, J. (1966) Acetyl transfer in arylamine metabolism. *Biochem. J.*, **100**, 745–53.

Garner, R.C., Martin, C.N. and Clayson, D.B. (1984) Carcinogenic aromatic amines and related compounds, in *Chemical Carcinogens*, 2nd edn, (ed. C.E. Searle), ACS Monograph 182, American Chemical Society, Washington DC, pp. 175–276.

Glowinski, I.B., Fysh, J.M., Vaught, J.B. *et al.* (1980) Evidence for common genetic control of arylhydroxamic acid *N*,*O*-acyltransferase and *N*-acetyltransferase of rabbit liver. *J. Biol. Chem.*, **255**, 7883–90.

King, C.M. (1974) Mechanism of reaction, tissue distribution and inhibition of arylhydroxamic acid acyltransferase. *Cancer Res.*, **34**, 1503–16.

King, C.M. and Allaben, W.T. (1980) Arylhydroxamic acid acyltransferase, in *Enzymatic Basis of Detoxication* (ed. W.B. Jakoby), Academic Press, New York, pp. 187–97.

King, C.M. and Phillips, B. (1968) Enzyme-catalyzed reactions of the carcinogen *N*-hydroxy-2-fluorenylacetamide with nucleic acid. *Science*, **159**, 1351–3.

Kriek, E. (1965) On the interaction of *N*-2-fluorenylhydroxylamine with nucleic acids *in vitro*. *Biochem. Biophys. Res. Commun.*, **20**, 793–9.

Land, S.J., Zukowski, K., Lee, M-S. *et al.* (1989) Metabolism of aromatic amines: relationships of *N*-acetylation, *O*-acetylation, *N*,*O*-acetyltransfer and deacetylation in human liver and bladder. *Carcinogenesis*, **10**, 727–31.

Mattano, S.S., Land, S.J., King, C.M. and Weber, W.W. (1989) Purification and biochemical characterization of hepatic arylamine *N*-acetyltransferase from rapid and slow acetylator mice: identity with arylhydroxamic acid *N*,*O*-acyltransferase and arylamine *O*-acetyltransferase. *Mol. Pharmacol.*, **35**, 599–609.

Shirai, T., Fysh, J.M., Lee, M.-S. *et al.* (1981) *N*-Hydroxy-*N*-acylarylamines: relationship of metabolic activation to biological response in the liver and mammary gland of the female CD rat. *Cancer Res.*, **41**, 4346–53.

Weeks, C.E., Allaben, W.T., Tresp, N.M. *et al.* (1980) Effects of structure of *N*-acyl-*N*-2-fluorenylhydroxylamines on arylhydroxamic acid acyltransferase, sulfotransferase and deacylase activities, and on mutations in *Salmonella typhimurium* TA1538. *Cancer Res.*, **40**, 1204–11.

PART 4

Bioactivation of Nitrogenous Compounds and Cell Toxicity

17

Metabolism and activation of nitrosamines catalysed by cytochrome *P*-450 isoenzymes

C.S. Yang, T. Smith, H. Ishizaki, J.S.H. Yoo and J-Y. Hong

Department of Chemical Biology and Pharmacognosy, College of Pharmacy, Rutgers University, Piscataway, NJ, USA

1. A large number of nitrosamines are metabolized by cytochromes *P*-450, among which P450IIE1 has received much attention because of its role in the metabolic activation of *N*-nitrosodimethylamine (NDMA). This enzyme exists in many animal species and is inducible by fasting, diabetes, as well as by exposure to ethanol, acetone and other chemicals.
2. P450IIE1 represents the low K_m form of NDMA demethylase, and is the major enzyme catalysing the metabolic activation and denitrosation of this carcinogen. The mechanisms are discussed.
3. The substrate specificity and alkyl group selectivity in the metabolism of nitrosamines by *P*-450 isoenzymes are illustrated.
4. The metabolism of 4-(*N*-nitrosomethylamino)-1-(3-pyridyl)-1-butanone (NNK) is discussed to illustrate the metabolism of carcinogenic nitrosamines in non-hepatic target tissues.
5. The relationship between metabolism and carcinogenicity of nitrosamines is discussed.

17.1 INTRODUCTION

Nitrosamines are a group of nitrogenous compounds that are widely distributed in the environment and can also be synthesized endogenously (Bartsch and Montesano, 1984). Many of these compounds are potent carcinogens showing remarkable tissue and species specificity. Since the discovery of the carcinogenicity of NDMA, the metabolism of nitrosamines has been studied extensively (Magee and Barnes, 1967; Lai and Arcos, 1980). Although α-hydroxylation has been shown many years ago to be a key step in the activation of nitrosamines the enzymatic mechanisms of the metabolic activation of some of these compounds are not clearly understood. The lack of such information has hindered our progress in understanding the tissue and species specific carcinogenicity of nitrosamines. In this review, the enzymes and mechanisms involved in the activation of NDMA and other nitrosamines are discussed.

17.2 ENZYMOLOGY OF NDMA METABOLISM

The involvement of cytochrome *P*-450 in *N*-nitrosodimethylamine (NDMA) metabolism was demonstrated by various investigators (Czygan *et al.*, 1973; Lotikar *et al.*, 1975; Guengerich *et al.*, 1982). However, the role of *P*-450 in the metabolic activation of NDMA *in vivo* and *in vitro* has been questioned (Lai and Arcos, 1980; Yoo and Yang, 1985; Yoo *et al.*, 1987) because of the following observations: (1) NDMA demethylase is not induced by classical inducers such as phenobarbital and 3-methyl-cholanthrene. (2) Activity is not inhibited by well-known *P*-450 inhibitors such as metyrapone or SKF-525A. (3) Multiple K_m values have been determined for NDMA demethylase, and unrealistically high K_m values (>100 mM) were observed with phenobarbital-induced microsomes. (4) Rat liver microsomes and certain purified *P*-450 preparations were reported to be not effective in the activation of NDMA to a mutagen.

We initiated our work in 1980 with the hypothesis that the multiplicity of K_m values for NDMA demethylation is due to the catalytic activity of multiple forms of *P*-450; the *P*-450 species showing the lowest K_m and highest V_{max} most likely representing the enzyme responsible for activation of this carcinogen. In an effort to identify this key *P*-450 isoenzyme for the metabolism of NDMA, we re-examined the kinetic parameters for NDMA demethylation and studied the induction of the low K_m enzyme form. With uninduced rat liver microsomes, we observed a much lower K_m value of 50–70 μM (Peng *et al.*, 1982; Tu *et al.*, 1983), in addition to K_m values of 0.3–0.5 mM and 30–50 mM, which corresponded to the NDMA demethylases I and II, respectively (Lai and Arcos, 1980; Lake *et al.*, 1976). This low K_m

form of activity was induced by pretreatment of rats with ethanol, acetone, isopropanol, pyrazole, and other chemicals as well as by fasting and diabetes (Peng *et al.*, 1982, 1983; Tu *et al.*, 1981, 1983; Tu and Yang, 1983). However, it was not inducible by classical *P*-450 inducers such as phenobarbital and 3-methylcholanthrene (Hong *et al.*, 1987a). This enzyme, *P*-450ac, was purified from acetone-induced rat liver microsomes (Patten *et al.*, 1986). It is probably identical to *P*-450j purified from isoniazid-induced rat liver microsomes (Ryan *et al.*, 1985) and is orthologous to rabbit P450LM3a (Koop *et al.*, 1982, 1985). Similar orthologues probably also exist in humans (Wrighton *et al.*, 1986; Yoo *et al.*, 1988), mice, hamsters, guinea pigs, and other animal species (Yang *et al.*, 1985a; Hong *et al.*, 1989). In this review, we shall use the systematic name P450IIE1 for this type of *P*-450 in all species.

The alcohol-inducible P450LM3a (P450IIE1) is most efficient, among six purified rabbit liver *P*-450 forms, in catalysing the demethylation and denitrosation of NDMA (Yang *et al.*, 1985b). Some of the results are summarized in Table 17.1. Studies with purified rat liver *P*-450 isoenzymes in our and other laboratories also demonstrated that P450IIE1 is more active than other forms in catalysing the metabolism of NDMA (Tu and Yang, 1985; Patten *et al.*, 1986; Levin *et al.*, 1986). Other *P*-450 forms show substantial activities only at high substrate concentrations, reflecting high K_m values (Yang *et al.*, 1985b; Tu and Yang, 1985). The results are consistent with the concept that the multiple K_m values for microsomal

Table 17.1 Metabolism of nitrosamines by rabbit liver *P*-450 isoenzymes[a]

N-*Nitrosamine*	(*mM*)	Cytochrome P-450 isoenzymes					
		2	3a	3b	3c	4	6
		Turnover number (min^{-1})					
NDMA	(4)	0.10	5.89	0.03	0.08	<0.03	0.31
NDMA	(100)	2.21	6.71	1.45	0.79	2.01	4.35
NEMA	(4)	0.07	0.94	—	—	0.07	—
NBMA	(4)	1.73	0.23	—	—	0.10	—
NMBzA	(4)	0.50	0.28	—	—	0.08	—
NMA	(4)	5.57	2.51	—	—	1.57	—
NPYR	(5)	1.11	6.13	—	—	0.93	2.97
NPYR	(20)	2.50	7.05	2.22	1.48	3.14	7.31
NdiMMOR	(1)	1.63	0.25	0.04	0.04	0.04	0.04

[a] Results for NDMA, NEMA, NBMA, NMA and NMBzA are from Yang *et al.* (1985b); data for NPYR are from McCoy and Koop (1988) and those for NdiMMOR (*N*-nitroso-2,6-dimethylmorpholine) are from Kokkinakis *et al.* (1985). Dashed lines indicate that studies were not performed.

NDMA demethylase are due to the catalytic activity of multiple forms of P-450 in microsomes. The P450IIE1-dependent NDMA demethylase was inhibited by non-classical inhibitors such as 2-phenylethylamine, 3-amino-1,2,4-triazole, and pyrazole, but not effectively by the well-known P-450 inhibitor SKF-525A (Yang *et al.*, 1985b; Tu and Yang, 1985).

17.3 KINETICS AND MECHANISMS OF NDMA METABOLISM

Although, in a reconstituted NDMA demethylase system, P450IIE1 displays K_m values of 0.35 mM or higher (Patten *et al.*, 1986), this isoenzyme is believed to be the species responsible for the low K_m in NDMA demethylation by liver microsomes. The higher K_m values in the reconstituted systems are probably due to the assay conditions used. The K_m of NDMA demethylase is decreased in the presence of cytochrome b_5 (Patten *et al.*, 1986) and increases in the presence of glycerol serving as a competitive inhibitor. Both the reconstituted system and in microsomes, K_m values are increased in the presence of solvents such as acetone, ethanol, and dimethyl sulphoxide as well as by unknown inhibitors introduced into the incubation mixtures by the NADPH generating system (Yoo *et al.*, 1987). Considering these factors, NDMA demethylase K_m values of 15–22 μM were recently obtained using various amounts of acetone-induced microsomes suggesting correlation between K_m value and protein concentration. K_m values ranging from 16 to 24 μM were also obtained using different preparations of control, acetone-induced and fasting-induced microsomes. These values are close to the reported K_m values (8–25 μM) for NDMA metabolism in perfused liver, liver slices, and isolated hepatocytes (Skipper *et al.*, 1983; Hauber *et al.*, 1984; Swann, 1984).

As demonstrated in previous studies, the denitrosation and demethylation reactions are catalysed by the same P-450 isoenzymes and seem to involve a common intermediate; the ratio of the catalytic rates for these two reactions is about 1 : 10 under various inhibitory and induction conditions (Lorr *et al.*, 1982; Yang *et al.*, 1985b; Patten *et al.*, 1986). A proposed mechanism for NDMA demethylation and denitrosation is shown in Fig. 17.1 (Wade *et al.*, 1987). The α-nitrosamino radical is proposed as a common intermediate for both reactions (Wade *et al.*, 1987; Haussman and Werringloer, 1987). Recombination of this radical with $\cdot$OH at the active site of P450IIE1 would lead to the hydroxylation and demethylation (activation) pathway, whereas fragmentation of this radical species would produce an imine and nitric oxide which is subsequently oxidized to nitrite. The finding that superoxide dismutase partially inhibits formation of nitrite but not formaldehyde (Tu and Yang, 1985) suggests the involvement of

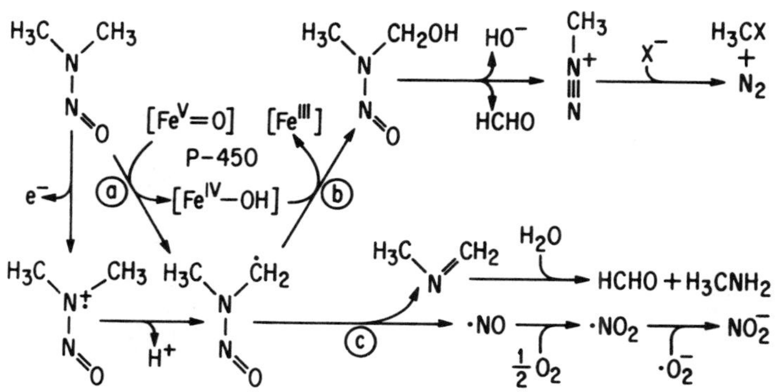

Figure 17.1 Proposed mechanism for the demethylation and denitrosation of NDMA (from Wade *et al.*, 1987).

superoxide in the oxidation of nitric oxide to nitrite. The imine would be hydrolysed to formaldehyde and methylamine. The latter should be formed in equimolar amounts to nitrite, as has been indeed observed (Keefer *et al.*, 1987). N-Oxidation of the unfragmented NDMA was not observed.

Antibodies against P450IIE1 inhibit NDMA demethylase activity in both control and acetone-induced microsomes and impair the denitrosation reaction to about the same extent (Fig. 17.2). This result is different from that of Amelizad *et al.* (1988) and is consistent with the proposal shown in Fig. 17.1. The results also show that the antibodies inhibit the reaction almost to completion, reflecting the key role of P450IIE1 in NDMA metabolism.

17.4 THE ROLE OF CYTOCHROME P450IIE1 IN THE ACTIVATION OF NDMA

The key role of NDMA demethylase in the activation of NDMA was demonstrated in several different systems. (1) P450IIE1 was more efficient than other *P*-450 species in the activation of NDMA to a mutagen in Chinese Hamster V79 cells (Yoo and Yang, 1985). At high NDMA concentrations, P450IIB1, which displays a high K_m value, also activates this mutagen. Microsomes and S-9 fractions, containing the same amount of P450IIE1, produce the same extent of mutagenicity. Many of the previous confusions concerning the role of *P*-450 in the activation of NDMA might be due to the use of inappropriate enzyme inducers or other inadequate experimental conditions, e.g. the presence of dimethyl sulphoxide. (2) Some of the

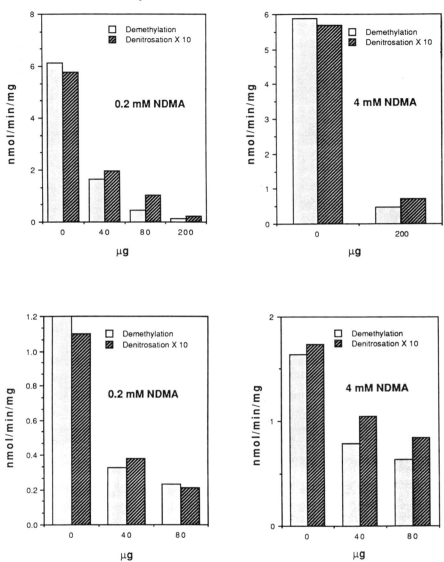

Figure 17.2 Inhibition of demethylation and denitrosation of NDMA by antibodies against P450IIE1. NDMA concentrations were 0.2 and 4.0 mM, respectively. Reactions were catalysed by acetone-induced rat liver microsomes (upper panels) or control microsomes (lower panels). Reaction rates in the presence or absence of antibodies are expressed in nmol/min/mg protein, except that denitrosation rates are at one-tenth of the scale.

previously reported species and age differences in rats and hamsters concerning the ability to activate NDMA could be interpreted in terms of the quantity of P450IIE1 present in the microsomes (Yoo *et al.*, 1987). (3) Pretreatment of rats with ethanol or acetone increases both NDMA-induced methylation of DNA *in vivo* and hepatotoxicity. However, enhancement is observed only when high doses of NDMA (>25 mg/kg body weight) are given to the rats (Hong and Yang, 1985; Lorr *et al.*, 1984), suggesting that there is an amount of P450IIE1 in untreated rats sufficient to metabolize low doses of NDMA. Alternatively, the result may suggest the involvement of other factors in the metabolic activation of NDMA.

Many previously observed effects on NDMA carcinogenicity can also be interpreted in terms of the catalytic activity of P450IIE1. For example, Keefer and Lijinsky (1973) observed that deuteration of NDMA decreased its carcinogenicity. Studies on the *in vivo* metabolic transformation of NDMA and [^{2}H$_6$]NDMA to CO_2 showed a large isotope effect when these two compounds were administered simultaneously and a small isotope effect when they were administered separately. The results are consistent with the observation that deuteration increases the K_m for NDMA demethylation (mediated by P450IIE1) but has no effect on the V_{max} value (Wade *et al.*, 1987). The fact that ethanol is a competitive inhibitor of NDMA metabolism is consistent with the report that ethanol decreases NDMA-induced hepatocarcinogenicity (Gricuite *et al.*, 1981).

17.5 REGULATION AND FUNCTIONS OF CYTOCHROME P450IIE1

The cDNAs and the complete gene sequence for rat and human P450IIE1 have been determined (Song *et al.*, 1986; Umeno *et al.*, 1988a,b). The human enzyme shares 78% amino acid similarity with the rat enzyme and immunochemically cross-reacts with antibodies prepared against the rat haemoprotein (Yoo *et al.*, 1988). P450IIE1 is inducible by a variety of environmental chemicals and by metabolic conditions such as fasting and diabetes (Fig. 17.3). The induction of NDMA demethylase activity under different conditions is due to the increase in quantity of this enzyme rather than post-translational modifications. The induction of P450IIE1 by acetone as well as by isopropanol, pyrazole, and 4-methylpyrazole is not accompanied by elevation of the P450IIE1 mRNA level (Song *et al.*, 1986; Hong *et al.*, 1987a,b). Acetone treatment may result in stabilization of the P450IIE1 protein (Song *et al.*, 1987). The induction of P450IIE1 by fasting and diabetes is accompanied by an elevation of P450IIE1 mRNA (Hong *et al.*, 1987a,b; Dong *et al.*, 1988). Nuclear run-on transcriptional analysis

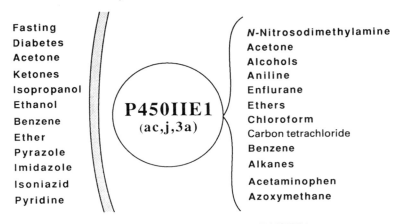

Fasting
Diabetes
Acetone
Ketones
Isopropanol
Ethanol
Benzene
Ether
Pyrazole
Imidazole
Isoniazid
Pyridine

P450IIE1
(ac,j,3a)

N-Nitrosodimethylamine
Acetone
Alcohols
Aniline
Enflurane
Ethers
Chloroform
Carbon tetrachloride
Benzene
Alkanes
Acetaminophen
Azoxymethane

Figure 17.3 Inducers and substrates for cytochrome P450IIE1.

failed to demonstrate an increased rate of transcription in the liver nuclei of diabetic rats, suggesting that P450IIE1 mRNA stabilization may be a mechanism of the induction (Song *et al.*, 1987). However, cycloheximide and actinomycin D inhibit induction of NDMA demethylase by acetone and fasting (Tu and Yang, 1983; and unpublished results) suggesting that other mechanisms may also be involved. P450IIE1 is regulated developmentally; transcription activation after birth has been demonstrated (Song *et al.*, 1986). In several mouse strains, the kidney P450IIE1 (or a similar enzyme) is regulated by testosterone (Hong *et al.*, 1989). The higher P450IIE1-associated NDMA demethylase activity found in male mice as compared to female mice might be correlated with the higher NDMA toxicity observed in the males (Mohla *et al.*, 1981).

P450IIE1 also catalyses the metabolism of other important environmental chemicals such as diethyl ether (Brady *et al.*, 1988), enflurane (Pantuck *et al.*, 1987), acetaminophen (Morgan *et al.*, 1983), carbon tetrachloride (Johansson and Ingelman-Sundberg, 1985), benzene (Johansson and Ingelman-Sundberg, 1988), alcohols (Morgan *et al.*, 1982), aniline (Patten *et al.*, 1986), *p*-nitrophenol (Koop, 1986), and other chemicals (Fig. 17.3). It is worth noting that many compounds are both substrates and inducers of P450IIE1. In addition, one substrate can serve as a competitive inhibitor of metabolism of another one. The understanding of the induction and function of P450IIE1 helps to explain, at least in part, many of the previously observed interactions between chemicals, such as the potentiation of carbon tetrachloride toxicity by alcohol (Traiger and Plaa, 1971) and the inhibition of nitrosamine metabolism *in vivo* by ethanol or diethyl ether.

17.6 SUBSTRATE SPECIFICITY AND ALKYL GROUP SELECTIVITY IN THE METABOLISM OF OTHER NITROSAMINES

Fasting or treatment of rats with P450IIE1-inducers enhances microsomal demethylase activities with several nitrosamines (Peng *et al.*, 1982; Tu and Yang, 1983; Tu *et al.*, 1983). However, this does not mean that P450IIE1 is the enzyme responsible for the metabolic activation of all types of nitrosamines. The enzyme and substrate specificities in the metabolism of various nitrosamines, the structures of which are shown in Fig. 17.4, are summarized in Table 17.1. Because of structural similarities, *N*-nitroso-ethylmethylamine (NEMA) and *N*-nitrosodiethylamine (NDEA) are perhaps preferentially metabolized by P450IIE1. P450IIE1 is also more active than other *P*-450 forms in catalysing α-oxygenation of *N*-nitroso-pyrrolidine (NPYR) (McCoy and Koop, 1988). With nitrosamines bearing larger alkyl chains, the situation can be very different. For example, *N*-nitroso-2,6-dimethylmorpholine (NdiMMOR) is metabolized more efficiently by the phenobarbital-inducible LM2 (IIB4) than by LM3a (IIE1) and other *P*-450 isoenzymes from rabbit liver microsomes (Kokkinakis *et al.*, 1985). In the case of *N*-nitrosobutylmethylamine (NBMA), *N*-nitroso-methylbenzylamine (NMBzA), and *N*-nitrosomethylaniline (NMA), demethylation is more effectively catalysed by the LM2 than by the LM3a isoenzyme. Furthermore, the phenobarbital-inducible rat P450IIB1 is more active in oxygenating the butyl and pentyl groups of NBMA and

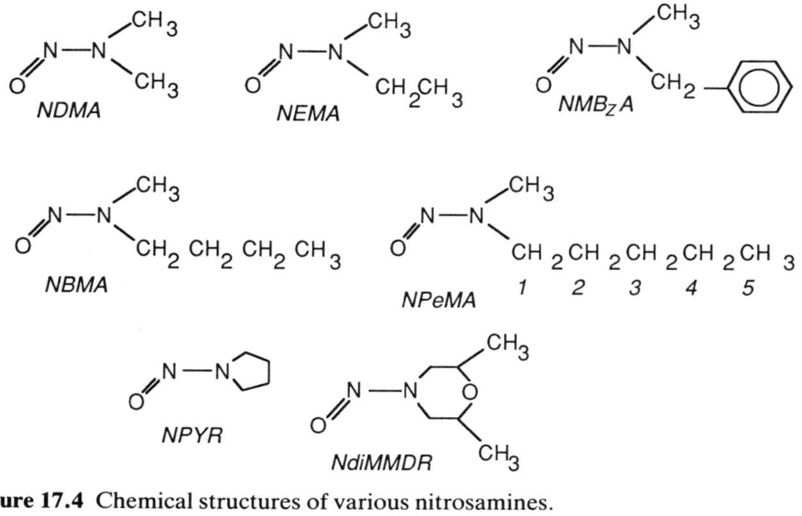

Figure 17.4 Chemical structures of various nitrosamines.

N-nitrosopentylmethylamine (NPeMA), respectively (Lee *et al.*, 1989; Ji *et al.*, 1989). Debutylation and depentylation of these compounds leads to the formation of methyldiazonium, which is likely to be the reactive species for carcinogenesis.

In the metabolism of asymmetrical N-nitrosodialkylamines, a key question concerns the selectivity of oxidative attack on the different alkyl groups. This is determined by the molar ratio of the enzymes in a specific tissue and the substrate specificities of the enzymes involved. For example, if P450IIE1 is the major enzyme form, demethylation is favoured over debutylation, since the butyl group is likely to bind to a hydrophobic pocket in the active site of P450IIE1, leaving the methyl group at the oxygenation site. Accordingly, the K_m for debutylation is higher than that for demethylation, and only the latter reaction is markedly decreased by the presence of cytochrome b_5 (Lee *et al.*, 1989).

In collaboration with other laboratories stereoselectivity in the metabolism of NPeMA by P450IIB1 was studied. P450IIB1 catalyses α-oxygenation of the pentyl and methyl groups of NPeMA at a ratio of 2:1. It also oxygenates other carbon atoms, the rates of formation of the 2-HO, 3-HO, 4-HO, and 5-HO metabolites accounting for only 0.6, 1.5, 26.5 and 2.4%, respectively, of the rate of pentaldehyde formation. The rate of denitrosation is 25% of that of depentylation. Similarly, N-nitroso-dibutylamine (NDBA) is activated by a presumably P-450-catalysed α-oxygenation pathway, leading to the formation of a butylating species. In addition, 2-HO, 3-HO, and 4-HO derivatives of NDBA are formed, and glucuronide conjugates were isolated from the urine (Okada, 1983).

17.7 METABOLISM OF 4-(N-NITROSO-METHYLAMINO)-1-(3-PYRIDYL)-1-BUTANONE (NNK)

Even though the P-450 concentration in the liver is much higher than in non-hepatic tissues, many nitrosamines can be effectively activated in non-hepatic target tissues. For example, NNK, a potent tobacco-specific lung carcinogen, is metabolized efficiently by lung microsomes. The pathways of metabolism, previously established by Hecht *et al.* (1980), are summarized in Fig. 17.5. In Sprague–Dawley rats liver microsomes are as active as lung microsomes in metabolizing NNK to a keto alcohol derivative, but are 200-times more active in the formation of NNal, a reductive product (Table 17.2). NNK N-oxide is formed in lung but not in liver microsomes. Nasal microsomes are about 100-times more active than liver and lung microsomes in transforming NNK to keto alcohol. Nasal microsomes are also active in the conversion of NNK into many other metabolites except

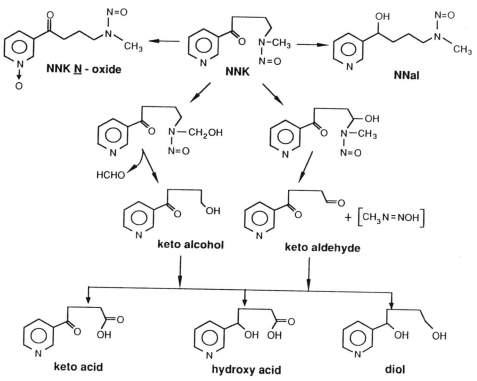

Figure 17.5 Metabolic pathway of NNK (modified from Hecht *et al.*, 1980).

NNal. With A/J mice, liver and lung microsomes show similar activities in the formation of keto alcohol. The lung has lower activity in catalysing conversion of NNK to NNal, but has much higher activity for the formation of NNK N-oxide (Table 17.2). With mouse lung microsomes, a K_m value of 5 μM and V_{max} of 58 pmol/min/mg protein were measured both for the formation of keto alcohol and NNK N-oxide, but a much higher K_m was observed for the formation of NNal, suggesting the involvement of a different enzyme.

The involvement of P450IIB1 in the metabolism of NNK was suggested by Devereux *et al.* (1988) and confirmed in our laboratory. Thus, antibodies against P450IIB1 inhibit keto alcohol formation from NNK by 40% in rat lung microsomes. Although to a smaller extent they also inhibit the formation of NNal and NNK N-oxide. With mouse lung microsomes, anti-P450IIB1 and anti-P450IA1 sera inhibit keto alcohol formation by 25 and 15%, respectively. They also inhibit formation of a metabolite characterized

Table 17.2 Metabolism of NNK in mice and rats[a]

Microsomes	NNal	Keto alcohol	NNK N-oxide	Diol pmol/min/mg	NNal N-oxide	Keto acid[b]	Hydroxy acid
Rat							
Liver	950.1	5.9	—	—	—	—	—
Lung	5.5	3.9	2.9	—	—	—	—
Nasal	—	451.0	7.7	15.5	7.5	281.7	6.6
Rat, acetone treated							
Nasal	—	150.3	2.8	0.9	—	11.6	0.7
Mouse							
Liver	65.5	50.0	—	—	—	—	—
Lung	14.7	37.1	33.3	—	—	2.0	—

[a] The microsomes (0.1–0.4 mg protein) were incubated with 10 or 50 μM [5-^{3}H]NNK in the presence of an NADPH generating system for 20 or 30 min. The metabolites were separated by reverse-phase HPLC. Abbreviations for the metabolites are as given by Hecht *et al.* (1980): NNal, 4-(methylnitrosamino)-1-(3-pyridyl)-1-butanol; keto alcohol, 4-oxo-4-(3-pyridyl-N-oxide)-1-butanol; NNK N-oxide, 4-(methylnitrosamino)-1-(3-pyridyl-N-oxide)-1-butanone; NNal-N-oxide, 4-(methylnitrosamino)-1-(3-pyridyl-N-oxide)-1-butanol; Keto acid, 4-oxo-4-(3-pyridyl)butyric acid; hydroxy acid, 4-hydroxy-4-(3-pyrridyl)butyric acid; Diol, 4-hydroxy-4-(3-pyridyl)-1-butanol.
[b] Identity remains to be established.

by the retention time of the keto acid by 19 and 23%, respectively. Anti-P450IA2 serum shows no inhibitory effect. The results suggest that more than half of the activity is due to other enzymes, and their identities remain to be established. Cyclo-oxygenase inhibitors, such as indomethacin and aspirin, inhibit the metabolism of NNK when present at concentrations of 100 and 300 μM, respectively. The N-oxidation pathway appears to be inhibited to the greatest extent. More work is needed to study the identities of the non-hepatic enzymes responsible for the metabolism of NNK and other nitrosamines.

17.8 ORGAN SPECIFICITY IN NITROSAMINE METABOLISM AND CARCINOGENESIS

The remarkable organ specificity of nitrosamine carcinogenesis can in part be interpreted in terms of the distribution of the activating enzymes in specific tissues. For example, the high capacity for NNK metabolism in rat nasal and mouse lung microsomes may be responsible for the fact that these organs represent main targets for NNK carcinogenesis.

The metabolism and carcinogenicity of different nitrosamines in the liver and oesophagus has been studied by many investigators. As summarized by Kleihues *et al.* (1987), the liver is much more effective than the oesophagus in activating NDMA, NEMA, and N-nitrosopropylmethylamine to a methylating agent, as determined by the formation of 7-methylguanine, but the oesophagus is more effective than the liver in the activation of NMBzA, NBMA, and NPeMA. If any, the oesophagus contains very low amounts of P450IIE1, and this is likely to be the reason for the low activity of this tissue towards NDMA and NEMA. Considering the enzymology of oesophageal metabolism of NMBzA, NBMA and NPeMA it is easy to understand why NDMA is a liver but not an oesophageal carcinogen, whereas NMBzA is an oesophageal carcinogen. However, the fact that NMBzA is activated in the liver but does not induce hepatocarcinogenesis suggests that factors other than metabolism must be involved.

Although cytochrome *P-450* enzymes are responsible for metabolism of the major portion of the known nitrosamines, they are not responsible for metabolism of all types of nitrosamines. For example, prostaglandin H synthetase has been suggested to be involved in the metabolism of NDEA by pulmonary endocrine cells (Schuller *et al.*, 1987). *P-450*-independent oxidative pathways may also be important in the metabolism of other nitrosamines, especially in non-hepatic tissues. A well-established *P-450*-independent non-oxidative pathway is illustrated by the activation of N-nitrosodiethanolamine and other β-hydroxyalkylnitrosamines (Michejda *et al.*, 1987). These compounds are refractory to microsomal oxidation but activated by conjugation with sulphate.

The aforementioned examples show that nitrosamines can be metabolically activated by reactions catalysed by different species of *P*-450 and other enzymes. Differences in organ distribution of these enzymes represent a key factor in determining the tissue specificity in carcinogenesis.

REFERENCES

Amelizad, A., Appel, K.E., Oesch, F. and Hildebrandt, A.G. (1988) Effect of antibodies against cytochrome *P*-450 on demethylation and denitrosation of *N*-nitrosodimethylamine and *N*-nitrosomethylaniline. *Cancer Res.*, **114**, 380–4.

Bartsch, H. and Montesano, R. (1984) Relevance of nitrosamines to human cancer. *Carcinogenesis*, **5**, 1381–93.

Brady, J.F., Lee M.J., Li, M. *et al.* (1988) Diethyl ether as a substrate for acetone/ethanol-inducible cytochrome *P*-450 and as an inducer for cytochrome *P*-450. *Mol. Pharmacol.*, **33**, 148–54.

Czygan, P., Greim, H., Garro, A.J. *et al.* (1973) Microsomal metabolism of dimethylnitrosamine and the cytochrome *P*-450 dependency of its activation to a mutagen. *Cancer Res.*, **33**, 2983–6.

Devereux, T.R., Anderson, M.W. and Belinsky, S.A. (1988) Factors regulating activation and DNA alkylation by 4-(*N*-methyl-*N*-nitrosamino)-1-(3-pyridyl)-1-butanone and nitrosodimethylamine in rat lung and isolated lung cells, and the relationship to carcinogenicity. *Cancer Res.*, **48**, 4215–21.

Dong, Z., Hong, J.Y., Ma, Q. *et al.* (1988) Mechanisms of induction of cytochrome *P*-450ac (*P*-450j) in chemically induced and spontaneously diabetic rats. *Arch. Biochem. Biophys.*, **263**, 29–35.

Griciute, L., Castegnaro, M. and Bereziat, J.C. (1981) Influence of ethyl alcohol on carcinogenesis with *N*-nitrosodimethylamine. *Cancer Lett.*, **13**, 345–52.

Guengerich, F.P., Dannan, G.A., Wright, S.T. *et al.* (1982) Purification and characterization of liver microsomal cytochrome *P*-450: electrophoretic, spectral, catalytic, and immunochemical properties and inducibility of eight isozymes isolated from rat treated with phenobarbital or β-naphthoflavone. *Biochemistry*, **21**, 6019–30.

Hauber, G., Frommberger, R., Remmer, H. and Schwenk, M. (1984) Metabolism of low concentrations of *N*-nitrosodimethylamine in isolated liver cells of the guinea pig. *Cancer Res.*, **44**, 1343–6.

Haussmann, H-J. and Werringloer, J. (1987) On the mechanism and the control of the denitrosation of *N*-nitrosomethylamine, *N-Nitroso Compounds: Relevance to Human Cancer* (eds I.K. O'Neill *et al.*), IRAC, Lyon, pp. 109–12.

Hecht, S.S., Young, R. and Chen, C.B. (1980) Metabolism in the F344 rat of 4-(*N*-methyl-*N*-nitrosamino)-1-(3-pyridyl)-1-butanone, a tobacco-specific carcinogen. *Cancer Res.*, **40**, 4144–50.

Hong, J. and Yang, C.S. (1985) The nature of microsomal *N*-nitrosodimethylamine demethylase and its role in carcinogen activation. *Carcinogenesis*, **6**, 1805–9.

Hong, J., Pan, J., Gonzalez, F.J. *et al.* (1987a) The induction of a specific form of cytochrome *P*-450 (*P*-450j) by fasting. *Biochem. Biophys. Res. Commun.*, **142**, 1077–83.

Hong, J., Pan, J., Dong, Z. and Yang, C.S. (1987b) Regulation of N-nitroso-dimethylamine demethylase in rat liver and kidney. *Cancer Res.*, **47**, 5948–53.

Hong, J-Y., Pan, J., Ning, S. and Yang, C.S. (1989) Molecular basis for the sex-related difference in renal N-nitrosodimethylamine demethylase in C3H/HeJ mice. *Cancer Res.*, **49**, 2973–9.

Ji, C., Mirvish, S.S. Nickols, J. *et al.* (1989) Formation of hydroxy derivatives, aldehydes and nitrite from N-nitrosomethyl-n-amylamine by rat liver microsomes and by purified cytochrome *P*-450 IIB1. *Cancer Res.*, **2**.

Johansson, I. and Ingleman-Sundberg, M. (1985) Carbon tetrachloride-induced lipid peroxidation dependent on an ethanol-inducible form of rabbit liver microsomal cytochrome *P*-450. *FEBS Lett.*, **183**, 265–9.

Johansson, I. and Ingleman-Sundberg, M. (1988) Benzene metabolism by ethanol, acetone, and benzene-inducible cytochrome *P*-450 (IIE1) in rat and rabbit liver microsomes. *Cancer Res.*, **48**, 5387–90.

Keefer, L.K. and Lijinsky, W. (1973) Deuterium isotope effect on the carcinogenicity of dimethylnitrosamine in rat liver. *J. Natl Cancer Inst.*, **51**, 299–302.

Keefer, L.K., Anjo, T., Wade, D. *et al.* (1987) Concurrent generation of methyl-amine and nitrite during denitrosation of N-nitrosodimethylamine by rat liver microsomes. *Cancer Res.*, **47**, 992–8.

Kleihues, P., von Hofe, E., Schmerold, I. *et al.* (1987) Organ specificity, metabolism and reaction with DNA of aliphatic nitrosomethylalkylamines. *Relevance of N-Nitroso Compounds to Human Cancer: Exposures and Mechanism* (eds H. Bartsch, I.K. O'Neill and R. Schulte-Hermann), IARC, Lyon, pp. 49–54.

Kokkinakis, D.M., Koop, D.R., Scarpelli, D.G. *et al.* (1985) Metabolism of N-nitroso-2,6-dimethylmorpholine by isozymes of rabbit liver microsomal cytochrome P450. *Cancer Res.*, **45**, 619–24.

Koop, D.R. (1986) Hydroxylation of p-nitrophenol by rabbit ethanol-inducible cytochrome *P*-450 isozyme 3a. *Mol. Pharmacol.*, **29**, 399–404.

Koop, D.R., Morgan, E.T., Tarr, G.E. and Coon, M.J. (1982) Purification and characterization of a unique isozyme of cytochrome *P*-450 from liver microsomes of ethanol-treated rabbits. *J. Biol. Chem.*, **257**, 8472–80.

Koop, D.R., Crump, B.L., Nordblom, G.D. and Coon, M.J. (1985) Immuno-chemical evidence for induction of the alcohol-oxidizing cytochrome *P*-450 of rabbit liver microsomes by diverse agents: ethanol, imidazole, trichloroethylene, acetone, pyrazole, and isoniazid. *Biochemistry*, **82**, 4065–9.

Lai, D.Y. and Arcos, J.D. (1980) Dialkylnitrosamine bioactivation and carcino-genesis. *Life Sci.*, **27**, 2149–65.

Lake, B.G., Phillips, J.C., Heading, C.E. and Gangolli, S.D. (1976) Studies on the *in vitro* metabolism of dimethylnitrosamine by rat liver. *Toxicology*, **5**, 297–309.

Lee, M.J., Ishizaki, H., Brady, J.F. and Yang, C.S. (1989) Substrate specificity and alkyl group selectivity in the metabolism of N-nitrosodialkylamines. *Cancer Res.*, **49**, 1470–4.

Levin, W., Thomas, P.E., Oldfield, N. and Ryan, D.E. (1986) N-Demethylation of N-nitrosodimethylamine by purified rat hepatic microsomal cytochrome *P*-450: isozyme specificity and role of cytochrome b_5. *Arch. Biochem. Biophys.*, **248**, 158–65.

Lorr, N.A., Tu, Y.Y. and Yang, C.S. (1982) The nature of nitrosamine denitrosation by rat liver microsomes. *Carcinogenesis*, **3**, 1039–43.

Lorr, N.A., Miller, K.W., Chung, H.R. and Yang, C.S. (1984) Potentiation of the hepatotoxicity of N-nitrosodimethylamine by fasting, diabetes, acetone and isopropanol. *Toxicol. Appl. Pharmacol.*, **73**, 423–31.

Lotlikar, P.D., Baldy, Jr, W.J. and Dwyer, E.N. (1975) Dimethylnitrosamine demethylation by reconstituted liver microsomal cytochrome *P*-450 enzyme system. *Biochem. J.*, **152**, 705–8.

Magee, P.N. and Barnes, J.M. (1967) Carcinogenic nitroso compounds. *Adv. Cancer Res.*, **10**, 163–246.

McCoy, G.D. and Koop, D.R. (1988) Reconstitution of rabbit liver microsomal *N*-nitrosopyrrolidine α-hydroxylase activity. *Cancer Res.*, **48**, 3987–92.

Michejda, C.J., Koepke, S.R., Kroeger-Koepke, M.B. and Bosan, W. (1987) Recent findings on the metabolism of β-hydroxyalkylnitrosamines, in *N-Nitroso Compounds to Human Cancer: Exposure and Mechanisms* (eds H. Bartsch, I.K. O'Neill and R. Schulte-Hermann), IARC Publications, Lyon, **84**, pp. 77–82.

Mohla, S., Ampy, F.R., Sanders, K.J. and Criss, W.E. (1981) Hormonal regulation of the metabolism of carcinogens in renal tissue of BALB/c mice. *Cancer. Res.*, **41**, 3821–3.

Morgan, E.T., Koop, D.R. and Coon, M.J. (1982) Catalytic activity of cytochrome *P*-450 isozyme 3a isolated from liver microsomes of ethanol-treated rabbits. *J. Biol. Chem.*, **257**, 13951–7.

Morgan, E.T., Koop, D.R. and Coon, M.J. (1983) Comparison of six rabbit liver cytochrome *P*-450 isozymes in formation of a reactive metabolite of acetaminophen. *Biochem. Biophys. Res. Commun.*, **112**, 8–13.

Okada, M. (1983) Comparative metabolism of *N*-nitrosamines in relation to their organ and species specificity, in *N-Nitroso Compounds: Occurrence, Biological Effects and Relevance to Human Cancer* (eds I.K. O'Neill, R. C. von Borstel, C.T. Miller, J. Long and H. Bartsch), IARC Publications, Lyon, **57**, pp. 401–9.

Pantuck, E.J., Pantuck, C.B. and Conney, A.H. (1987) Effect of streptozotocin-induced diabetes in the rat on the metabolism of fluorinated volatile anesthetics. *Anesthesiology*, **66**, 24–8.

Patten, C., Ning, S.M., Lu, A.Y.H. and Yang, C.S. (1986) Acetone-inducible cytochrome *P*-450: purification, catalytic activity and interaction with cytochrome b_5. *Arch. Biochem. Biophys.*, **251**. 629–38.

Peng, R., Tennant, P., Lorr, N.A. and Yang, C.S. (1983) Alterations of microsomal monooxygenase system and carcinogen metabolism by streptozotocin-induced diabetes in rats. *Carcinogenesis*, **4**, 703–8.

Peng, R., Tu, Y.Y. and Yang, C.S. (1982) The induction and competitive inhibition of a high affinity microsomal nitrosodimethylamine demethylase by ethanol. *Carcinogenesis*, **3**, 1457–61.

Ryan, D.E., Ramanthan, L., Iida, S. *et al.* (1985) Characterization of a major form of rat hepatic microsomal cytochrome *P*-450 induced by isoniazid. *J. Biol. Chem.*, **260**, 6385–93.

Schuller, H.M., Falzon, M., Gazdar, A.F. and Hegedus, T. (1987) Cell type-specific differences in metabolic activation of *N*-nitrosodiethylamine by human lung cancer cell lines, in *N-Nitroso Compounds to Human Cancer: Exposures and Mechanisms* (eds H. Bartsch, I.K. O'Neill and R. Schulte-Hermann), IARC Publications, Lyon, **84**, pp. 138–40.

Skipper, P.L., Tomera, J.F., Wishnok, J.S. *et al.* (1983) Pharmacokinetics model for *N*-nitrosodimethylamine based on Michaelis–Menten constants determined with isolated perfused rat liver. *Cancer Res.*, **43**, 4786–90.

Song, B.J., Gelboin, H.V., Park, S.S. *et al.* (1986) Complementary DNA and protein sequence of ethanol-inducible rat and human *P*-450: transcriptional and post-transcriptional regulation of the rat enzyme. *J. Biol. Chem.*, **261**, 16689–97.

Song, B.J., Matsunaga, T., Hardwick, J.P. *et al.* (1987) Stabilization of cytochrome *P*-450j mRNA in the diabetic rat. *Mol. Endocrinol.*, **1**, 542–7.

Swann, P.F. (1984) Effect of ethanol on nitrosamine metabolism and distribution. Implications for the role of nitrosamines in human cancer and for the influence of alcohol consumption on cancer incidence, in *N-Nitroso Compounds: Occurrence, Biological Effects and Relevance to Human Cancer* (eds. I.K. O'Neill, R.C. von Borstel, C.T. Miller, J. Long and H. Bartsch), IARC Sci. Publ., Lyon, 501–12.

Traiger, G.J. and Plaa, G.L.P. (1971) Differences in the potentiation of carbon tetrachloride in rats by ethanol and isopropanol pretreatment. *Toxicol. Appl. Pharmacol.*, **20**, 105–12.

Tu, Y.Y. and Yang, C.S. (1983) A high affinity nitrosamine dealkylase system in rat liver microsomes and its induction by fasting. *Cancer Res.*, **43**, 623–9.

Tu, Y.Y. and Yang, C.S. (1985) Demethylation and denitrosation of nitrosamines by cytochrome *P*-450 isozymes. *Arch. Biochem. Biophys.*, **242**, 32–40.

Tu, Y.Y., Sonnenberg, J., Lewis, K.F. and Yang, C.S. (1981) Pyrazole-induced cytochrome *P*-450 in rat liver microsomes: an isozyme with high affinity for dimethylnitrosamine. *Biochem. Biophys. Res. Commun.*, **103**, 905–12.

Tu, Y.Y., Peng, R.X., Chang, Z.F. and Yang, C.S. (1983) Induction of a high affinity nitrosamine demethylase in rat liver microsomes by acetone and isopropanol. *Chem. Biol. Interact.*, **44**, 247–60.

Umeno, M., McBride, O.W., Yang, C.S. *et al.* (1988a) Human ethanol-inducible P450IIE1: complete gene sequence, promoter characterization, chromosome mapping, and cDNA-directed expression. *Biochemistry*, **27**, 9006.

Umeno, M., Song, B.J., Kozak, C. *et al.* (1988b) The rat P450IIE1 gene: complete intron and exon sequence, chromosome mapping, and correlation of developmental expression with specific 5' cytosine demethylation. *J. Biol. Chem.*, **263**, 4956–62.

Wade, D., Yang, C.S., Metral, C.J. *et al.* (1987) Deuterium isotope effect on denitrosation and demethylation of *N*-nitrosodimethylamine by rat liver microsomes. *Cancer Res.*, **47**, 3373–7.

Wrighton, S.A., Thomas, P.E., Molowa, D.T. *et al.* (1986) Characterization of ethanol-inducible human liver *N*-nitrosodimethylamine demethylase. *Biochemistry*, **25**, 6731–5.

Yang, C.S., Koop, D.R., Wang, T. and Coon, M.J. (1985a) Immunochemical studies on the metabolism of nitrosamines by ethanol-inducible cytochrome *P*-450. *Biochem. Biophys. Res. Commun.*, **128**, 1007–13.

Yang, C.S., Tu, Y.Y., Koop, D.R. and Coon, M.J. (1985b) Metabolism of nitrosamines by purified rabbit liver microsomal *P*-450 isozymes. *Cancer Res.*, **45**, 1140–5.

Yoo, J.S.H., Guengerich, F.P. and Yang, C.S. (1988) Metabolism of *N*-nitrosodialkylamines by human liver microsomes. *Cancer Res.*, **48**, 1499–504.

Yoo, J.S.H., Ning, S.M., Patten, C. and Yang, C.S. (1987) Metabolism and activation of *N*-nitrosodimethylamine by hamster and rat microsomes: a comparative study with weaning and adult animals. *Cancer Res.*, **47**, 992–8.

Yoo, J.S.H. and Yang, C.S. (1985) Enzyme specificity in the metabolic activation of *N*-nitrosodimethylamine to a mutagen for Chinese hamster V79 cells. *Cancer Res.*, **45**, 5569–74.

18

Cytochrome *P*-450 catalysed activation of carcinogenic aromatic amines and amides

P.D. Lotlikar

Fels Institute for Cancer Research and Molecular Biology,
Temple University School of Medicine, Philadelphia, PA 19140, USA

1. Metabolic *N*-hydroxylation and ring-hydroxylations are considered as activation and inactivation steps respectively in the carcinogenesis by several aromatic amines and amides.
2. Both *N*- and ring-hydroxylations of 4-aminobiphenyl, 2-aminostilbene and their corresponding amides, and 2-aminoanthracene, 4,4'-methylene-bis-2-chloroaniline and *N*,*N*'-diacetylbenzidine by various mammalian hepatic microsomes are reviewed.
3. Evidence for participation of the cytochrome *P*-450 enzyme system in *N*- and *C*-oxidation of various aromatic amines is presented.
4. Kinetic parameters for *N*- and *C*-oxidation of aromatic amines and amides by hepatic microsomes and several constitutive forms of purified cytochromes *P*-450 from various mammalian species are also presented.

18.1 INTRODUCTION

Aromatic amines and amides need to be metabolized in any animal species before they are carcinogenic in that species, because large amounts of compounds are required for tumour induction and tumours appear at sites distant from those of administration (Miller and Miller, 1969, 1981). Epidemiological data have demonstrated that aromatic amines such as 2-naphthylamine, 4-aminobiphenyl and benzidine (Fig. 18.1) are responsible for the induction of bladder cancer among industrial workers (Scott, 1962).

Pioneering studies on the aromatic amine toxicity, carcinogenicity and metabolism have been carried out by several groups in three continents: Boyland at the Chester Beatty Research Institute, London; Gutmann at the University of Minnesota, Minneapolis, Minnesota; Kiese, at the University of Munich, Munich, Germany; Elizabeth and James Miller at the McArdle Laboratory for Cancer Research, University of Wisconsin, Madison, Wisconsin; Terayama at the University of Tokyo, Tokyo, Japan; and Elizabeth and John Weisburger at the National Cancer Institute, Bethesda, Maryland, USA. In addition to the earlier reviews on the toxicity, carcinogenicity and metabolism of aromatic amines and amides (Kiese, 1966; Miller and Miller, 1969, 1981; Weisburger and Weisburger, 1973; Kriek, 1974; Irving, 1979; Lotlikar and Hong, 1981; Garner *et al.*, 1984; Lotlikar, 1985), excellent reviews on the role of cytochromes *P*-450 in the metabolic activation and inactivation of these compounds have appeared recently (Guengerich *et al.*, 1987; Kadlubar and Hammons, 1987; Guengerich, 1988).

2-Acetylaminofluorene (AAF) has been used as a model compound by several laboratories in studying the mechanism of carcinogenesis by chemicals in general and aromatic amines in particular. AAF is readily ring-hydroxylated at several positions including 1-, 3-, 5-, 7- and 9 (Fig. 18.2). These ring-hydroxylation reactions are mediated via a hepatic microsomal enzyme system requiring NADPH and O_2 (Booth and Boyland, 1957; Seal

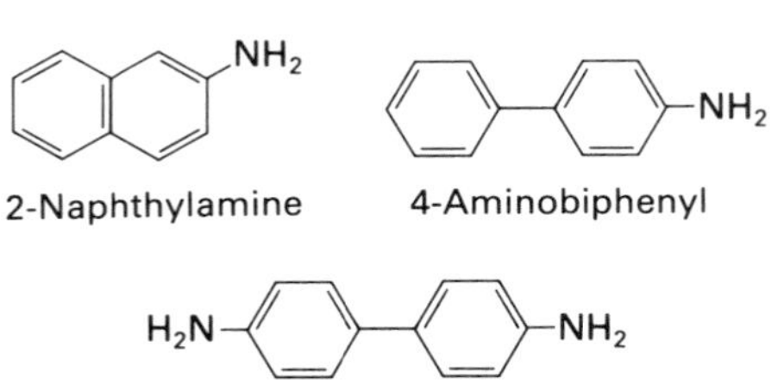

2-Naphthylamine   4-Aminobiphenyl

Benzidine

Figure 18.1 Human bladder carcinogens.

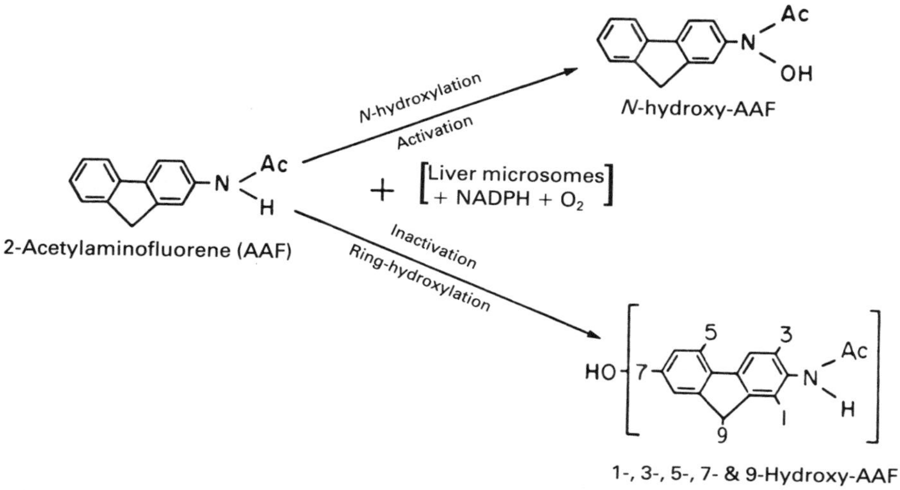

Figure 18.2 Metabolic hydroxylation of 2-acetylaminofluorene.

and Gutmann, 1959; Cramer *et al.*, 1960a; Benkert *et al.*, 1975). These non-carcinogenic monophenols are excreted primarily as glucuronide and sulphate conjugates in the urine and bile (Garner *et al.*, 1984 and references therein). Since the discovery of *N*-hydroxy-AAF glucuronide conjugate as a metabolite of AAF in the rat urine (Cramer *et al.*, 1960b), there is unequivocal evidence that *N*-hydroxylation is an activation step, whereas ring-hydroxylation is an inactivation step in the carcinogenesis by AAF and several other aromatic amines and amides (Miller and Miller, 1969, 1981).

18.2 AAF *N*-HYDROXYLATION BY LIVER MICROSOMES

Irving (1962) was the first to demonstrate AAF *N*-hydroxylation with rabbit liver microsomes in the presence of NADPH and O_2. Since then, several other laboratories have shown AAF *N*- and ring-hydroxylations with liver microsomes from cat, chicken, dog, hamster, mouse and rabbit (Table 18.1). In earlier studies (Irving, 1964; Lotlikar *et al.*, 1967), AAF *N*-hydroxylation could not be detected with guinea-pig liver microsomes. This lack of *N*-hydroxylation could explain the lack of AAF carcinogenicity in this species. However, later studies (Gutmann and Bell, 1977; Razzouk *et al.*, 1980b) demonstrated AAF *N*-hydroxylation with guinea-pig liver microsomes, though to a much smaller extent than with those from other

Table 18.1 AAF *N*- and ring-hydroxylation by control liver microsomes

Species	Oxidation N–	C–	References
Cat, chicken, dog, hamster, mouse, rabbit and rat	+	+	Irving (1962, 1964); Lotlikar *et al.* (1967); Matsushima *et al.* (1972); Thorgeirsson *et al.* (1973); Malejka-Giganti *et al.* (1978); Razzouk *et al.* (1980a, 1982); Aström *et al.* (1983); McManus *et al.* (1983b, 1984)
Guinea pig	+		Gutmann and Bell (1977); Razzouk *et al.* (1980b)
Monkey	+	+	Thorgeirsson *et al.* (1978); Razzouk *et al.* (1980b)
Human	+	+	Enomoto and Sato (1967); Dybig *et al.* (1979); McManus *et al.* (1983a); Shimada *et al.* (1989)

rodent species. Enomoto and Sato (1967) were the first to show AAF
N-hydroxylation with human liver microsomes. Detailed studies on AAF
N- and ring-hydroxylations by liver microsomes from the monkey and
humans have been reported by other laboratories (Table 18.1).

18.3 *N*-HYDROXYLATION OF VARIOUS AROMATIC AMIDES

In addition to AAF, data on *N*-hydroxylation of several other aromatic
amides by liver microsomes from different species have been summarized
in Table 18.2.

Table 18.2 *N*-Hydroxylation of aromatic amides by control liver microsomes from
various species

Aromatic amide	Species	References
4-Acetylaminobiphenyl (AABP)	Dog, rabbit	Booth and Boyland (1964); Brill and Radomski (1971)
3-AAF	Rat	Gutmann and Bell (1977)
7-Fluoro-AAF	Rabbit	Irving (1964)
2-Acetylaminonaphthalene (AAN)	Dog, rabbit	Brill and Radomski (1971)
4-Acetylaminostilbene (AAS)	Hamster, rat	Baldwin and Smith (1965); Gammans *et al.* (1977)
N-Acetylbenzidine (ABZ)	Rabbit	Booth and Boyland (1964)
N,N'-Diacetylbenzidine (DBZ)	Hamster, mouse, rat	Morton *et al.* (1979)

Some of the quantitative aspects of these studies need emphasis. Rate of *N*-hydroxylation of 2-AAF is 40 to 50 times higher than that of 3-AAF. The inability of rat liver microsomes to *N*-hydroxylate 3-AAF at a rate comparable to that of 2-AAF thus explains the lack of carcinogenic potency of the former (Gutmann and Bell, 1977). With rabbit liver microsomes, 7-fluoro-AAF yields 2 to 3 times more *N*-hydroxy metabolite than AAF as a substrate (Irving, 1964). Hamster liver microsomes catalyse *N*-hydroxylation of 4-acetylaminostilbene (AAS) and its 4'-chloro-, 4'-bromo- and 4'-fluoro analogues. The rate of *N*-hydroxylation of the 4'-bromo-analogue is about 16% of that of AAS, whereas the rates of *N*-hydroxylation of the 4'-chloro and 4'-fluoro analogues are only 3% of that of AAS (Gammons *et al.*, 1977).

18.4 *N*-OXIDATION OF AROMATIC AMINES

N-Oxidation of various aromatic amines such as 2-aminoanthracene, 4-aminobiphenyl (ABP), 2-aminofluorene (AF), 2-aminonaphthalene and MOCA has been observed with human liver microsomes in addition to liver microsomes from different species (Table 18.3). Unlike *N*-oxidation of 2-aminonaphthalene, *N*-oxidation of 1-aminonaphthalene could not be detected with liver microsomes from either dog, rat or humans. In contrast to *N*-oxidation, *C*-oxydation of these two amines could be easily detected with hepatic microsomes from all the species examined (Hammons *et al.*, 1985). These data are consistent with the known carcinogenicity of 2-amino-naphthalene and with the failure of 1-aminonaphthalene to induce tumours in experimental animals (Garner *et al.*, 1984 and references therein). *N*-Oxidation of 2-aminoanthracene has been demonstrated with human liver microsomes by an indirect way, namely *umu* gene response in a bacterial mutation assay (Shimada *et al.*, 1989).

18.4 The role of hepatic microsomal cytochrome *P*-450 in *N*- and *C*-oxidation of aromatic amines and amides

Several lines of evidence for participation of the liver microsomal cytochrome *P*-450 enzyme system in the *N*- and *C*-oxidation of aromatic amines are summarized in Table 18.4.

Studies from our laboratory (Lotlikar *et al.*, 1973) ruled out the possibility of the participation of mixed-function amine oxidase in AAF *N*-hydroxylation by hamster liver microsomes. The role of cytochrome *P*-450 in *N*-oxidation of aromatic amines was uncertain (Weisburger and

Table 18.3 Cytochrome *P*-450-mediated *N*-and ring-hydroxylation of aromatic amines by control liver microsomes

Compound	Species	Oxidation		References
		N–	C–	
2-Aminoanthracene (AA)	Human	+		Shimada *et al.* (1989)
4-Aminoazobenzene (AB)	Hamster, guinea pig, mouse, rabbit, rat	+		Kadlubar *et al.* (1976)
4-Aminobiphenyl (ABP)	Dog, guinea pig, hamster, human, mouse, rabbit, rat	+	+	Uehleke (1963); Booth and Boyland (1964); Brill and Radomski (1971); McMahon *et al.* (1980); Hammons *et al.* (1985); Butler *et al.* (1989); Shimada *et al.* (1989)
2-Aminofluorene (AF)	Dog, guinea pig, hamster, human, monkey, rat	+	+	Uehleke (1961, 1963); Kiese *et al.* (1966); Razzouk *et al.* (1980a, b); Frederick *et al.* (1982); Hammons *et al.* (1985); Shimada *et al.* (1989)
2-Aminonaphthalene (AN)	Dog, human, rabbit, rat	+	+	Uehleke (1963); Brill and Radomski (1971); Hammons *et al.* (1985)
4-Aminostilbene (AS)	Rat	+		Uehleke (1963)
4,4'-Methylene-bis-2-chloro-aniline (MOCA)	Human, rat	+	+	Morton *et al.* (1988); Butler *et al.* (1989)

Table 18.4 Evidence for participation of the cytochrome *P-450* enzyme system in *N*- and *C*-oxidation of aromatic amines and amides by control liver microsomes

Evidence	Substrate	Species	References
1. CO inhibition	AAF	Hamster, guinea pig, mouse, rat	Thorgeirsson *et al.* (1973); Lotlikar and Zaleski (1974); Gutmann and Bell (1973, 1977); Malejka-Giganti *et al.* (1978)
2. Substrate binding	AB	Rat	Kadlubar *et al.* (1976)
	ABP	Rat	McMahon *et al.* (1980)
	AAF	Guinea pig, hamster, rat	Gutmann and Bell (1977); Malejka-Giganti *et al.* (1978); Lotlikar *et al.* (1978)
3. NADPH–cytochrome *P-450* reductase antibody inhibition	AAF	Hamster, rat	Thorgeirsson *et al.* (1973); Kawajiri *et al.* (1983)
4. DPEA inhibition	AB	Rat	Kadlubar *et al.* (1976)
	ABP	Rat	McMahon *et al.* (1980)
	AF	Dog, human, pig, rat	Frederick *et al.* (1982)
	AAF	Human, rabbit, rat	McManus *et al.* (1984); Malejka-Giganti *et al.* (1985); Shimada *et al.* (1989)
5. ANF inhibition	ABP	Human	Butler *et al.* (1989); Shimada *et al.* (1989)
	AF	Guinea pig, human	Razzouk *et al.* (1980c); Shimada *et al.* (1989)

Weisburger, 1973) until 1973 when two laboratories showed that AAF N-hydroxylation by liver microsomes from several species was inhibited to a great extent in the presence of $CO-O_2$ mixtures (Gutmann and Bell, 1973; Thorgeirsson *et al.*, 1973). Subsequently, differences in the inhibition of AAF N- and ring-hydroxylations by CO were demonstrated, suggesting the involvement of multiple forms of cytochrome P-450 (Lotlikar and Zaleski, 1974).

Several laboratories have also shown AAF binding to cytochrome P-450 giving a characteristic type I binding spectrum. Such a binding phenomenon also suggests participation of the cytochrome P-450 enzyme system in the oxidation of AAF.

Inhibition of AAF N-hydroxylation by liver microsomes in the presence of antibodies prepared against NADPH–cytochrome P-450 reductase gave additional proof that AAF N-hydroxylation is indeed mediated via the cytochrome P-450 enzyme system (Thorgeirsson *et al.*, 1973). DPEA, a specific inhibitor of cytochrome P-450-mediated reactions, was first shown to inhibit AB N-oxidation by rat liver microsomes (Kadlubar *et al.*, 1976). Later studies demonstrated DPEA inhibition of N-oxidation of both ABP and AF by liver microsomes from several species (McMahon *et al.*, 1980; Frederick *et al.*, 1982). Another inhibitor widely used for demonstrating the role of cytochrome P-450 in drug oxidations is ANF. N-Oxidation of AAF, ABP and AF is inhibited by the presence of this agent (Table 18.4).

18.5 KINETIC STUDIES WITH MICROSOMAL PREPARATIONS

18.5.1 AAF N-Hydroxylation

The kinetic characteristics of AAF N-hydroxylation by control liver microsomes from various species are presented in Table 18.5.

Large differences seen in the K_m values for AAF in rat liver microsomes may be due to strain differences, even though much smaller differences in the rates of N-hydroxylation were observed in those two strains which are sensitive to AAF carcinogenesis (Razzouk *et al.*, 1980a; McManus *et al.*, 1983b). Comparative study with Sprague–Dawley rats indicates that the rate of AAF N-hydroxylation is not a limiting factor in the resistance of the cotton rat to AAF-induced carcinogenesis (Schut and Thorgeirsson, 1978). K_m values for AAF in microsomes from various species indicate a wide range from 0.03 μM for the rat to 8.6 μM for the monkey (Table 18.5). Hamster liver microsomes show the highest catalytic activity for AAF N-hydroxylation. There are no differences either in the K_m values or the catalytic activities with the non-inducible DBA mouse strain and the hydrocarbon-

Table 18.5 Kinetic constants for AAF N-hydroxylation by control liver microsomes

Species	K_m (μm)	AAF N-*hydroxylation* V_{max} (*pmol/min/mg protein*)	References
Guinea pig	1.8 ± 0.4	9.5 ± 0.7	Razzouk *et al.* (1980b)
Hamster	0.93 ± 0.15	140 ± 7.2	Razzouk *et al.* (1980a)
Human	1.63 ± 0.59	61 ± 14	McManus *et al.* (1983a)
Monkey (Rhesus)	8.6 ± 1.4	12.8 ± 1.2	Razzouk *et al.* (1980b)
Mouse			
DBA	0.47 ± 0.10	86 ± 5.7	Razzouk *et al.* (1982)
C57BL	0.33 ± 0.05	59 ± 4.5	Razzouk *et al.* (1982)
Rabbit	0.36 ± 0.03	46 ± 2.0	McManus *et al.* (1984)
Rat			
Wistar	0.53 ± 0.02	13.2 ± 0.4	Razzouk *et al.* (1980a)
Sprague–Dawley	0.03 ± 0.01	3.6 ± 1.6	McManus *et al.* (1983b)

inducible C57BL mouse strain. Human liver microsomes show intermediate catalytic activities among the species reviewed. Detailed kinetic data on AAF oxidations with rat, rabbit and human liver microsomes indicate that AAF N-hydroxylation is carried out by a single cytochrome *P*-450 species (McManus *et al.*, 1983a,b, 1984).

18.5.2 AAF ring-hydroxylation

Like N-hydroxylation, detailed studies on AAF ring-hydroxylations were also carried out with human, rabbit and rat liver microsomes. Kinetic parameters of AAF 7-hydroxylation are summarized in Table 18.6. The data indicate that 7-hydroxylation of AAF by hepatic microsomes from all three

Table 18.6 Kinetic constants for 7-hydroxylation of AAF by control liver microsomes

Species	K_{m1} (μM)	K_{m2}	AAF 7-*hydroxylation* V_{max1} (*pmol/min/mg protein*)	V_{max2}	References
Human	0.69 ± 0.29	75 ± 26	30 ± 8	286 ± 124	McManus *et al.* (1983a)
Rabbit	0.53 ± 0.07	253 ± 73	938 ± 47	3610 ± 531	McManus *et al.* (1984)
Rat	0.05 ± 0.02	103 ± 16	3.5 ± 1.4	1351 ± 76	McManus *et al.* (1983b)

species is mediated by at least two different cytochrome *P*-450 species, one species displaying a high affinity for AAF with relatively low metabolic capacity and a second one indicating a low affinity and high capacity. Similar biphasic kinetics suggesting participation of at least two *P*-450 species were observed for the 5-, 3- and 1-hydroxylation of AAF with rat liver microsomes (McManus *et al.*, 1983b) and for 5-hydroxylation with human liver (McManus *et al.*, 1984). Even though formation of 9-hydroxy-AAF represents a major metabolic pathway as compared with 7-hydroxylation, kinetic studies were inadequate for describing the rate of 9-hydroxy-AAF formation.

18.5.3 *N*-Oxidation of various amines and amides

The kinetic characteristics of aromatic amine and amide *N*-oxidation by hepatic microsomes from various species are presented in Table 18.7.

In contrast to AAF (Table 18.5), the rate of *N*-oxidation with AF is two orders of magnitude higher with both guinea-pig and monkey liver microsomes. The rate of *N*-oxidation of ABP with human liver microsomes is intermediate to AF and AN, whereas the rates of *N*-oxidation of AF and MOCA are similar.

It is very striking that the K_m values for AAS (Table 18.7) are three orders of magnitude higher than those reported for AAF (Table 18.5) with hamster liver microsomes.

Table 18.7 Kinetic parameters for *N*-oxidation of various aromatic amines and amides by control liver microsomes

Substrate	Species	N-Oxidation		References
		K_m (μM)	V_{max} (nmol/min/mg protein)	
AF	Guinea pig	0.95 ± 0.2	1.03 ± 0.06	Razzouk *et al.* (1980b)
	Monkey	3.3 ± 1.1	1.12 ± 0.21	Razzouk *et al.* (1980b)
	Dog	ND[b]	1.0 ± 0.6[a]	Frederick *et al.* (1982)
	Human	ND	1.2 ± 0.6[a]	Hammons *et al.* (1985)
	Rat	ND	1.6 ± 0.9[a]	
AAS	Hamster	2800 ± 200	2.87 ± 0.08	Gammans *et al.* (1977)
AN	Dog	ND	0.24 ± 0.01[a]	Hammons *et al.* (1985)
	Human	ND	$0.14(0.01-0.35)$[a]	Hammons *et al.* (1985)
	Rat	ND	0.12 ± 0.03[a]	Hammons *et al.* (1985)
ABP	Human	ND	$0.55(0.02-1.81)$[a]	Butler *et al.* (1989)
MOCA	Human	ND	$1.40(0.29-2.66)$[a]	Butler *et al.* (1989)

[a] These are initial rates of *N*-oxidation.
[b] ND denotes not determined.

18.6 RESOLUTION AND RECONSTITUTION OF THE CYTOCHROME *P*-450 ENZYME SYSTEM

The mammalian hepatic microsomal cytochrome *P*-450 enzyme system has been solubilized and resolved into three components: cytochrome *P*-450, NADPH–cytochrome *P*-450 reductase and phospholipid. All three factors are required for the oxidation of many physiological and xenobiotic compounds (Lu and Levin, 1974).

Earlier reconstitution studies with hamster (Lotlikar *et al.*, 1974), rat (Lotlikar and Zaleski, 1975) and mouse (Lotlikar and Wang, 1982) liver preparations demonstrated that even though both *P*-450 and reductase fractions are required for AAF *N*- and ring-hydroxylation, the degree of hydroxylation is determined by the source of cytochrome *P*-450. Lipid requirement could not be demonstrated in these studies because of incomplete removal of Triton X-100 from the cytochrome *P*-450 fractions. Later reconstitution studies, however, have shown requirement of lipid when cholate-solubilized cytochrome *P*-450 fraction was used (Lotlikar and Hong, 1981).

18.6.1 *N*-Oxidation of aromatic amines and amides with purified specific cytochrome *P*-450 forms

Several mammalian hepatic microsomal cytochrome *P*-450 species purified by affinity chromatography have now been used in reconstitution studies on *N*-oxidation of various aromatic amines and amides (Table 18.8). There are marked differences in the *N*-oxidation rates of various amines with different purified cytochromes *P*-450. Rates of AF and ABP *N*-oxidation with the rat cytochromes indicated P-450_d to exhibit the highest activity followed by P-450_c. In contrast to AF and ABP *N*-oxidation, rat P-450_b showed highest activity for the *N*-oxidation of MOCA followed by P-450_e. *N*-Oxidation of 2-AN could be seen only with rat P-450_d. No *N*-oxidation of 1-AN could be detected with any of the purified rat cytochromes *P*-450, even though these cytochromes catalyse ring oxidation of both substrates. Rat P-450_d, P-450_c and P-450_b also mediate ring hydroxylation of AF and 2-AN. Cytochrome P-450_h (P-450_{UT-A}), a major cytochrome *P*-450 in the uninduced male rat, displays appreciable ABP *N*-oxidase activity. Another constitutive enzyme P-450_a (P-450_{UT-F}), also catalyses ABP *N*-oxidation to an appreciable extent in the rat. It appears that the ratio of activation (*N*-oxidation) to inactivation (ring-oxidations) with all purified cytochromes *P*-450 is higher for ABP than for MOCA. It is, therefore, speculated that MOCA would have a lower carcinogenic potential than ABP (Butler *et al.*, 1989).

AAF *N*-hydroxylation is primarily carried out by cytochrome *P*-450 LM_4 in the rabbit, whereas it is mediated via P-450_d in the rat.

Table 18.8 Rates of N-oxidation of aromatic amines and amides by constitutive hepatic cytochromes P-450 from various species

Substrate	Species	Cytochrome P-450	Isoenzyme nmol/min/nmol P-450	References
AA	Human	$P\text{-}450_{PA}$		Shimada et al. (1989)
AAF	Rabbit	LM_4	2.1	Johnson et al. (1980)
		LM_4	0.78	McManus et al. (1984)
	Rat	$P\text{-}450_d$	1.4	Aström and Depierre
		$P\text{-}450_c$	1.1	(1985); Goldstein et al. 1986
	Human	$P\text{-}450_{PA}$		Shimada et al. (1989)
		$P\text{-}450_c$	5.2	
		$P\text{-}450_d$	7.4	
AF	Rat	$P\text{-}450_b$	1.06	Hammons et al. (1985)
		$P\text{-}450_h$	0.54	
		$P\text{-}450_{UT\text{-}H}$	0.44	
	Dog	$P\text{-}450_{II}$	0.76	Hammons et al. (1985)
		$P\text{-}450_I$	0.38	
	Human	$P\text{-}450_{PA}$		Shimada et al. (1989)
ABP	Rat	$P\text{-}450_d$	13.6	Butler et al. (1989)
		$P\text{-}450_c$	4.4	
		$P\text{-}450_a$	2.7	
		$P\text{-}450_h$	2.5	
		$P\text{-}450_b$	2.1	
		$P\text{-}450_{UT\text{-}H}$	0.32	
		$P\text{-}450_I$	0.26	
	Human	$P\text{-}450_{PA}$		Shimada et al. (1989)
AN	Rat	$P\text{-}450_d$	2.5	Hammons et al. (1985)
MOCA	Rat	$P\text{-}450_b$	8.9	Butler et al. (1989)
		$P\text{-}450_e$	6.6	
		$P\text{-}450_d$	2.7	
		$P\text{-}450_h$	2.1	
		$P\text{-}450_c$	1.8	
		$P\text{-}450_a$	0.45	
		$P\text{-}450_{UT\text{-}H}$	0.25	
		$P\text{-}450_I$	0.11	

$P\text{-}450_{PA}$ highly purified from human liver microsomes activates a series of aromatic amines such as AA, AF and ABP as well as amides such as AAF (Table 18.8); this cytochrome appears to be the human orthologue of rat $P\text{-}450_d$ and rabbit P-450 LM4. Activation was assayed on the basis of bacterial mutations induced through N-oxidation of aromatic amines and amides. The *umu* gene expression induced by different aromatic amines, such as AA, AF and ABP, was well correlated with phenacetin O-deethylase activity in different human liver microsomal preparations (Shimada *et al.*, 1989). However, benzidine, 1-NA, 2-NA, MOCA and *N*-acetylbenzidine did not show appreciable *umu* gene response with human liver microsomes.

18.7 CONCLUSIONS

It should be pointed out that several studies with human liver microsomes show a great variability in cytochrome *P*-450 profiles, suggesting large differences in carcinogen metabolism (Butler *et al.*, 1989; Shimada *et al.*, 1989). One factor for these differences appears to be a genetic polymorphism. Since there exist similarities in drug and carcinogen metabolism, it may be possible to phenotype humans by their rate of drug and carcinogen metabolism, as has been suggested previously (Gelboin, 1983). Recent (Butler *et al.*, 1989; Shimada *et al.*, 1989) and future studies of this kind thus would enable us to predict and categorize human populations at high risk for cancer induced by aromatic amines and other chemicals.

REFERENCES

Aström, A. and DePierre, J. (1985) Metabolism of 2-acetylaminofluorene by eight different forms of cytochrome *P*-450 isolated from rat liver. *Carcinogenesis (Lond.)*, **6**, 113–20.

Aström, A., Meijer, J. and DePierre, J.W. (1983) Characterization of the microsomal cytochrome *P*-450 species induced in rat liver by 2-acetylaminofluorene. *Cancer Res.*, **43**, 342–8.

Baldwin, R.W. and Smith, W.R.D. (1965) *N*-Hydroxylation in aminostilbene carcinogenesis. *Br. J. Cancer*, **19**, 433–43.

Benkert, K., Fries, W., Kiese, M. and Lenk, W. (1975) *N*-(9-hydroxy-9H-fluoren-2yl)acetamide and *N*-(9-oxo-9H-fluoren-2yl)acetamide: metabolites of *N*-(9H-fluoren-2yl)acetamide. *Biochem. Pharmacol.*, **24**, 1375–80.

Booth, J. and Boyland, E. (1957) The biochemistry of aromatic amines. 3. Enzymic hydroxylation by rat liver microsomes. *Biochem. J.*, **66**, 73–8.

Booth, J. and Boyland, E. (1964) The biochemistry of aromatic amines. 10. Enzymic *N*-hydroxylation of arylamines and conversion of arylhydroxylamines into *O*-aminophenols. *Biochem. J.*, **91**, 362–9.

Brill, E. and Radomski, J.L. (1971) Comparison of the *in vitro* and *in vivo* *N*-oxidation of the carcinogenic aromatic amines. *Xenobiotica*, **1**, 347–8.

Butler, M.A., Guengerich, F.P. and Kadlubar, F.F. (1989) Metabolic oxidation of the carcinogens 4-aminobiphenyl and 4,4'-methylene-bis(2-chloroaniline) by human hepatic microsomes and by purified rat hepatic cytochrome *P*-450 monooxygenases. *Cancer Res.*, **49**, 25–31.

Cramer, J.W., Miller, J.A. and Miller, E.C. (1960a) The hydroxylation of the carcinogen 2-acetylaminofluorene by rat liver: stimulation by pretreatment *in vivo* with 3-methylcholanthrene. *J. Biol. Chem.*, **235**, 250–6.

Cramer, J.W., Miller, J.A. and Miller, E.C. (1960b) N-hydroxylation: a new metabolic reaction observed in the rat with the carcinogen 2-acetylaminofluorene. *J. Biol. Chem.*, **235**, 885–8.

Dybing, E., Bahr, C.V., Aune, T. *et al.* (1979) *In vitro* metabolism and activation of carcinogenic aromatic amines by subcellular fractions of human liver. *Cancer Res.*, **39**, 4206–11.

Enomoto, M. and Sato, K. (1967) N-Hydroxylation of the carcinogen 2-acetyl-aminofluorene by human liver tissue *in vitro*. *Life Sci.*, **6**, 881–7.

Frederick, C.B., Mays, J.B., Ziegler, D.M. *et al.* (1982) Cytochrome *P*-450 and flavin-containing monooxygenase catalyze formation of the carcinogen N-hydroxy-2-aminofluorene and its covalent binding to nuclear DNA. *Cancer Res.*, **42**, 2671–7.

Gammans, R.E., Sehon, R.D., Anders, M.W. and Hanna, P.E. (1977) Microsomal N-hydroxylation of trans-4'-halo-4-acetamidostilbenes. *Drug Metab. Dispos.*, **5**, 310–16.

Garner, R.C., Martin, C.N. and Clayson, D.B. (1984) Carcinogenic aromatic amines and related compounds, in *Chemical Carcinogens* (ed. C.E. Searle), Vol. 1. American Chemical Society, Washington, pp. 175–276.

Gelboin, H.V. (1983) Carcinogens, drugs and cytochromes *P*-450. *N. Engl. J. Med.*, **309**, 105–7.

Goldstein, J.A., Weaver, R. and Sundheimer, D.W. (1986) Metabolism of 2-acetylaminofluorene by two 3-methylcholanthrene-inducible forms of rat liver cytochromes *P*-450. *Cancer Res.*, **44**, 3768–71.

Guengerich, F.P. (1988) Roles of cytochrome *P*-450 enzymes in chemical carcinogenesis and cancer chemotherapy. *Cancer Res.*, **48**, 2946–54.

Guengerich, F.P., Butler, M.A., MacDonald, T.L. and Kadlubar, F.F. (1987) Oxidation of carcinogenic aryl amines by cytochrome *P*-450 enzymes, in *Carcinogenic and Mutagenic Responses to Aromatic Amines and Nitroarenes* Elsevier, New York, pp. 89–95.

Gutmann, H.R. and Bell, P. (1973) Specificity of N-hydroxylation of arylamides. *Fed. Proc.*, **32**, 665.

Gutmann, H.R. and Bell, P. (1977) N-Hydroxylation of arylamides by the rat and guinea pig. Evidence for substrate specificity and participation of cytochrome *P*-450. *Biochim. Biophys. Acta*, **498**, 229–43.

Hammons, G.J., Guengerich, F.P., Weis, C.C. *et al.* (1985) Metabolic oxidation of carcinogenic arylamines by rat, dog and human hepatic microsomes and by purified flavin-containing and cytochrome *P*-450 monooxygenases. *Cancer Res.*, **45**, 3578–85.

Irving, C.C. (1962) N-Hydroxylation of the carcinogen 2-acetylaminofluorene by rabbit liver microsomes. *Biochim. Biophys. Acta*, **65**, 564–6.

Irving, C.C. (1964) Enzymatic N-hydroxylation of the carcinogen 2-acetylamino-fluorene and the metabolism of N-hydroxy-2-acetylaminofluorene-9-[14]C *in vitro*. *J. Biol. Chem.*, **239**, 1589–96.

Irving, C.C. (1979) Species and tissue variations in the metabolic activation of aromatic amines, in *Carcinogens: Identification and Mechanisms of Action* (eds A.C. Griffin and C.R. Shaw), Raven Press, New York, pp. 211–27.

Johnson, E.F., Levitt, D.S., Muller-Eberhard, U. and Thorgeirsson, S.S. (1980) Divergent pathways of carcinogen metabolism: metabolism of 2-acetylamino-fluorene by multiple forms of cytochrome *P*-450. *Cancer Res.*, **40**, 4456–60.

Kadlubar, F.F. and Hammons, G.J. (1987) The role of cytochrome *P*-450 in the metabolism of chemical carcinogens, in *Mammalian Cytochromes P-450* (ed. F.P. Guengerich), CRC Press, Boca Raton, pp. 81–130.

Kadlubar, F.F., Miller, J.A. and Miller, E.C. (1976) Microsomal *N*-oxidation of the hepatocarcinogen *N*-methyl-4-aminoazobenzene and the reactivity of *N*-hydroxy-*N*-methyl-4-aminoazobenzene. *Cancer Res.*, **36**, 1196–206.

Kawajiri, K., Yonekawa, H., Gotoh, O. *et al.* (1983) Contributions of two inducible forms of cytochrome *P*-450 in rat liver microsomes to the metabolic activation of various chemical carcinogens. *Cancer Res.*, **43**, 819–23.

Kiese, M. (1966) The biochemical production of ferrihemoglobin – forming derivatives from aromatic amines and mechanisms of ferrihemoglobin formation. *Pharmacol. Rev.*, **18**, 1091–161.

Kiese, M., Renner, G. and Wiedemann, I. (1966) *N*-Hydroxylation of 2-amino-fluorene in the guinea pig and by guinea pig liver microsomes *in vitro*, Naunyn-Schmiedeberg's. *Arch. Exp. Pathol. Pharmacol.*, **252**, 418–23.

Kriek, E. (1974) Carcinogenesis by aromatic amines. *Biochim. Biophys. Acta*, **355**, 177–203.

Lotlikar, P.D. (1985) Enzymatic *N*-hydroxylation of aromatic amides, in *Biological Oxidation of Nitrogen in Organic Molecules* (eds J.W. Gorrod and L.A. Damani), Ellis Horwood Limited, Chichester, pp. 163–74.

Lotlikar, P.D. and Hong, Y.S. (1981) Microsomal *N*- and *C*-oxidations of carcinogenic aromatic amines and amides, in *Carcinogenic and Mutagenic N-Substituted Aryl Compounds. J. Natl Cancer Inst. Monograph*, no. 58, pp. 101–7.

Lotlikar, P.D. and Wang, T.F. (1982) Ring- and *N*-hydroxylation of 2-acetylamino-fluorene by reconstituted mouse liver microsomal cytochrome *P*-450 enzyme system. *Toxicol. Lett.*, **11**, 173–9.

Lotlikar, P.D. and Zaleski, K. (1974) Inhibitory effect of carbon monoxide on the *N*- and ring-hydroxylation of 2-acetamidofluorene by hamster hepatic microsomal preparations. *Biochem. J.*, **144**, 427–30.

Lotlikar, P.D. and Zaleski, K. (1975) Ring- and *N*-hydroxylation of 2-acetamido-fluorene by rat liver reconstituted cytochrome *P*-450 enzyme system. *Biochem. J.*, **150**, 561–4.

Lotlikar, P.D., Enomoto, M., Miller, J.A. and Miller, E.C. (1967) Species variation in the *N*- and ring-hydroxylation of 2-acetylaminofluorene and effects of 3-methyl-cholanthrene pretreatment. *Proc. Soc. Exp. Biol. Med.*, **125**, 341–6.

Lotlikar, P.D., Wertman, K. and Luha, L. (1973) Role of mixed-function amine oxidase in *N*-hydroxylation of 2-acetamidofluorene by hamster liver microsomal preparations. *Biochem. J.*, **136**, 1137–40.

Lotlikar, P.D., Luha, L. and Zaleski, K. (1974) Reconstituted hamster liver microsomal enzyme system for *N*-hydroxylation of the carcinogen 2-acetyl-aminofluorene. *Biochem. Biophys. Res. Commun.*, **59**, 1349–55.

Lotlikar, P.D., Hong, Y.S. and Baldy, W.J. Jr (1978) Cytochrome *P*-450 dependent *N*-hydroxylation of 2-acetylaminofluorene, in *Biological Oxidation of Nitrogen* (ed. J.W. Gorrod), Elsevier, Biomedical Press, Amsterdam, pp. 185–93.

Lu, A.Y. and Levin, W. (1974) The resolution and reconstitution of the liver microsomal hydroxylation system. *Biochim. Biophys. Acta*, **344**, 205–40.

Malejka-Giganti, D., McIver, R.C., Glasebrook, A.L. and Gutmann, H.R. (1978) Induction of microsomal *N*-hydroxylation of *N*-2-fluorenylacetamide in rat liver. *Biochem. Pharmacol.*, **27**, 61–9.

Malejka-Giganti, D., Decker, R.W., Ritter, C.L. and Polovina, M. (1985) Microsomal metabolism of the carcinogen, *N*-2 fluorenylacetamide, by the mammary gland and liver of female rats. I. Ring- and *N*-hydroxylations of *N*-2-fluorenylacetamide. *Carcinogenesis (Lond.)*, **6**, 95–103.

Matsushima, T., Grantham, P.H., Weisburger, E.K. and Weisburger, J.H. (1972)

Phenobarbital-mediated increase in ring- and N-hydroxylation of the carcinogen N-2-fluorenylacetamide, and decrease in amounts bound to liver deoxyribonucleic acid. *Biochem. Pharmacol.*, **21**, 2043–51.

McMahon, R.E., Turner, J.C. and Whitaker, G.W. (1980) The N-hydroxylation and ring-hydroxylation of 4-aminobiphenyl *in vitro* by hepatic mono-oxygenases from rat, mouse, hamster, rabbit and guinea pig. *Xenobiotica*, **10**, 469–81.

McManus, M.E., Minchin, R.F., Sanderson, N. *et al.* (1984) Metabolic processing of 2-acetylaminofluorene by microsomes and six highly purified cytochrome P-450 forms from rabbit liver. *Carcinogenesis (Lond.)*, **5**, 1717–23.

McManus, M.E., Minchin, R.F., Sanderson, M.D. *et al.* (1983a) Kinetic evidence for the involvement of multiple forms of human liver cytochrome P-450 in the metabolism of acetylaminofluorene. *Carcinogenesis (Lond.)*, **4**, 693–8.

McManus, M.E., Minchin, R.F., Sanderson, N. *et al.* (1983b) Kinetics of N- and C-hydroxylations of 2-acetylaminofluorene in male Sprague–Dawley rat liver microsomes; implications for carcinogenesis. *Cancer Res.*, **43**, 3720–4.

Miller, J.A. and Miller, E.C. (1969) The metabolic activation of carcinogenic aromatic amines and amides. *Prog. Exp. Tumor Res.*, **11**, 273–301.

Miller, E.C. and Miller, J.A. (1981) Searches for ultimate chemical carcinogens and their reactions with cellular macromolecules. *Cancer*, **47**, 2327–45.

Morton, K.D., King, C.M. and Baetcke, K.P. (1979) Metabolism of benzidine to N-hydroxy-N,N'-diacetylbenzidine and subsequent nucleic acid binding and mutagenicity. *Cancer Res.*, **39**, 3107–13.

Morton, K.C., Lee, M.S., Siedlik, P. and Chapman, R. (1988) Metabolism of 4,4'-methylene-bis-2-chloraniline (MOCA) by rats *in vivo* and formation of N-hydroxy MOCA by rat and human liver microsomes. *Carcinogenesis* (Lond.), **9**, 731–9.

Ràzzouk, C., Batardy-Grégoire, M. and Roberfroid, M. (1982) Genetic differences in the enzymic properties of the aromatic hydrocarbon inducible N-hydroxylation of 2-acetylaminoafluorene in mouse liver. *Carcinogenesis (Lond.)*, **3**, 1325–9.

Razzouk, C., Mercier, M. and Roberfroid, M. (1980a) Induction, activation and inhibition of hamster and rat liver microsomal arylamide and arylamine N-hydroxylase. *Cancer Res.*, **40**, 3540–6.

Razzouk, C., Mercier, M. and Roberfroid, M. (1980b) Biochemical basis for the resistance of guinea pig and monkey to the carcinogenic effects of arylamines and arylamides. *Xenobiotica*, **10**, 565–71.

Razzouk, C., Mercier, M. and Roberfroid, M. (1980c) Characterization of the guinea pig liver microsomal 2-fluorenylamine and N-2-fluorenylacetamide and N-hydroxylase. *Cancer Lett.*, **9**, 123–31.

Schut, H.A.J. and Thorgeirsson, S.S. (1978) *In vitro* metabolism and mutagenic activation of 2-acetylaminofluorene by subcellular liver fractions from cotton rats. *Cancer Res.*, **38**, 2501–7.

Scott, T.S. (1962) *Carcinogenic and Chronic Toxic Hazards of Aromatic Amines*, Elsevier, Amsterdam.

Seal, U.S. and Gutmann, H.R. (1959) The metabolism of the carcinogen N-(2-fluorenyl)acetamide by liver cell fractions. *J. Biol. Chem.*, **234**, 648–54.

Shimada, T., Iwasaki, M., Martin, M.V. and Guengerich, F.P. (1989) Human liver microsomal cytochrome P-450 enzymes involved in the bioactivation of procarcinogens detected by *umu* gene response in *Salmonella typhimurium* TA 1535/pSK 1002, *Cancer Res.*, **49**, 3218–28.

Thorgeirsson, S.S., Jollow, D.J., Sasame, H.A. *et al.* (1973) The role of cytochrome P-450 in N-hydroxylation of 2-acetylaminofluorene. *Mol. Pharmacol.*, **9**, 398–404.

Thorgeirsson, S.S., Sakai, S. and Adamson, R.H. (1978) Induction of mono-oxygenases in Rhesus monkeys by 3-methylcholanthrene: metabolism and mutagenic activation of *N*-2-acetylaminofluorene and benzo[a]pyrene. *J. Natl Cancer Inst.*, **60**, 365–9.

Uehleke, H. (1961) *N*-hydroxylierung von 2-Aminofluorene durch Leber-mikrosomen. *Experientia*, **17**, 557.

Uehleke, H. (1963) *N*-hydroxylation of carcinogenic amines *in vivo* and *in vitro* with liver microsomes. *Biochem. Pharmacol.*, **12**, 219–21.

Weisburger, J.D. and Weisburger, E.K. (1973) Biochemical formation and pharmacological, toxicological and pathological properties of hydroxylamines and hydroxamic acids. *Pharmacol. Rev.*, **25**, 1–66.

19

Comparative biochemistry of cytochrome *P*-450 species responsible for the activation of mutagenic food-derived heterocyclic amines

T. Kamataki, M. Kitada, M. Komori, K. Ohta, T. Uchida, O. Kikuchi, M. Taneda and H. Fukuta

Division of Analytical Biochemistry, Faculty of Pharmaceutical Sciences, Hokkaido University, N12 W6 060 Japan

1. The mutagen-producing activities of liver microsomes from rats, beagle dogs, crab-eating monkeys and humans were compared using various heterocyclic amines as substrates. Liver microsomes from monkeys and humans activated IQ efficiently.
2. The abilities of adult and fetal human livers to activate promutagens were investigated. *P*-450HFLa and the recently purified *P*-450HFLb were found to be responsible for the activation of IQ.
3. *P*-450-D2 and *P*-450-D3 purified from liver microsomes from PCB-treated beagle dogs were capable of activating heterocyclic amine promutagens.
4. Two cDNA clones, λDah1 and λDah2, encoding cytochrome *P*-450, were isolated, and the sequences analysed and compared with corresponding forms of cytochrome *P*-450 from other animal species.
5. A cDNA clone, λMKah1, encoding cytochrome *P*-450 corresponding to rat *P*-450c, was isolated and the sequence analysed and compared with the forms from other animal species.

19.1 INTRODUCTION

Numerous promutagens have been isolated from the pyrolysates of proteins, amino acids and other constituents of foods. Extensive studies have shown that most of these agents are carcinogenic. The mechanisms involved in the induction of tumours by these compounds have also been examined (Sugimura and Sato, 1983; Matsukura *et al.*, 1981). Trp-P-2 (3-amino-1-methyl-5H-pyrido[4,3-b]indole), Glu-P-1 (2-amino-6-methyl-dipyrido[1,2-a: 3′,2′-d]imidazole) and IQ (2-amino-3-methylimidazo-[4,5-f]quinoline) are heterocyclic amines which are hydroxylated at the *N*-positions by cytochrome *P*-450 as a common activation step in mutagenicity (Ishii *et al.*, 1980, 1981; Yamazoe *et al.*, 1983; Kato, 1986). Despite these studies, few data have been reported on species differences in the activation of promutagens, especially with regard to beagle dogs, monkeys and humans. Thus, this review focuses on comparison of the mutagen-producing activities of liver microsomes and purified preparations of cytochrome *P*-450 from rats, beagle dogs, monkeys and humans using heterocyclic amines as the substrates. The results of molecular cloning and expression in yeast of certain forms of cytochrome *P*-450 from beagle dogs and monkeys will also be discussed.

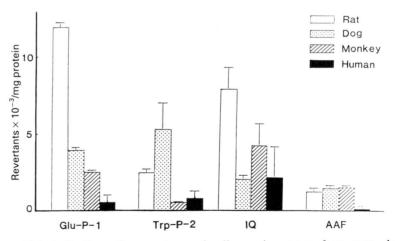

Figure 19.1 Activation of promutagens by liver microsomes from rats, dogs, monkeys and humans. Mutagenicity test was carried out with the Ames *Salmonella typhimurium* test strain TA98. The concentration of most promutagens used was 0.8 mM; the concentration of IQ was 0.2 mM. Microsomal protein added to the reaction mixtures was 150 μg in the assays with Glu-P-1 and IQ and 20 and 50 mg, respectively, in the assays with Trp-P-2 and AAF. The values represent the means ± SD of three to eight liver samples.

19.2 ACTIVATION OF MUTAGENIC FOOD-DERIVED HETEROCYCLIC AMINES BY LIVER MICROSOMES

The mutagen-producing activities of liver microsomes from rats, beagle dogs, monkeys and humans were compared using Glu-P-1, Trp-P-2 and IQ (Fig. 19.1). As can be seen, liver microsomes from rats, beagle dogs, monkeys and humans show highest catalytic capacities for the activation of Glu-P-1, Trp-P-2 and IQ, respectively. IQ is efficiently activated by liver microsomes from monkeys and humans.

The effect of pretreatment of rats, beagle dogs and monkeys with PCB (polychlorinated biphenyl, KC-500) on the mutagen-producing activities was examined. The enzyme(s) responsible for activation of Trp-P-2 is strongly induced in rats and monkeys, whereas induction in beagle dogs is relatively weak. These results do not indicate that enzyme(s) in the beagle dog are generally not inducible by PCB. In fact, pretreatment of the animals markedly induces activity for mutagen production from IQ.

19.3 ACTIVATION OF MUTAGENIC HETEROCYCLIC AMINES BY HUMAN FETAL LIVER CYTOCHROME *P*-450

Unlike in rats and rabbits, human liver contains cytochrome *P*-450 even during fetal life (Juchau *et al.*, 1980). Since knowledge of the toxicological significance of fetal cytochrome *P*-450 seems important, the mutagen-producing activities of fetal and adult human livers were compared. Fetal and adult livers were found to activate IQ, MeIQ (2-amino-3,4-

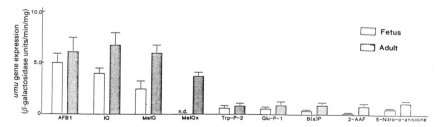

Figure 19.2 Activation of various promutagens by human fetal and adult livers. Mutagenicity test was carried out by measuring the induction of *umu* gene in *Salmonella typhimurium* TA 1535/pSK1002. The concentration of promutagens used was 10 μM. The values represent the means ± SD of five to six preparations, n.d. = not detectable.

dimethylimidazo[4,5-f]quinoline) and AFB_1 (aflatoxin B_1) (Fig. 19.2). It is noteworthy that MeIQx (2-amino-3,8-dimethylimidazo[4.5-f]-quinoxaline) was activated efficiently by adult but not by fetal livers.

IQ has been reported to be activated by rat P-450d and P-450c (Yamazoe *et al.*, 1984; Shimada and Okuda, 1988). 7,8-Benzoflavone is known as a representative inhibitor of these forms of cytochrome P-450. If a cytochrome P-450 species corresponding to these forms is responsible for the activation of IQ in human adult and fetal liver, 7,8-benzoflavone should inhibit activation of IQ. The results show that 7,8-benzoflavone inhibits activation in adult livers more strongly than in fetal livers (Table 19.1), suggesting the possibility that there exists an enzyme(s) in fetal liver capable of activating IQ, that is distinct from that in adult liver.

It has been previously reported that P-450HFLa, the major form of cytochrome P-450 in human fetal liver, is capable of activating various promutagens including pyrolysis products (Kitada *et al.*, unpublished observations, 1989). Thus, we looked for a possible correlation between the capacity of human fetal liver to activate IQ and the amount of P-450HFLa available. The results support the idea that the activation of IQ seen in human fetal liver is, in part, accounted for by P-450HFLa. As can be seen in Fig. 19.3, the capacity to activate IQ as well as Glu-P-1 and Trp-P-2 by the individual adult human liver microsomal preparations is in good correlation with the amount of cytochrome P-450 cross-reacting with antibodies to rat P-450d. As shown above, 7,8-benzoflavone inhibits activation of IQ by fetal livers. Since this activity is also inhibited by antibodies to rat P-450d, we started to purify another form of cytochrome P-450 cross-reacting with the antibodies. The purified preparation, designated P-450HFLb, is also capable of activating IQ. Activity is enhanced by the addition of cytochrome b_5 purified from phenobarbital-treated rats. Most interestingly, this enzyme preparation does not show detectable activity when Glu-P-1 is employed

Table 19.1 Effect of 7,8-benzoflavone on activation of mutagenic IQ in human fetal and adult livers

Addition	*β-Galactosidase activity (units/mg of protein)*			
	Fetus		*Adult*	
None	3.5	(100)	5.1	(100)
7,8-Benzoflavone				
100 μM	2.8	(79)	1.3	(26)
200 μM	2.5	(72)	0.8	(15)

Values represent the means of duplicate determinations. Figures in parentheses indicate percentage of control.

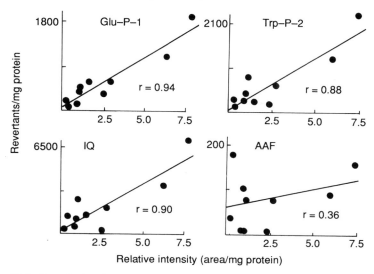

Figure 19.3 Correlation between the amount of cytochrome *P*-450 cross-reacting with anti-*P*-450d antibodies and the extent of activation of mutagens in human liver microsomes. The amount of human cytochrome *P*-450 immunochemically related to *P*-450d was quantified by Western blot analysis and PAP-staining. Activation of the mutagens was measured by using the Ames assay. The values are expressed as relative intensity of staining per mg of microsomal protein.

as the substrate. The precise properties of *P*-450HFLb are not known as yet. However, efforts to purify a form of cytochrome *P*-450 from liver microsomes of adults analogous to *P*-450HFLb were unsuccessful, suggesting that *P*-450HFLb may be absent in adult human livers.

19.4 PURIFICATION AND PROPERTIES OF P-450-*D2* AND P-450-*D3* FROM LIVER MICROSOMES OF PCB (KC-500)-TREATED DOGS

P-450-D2 and *P*-450-D3 were purified from liver microsomes of PCB-treated beagle dogs. These forms of cytochrome *P*-450 were distinct from each other in molecular weight and spectral properties. The molecular weights of *P*-450-D2 and *P*-450-D3 were determined to be 50 000 and 54 000 Da, respectively, as judged from sodium dodecyl sulphate (SDS)–polyacrylamide gel electrophoresis. The oxidized absolute spectra of *P*-450-D2 and *P*-450-D3 show absorption peaks at 394 and 416 nm, respectively, indicating that *P*-450-D2 is in the high-spin state, whereas *P*-450-D3

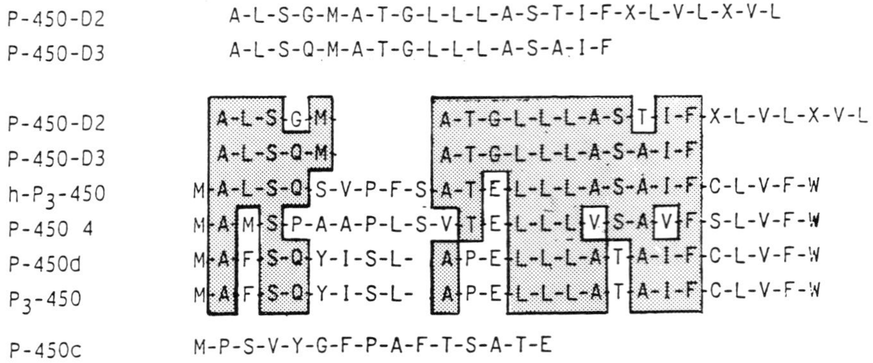

```
P-450-D2      A-L-S-G-M-A-T-G-L-L-L-A-S-T-I-F-X-L-V-L-X-V-L
P-450-D3      A-L-S-Q-M-A-T-G-L-L-L-A-S-A-I-F
```

```
P-450-D2      A-L-S-G-M-            A-T-G-L-L-L-A-S-T-I-F-X-L-V-L-X-V-L
P-450-D3      A-L-S-Q-M-            A-T-G-L-L-L-A-S-A-I-F
h-P₃-450    M-A-L-S-Q-S-V-P-F-S-A-T-E-L-L-L-A-S-A-I-F-C-L-V-F-W
P-450 4     M-A-M-S-P-A-A-P-L-S-V-T-E-L-L-L-V-S-A-V-F-S-L-V-F-W
P-450d      M-A-F-S-Q-Y-I-S-L-  A-P-E-L-L-L-A-T-A-I-F-C-L-V-F-W
P₃-450      M-A-F-S-Q-Y-I-S-L-  A-P-E-L-L-L-A-T-A-I-F-C-L-V-F-W
```

```
P-450c        M-P-S-V-Y-G-F-P-A-F-T-S-A-T-E
```

Figure 19.4 Comparison of *N*-terminal amino acid sequences of *P*-450-D2 and *P*-450-D3 with those of corresponding cytochrome forms of other species. Samples (500 pmol) of purified *P*-450-D2 and *P*-450-D3 were subjected to automated liquid-phase sequencing. Published data for amino acid sequences were taken as follows: h-P_3-450, Jaiswal *et al.* (1986); *P*-450 form 4, Okino *et al.* (1985); *P*-450c and *P*-450d, Haniu *et al.* (1984); P_3-450, Kimura *et al.* (1984). Residues in boxes indicate amino acids homologous to residues in *P*-450-D3.

exists in the low-spin state. The *N*-terminal amino acid sequences of these two forms of cytochrome *P*-450 were found to be highly homologous but not identical: there are only two amino acid replacements in the first 16-amino acid region (Fig. 19.4). In addition, the *N*-terminal domains of these cytochromes appeared to be similar to those of rat *P*-450d (Haniu *et al.*, 1984), human P_3-450 (Jaiswall *et al.*, 1986) and mouse P_3-450 (Kimura *et al.*, 1984), provided that four amino acid deletions are tentatively inserted

Table 19.2 Drug metabolizing enzyme activities of *P*-450-D2 and *P*-450-D3

Substrate	High-spin type		Low-spin type	
	P-450-D2	*P-450d*	*P-450-D3*	*P-450c*
Benzphetamine	39.2	23.1	48.6	19.6
Aminopyrine	38.3	30.1	45.2	20.4
7-Ethoxycoumarin	4.38	1.60	4.36	123
Aniline	0.35	5.51	0.58	0.70
p-Propoxyaniline	5.16	4.81	6.30	1.49
Benzo(a)pyrene	0.18	0.04	0.83	6.45

Numbers represent activities as expressed in nmol of product formed/min per nmol of cytochrome *P*-450.

Table 19.3 Activation of promutagens by *P*-450-D2 and *P*-450-D3 in a reconstituted system

Promutagens	Concentration (µM)	Test strain	High-spin type		Low-spin type	
			P-450-D2	*P*-450d	*P*-450-D3	*P*-450c
				(revertants $\times 10^{-3}$/nmol P-450)		
Trp-P-2	500	TA98	167	200	25	116
Glu-P-1	500	TA98	83	110	37	56
Aflatoxin Bl	2	TA100	4.6	8.8	0.6	3.7
Benzo(a)pyrene	40	TA100	1.0	1.7	1.3	4.7

between the fifth and sixth amino acids starting from the *N*-terminus. Both *P*-450-D2 and *P*-450-D3 mediate activation of Trp-P-2 and Glu-P-1 to different extents. *P*-450-D3 was expected to show catalytic properties similar to those of rat *P*-450c. The results shown in Tables 19.2 and 19.3, however, indicated that *P*-450-D3 is not necessarily homologous to rat *P*-450c. *P*-450-D2 shows higher capacity for the activation of Trp-P-2 and Glu-P-1 than does *P*-450-D3.

19.5 MOLECULAR CLONING AND EXPRESSION OF DOG LIVER CYTOCHROME *P*-450

Two cDNA clones, namely λDah1 and λDah2, encoding cytochrome *P*-450 probably corresponding to rat *P*-450c and *P*-450d were isolated from a cDNA library constructed from liver RNA of PCB-treated beagle dogs. λDah1 shows a chain length of 2394 bp and contains the entire coding region for 524 amino acid residues. λDah2 shows a length of 1623 bp and has an open reading frame consisting of 503 amino acid residues, but lacks the translational initiation codon. The *N*-terminal amino acid sequence of *P*-450(Dah2) is identical to that of *P*-450-D2 (Fig. 19.5). Analysing the

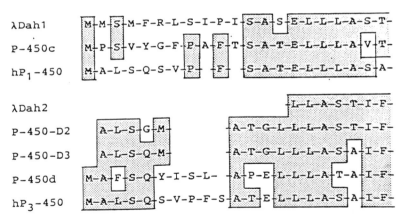

Figure 19.5 Comparison of the *N*-terminal amino acid sequences of λDah1 protein and λDah2 protein with those of corresponding forms of cytochrome *P*-450. Published data for amino acid sequences were obtained from the following sources. *P*-450c, Yabusaki *et al.* (1984); *P*-450-D3; Ohta *et al.*, (1989a); *P*-450d, Kawajiri *et al.* (1984); h-P_3-450, Quattrochi *et al.* (1986). Identical residues are indicated by shaded areas.

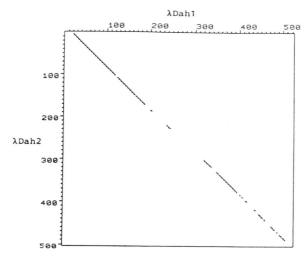

Figure 19.6 Dot matrix alignment of the deduced amino acid sequences of λDah1 and λDah2. The window size for a spot on the matrix was set at 80% similarity between 10 consecutive residues. The diagonal line indicates significant similarity.

extent of sequence homology it is concluded that λDah1 and λDah2 belonged to the *P*-450IA1 and *P*-450IA2 subfamilies, respectively. Dot matrix analysis of the deduced amino acid sequences for λDah1 and λDah2 (Fig. 19.6) shows that high identity of the sequences between these two clones exists at the *N*- and *C*-terminal regions, respectively.

Using fragments of cDNA in the 3'-non-coding region of λDah1 and λDah2 as the probes, the expression in organs of untreated and PCB-treated beagle dogs was measured by Northern blot analysis. The mRNAs for *P*-450(Dah1) and *P*-450(Dah2) were not detectable in most tissues from untreated beagle dogs, but small amounts of mRNA for *P*-450(Dah2) were present in the liver. PCB induced the expression of both mRNAs in all tissues, especially in the kidney.

An expression plasmid (pDC-1) containing the entire coding region of λDah1 was constructed with a vector plasmid pAAH5. The yeast *Saccharomyces cerevisiae* AH22 was transformed with the expression plasmid. The transformant was judged to express mRNA coding for *P*-450(Dah1). Western blot analysis also confirmed expression of *P*-450(Dah1) in the yeast. Results of preliminary experiments have shown that *P*-450(Dah1) expressed in yeast is capable of activating IQ, MeIQx and Glu-P-1.

19.6 MOLECULAR CLONING AND EXPRESSION OF MONKEY LIVER CYTOCHROME *P*-450

A cDNA clone, λMKah1, was isolated from a cDNA library in λgt11 constructed from liver RNA of PCB-treated crab-eating monkeys. λMKah1 shows a length of 2455 bp and contains the entire coding region for 512 amino acid residues. Comparison of the cDNA sequence of λMKah1 with those of cytochrome *P*-450 forms in the *P*-450IA1 and *P*-450IA2 subfamily from various animal species indicates that *P*-450(MKah1) belongs to the *P*-450IA1 enzyme family and is highly homologous to human cytochrome *P*-450 (Table 19.4). Based on the nucleotide sequences, the human and monkey *P*-450 was calculated to be 95% identical. Amino acid replacements were found to be randomly distributed over the entire peptide chains. In addition to this, interesting interspecies homologies or differences were found in the nucleotide sequences around the stop codon on each clone. As shown in Fig. 19.7, cDNAs coding for human, monkey and rabbit cyto-

Table 19.4 Comparison of nucleotide and deduced amino acid sequences of λMKahl with those of other *P*-450IA1 genes

	Human	*Dog*	*Rabbit*	*Rat*[a]	*Mouse*[b]
Nucleotide	95	79	69	69	70
Coding	95	84	81	81	82
3'-Non-coding	94	69	53	56	57
Amino acid	94	82	77	78	80

Values indicates percentage identity.
[a] Yabusaki *et al.* (1984).
[b] Kimura *et al.* (1984).

Monkey	CGC TCT ░TAG░GTG CTT GAG AGC CCT GAG GCC TAG
Human	CGC TCT ░TAG░GTG CTT GAG AGC CCT GAG GCC TAG
Dog	CGC ACT GAG GGG GCT GAG ACC CCT GCA GCC ░TAG░
Rabbit	CGC TTC GAG GCC ░TAG░ACT GTG CCC CGG GCC TAA
Rat	CGG TCT TCT GGT CCT CAG CAT CTC CAG GCT ░TAG░
Mouse	CGG TCT TCT GGT CCT CAG CAT CTT CTG CCT ░TAG░

Figure 19.7 Comparison of regions in the nucleotide sequences of various *P*-450IA1 genes located close to the stop codons. Known cDNA sequences were obtained from the following sources: rat *P*-450c, Yabusaki *et al.* (1984); mouse P_1-450: Kimura *et al.* (1984). Nucleotides in boxes indicate stop codons used in each DNA sequence.

chrome *P*-450 possess two stop codons, one of which is seen at corresponding sites in the cDNA of other animal species. Another stop codon appears in these species to shorten the peptide at the *C*-terminal region. At present, there is no clear explanation for the deletion of the *C*-terminal peptide during the evolutionary process, although it has been suggested that the *C*-terminal peptide region is not essential in supporting enzyme activity.

19.7 CONCLUSIONS

The strategy of the experiments described was to develop a method serving to predict the carcinogenic risk of certain compounds in humans.

The results reviewed that mutagen-producing activities vary widely among the individual animal species. A close correlation between humans and other species could not be found. Taking into account the large variations in activity of human livers, it appears that the situation in monkeys resembles more closely that in humans, as compared to other species (Komori *et al.*, 1988; Ohta *et al.*, 1989c). Examination, on a molecular basis, of similarities between corresponding cytochrome *P*-450 forms from various animal species indicates that enzyme from beagle dogs has many characteristic properties. Two forms of cytochrome *P*-450, termed *P*-450-D2 and *P*-450-D3, exhibiting very similar catalytic properties were purified. *P*-450-D2 and *P*-450-D3 also show high similarity in their *N*-terminal amino acid sequences. The results show for the first time that there are two forms in the *P*-450IA family characterized by very similar features. As shown in Fig. 19.4, homology of *P*-450-D2 and *P*-450-D3 to cytochrome *P*-450 forms from other animal species cannot be constructed unless deletion is assumed of four residues between the fifth and sixth amino acids starting from the *N*-terminus. Both the nucleotide and the amino acid sequences of the monkey cytochrome *P*-450(MKah1) show that this form of cytochrome *P*-450 is highly homologous to human cytochrome P_3-450. At present, the functions of the monkey and human cytochrome *P*-450 species have not been clarified. Attempts to purify the former cytochrome *P*-450 isoenzyme from liver microsomes of PCB-treated crab-eating monkeys were unsuccessful. However, it may be possible to compare the catalytic properties using preparations of cytochrome *P*-450 expressed in yeast.

Human fetal livers contain at least two forms of cytochrome *P*-450, *P*-450HFLa and *P*-450HFLb. A cDNA clone, λHFL33, probably coding for *P*-450HFLa (Komori *et al.*, 1989a) was obtained. The nucleotide sequence of this clone (Komori *et al.*, 1989b) is highly similar to that of NF 25 (Beaune *et al.*, 1986) isolated from a cDNA library constructed with RNA from adult human liver. Using synthetic oligo DNA fragments as probes, which are expected to hybridize specifically with mRNA from *P*-450(HFL33) and *P*-450(NF25), the latter two cytochrome *P*-450 species were indeed expressed both in fetuses and in adults. As mentioned above, *P*-450HFLb could not be detected in adult human livers. Therefore, both *P*-450HFLa and *P*-450HFLb might be expressed only during fetal life in humans. Based on these results it is concluded that functional properties of cytochrome *P*-450 in human fetuses cannot be deduced solely from studies on cyto-chrome *P*-450 in adult livers.

REFERENCES

Beaune, P.H., Umbenhauer, D.R., Bork, R.W. *et al.* (1986) Isolation and sequence determination of cDNA clone related to human cytochrome *P*-450 nifedipine oxidase. *Proc. Natl. Acad. Sci. USA*, **83**, 8064–8.

Haniu, M., Ryan, D.E., Iida, S. *et al.* (1984) NH_2-Terminal sequence analyses of four rat hepatic microsomal cytochrome *P*-450. *Arch. Biochem. Biophys.*, **235**, 304–11.

Ishii, K., Yamazoe, Y., Kamataki, T. and Kato, R. (1980) Metabolic activation of mutagenic tryptophan pyrolysis products by rat liver microsomes. *Cancer Res.*, **40**, 2596–600.

Ishii, K., Yamazoe, Y., Kamatkai, T. and Kato, R. (1981) Metabolic activation of glutamic acid pyrolysis products, 2-amino-6-methyldipyrido[1,2-a: 3′,2′-d]-imidazole and 2-aminodipyrido[1,2-a: 3′,2′d]imidozole, by purified cytochrome *P*-450. *Chem. Biol. Interact.* **38**, 1–13.

Jaiswall, K., Nebert, D.W. and Gonzalez, F.T. (1986) Human P_3-450: cDNA and complete amino acid sequence. *Nucleic Acids Res.*, **14**, 6773–4.

Juchau, M.R., Chao, S.T. and Omiecinski, C.J. (1980) Drug metabolism by the human fetus. *Clin. Pharmacokinet.* **5**, 320–59.

Kato, R. (1986) Metabolic activation of mutagenic heterocyclic aromatic amines from protein pyrolysates, *CRC Crit. Rev. Toxicol.*, **16**, 307–48.

Kawajiri, K., Gotoh, O., Sogawa, K. *et al.* (1984) Coding nucleotide sequence of 3-methylcholanthrene-inducible cytochrome *P*-450d cDNA from rat liver, *Proc. Natl. Acad. Sci. USA*, **81**, 1649–53.

Kimura, S., Gonzalez, F.J. and Nebert, D.W. (1984) The murine Ah locus comparison of the complete cytochrome P_1-450 and P_3-450 cDNA nucleotide and amino acid sequences. *J. Biol. Chem.*, **259**, 10705–13.

Komori, M., Hashizume, T., Ohi, H. *et al.* (1988) Cytochrome *P*-450 in human liver microsomes: high-performance liquid chromatographic isolation of three forms and their characterization. *J. Biochem. (Tokyo)*, **104**, 912–16.

Komori, M., Nishio, K., Ohi, H. *et al.* (1989a) Molecular cloning and sequence analysis of cDNA containing the entire coding region for human fetal liver cytochrome *P*-450. *J. Biochem. (Tokyo)*, **105**, 161–3.

Komori, M., Nishio, K., Fujitani, T. *et al.* (1989b) Isolation of a new human fetal liver cytochrome *P*-450 cDNA clone: evidence for expression of a limited number of forms of cytochrome *P*-450 in human fetal livers. *Arch. Biochem. Biophys.*, **273**, 219–25.

Matsukura, N., Kawachi, T., Morio, K. *et al.* (1981) Carcinogenicity in mice of mutagenic compounds from a tryptphan pyrolysate. *Science*, **213**, 346–7.

Ohta, K., Motoya, M., Komori, M. *et al.* (1989a) A novel form of cytochrome *P*-450 in beagle dogs: *P*-450-D3 is a low spin form of cytochrome *P*-450 but with catalytic and structural properties similar to PCB P-448-H (*P*-450d). *Biochem. Pharmacol.,* **38**, 91–6.

Ohta, K., Kitada, M., Hashizume, T. *et al.* (1989c) Purification of cytochrome *P*-450 from polychlorinated biphenyl-treated crab-eating monkeys: high homology to a form of human cytochrome *P*-450, *Biochim. Biophys. Acta,* **996**, 142–5.

Okino, S.T., Quattrochi, L.C., Barnes, H.J. *et al.* (1985) Cloning and characterization of cDNA encoding 2,3,7,8-tetrachlorodibenzo-*p*-dioxin-inducible rabbit mRNAs for cytochrome *P*-450 isozyme 4 and 6. *Proc. Natl. Acad. Sci. USA,* **82**, 5310–14.

Quattrochi, L.C., Pendurthi, U.R., Okino, S.T. *et al.* (1986) Human cytochrome *P*-450 4 mRNA and gene: part of a multigene family that contains Alu sequences in its mRNA. *Proc. Natl. Acad. Sci. USA,* **83**, 6731–5.

Shimada, T. and Okuda, Y. (1988) Metabolic activation of environmental carcinogens and mutagens by human liver microsomes. Role of cytochrome *P*-450 homologous to a 3-methylcholanthrene-inducible isozyme in rat liver. *Biochem. Pharmacol.,* **37**, 459–65.

Sugimura, T. and Sato, S. (1983) Mutagens-carcinogens in foods, *Cancer Res.,* **43**, 2415–21.

Yabusaki, Y., Shimizu, M., Murakami, H. *et al.* (1984) Nucleotide sequence of a full-length cDNA coding for 3-methylcholanthrene-induced rat liver *P*-450MC, *Nucleic Acids Res.,* **12**, 2929–38.

Yamazoe, Y., Shimada, M., Maeda, K. *et al.* (1984) Specificity of four forms of cytochrome *P*-450 in the metabolic activation of several aromatic amines and benzo(a)pyrene. *Xenobiotica,* **7**, 549–52.

Yamazoe, Y., Shimada, M. Kamataki, T. and Kato, R. (1983) Microsomal activation of 2-amino-3-methyl imidazo[4,5-f]quinoline, a pyrolysate of sardine and beef extracts, to a mutagenic intermediate. *Cancer Res.,* **43**, 5768–74.

20

Specificity and inducibility of cytochrome *P-450* catalysing the activation of food-derived mutagenic heterocyclic amines

A.R. Boobis, D. Sesardic, B.P. Murray, R.J. Edwards and D.S. Davies

Department of Clinical Pharmacology, Royal Postgraduate Medical School, Ducane Road, London W12 0NN, UK

1. During the normal cooking of meat highly mutagenic heterocyclic aromatic amines, such as MeIQx, DiMeIQx, MeIQ and IQ, are produced from endogenous constituents. The activation of these compounds requires *N*-hydroxylation, catalysed by cytochrome *P-450*, prior to the manifestation of any genotoxic effects.

2. The activation of these compounds to mutagenic intermediates by hepatic microsomal fractions in the Ames *Salmonella* test is increased considerably by pretreatment of rats or rabbits with inducers of the P-450IA subfamily of cytochrome *P-450*.

3. Following the selective destruction of P-450IA2 in 3-methylcholanthrene-induced rats by the *in vivo* administration of CCl_4 the mutagenic activation of MeIQx and related amines is decreased by 50%, but remains well above values with uninduced animals. This suggests that both P-450IA1 and P-450IA2 are involved in the activation of these amines.

4. An inhibitory antibody specific to P-450IA2 in the rat was raised against a synthetic peptide corresponding to a sequence unique to this isoenzyme. This antibody inhibited approx 50% of the activation of MeIQx and related amines, whereas a polyclonal antibody that cross-reacts with both P-450IA1 and P-450IA2 almost completely inhibited the mutagenic activation of these compounds.

5. Unlike the situation in the rat, in the rabbit the hepatic activation of MeIQx and related amines is completely inhibited by a specific monoclonal antibody to P-450IA2.

6. The methylxanthine furafylline has been shown to be a selective and potent inhibitor of P-450IA2 in man. This compound almost completely inhibited the mutagenic activation of MeIQx and related amines by

microsomal fractions of human liver, demonstrating that in this tissue P-450IA2 is almost exclusively involved in the activation of these dietary mutagens.

20.1 INTRODUCTION

In all recent surveys of the aetiology of human cancer, environment has been identified as a major factor (Wynder and Gori, 1977; Higginson and Muir, 1979; Doll and Peto, 1981). In most studies, diet has proven to be the single most important determinant, associated with more cancer deaths than even cigarette smoking. Indeed, in some studies as many as 35% of all cancer deaths have been attributed to dietary factors which exclude alcohol and food additives (Doll and Peto, 1981). There are numerous mechanisms whereby diet might influence the incidence of cancer. However, the identification of mutagenic chemicals in food formed during its processing and/or cooking (Sugimura *et al.*, 1977) has focussed considerable attention on the possible role of these compounds in human cancer. Recently, a number of extremely potent mutagens arising from the Maillard reaction, involving sugars, amino acids and creatine, have been isolated from cooked meat. These are all heterocyclic aromatic amines amongst which the most potent and abundant are 2-amino-3,8-dimethylimidazo[4,5-f]quinoxaline (MeIQx) and 2-amino-3,4,8-trimethylimidazo[4,5-f]quinoxaline (DiMeIQx) (Felton *et al.*, 1986). Smaller amounts of 2-amino-3-methylimidazo[4,5-f]-quinoline (IQ) and 2-amino-3,4-dimethylimidazo[4,5-f]quinoline (MeIQ) are also found, together with the less mutagenic but more abundant 2-amino-1-methyl-6-phenylimidazo[4,5-b]pyridine (PhIP) (Felton *et al.*, 1986). All these compounds are extremely potent mutagens in the Ames test when activated by mammalian hepatic subcellular fractions. In addition, they have been shown to be carcinogenic in rodent species (Sugimura, 1988). Thus, it is essential to assess the risk that these compounds pose to man.

In common with many carcinogenic compounds, the heterocyclic aromatic amines found in cooked food are not mutagenic unless metabolically activated by the cytochrome *P*-450-dependent mixed function oxidase system, via *N*-hydroxylation of an exocyclic amino group (Yamazoe *et al.*, 1988). As a consequence, susceptibility to the carcinogenicity of these compounds will to some extent be determined by the activity of the enzymes responsible for their activation. It is thus of some importance to establish which isoenzymes of cytochrome *P*-450 are responsible for the activation of these amines and to determine to what extent their activity varies between species and amongst individuals.

20.2 THE ROLE OF THE P-450IA SUBFAMILY IN THE ACTIVATION OF FOOD MUTAGENS AND OTHER AROMATIC AMINES IN VARIOUS SPECIES

The metabolic activation of food mutagens such as MeIQx by hepatic microsomal fractions from rats is affected by pretreatment of the animals with classical inducers of mono-oxygenase activity (Gooderham *et al.*, 1988). Thus, the ability to activate MeIQx is considerably increased by prior treatment of the animals with inducers of the P-450IA subfamily, such as 3-methylcholanthrene (MC) or isosafrole (ISF), but not by acetone (an inducer of P-450IIE1), pregnenolone-16α-carbonitrile (an inducer of P-450IIIA) or clofibrate (an inducer of P-450IVA) (Fig. 20.1). Phenobarbitone, an inducer of predominantly P-450IIB1 and IIB2 causes a modest increase in the activation of MeIQx. Thus, the major inducible isoenzymes responsible for the activation of these aromatic amines belong to

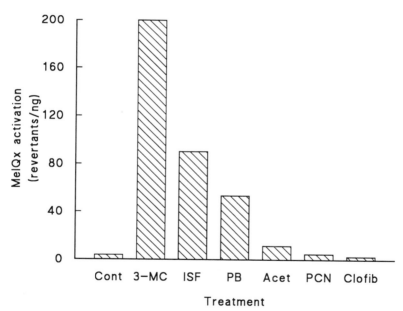

Figure 20.1 Effects of pretreatment of rats with inducers of different isoenzymes of cytochrome *P*-450 on the hepatic microsomal activation of MeIQx to a mutagen in the Ames/*Salmonella* test. Rats were either left untreated (Cont), or they were treated with 3-methylcholanthrene (3-MC), isosafrole (ISF), phenobarbitone (PB), acetone (Acet), pregnenolone-16α-carbonitrile (PCN) or clofibrate (Clofib). The treatment regimens were such that they caused maximum induction of hepatic *P*-450.

the hydrocarbon-inducible subfamily. There are two polycyclic aromatic hydrocarbon (PAH)-inducible forms of cytochrome P-450 in mammalian species, products of the *CYPIA1* and *CYPIA2* genes, respectively (Nebert *et al.*, 1989). Both MC and ISF induce these two isoenzymes in the liver of rats (Sesardic *et al.*, 1990b).

The contribution of these two isoenzymes to the activation of aromatic amines was first investigated by selectively destroying the hepatic content of P-450IA2. This was accomplished by treating MC-induced rats with a high dose of carbon tetrachloride up to 6 hours before killing the animals (Sesardic *et al.*, 1989). This treatment results in the destruction of essentially all of the P-450IA2 in the liver within 6 hours whilst having no effect on the hepatic content of P-450IA1 (Table 20.1). There was a parallel loss of activities dependent on P-450IA2, such as the high affinity component of phenacetin *O*-de-ethylase (POD) whereas activities dependent on P-450IA1, such as aryl hydrocarbon (benzo[*a*]pyrene) hydroxylase (AHH), remained unaltered. Carbon tetrachloride treatment of MC-induced rats reduced the activation of MeIQx by hepatic microsomal fractions by approximately 50%, although the activity remaining was still above that in control animals. This suggests that both P-450IA1 and IA2 are responsible for the activation of MeIQx in the liver of hydrocarbon-induced rats. Similar results have been reported for the *N*-hydroxylation of 2-acetylamino-fluorene (2-AAF) in this species (Åström and DePierre, 1985).

Although the selective destruction of P-450IA2 by carbon tetrachloride is a useful means of investigating the contribution of this isoenzyme to the oxidation of a compound, there are obviously considerable limitations in this approach. An alternative, more widely applicable technique is that of inhibition by specific antibodies. One of the difficulties in raising such antibodies against isoenzymes of cytochrome *P*-450 is the high degree of homology that exists between members of the same subfamily (Nebert *et al.*, 1989). Thus, polyclonal antibodies raised against a purified isoenzyme tend to cross-react with other, closely related, isoenzymes (Edwards *et al.*, 1990). Whilst many monoclonal antibodies are much more specific, they suffer from the disadvantage that such antibodies are often not inhibitory (Thomas *et al.*, 1984), and it is difficult to ensure that such antibodies will have this property during the initial screening process. An alternative approach has been adopted to produce specific inhibitory antibodies. This method combines the specificity associated with monoclonal antibodies with the ease of production of polyclonal antibodies. The specificity of the antibodies is ensured by raising them against peptides corresponding to a unique sequence of the isoenzyme of interest (Edwards *et al.*, 1988). However, to be inhibitory, the sequence against which an antibody is raised must also be involved in the functional activity of the protein.

Table 20.1 Effect of treatment of 3-methylcholanthrene (3-MC)-induced rats with CCl_4 *in vivo* on hepatic microsomal P-450IA content and mono-oxygenase activities. Rats were treated with 3-MC 60 h before administration of CCl_4. Animals were then killed at the times shown after CCl_4 administration. Values are mean $\pm$ SD ($n = 3$).

| Time after CCl_4 (h) | Specific content (pmol/mg) | | POD activity (pmol/min/mg of protein) | AHH activity (nmol/min/mg of protein) | MeIQx activation (revertants/ng) |
	P-450IA1	P-450IA2			
Untreated					
0	< 0.5	9 ± 1	20 ± 4	0.9 ± 0.1	4.0 ± 1.1
3-MC					
0	350 ± 90	125 ± 10	1250 ± 50	9.5 ± 2.0	71.4 ± 0.6
1	310 ± 78	37 ± 8	360 ± 40	9.1 ± 0.8	53.7 ± 1.7
3	544 ± 75	22 ± 6	347 ± 33	9.3 ± 1.2	49.0 ± 5.6
6	428 ± 54	5 ± 3	49 ± 20	7.9 ± 0.4	41.5 ± 5.0

A model of the three-dimensional structure of mammalian cytochromes *P*-450 has been produced (Edwards *et al.*, 1989), based on the crystallographic structure of P-450cam and alignment of the mammalian cytochromes *P*-450. This, together with predictions of α-helical structures, loop and turn regions, has enabled the identification of surface areas of mammalian cytochromes *P*-450 (Edwards *et al.*, 1990). One of these regions was considered almost certainly to be involved in the function of the protein, by virtue of the fact that chemical modification of a single amino acid located within this sequence resulted in reduction of enzyme activity (Parkinson *et al.*, 1986). The corresponding peptide was synthesized and coupled to a carrier protein which was then used to immunize rabbits (Edwards *et al.*, 1990). Antiserum was obtained and this had the predicted specificity, reacting only with P-450IA2, unlike a polyclonal antibody against the native protein, which recognized P-450IA1 as well as P-450IA2 (Edwards *et al.*, 1990). The polyclonal antibody against the purified isoenzyme was inhibitory, as expected, and when the activation of aromatic amines such as MeIQx and 2-AAF in the fluctuation test was studied in the presence of this antibody, almost complete inhibition of mutagenicity was obtained. In contrast, the activation of 2-AAF and MeIQx was inhibited by only 50% by the specific antipeptide antibody, confirming the results of the experiments with carbon tetrachloride that both P-450IA1 and IA2 contribute to the activation of these compounds in the rat.

In marked contrast to the results obtained in rat liver, the activation of aromatic amines by rabbit liver appears to be catalysed exclusively by P-450IA2. When a specific inhibitory monoclonal antibody against this isoenzyme (Boobis *et al.*, 1985) was used in the Ames test with hepatic microsomes from hydrocarbon-treated animals, virtually complete inhibition of the activation of the amines (IQ, MeIQ, MeIQx, DiMeIQx) was obtained, with almost identical inhibition profiles (Fig. 20.2). This would suggest that the activation of all four amines is catalysed by the same isoenzyme and that this is exclusively P-450IA2 in rabbit liver. Further, activation of these amines by liver microsomes from untreated rabbits was also almost completely inhibited by this antibody, demonstrating that not only in induced animals but also in control animals P-450IA2 is responsible for the activation of these amines.

In extrahepatic tissues such as the kidney, activation is also inhibited by the anti-P-450IA2 antibody in the rabbit. Indeed, there is a good correlation between the immunodetectable content of P-450IA2 in liver and kidney and the activation of MeIQx by these tissues, in both untreated and MC-treated rabbits. This suggests that unlike the rat, P-450IA1 plays little role in the activation of aromatic amines in the rabbit. Thus, it is difficult to extrapolate from one species to another with respect to the specificity of activation of these compounds.

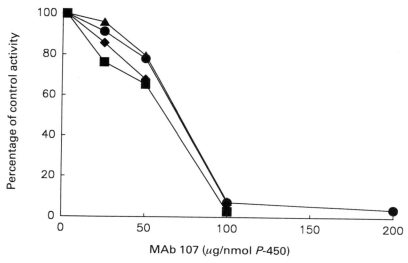

Figure 20.2 Effect of a specific monoclonal antibody against rabbit P-450IA2 (MAb 107) on the mutagenic activation of (●) MeIQx, (▲) DiMeIQx, (■) IQ and (◆) MeIQ by hepatic microsomal fractions from 3-methylcholanthrene-treated rabbits. Mutagenic activity in the Ames/*Salmonella* test has been expressed as a percentage of activity in the presence of a null monoclonal antibody as a negative control.

Of obvious interest is the question of whether the amines are activated by human tissues. In the Ames test, human liver microsomes are extremely active in converting MeIQx and the other aromatic amines in food into mutagenic species (Murray *et al.*, 1988). Indeed, human liver microsomes are more active than those from rat or rabbit in activating aromatic amines to genotoxic intermediates. Human liver microsomes contain an isoenzyme which is the orthologue of P-450IA2 in the rat (Sesardic *et al.*, 1988; Butler *et al.*, 1989). The content of this isoenzyme in human liver is greater than that in control rat liver and is inducible by cigarette smoking (Sesardic *et al.*, 1988). There is an approximate three-fold increase in both P-450IA2 and the *O*-de-ethylation of phenacetin in human liver in cigarette smokers compared to non-smokers. The *O*-de-ethylation of phenacetin is highly significantly correlated with the immunodetectable content of P-450IA2 in human liver samples. Thus, as in the rat (Sesardic *et al.*, 1990b), the *O*-de-ethylation of phenacetin is inducible by PAHs and is catalysed by P-450IA2 (Sesardic *et al.*, 1988).

Whilst, ideally, the contribution of P-450IA2 to metabolic activities should be investigated using inhibitory antibodies, such antibodies are still under development in our laboratory. However, it has been reported that the xanthine analogue furafylline, which was developed as a possible

replacement for theophylline in the treatment of asthma, inhibits the
N-demethylation of caffeine (Tarrus *et al.*, 1987), a reaction which is also
inducible by cigarette smoking in man (Campbell *et al.*, 1987). The effects of
furafylline were investigated on the O-de-ethylation of phenacetin by
microsomal fractions of human liver. Furafylline was an extremely potent
inhibitor of this reaction, but did not inhibit the 4-hydroxylation of
debrisoquine, which is catalysed by a different isoenzyme, P-450IID6
(Gonzalez *et al.*, 1988). Thus, furafylline has the potential to serve as a
potent inhibitor of the activity of human P-450IA2. However, it was first
necessary to establish its selectivity.

As there is no detectable P-450IA1 in human liver samples (Sesardic *et al.*,
1988), it was not possible to determine the specificity of furafylline with
respect to these two members of the same subfamily of P-450 in such
samples. However, in the placenta of cigarette smoking women, AHH
activity and mRNA for P-450IA1 are elevated (Pasanen *et al.*, 1989). This
has been attributed to the induction of P-450IA1 in this tissue, similar to the
effects of induction in extrahepatic tissues of other species (Goldstein and
Linko, 1984). The O-de-ethylation of phenacetin in placenta was also
induced by cigarette smoking (Table 20.2) and there was a highly significant
correlation between AHH and POD activity ($r_s = 0.755$, $P < 0.02$). Thus,
as in the rat, POD activity of extrahepatic tissues in man is catalysed by
P-450IA1 and not by P-450IA2, which is not expressed. Thus, in smokers the
placental activity of either AHH or POD can be used as a measure of
P-450IA1 in man.

Furafylline, even at concentrations of > 10 μM, was without effect on
either activity in the placenta but inhibited hepatic high affinity POD with an
IC50 value of 0.05 μM. Additional studies using marker substrates for other
isoenzymes of P-450 in human liver confirmed the high degree of selectivity
of furafylline for P-450IA2 in man (Sesardic *et al.*, 1990a). The effects of
furafylline were then investigated on the activation of MeIQx and other

Table 20.2 Effect of cigarette smoking on placental aryl hydrocarbon
hydroxylase (AHH) and phenacetin O-de-ethylase (POD) activities. Values
are mean $\pm$ SD. Smokers were defined as those subjects with plasma cotinine
levels of $\geqslant 20$ ng/ml (Pasanen *et al.*, 1988). Both AHH and POD activities
of smokers were significantly greater than in non-smokers ($P < 0.025$,
Mann–Whitney test)

Smoking status	AHH activity (pmol/h/mg of protein)	POD activity (pmol/h/mg of protein)
Smokers (5)	481 ± 425	114 ± 143
Non-smokers (6)	42.9 ± 33.1	3.75 ± 2.43

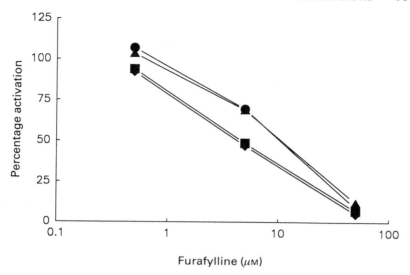

Figure 20.3 Effects of furafylline on the mutagenic activation of (●) MeIQx, (▲) DiMeIQx, (■) IQ and (◆) MeIQ by hepatic microsomal fractions from human liver. Mutagenic activity in the Ames/*Salmonella* test has been expressed as a percentage of control activity in the presence of vehicle alone.

aromatic amines by microsomal fractions of human liver in the Ames test. Furafylline almost completely inhibited the activation of these amines (Fig. 20.3), demonstrating the virtually exclusive involvement of P-450IA2 in this reaction. This does not exclude the possible contribution of P-450IA1 to the activation of the food mutagens in extrahepatic tissues in man, and studies are currently in progress to investigate this aspect of the activity of the PAH-inducible isoenyzmes.

20.3 CONCLUSIONS

The activation of aromatic amines such as MeIQx depends on *N*-hydroxylation (Yamazoe *et al.*, 1988), and the mutagenicity of these compounds in an Ames test can be used as a convenient measure of this activity. Mutagenic activation is increased following treatment of rats and rabbits with PAHs (Murray *et al.*, 1988) and almost certainly in man following cigarette smoking. Whilst in the rat, both P-450IA1 and IA2 contribute to the activation of the aromatic amines, in the rabbit only P-450IA2 catalyses this reaction. In man, P-450IA2 in liver is responsible for mutagen activation, but in the absence of any P-450IA1 in this tissue it is not possible to determine the possible role of this isoenzyme in mutagen activation at present.

The importance of inducible isoenzymes of *P*-450 in the activation of mutagenic amines in man has yet to be determined, and the role that they might play in the aetiology of cancer will be even more difficult to establish. However, should these amines prove to be risk factors for human cancer, it is very likely that the activity of the isoenzymes responsible for their activation will contribute to interindividual differences in susceptibility and, further, the tissue-dependent regulation of the members of the P-450IA subfamily of cytochrome *P*-450 is likely to prove a major determinant in the target organ effects of these compounds.

REFERENCES

Åström, A. and DePierre, J.W. (1985) Metabolism of 2-acetylaminofluorene by eight different forms of cytochrome *P*-450 isolated from rat liver. *Carcinogenesis,* **6**, 113–20.

Boobis, A.R., Davies, D.S., McQuade, J. and Sesardic, D. (1985) Specificity of isozymes of cytochrome *P*-450 determined by inhibitory monoclonal antibodies, *Br. J. Pharmacol.,* **86**, 511P.

Butler, M.A., Iwasaki, M., Guengerich, F.P. and Kadlubar, F.F. (1989) Human cytochrome *P*-450$_{PA}$ (*P*-450IA2), the phenacetin *O*-deethylase, is primarily responsible for the hepatic 3-demethylation of caffeine and *N*-oxidation of carcinogenic arylamines. *Proc. Natl. Acad. Sci. USA*, **86**, 7696–700.

Campbell, M.E., Grant, D.M., Inaba, T. and Kalow, W. (1987) Biotransformation of caffeine, paraxanthine, theophylline, and theobromine by polycyclic hydrocarbon-inducible cytochrome(s) *P*-450 in human liver microsomes. *Drug Metab. Dispos.,* **15**, 237–49.

Doll, R. and Peto, R. (1981) The causes of cancer: quantitative estimates of avoidable risks of cancer in the United States today. *J. Natl. Cancer Inst.,* **66**, 1191–308.

Edwards, R.J., Murray, B.P., Boobis, A.R. and Davies, D.S. (1989) Identification and location of alpha-helices in mammalian cytochromes *P*-450. *Biochemistry,* **28**, 3762–70.

Edwards, R.J., Singleton, A.M., Murray, B.P. *et al.* (1990) An anti-peptide antibody targeted to a specific region of rat cytochrome P-450IA2 inhibits enzyme activity. *Biochem.,* **266**, 497–504.

Edwards, R.J., Singleton, A.M., Sesardic, D. *et al.* (1988) Antibodies to a synthetic peptide that react specifically with a common surface region on two hydrocarbon-inducible isoenzymes of cytochrome *P*-450 in the rat. *Biochem. Pharmacol.,* **37**, 3735–41.

Felton, J.S., Knize, M.G., Shen, N.H. *et al.* (1986) Identification of the mutagens in cooked beef. *Environ. Health Perspect.,* **67**, 17–24.

Goldstein, J.A. and Linko, P. (1984) Differential induction of two 2,3,7,8-terachlorodibenzo-*p*-dioxin inducible forms of cytochrome *P*-450 in extrahepatic versus hepatic tissues. *Mol. Pharmacol.,* **25**, 185–91.

Gonzalez, F.J., Skoda, R., Kimura, S. *et al.* (1988) Characterization of the common genetic defect in humans deficient in debrisoquine metabolism. *Nature*, **331**, 442–6.

Gooderham, N.J., Rice, J.C., Boobis, A.R. and Davies, D.S. (1988) Differences between hepatic and extrahepatic tissues in the activation of MeIQx to a mutagen in the Ames/*Salmonella* test, *Human Toxicol.*, **7**, 79.

Higginson, J. and Muir, C.S. (1979) Environmental carcinogens: misconceptions and limitations of cancer control. *J. Natl Cancer Inst.*, **63**, 1291–8.

Murray, B.P., Boobis, A.R., De La Torre, R. *et al.* (1988) Inhibition of dietary mutagen activation by methylxanthines. *Biochem. Soc. Trans.*, **16**, 620–1.

Nebert, D.W., Nelson, D.R., Adesnik, M. *et al.* (1989) The *P*-450 superfamily: updated listing of all genes and recommended nomenclature for the chromosomal loci. *DNA*, **8**, 1–13.

Pasanen, M., Stenbäck, F., Taskinen, T. *et al.* (1989) Human placental 7-ethoxy-resorufin *O*-deethylase activity: inhibition studies and immunological and molecular biological detection of *P*-450IA1, in *Cytochrome P-450: Biochemistry and Biophysics* (ed. I. Schuster), Taylor and Francis, London, pp. 532–5.

Parkinson, A., Ryan, D.E., Thomas, P.E. *et al.* (1986) Chemical modification and inactivation of rat liver microsomal cytochrome *P*-450c by 2-bromo-4'-nitro-acetophenone. *J. Biol. Chem.*, **261**, 11478–86.

Sesardic, D., Boobis, A.R., Edwards, R.J. and Davies, D.S. (1988) A form of cytochrome *P*-450 in man, orthologous to form d in the rat, catalyses the *O*-deethylation of phenacetin and is inducible by cigarette smoking. *Br. J. Clin. Pharmacol.*, **26**, 363–72.

Sesardic, D., Boobis, A.R., Murray, B.P. *et al.* (1990a) Furafylline is a potent and selective inhibitor of cytochrome *P*-450IA2 in man, *Br. J. Clin. Pharmacol.*, **29**, 651–63.

Sesardic, D., Edwards, R.J., Davies, D.S. *et al.* (1990b) High affinity phenacetin *O*-deethylase is catalysed specifically by cytochrome *P*-450d (P-450IA2) in the liver of the rat. *Biochem. Pharmacol.*, **39**, 489–98.

Sesardic, D., Rich, K.J., Edwards, R.J. *et al.* (1989) Selective destruction of cytochrome *P*-450d and associate monooxygenase activity by carbon tetrachloride in the rat. *Xenobiotica*, **19**, 795–811.

Sugimura, T. (1988) New environmental carcinogens in daily life. *Trends in Pharmacol. Sci.*, Elsevier, Cambridge, **9**, 205–9.

Sugimura, T., Kawachi, T., Nagao, M. *et al.* (1977) Mutagenic principle(s) in tryptophan and phenylalanine pyrolysis products. *Proc. Japan Acad.*, **53**, 58–61.

Tarrus, E., Cami, J., Roberts, D.J. *et al.* (1987) Accumulation of caffeine in healthy volunteers treated with furafylline. *Br. J. Clin. Pharmacol.*, **23**, 9–18.

Thomas, P.E., Reik, L.M., Ryan, D.E. and Levin, W. (1984) Characterization of nine monoclonal antibodies against rat hepatic cytochrome P-450c. Delineation of at least five spatially distinct epitopes. *J. Biol. Chem.*, **259**, 3890–9.

Wynder, E.L. and Gori, G.B. (1977) Contribution of the environment to cancer incidence: an epidemiological exercise. *J. Natl. Cancer Inst.*, **58**, 825–32.

Yamazoe, Y., Abu-Zeid, M., Manabe, S. *et al.* (1988) Metabolic activation of a protein pyrolysate promutagen 2-amino-3,8-dimethylimidazo[4,5-f]quinoxaline by rat liver microsomes and purified cytochrome *P*-450. *Carcinogenesis*, **9**, 105–9.

21

Monoamine oxidase-mediated activation of MPTP and related compounds

V. Glover and M. Sandler

Department of Chemical Pathology, Queen Charlotte's and Chelsea Hospital, Goldhawk Road, London W6 0XG, UK

1. The discovery that 1-methyl-4-phenyl-1,2,3,6-tetrahydropyridine (MPTP) produces a model of Parkinson's disease in man and primates suggests that related compounds might be responsible for the idiopathic disease.
2. The finding that MPTP is metabolized to its toxic metabolite, 1-methyl-4-phenylpyridinium (MPP^+), by monoamine oxidase (MAO B) has focused attention on the possible role of this enzyme in the pathogenesis of Parkinson's disease.
3. Recent developments include studies of alternative mechanisms of MPTP metabolism, including *N*-oxidation, possible toxic mechanisms, the histochemical localization of MAO, and the finding of new toxic analogues of MPTP, including some which may be endogenous; certain of these analogues are partially or totally activated by MAO A rather than B.
4. Other relevant discoveries include an endogenous MAO B inhibitor and the fact that (−)deprenyl, a selective MAO B inhibitor, prolongs life in rats and appears to arrest the progress of idiopathic Parkinson's disease in man.

21.1 INTRODUCTION

The question why some individuals and not others get idiopathic Parkinson's disease, which involves a selective degeneration of dopamine-containing cells in the substantia nigra, is still unanswered. The disease incidence in identical twins has shown no higher concordance than between random members of the population (Ward *et al.*, 1983). The pattern of incidence has remained reasonably constant over time, and although there is some population clustering, it is insufficient to point to a clear environmental cause (Schoenberg, 1987). Other epidemiological findings amount to little. There is some evidence for an inverse relationship between smoking and Parkinson's disease (Tanner *et al.*, 1987), but the reason for this finding is unclear.

When striatal dopamine falls to less than 20% of normal values, clinical signs of the illness begin to appear; it is conceivable that some members of the population are protected by a richer endowment of the monoamine. The initial insult, giving rise to subclinical damage, may only make its presence known when normal age-related neuronal attrition results in the onset of disease. There is good evidence that striatal dopamine falls substantially in normal individuals after about 60 years of age (Calne and Peppard, 1987). However, it is still surprising that this phenomenon is not under genetic control. With Parkinson's disease, it seems possible that the traditional genetic/environmental way of thinking is incomplete and that some additional new concept, such as nigral cell destruction by an endogenously generated agent, will eventually be introduced.

The discovery that MPTP (1-methyl-4-phenyl-1,2,3,6-tetrahydro-pyridine) can cause a selective destruction of nigrostriatal neurons in primates, providing the best model to date of Parkinson's disease (Kopin and Markey, 1988), has led to an explosion of research on its mode of action. The MPTP phenomenon seems likely to provide new understanding of the idiopathic disease, and may help us to identify exogenous or endogenous toxins with similar properties.

21.2 MPTP AND PARKINSON'S DISEASE

MPTP came to light when a young drug abuser self-administered what he considered to be a pethidine-like drug (which he had synthesized himself) and developed a syndrome indistinguishable from advanced Parkinson's disease; the active chemical principle was promptly isolated and identified (Davis *et al.*, 1979). Rats are resistant to the action of MPTP and this finding misled the early investigators who thought they were on the wrong track, until species differences became apparent and primates were shown to be highly susceptible (Burns *et al.*, 1983). In mice, nigrostriatal dopamine depletion but not cell death is observed (Heikkila *et al.*, 1984). The clinical syndrome in man is clinically indistinguishable from the idiopathic disease, except for one major feature; idiopathic Parkinson's disease usually develops over years, often so slowly that the patient cannot recall the onset of the first symptoms. In contrast, MPTP-induced parkinsonism usually develops in 5–15 days. However, like the idiopathic disease it does cause a selective destruction of the substantia nigra and its projections, leaving other dopaminergic regions, like the nucleus accumbens, largely unaffected. These findings raise at least two questions. Why are primates more susceptible than rodents, and why does MPTP selectively attack some dopaminergic regions and not others? One possible answer to both questions could be linked with neuromelanin (Langston, 1988), which is present in high concentration in the substantia nigra, and also generally present in primates but not in rodents. Beagle dogs, which have neuromelanin in their substantia nigra, are also susceptible to MPTP (Kopin and Markey, 1988). Pigmented leeches are more susceptible than non-pigmented leeches. However, other factors may also play a part and, if neuromelanin is important in the story, it is not clear how.

21.3 MAO B AND MPTP

An early discovery was that MPTP itself is only toxic after enzymatic activation by MAO B, which catalyses its transformation into the neurotoxic agent, MPP⁺ (Fig. 21.1, Salach *et al.*, 1984). If the selective MAO B inhibitor, (−)deprenyl, is given before MPTP, nigrostriatal damage can be averted in both primate (Cohen *et al.*, 1985) and mouse models (Heikkila *et al.*, 1984). The selective MAO A inhibitor, clorgyline, has no such

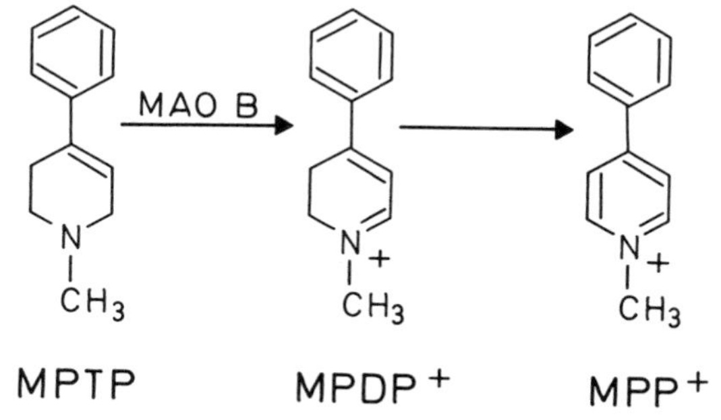

Figure 21.1 Metabolic conversion of MPTP to MPP⁺ by monoamine oxidase B.

protective effect. Both rodent and human brain enzyme are able selectively to catalyse this oxidation (Glover *et al.*, 1986).

Much work has been concentrated on trying to understand the exact localization of MAO A and B in rodent and primate brains, using histochemical and immunohistochemical methodology. While low concentrations of both forms of the enzyme are evenly spread throughout the brain, MAO A is particularly concentrated in noradrenergic neurons and MAO B in those containing 5-hydroxytryptamine (Westlund *et al.*, 1985, Willoughby *et al.*, 1988a,b). Dopaminergic neurons appear to contain low levels of MAO A and negligible activity of MAO B. Thus, most of the conversion of MPTP into MPP$^+$ by MAO B must occur outside the nigrostriatal tract. It probably occurs in the glial cells surrounding dopaminergic neurons, which are, in general, much richer in MAO B in primates than in rodents (Willoughby *et al.*, 1988a,b). However, it is important to emphasize that, although this may make some contribution to species differences in vulnerability, it is far from the whole story. The susceptibility to MPTP of different dopaminergic regions in the marmoset brain does not correspond with the local density of MAO B; nor does that of different mouse strains correlate with MAO B level. In rat brain, the ventral tegmental area has higher MAO B activity than the substantia nigra but does not show any special susceptibility to MPTP. Certainly, the overall distribution of MAO B does not explain the selective toxic effect of MPP$^+$ on the nigrostriatal tract, as the 5-HT cell bodies in the raphe nuclei are particularly rich in MAO B yet unaffected by MPTP.

21.4 THE MECHANISM OF MPTP TOXICITY

As MPP$^+$ is formed outside dopaminergic neurons, it has to be taken up by them to make possible their selective destruction. In fact, dopaminergic synaptosomes have a highly selective mechanism for concentrating MPP$^+$ via the dopamine uptake system (Jaritch *et al.*, 1985) (Fig. 21.2). This uptake process is present in both rat and human preparations, which seem to operate very similarly (Willoughby *et al.*, 1989). Such an uptake mechanism certainly helps to shed light on the selective toxicity of MPTP, but does not do so entirely. It does not account for species differences nor explain why only some dopaminergic systems are affected.

MPP$^+$ does bind with high affinity to neuromelanin (d'Amato *et al.*, 1986). This finding may account for the particular retention of MPP$^+$ in neuromelanin-containing cells. However, it is also possible that other characteristics of the cell that make it accumulate neuromelanin, possibly related to free-radical production, are the same as those that make it especially sensitive to MPP$^+$.

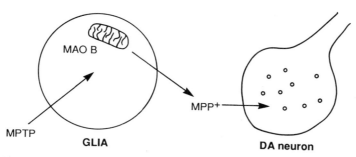

Figure 21.2 Model to show the separation of the bioactivation of MPTP by MAO B in glia from the site of action of MPP$^+$ in dopaminergic neurons.

There has been much speculation (well reviewed by Kopin and Markey, 1988) that the mechanism of the toxic effects of MPP$^+$ is linked with free radical production. MPP$^+$ has been shown to stimulate superoxide and hydroxyl radical formation during incubation of NADPH cytochrome P-450 reductase with NADPH and oxygen. However, ethanol, which is a good hydroxyl radical scavenger, potentiates the toxicity of MPTP. Other studies of the effects of antioxidants on the toxicity of MPTP have been contradictory. It has also been suggested that the degradation of MPTP to MPP$^+$ by MAO B, a reaction which also generates H_2O_2, is responsible for the toxicity. However, as this metabolic step occurs outside the dopaminergic system, it is hard to see on this model why the nigrostriatal system is specifically affected.

There has also been much recent work showing that MPP$^+$ inhibits mitochondrial oxidations. It is selectively accumulated by an energy-dependent process in mitochondria and selectively inhibits complex 1-catalysed NADH-linked oxidation of pyruvate or glutamate. It is of interest that complex 1 has recently been shown to be significantly low in the substantia nigra of patients with the idiopathic disease (Schapira *et al.*, 1989).

21.5 ANALOGUES OF MPTP AS SUBSTRATES FOR MAO A AND B

Several analogues of MPTP have now been shown to be substrates for MAO (Fig. 21.3). Compounds with fully saturated or unsaturated pyridine rings are not substrates. Minor substitutions on the phenyl ring can be tolerated, and PTP, the MPTP analogue lacking the *N*-methyl group, is also a substrate. Ethyl-MTP-carboxylate was the first MPTP analogue without a second ring which was still found to be a substrate. All these new substrates are also, at least partially, oxidized by MAO A.

Figure 21.4 shows some MPTP analogues which are neurotoxic in mice (Langston and Irwin, 1989). They all are tetrahydroxypyridines, but show

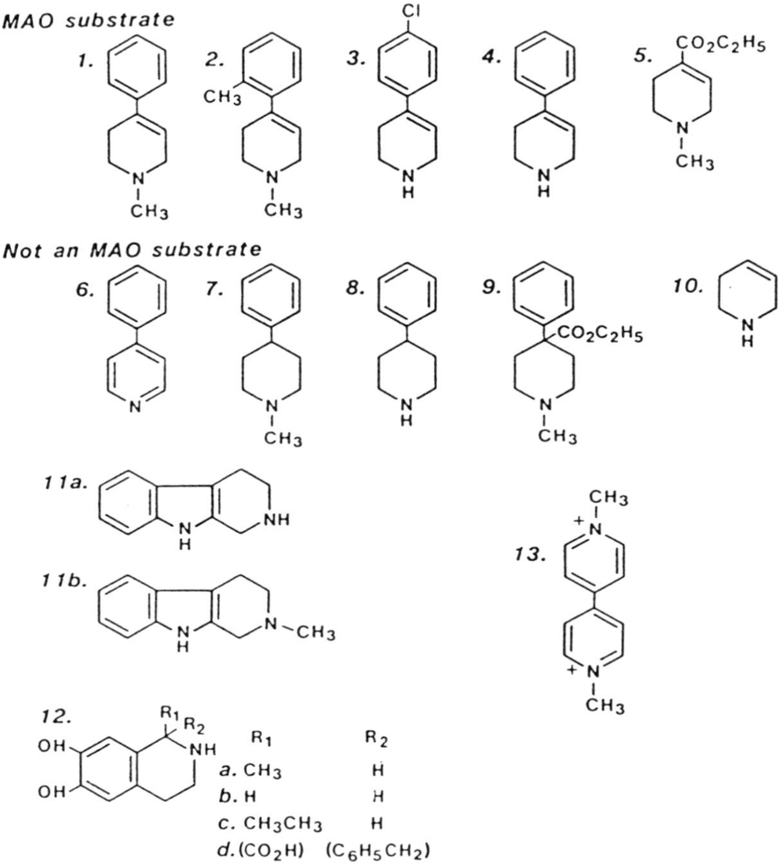

Figure 21.3 Structural requirements for a compound to act as a substrate for MAO. 1, 1-Methyl-4-phenyl-1,2,3,6-tetrahydropyridine (MPTP); 2, 1-methyl-4-(2-methyl-phenyl), 1,2,3,6-tetrahydropyridine (1 Me-MPTP); 3, 4-(*p*-chlorophenyl)-1,2,3,6-tetrahydropyridine (Cl-PTP); 4, 4-phenyl-1,2,3,6-tetrahydropyridine (PTP); 5, ethyl-1-methyl-1,2,3,6-tetrahydro-4-pyridine carboxylate (ethyl-MPTP-carboxylate); 6, 4-phenylpyridine; 7, 1-methyl-4-phenylpiperidine; 8, 4-phenyl-piperidine; 9, pethidine; 10, 1,2,3,6-tetrahydropyridine; 11a, tetrahydro-β-carboline; 11b, 2-N-methyl-tetrahydro-β-carboline; 12a, salsolinol; 12b, norsalsolinol; 12c, 1-methyl-salsolinol; 12d, 1-carboxy-1-benzyl-6,7-dihydroxy-tetrahydroisoquinoline; 13, paraquat.

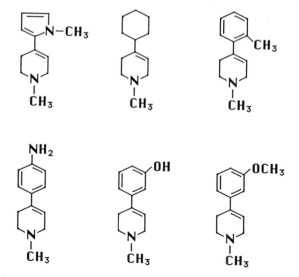

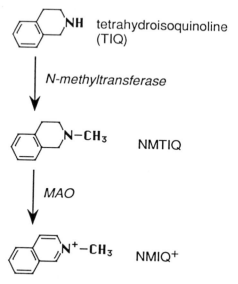

Figure 21.4 Structure of some MPTP analogues known to deplete striatal dopamine in mice.

Figure 21.5 Pathway of activation of tetrahydroisoquinoline to the *N*-methyliso-quinolinium ion (NMIQ$^+$), a possible endogenous analogue of MPP$^+$ (from Naoi *et al.*, 1989.)

variation in the phenyl ring. PTP, however, is not neurotoxic; thus the N-methyl substitution, although not necessary for metabolism by MAO, does seem to be a prerequisite for neurotoxicity. Other essential features for both activities appear to be the 4-5 double bond in the pyridine ring, and substitution at the 4-position.

With 2'-Me-MPTP and 2'-Et-MPTP, neurotoxicity cannot be prevented by deprenyl alone, but only after addition of the MAO A inhibitor, clorgyline, as well (Heikkila *et al.*, 1988). As several of the other analogues are also substrates of both MAO B and MAO A (Gibb *et al.*, 1987), it is important to consider the potential role of MAO A in the toxicity of MPTP analogues.

Of particular interest is the finding that tetrahydroisoquinoline (TIQ) can be N-methylated by human brain N-methyltransferase (Naoi *et al.*, 1989) to N-methyltetrahydroisoquinoline (NMTIQ) and subsequently oxidized by MAO to form the N-methylisoquinolinium ion (NMIQ$^+$) (Fig. 21.5). TIQ has been found in human brain (Niwa *et al.*, 1987) and has also been shown to be neurotoxic in mice (Ogawa *et al.*, 1989). This, then, is a potential endogenous pathway which could lead to an analogue of MPP$^+$. It is of note that NMTIQ fulfils the requirements for toxicity of MPTP analogues discussed by Langston and Irwin (1989).

21.6 OTHER ROUTES OF MPTP METABOLISM

It is relevant to consider other routes of MPTP metabolism as selective vulnerability to the toxin might depend on the balance of competing

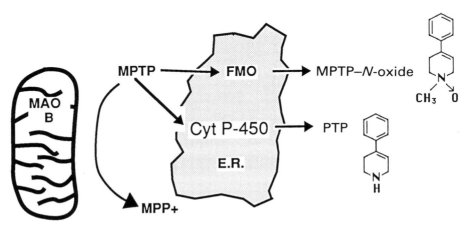

Figure 21.6 Alternative routes of MPTP metabolism.

pathways. Di Monte *et al.* (1988) have shown that in rat hepatocytes, cytochrome *P*-450 catalyses the demethylation of MPTP to form 4-phenyl-1,2,3,6-tetrahydropyridine (PTP), whereas the flavin containing mono-oxygenase (FMO) is responsible for the generation of MPTP-oxide (see Fig. 21.6). Even so, 90% of the metabolism of MPTP is by MAO B to form MPP^+, the other pathways being relatively very minor.

It remains possible that such other metabolic pathways are relatively more important in human brain or in the metabolism of MPTP analogues, including any responsible for the idiopathic disease. One retrospective epidemiological study which might be relevant in this context has shown a deficiency in the cytochrome *P*-450 system, catalysing the 4-hydroxylation of debrisoquine (Barbeau *et al.*, 1985).

21.7 MAO B IN HUMAN BRAIN AND PLATELETS

MAO B concentration in both human and animal brain is raised in old age (Strolin Benedetti and Dostert, 1989), a finding which may derive from a relative increase in the glial population. However, several studies of post-mortem human brains have failed to find any difference in MAO B activity between parkinsonian and control brains. In addition, the susceptibility of different strains of mice to MPTP does not correlate with their brain MAO B levels (Sonsalla and Heikkila, 1988). It thus seems unlikely that vulnerability to Parkinson's disease is increased by high brain MAO B activity. It remains possible, however, that a subgroup with low MAO B activity is protected. As mentioned above, several epidemiological studies indicate that smokers are less likely to suffer from the illness (Tanner *et al.*, 1987) and smoking has been linked with low platelet MAO activity (Littlewood *et al.*, 1984).

Platelet MAO B has been claimed by some to be raised in parkinsonian patients (Danielczyk *et al.*, 1988), although earlier studies failed to show this (Glover *et al.*, 1983). Whether a true finding or not, if brain levels do not mirror this rise, it is unlikely to be of pathological significance.

21.8 ISATIN, AN ENDOGENOUS MAO B INHIBITOR

It has recently been shown that the endogenous MAO inhibitory activity denominated tribulin derives substantially from the indole, isatin (Glover *et al.*, 1988). Although this compound was known previously and some of its pharmacological properties were recognized, it was not known to be generated endogenously. Isatin is present in both brain and tissues, with a quite distinct distribution; levels in the rat striatum are about 0.1 μg/g. Isatin is a potent inhibitor of MAO B (IC $50 = 3 \mu M$), and concentrations are significantly increased in conditions of stress.

Whether high isatin production plays some part in protection against Parkinson's disease is a question for the future.

21.9 EFFECTS OF (−)DEPRENYL ON LONGEVITY AND PARKINSON'S DISEASE

(−)Deprenyl has now been used for some years as an adjuvant in the treatment of idiopathic Parkinson's disease (Birkmayer *et al.*, 1975). Dopamine is predominantly metabolized by MAO B in the human striatum (Glover *et al.*, 1977, 1980) and this action is blocked by (−)deprenyl which 'smoothes out' end of dose akinesia (Yahr and Kaufmann, 1989) and seems to be a remarkably safe drug (Sandler and Glover, 1989). Unlike the non-selective MAO inhibitors used extensively for the treatment of depressive illness or, indeed, the selective MAO A inhibitor clorgyline, it does not have 'cheese effect' or potentiate the peripheral effects of L-dopa (Elsworth *et al.*, 1978). An unexpected retrospective finding (Birkmayer *et al.*, 1985) was that patients who had taken (−)deprenyl together with L-dopa plus a peripheral decarboxylase inhibitor, lived longer than those who received L-dopa plus decarboxylase inhibitor alone.

A parallel finding has recently been reported in aged rats. Animals given (−)deprenyl appear to live considerably longer than controls, almost doubling their expected life span (Knoll, 1988). Tetrud and Langston (1989) now report that in parkinsonian patients taking (−)deprenyl, deterioration from disease progression appears to be significantly slowed.

All these findings are remarkable. If true, they suggest that (−)deprenyl can both slow down nigrostriatal degeneration and senescence in general. Such effects might stem from increased levels of dopamine (Riederer and Youdim, 1986) or, perhaps, from blocking the production of a neurotoxic analogue of MPP^+. It is also possible that these effects of (−)deprenyl, if confirmed, derive from actions other than inhibition of MAO B. Knoll (1988), for example, has reported that the drug stimulates the induction of superoxide dismutase by 400%.

REFERENCES

Barbeau, A., Cloutier, T., Roy, M. *et al.*, (1985) Ecogenetics of Parkinson's disease: 4-hydroxylation of debrisoquine. *Lancet*, **ii**, 1213–16.

Birkmayer, W., Knoll, J., Riederer, P. *et al.* (1985) Increased life expectancy resulting from addition of 1-deprenyl to madopar treatment in Parkinson's disease: a long term study. *J. Neural Transm.*, **64**, 113–27.

Birkmayer, W., Riederer, P., Youdim, M.B.H. and Linauer, W. (1975) Potential of antiakinetic effect after L-dopa treatment by an inhibitor of MAO B, 1-deprenyl. *J. Neural Transm.*, **36**, 303–23.

Burns, R.S., Chiueh, C.C., Markey, S.P. *et al.* (1983) A primate model of parkinsonism: selective destruction of dopaminergic neurons in the pars compacta of the substantia nigra by *N*-methyl-4-phenyl-1,2,3,6-tetrahydropyridine. *Proc. Natl. Acad. Sci.*, USA, **80**, 4546–550.

Calne, D.B. and Peppard, R.F. (1987) Aging of the nigrostriatal pathway in humans. *Can. J. Neurol. Sci.*, **14**, 424–7.

Cohen, G., Pasik, P., Cohen, B. *et al.* (1985) Pargyline and deprenyl prevent the neurotoxicity of 1-methyl-4-phenyl-1,2,3,6-tetrahydropyridine (MPTP) in monkeys. *Eur. J. Pharmacol.* **106**, 209–10.

D'Amato, R.J., Lipman, Z.P. and Snyder, S.H. (1986) Selectivity of the parkinsonian neurotoxin MPTP: toxic metabolite MPP$^+$ binds to neuromelanin. *Science*, **231**, 987–9.

Danielczyk, M., Streifler, M., Konradi, C. *et al.* (1988) Platelet MAO-B activity and the psychopathology of Parkinson's disease, senile dementia and multi-infarct dementia. *Acta Psychiatr. Scand.*, **78**, 730–6.

Davis, G.C., Williams, A.C., Markey, S.P. *et al.* (1979) Chronic parkinsonism secondary to intravenous injection of meperidine analogues. *Psychiatr. Res.*, **1**, 249–54.

Di Monte, D., Shinka, T., Sandy, M.S. *et al.* (1988) Quantitative analysis of 1-methyl-4-phenyl-1,2,3,6-tetrahydropyridine metabolism in isolated rat hepatocytes. *Drug Metab. Dispos.*, **16**, 250–5.

Elsworth, J.D., Glover, V., Reynolds, G.P. *et al.* (1978) Deprenyl administration in man: a selective monoamine oxidase B inhibitor without the 'cheese effect'. *Psychopharmacology* (Berlin), **57**, 33–8.

Gibb, C., Willoughby, J., Glover, V. *et al.* (1987) Analogues of 1-methyl-4-phenyl-1,2,3,6-tetrahydropyridine as monoamine oxidase substrates: a second ring is not necessary. *Neurosci. Lett.*, **76**, 316–22.

Glover, V., Elsworth, J.D. and Sandler, M. (1980) Dopamine oxidation and its inhibition by (−)deprenyl in man. *J. Neural Transm. (Suppl)*, **16**, 163–72.

Glover, V., Gibb, C. and Sandler, M. (1986) Monoamine oxidase B (MAO B) is the major catalyst of 1-methyl-4-phenyl-1,2,3,6-tetrahydropyridine (MPTP) oxidation in human brain and other tissues. *Neurosci. Lett.*, **64**, 216–20.

Glover, V., Halket, J.M., Watkins, P.J. *et al.* (1988) Isatin: identity with the purified endogenous monoamine oxidase inhibitor tribulin. *J. Neurochem.*, **51**, 656–9.

Glover, V., Lees, A.J., Ward, C. *et al.* (1983) Platelet phenolsulphotransferase activity in Parkinson's disease. *J. Neural Transm.*, **57**, 95–102.

Glover, V., Sandler, M., Owen, F. and Riley, G.J. (1977) Dopamine is a mono-amine oxidase B substrate in man. *Nature*, **265**, 80–1.

Heikkila, R.E., Manzino, L., Cabbat, F.S. and Duvoisin, R.C. (1984) Protection against the dopaminergic neurotoxicity of 1-methyl-4-phenyl-1,2,5,6-tetra-hydropyridine by monoamine oxidase inhibitors. *Nature*, **311**, 467–9.

Heikkila, R.E., Kindt, M.V., Sonsalla, P.K. *et al.* (1988) Importance of monoamine oxidase A in the bioactivation of neurotoxic analogs of 1-methyl-4-phenyl-1,2,3,6-tetrahydropyridine. *Proc. Natl. Acad. Sci. USA*, **85**, 6172–6.

Jaritch, J.A., d'Amato, R.J., Strittmatter, S.M. and Snyder, S.H. (1985) Parkinsonism-inducing neurotoxin, *N*-methyl-4-phenyl-1,2,3,6-tetrahydro-pyridine: uptake of the metabolite *N*-methyl-4-phenylpyridinium by dopamine neurons explains selective toxicity. *Proc. Natl. Acad. Aci. USA*, **82**, 2173–7.

Knoll, J. (1988) The striatal dopamine dependency of life span in male rats: longevity study with (−)deprenyl. *Mech. Ageing Dev.*, **46**, 237–62.

Kopin, I.J. and Markey, S.P. (1988) MPTP toxicity: implications for research in Parkinson's disease. *Ann. Rev. Neurosci.*, **11**, 81–96.

Langston, J.W. (1988) Neuromelanin-containing neurons are selectively vulnerable in parkinsonism. *Trends Pharmacol. Sci.*, **9**, 347–8.

Langston, J.W. and Irwin, I. (1989) Pyridine toxins in *Drugs for the Treatment of Parkinson's Disease* (ed. D.B. Calne) Springer-Verlag, New York, pp. 205–26.

Littlewood, J.T., Glover, V., Sandler, M. *et al.* (1984) Migraine and cluster headache: links between platelet monoamine oxidase activity, smoking and personality. *Headache*, **24**, 30–4.

Naoi, M., Matsuura, S., Takahashi, T. and Nagatsu, T. (1989) A *N*-methyl-transferase in human brain catalyses *N*-methylation of 1,2,3,4-tetrahydro-isoquinoline into *N*-methyl-1,2,3,4-tetrahydroisoquinoline, a precursor of a dopaminergic neurotoxin, *N*-methylisoquinolinium ion. *Biochem. Biophys. Res. Commun.*, **161**, 1213–19.

Niwa, T., Takeda, N., Kaneda, N. *et al.* (1987) Presence of tetrahydroisoquinoline and 2-methyl-tetrahydroquinoline in parkinsonian and normal human brains. *Biochem. Biophys. Res. Commun.*, **144**, 1084–9.

Ogawa, M., Araki, M., Nagatsu, I. *et al.* (1989) The effect of 1,2,3,4-tetrahydro-isoquinoline (TIQ) on mesencephalic dopaminergic neurons in C57BL/6J mice: immunohistochemical studies – tyrosine hydroxylase. *Biogenic Amines*, **6**, 427–36.

Riederer, P. and Youdim, M.B.H. (1986) Monoamine oxidase activity and monoamine metabolism in brains of parkinsonian patients treated with 1-deprenyl. *J. Neurochem.*, **46**, 1359–65.

Salach, J.I., Singer, T.P., Castagnoli, N. Jr. and Trevor, A. (1984) Oxidation of the neurotoxic amine 1-methyl-4-phenyl-1,2,3,6-tetrahydropyridine (MPTP) by monoamine oxidases A and B and suicide inactivation of the enzymes by MPTP *Biochem. Biophys. Res. Commun.*, **125**, 831–5.

Sandler, M. and Glover, V. (1989) Monoamine oxidase inhibitors in Parkinson's disease, in *Drugs for the Treatment of Parkinson's Disease* (ed. D.B. Calne), Springer-Verlag, New York, pp. 411–31.

Schapira, A.H.V., Cooper, J.M., Dexter, D. *et al.* (1989) Mitochondrial complex 1 deficiency in Parkinson's disease. *Lancet*, **i**, 1269.

Schoenberg, B.S. (1987) Environmental risk factors for Parkinson's disease: the epidemiologic evidence. *Can. J. Neurol. Sci.*, **14**, 407–13.

Sonsalla, P.K. and Heikkila, R.E. (1988) Neurotoxic effects of 1-methyl-4-phenyl-1,2,3,6-tetrahydropyridine (MPTP) and methamphetamine in several strains of mice. *Prog. Neuropsychopharmacol. Biol. Psychiatr.* **12**, 345–54.

Strolin Benedetti, M. and Dostert, P. (1989) Monoamine oxidase, brain ageing and degenerative diseases. *Biochem. Pharmacol.*, **38**, 555–61.

Tanner, C.M., Chen, B., Wange, W.-Z. *et al.* (1987) Environmental factors in the etiology of Parkinson's disease. *Can. J. Neurol. Sci.*, **14**, 419–23.

Tetrud, J.W. and Langston, J.W. (1989) The effect of deprenyl (selegiline) on the natural history of Parkinson's disease. *Science*, **245**, 519–22.

Ward, C.D., Duvoisin, R.C. and Ince, S.E. (1983) Parkinson's disease in 65 pairs of twins and in a set of quadruplets. *Neurology*, **33**, 815–25.

Westlund, K.N., Denney, R.M., Kochersperger, L.M. *et al.* (1985) Distinct monoamine oxidase A and B populations in primate brain. *Science*, **230**, 181–3.

Willoughby, J., Cowburn, R.F., Hardy, J.A. *et al.* (1989) 1-Methyl-4-phenyl-pyridinium uptake by human and rat striatal synaptosomes. *J. Neurochem.*, **52**, 627–31.

Willoughby, J., Glover, V. and Sandler, M. (1988a) Histochemical localisation of monoamine oxidase A and B in rat brain. *J. Neural Transm.*, **74**, 29–42.

Willoughby, J., Glover, V., Sandler, M. *et al.* (1988b) Monoamine oxidase activity and distribution in marmoset brain: implications for MPTP toxicity. *Neurosci. Lett.*, **90**, 100–6.

Yahr, M.D. and Kaufmann, H. (1989) Clinical actions of L-deprenyl in Parkinson's disease, in *Drugs for the Treatment of Parkinson's Disease*, (ed. D.B. Calne), Springer-Verlag, New York, pp. 411–31.

22

Activation of aromatic amines by oxyhaemoglobin

P. Eyer

Walther-Straub-Institut für Pharmakologie und Toxikologie der Medizinischen Fakultät der Ludwig-Maximilians-Universität München, Nussbaumstrasse 26, D-8000 München 2, Federal Republic of Germany

1. Aromatic amines reacted directly with oxyhaemoglobin to afford ferrihaemoglobin. Arylamines with electron donating substituents were particularly active. In the series p-CH$_3$, p-S—CH$_3$, p-O—C$_2$H$_5$, p-NH$_2$, p-NH—C$_6$H$_5$, p-NH—CH$_3$, and p-N—(CH$_3$)$_2$ the ferrihaemoglobin forming activity increased in this order showing a good Hammett correlation with Hammett's σ_p^+.

2. The most active p-phenylenediamines, most of all N,N,N',N'-tetramethyl-p-phenylenediamine, produced ferrihaemoglobin in catalytic amounts, indicating aminyl radicals as the active molecules.

3. The stable Wurster's radical cations produced ferrihaemoglobin at rates greater than 10^3 M^{-1}/s, i.e. about three orders of magnitude faster than the parent phenylenediamines.

4. Wurster's blue radical cation formed an adduct with the SH groups of haemoglobin. With reduced glutathione (GSH) two different reactions were found: (1) a fast reaction with the respective quinone di-iminium dication which yielded both oxidized glutathione (GSSG) and 2-(glutathione-S-yl)-N,N,N',N'-tetramethyl-p-phenylenediamine at a ratio of $3:1$, and (2) a slow direct reaction which gave the parent amine and GSSG, presumably via GS* radicals. Reaction (2) gained significance only at GSH concentrations above 1 mM. In reaction (1) the velocity of the disproportionation controlled the overall reaction rates at GSH concentrations between 0.05 and 1 mM.

5. In contrast to the reactions of the homologous aminophenols with haemoglobin, free hydrogen peroxide was formed during ferrihaemoglobin formation by the arylamines. In the case of the phenylenediamines, catalase decreased the yield of ferrihaemoglobin by about 50%. However, in the case of aniline derivatives with electron-withdrawing substituents, catalase virtually prevented ferrihaemoglobin formation, indication for a differently operating reaction mechanism.

22.1 INTRODUCTION

It is generally accepted that aromatic amines belong to a group of compounds which produce ferrihaemoglobin only after biochemical transformation (for reviews see Kiese, 1966, 1974). Accordingly, aromatic amines are activated, primarily in the liver, to yield proximate reactive derivatives like aminophenols, N-hydroxyarylamines and N-hydroxy-N-arylacetamides. These compounds are then oxidized by oxyhaemoglobin to give the ultimate ferrihaemoglobin-forming species. Moreover, Mieyal *et al.* (1976) have shown that oxyhaemoglobin, both in reconstituted systems and in intact erythrocytes can also act as a mono-oxygenase-like catalyst of aniline hydroxylation to give mainly *p*-aminophenol. Similarly, Golly and Hlavica (1983) reported that 4-chloronitrosobenzene was formed in the reaction of oxyhaemoglobin with 4-chloroaniline in the presence of reduced pyridine nucleotides and cytochrome *c* reductase, indicating that oxy-haemoglobin can mediate N-oxygenations, in addition to the previous reports on aromatic carbon hydroxylations. These results show that the red cell has a dual function: it is a sensitive target for toxic actions of aromatic amines and plays a metabolic role in the biotransformation of aromatic amine derivatives (Eyer, 1983). Haemoglobin, therefore, has been recognized as a drug-metabolizing 'enzyme' (Mieyal *et al.*, 1976; Lenk and Riedl, 1989).

During the mutual redox-reactions of oxyhaemoglobin and the proximate reactive arylamine derivatives, macroscopically referred to as co-oxidation, radical intermediates have been detected: aminophenoxyl radicals (Eyer and Lengfelder, 1984) and phenylnitroxide radicals (Wahler *et al.*, 1959; Lenk and Riedl, 1989; Maples *et al.*, 1990). In addition, experiments reported by Lenk and Riedl (1989) strongly suggest intermediate formation of N-phenyl-N-acetylnitroxide radicals during the interaction of oxyhaemoglobin and aryl hydroxamic acids. These data indicate that oxyhaemoglobin is capable of one-electron abstraction reactions and also acts as an oxidase/peroxidase.

More recently, evidence has been accumulating that aromatic amines are activated by peroxidases, such as myeloperoxidase, horseradish peroxidase and prostaglandin H synthetase, to yield free arylaminyl radicals (Job and Dunford, 1976; Mason, 1982; Kalyanaraman and Sivarajah, 1984; O'Brien, 1984; Eling *et al.*, 1985; Lindquist *et al.*, 1985; McGirr and O'Brien, 1987). Hence, it was of interest to investigate whether oxyhaemoglobin is also capable of producing arylaminyl free radicals. To our knowledge, only a few reports exist on ferrihaemoglobin formation from pure oxyhaemoglobin by aromatic amines. Mieyal and Blumer (1976) and Sterzl (1984) observed enhanced autoxidation rates of oxyhaemoglobin in the presence of aniline, and Kiese (1974) reported on ferrihaemoglobin formation by durene-diamine, durenediamine semiquinone and duroquinone di-imine, of which he suggested the latter to be the active species.

This chapter reviews data on ferrihaemoglobin formation from purified human haemoglobin by 13 different arylamines, for which the reaction rates were correlated with Hammett constants. Phenylenediamine derivatives which were the most active compounds produced ferrihaemoglobin in catalytic amounts. Analysis of the reaction rates and products showed that aminyl radicals were the putative reactive intermediates which formed ferrihaemoglobin by three orders of magnitude faster than the parent amines.

Details of the experimental design are available in previous published reports (Michaelis and Granick, 1943; Eyer *et al.*, 1975; Eyer and Lengfelder, 1984; Störle, 1989).

22.2 FORMATION OF FERRIHAEMOGLOBIN BY AROMATIC AMINES

When purified human oxyhaemoglobin (5 mM) was incubated with a variety of aniline derivatives (5 mM each), formation of ferrihaemoglobin was observed as illustrated in Fig. 22.1. Of the compounds tested, aniline derivatives with strong electron-donating substituents were much more active than those with electron-withdrawing substituents (not shown in the figure). Since the rates of ferrihaemoglobin formation were found to be proportional to both oxyhaemoglobin and amine concentration, second order reactions were assumed. From the initial velocities of the reaction and after correction for autoxidation in the absence of amines, second order rate constants were calculated as listed in Table 22.1. It appears that a Hammett correlation exists only for the series of aniline derivatives with electron-donating substituents and a $+$I-effect, when Hammett σ_p^+ constants are plotted vs reaction rates (Fig. 22.2), whereas aniline derivatives with electron-withdrawing substituents scatter at random in the Hammett plot. Of the phenylenediamine derivatives, N,N,N',N'-tetramethyl-p-phenylenediamine (TMPD) was the most active compound which produced more than 90% ferrihaemoglobin within 10 min (Fig. 22.3). The second order rate constant of ferrihaemoglobin formation by TMPD was found to be $5.5 \, M^{-1} s^{-1}$, a value sixfold higher than found with N,N-dimethyl-p-phenylenediamine (DMPD).

The catalytic action of DMPD on ferrihaemoglobin formation was impressive, when low concentrations of DMPD (0.05 mM) were incubated with haemoglobin (3 mM). As shown in Fig. 22.4, ferrihaemoglobin

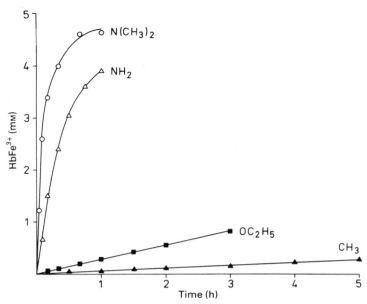

Figure 22.1 Ferrihaemoglobin formation by differently substituted aniline derivatives. Purified human haemoglobin (5 mM) reacted with various p-substituted aniline derivatives (5 mM each) in 0.2 M-sodium phosphate, pH 7.4, at 37°C under air. The data have been corrected for autoxidation of oxyhaemoglobin, i.e. 65 nM/s.

Table 22.1 Apparent second-order rate constants of ferrihaemoglobin formation by *para*-substituted aniline derivatives

Substituent	Hammett constant (σ_p^+)	$k \, (\mathrm{M^{-1} s^{-1}})$
$-\mathrm{N(CH_3)_2}$	-1.7	0.9
$-\mathrm{NHCH_3}$	?	0.8
$-\mathrm{NHC_6H_5}$	-1.4	0.25
$-\mathrm{NH_2}$	-1.31	0.11
$-\mathrm{OC_2H_5}$	-0.69	4×10^{-3}
$-\mathrm{SCH_3}$	-0.53	3×10^{-3}
$-\mathrm{CH_3}$	-0.31	6×10^{-4}
$-\mathrm{C_6H_5}$	-0.18	1×10^{-3}
$-\mathrm{H}$	0	7×10^{-4}
$-\mathrm{Cl}$	0.11	1×10^{-3}
$-\mathrm{COCH_3}$	0.5	3×10^{-4}
$-\mathrm{NO_2}$	0.79	1×10^{-3}

Purified human oxyhaemoglobin (5 mM) reacted with various aniline derivatives (5 mM, 4-aminobiphenyl 2.5 mM) in 0.2 M-sodium phosphate, pH 7.4, 37°C under air for up to 5 h. Initial velocities of ferrihaemoglobin formation were used to calculate the apparent second-order rate constants after correction for spontanous autoxidation of oxyhaemoglobin (i.e. 65 nM s^{-1}).

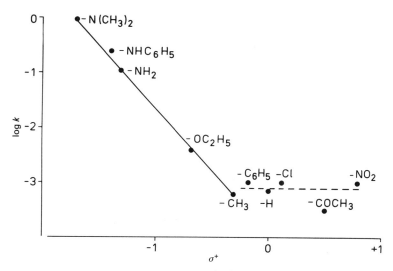

Figure 22.2 Second order rate constants ($M^{-1}s^{-1}$) of ferrihaemoglobin formation by aniline derivatives *vs* Hammett σ_P^+ constants. The rate constants from Table 22.1 were plotted vs σ_P^+ according to Hammett. The reaction constant ϱ for aniline derivatives with negative σ_P^+ values was -2.4. (For experimental conditions see Table 22.1.)

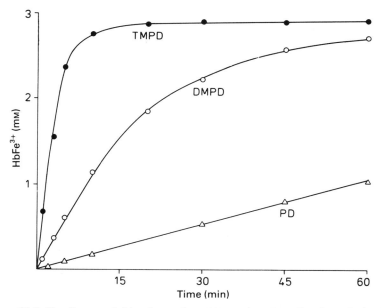

Figure 22.3 Ferrihaemoglobin formation by *p*-phenylenediamine derivatives. Purified human haemoglobin (3 mM) reacted with *p*-phenylenediamine (PD), *N*,*N*-dimethyl-*p*-phenylenediamine (DMPD), and *N*,*N*,*N'*,*N'*-tetramethyl-*p*-phenylenediamine (TMPD), respectively (1 mM each) in 0.2 M-sodium phosphate, pH 7.4, at 37°C under air.

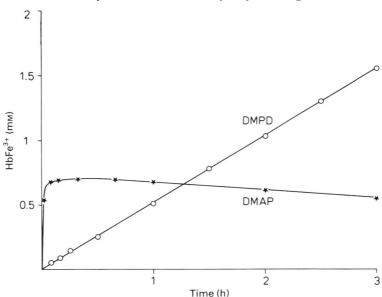

Figure 22.4 Comparison of ferrihaemoglobin formation by differently substituted *N*,*N*-dimethylaniline derivatives. Purified human haemoglobin (3 mM) reacted with *N*,*N*-dimethyl-*p*-aminophenol (DMAP) or *N*,*N*-dimethyl-*p*-phenylenediamine, 0.05 mM each, in 0.2 M-sodium phosphate, pH 7.4, at 37°C under air.

formation proceeded linearly with time and yielded more than 30 equivalents of ferrihaemoglobin in 3 h. In contrast, the phenolic analogue of DMPD, 4-dimethylaminophenol showed rapid cessation of its ferri-haemoglobin-forming activity. This compound was previously reported to disappear quickly from haemoglobin solutions by covalent binding to the SH groups in haemoglobin (Eyer *et al.*, 1974, 1983). Hence, it seemed of interest to study the fate of the phenylenediamines in solutions of haemoglobin.

22.3 FATE OF TMPD IN THE REACTION WITH HAEMOGLOBIN

TMPD (1 mM) was incubated with purified human haemoglobin (3 mM) or sodium phosphate buffer. Aliquots were extracted with ether to determine the TMPD concentration spectroscopically. As shown in Fig. 22.5, TMPD disappeared gradually from both incubates. Interestingly, TMPD decreased with a significant lag phase in the haemoglobin-containing solution, and the

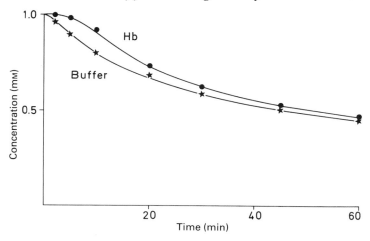

Figure 22.5 Disappearance of TMPD from haemoglobin and buffer solutions. TMPD (1 mM) reacted with human haemoglobin (3 mM) or 0.2 M-sodium phosphate buffer, pH 7.4, at 37°C under air. TMPD in ether extracts was determined spectroscopically (means of three experiments).

decay curve was retarded by about 5 min as compared to the curve obtained from the buffer incubate. Thus it appeared that the decrease in TMPD started only when ferrohaemoglobin had been markedly consumed. In a similar experiment TMPD (0.5 mM) reacted with haemoglobin (3 mM) for 3 h. After that time 0.29 mM-TMPD had been consumed and 0.31 mM of the reactive SH groups were blocked as revealed by SH group titration according to Boyer (1964). These data pointed to formation of oxidized TMPD, radical cation or quinone di-iminium dication, as the responsible metabolite that oxidizes ferrohaemoglobin and arylates the sulphydryl groups of haemoglobin.

22.4 FORMATION OF FERRIHAEMOGLOBIN BY THE *N,N*-DIMETHYL- AND *N,N,N',N'*-TETRA-METHYL-*p*-PHENYLENEDIAMINE RADICAL CATIONS

The synthetic radical cations, $DMPD^{+*}$ and $TMPD^{+*}$ immediately formed ferrihaemoglobin. To determine the reaction rates, ferrihaemoglobin formation was followed in a stopped-flow apparatus. Upon varying the concentrations of oxyhaemoglobin and the radical cations, a second order reaction was established with the following reaction constants: $DMPD^{+*}$,

5×10^3 M^{-1} s^{-1}; TMPD^{+*}, 1×10^3 M^{-1} s^{-1}. Identical constants were obtained when the rate of ferrihaemoglobin formation and the rate of disappearance of the radical cations was followed. Since oxidizing agents usually react with unliganded iron of haemoglobin, ferrihaemoglobin formation was also determined with deoxyhaemoglobin. With DMPD^{+*} a second order rate constant of 10^5 M^{-1} s^{-1} was found. All reactions were carried out in 0.2 M-sodium phosphate, pH 7.4, and 37°C. These data indicated that ferrihaemoglobin formation by the radical cations was about three orders of magnitude faster than by the parent phenylenediamines.

When purified ferrihaemoglobin was reacted with TMPD under an atmosphere of carbon monoxide, formation of TMPD^{+*} and carbon monoxide haemoglobin was observed spectroscopically. When ferrihaemoglobin reacted with TMPD^{+*} under carbon monoxide TMPD^{+*} gradually disappeared while carbon monoxide haemoglobin was formed concomitantly. When carbon monoxide was replaced by argon in these experiments, no significant spectral changes were observed within 10 min. Hence, the equilibrium of these redox reactions is in favour of ferrihaemoglobin and TMPD.

22.5 INFLUENCE OF SUPEROXIDE RADICALS AND HYDROGEN PEROXIDE ON FERRI-HAEMOGLOBIN FORMATION BY DMPD AND TMPD

Purified human haemoglobin (3 mM), free from catalase and superoxide dismutase (Eyer *et al.*, 1975), reacted with DMPD or TMPD, 0.1 mM each, in 0.2 M-sodium phosphate, pH 7.4, at 37°C under air. Ferrihaemoglobin formation was followed directly at 630 vs 690 nm in a dual wavelength mode. With both phenylenediamines, ferrihaemoglobin formation gained maximal speed after a short lag phase as shown in Fig. 22.6 (notice the different ordinate scale!). Addition of superoxide dismutase (600 U/ml) decreased the reaction rate by about 15% without influencing the lag phase. In contrast, addition of catalase (30 k/ml) abolished the lag phase and decreased the reaction rate by 40 to 50%. The presence of both, catalase and superoxide dismutase, had additive effects and suspended the lag phase. These data indicated that superoxide radical anions may be of minor importance for ferrihaemoglobin formation, whereas hydrogen peroxide apparently plays a major role. The acceleration of the reaction rates in the absence of catalase probably reflect a gradual build-up of hydrogen peroxide which itself produces ferrihaemoglobin. Addition of hydrogen peroxide (0.1 mM) to the phenylenediamines (0.1 mM each) did not lead to relevant oxidation.

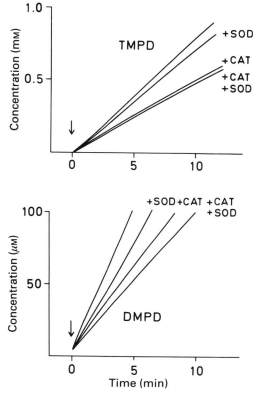

Figure 22.6 Influence of superoxide dismutase (SOD) and catalase (CAT) on ferrihaemoglobin formation by DMPD and TMPD. Purified human haemoglobin (3 mM) reacted with DMPD and TMPD, respectively, (0.1 mM each) in 0.2 M-sodium phosphate buffer, pH 7.4, 37°C under air in the absence or presence of superoxide dismutase (600 U/ml) and/or catalase (30 k/ml).

22.6 AUTOXIDATION OF DMPD AND TMPD AND THE INFLUENCE OF SUPEROXIDE RADICALS

Since DMPD and TMPD autoxidize readily to the radical cations which were assumed to be the active species responsible for ferrihaemoglobin formation, it was of interest to investigate also the kinetics of autoxidation. DMPD or TMPD (0.1 mM each) were allowed to react in 0.2 M-sodium phosphate buffer, pH 7.4 at 37°C under air, with radical detection at 550 and 611 nm, respectively. TMPD autoxidation showed an initial lag phase and gained maximal speed only after a few minutes (Fig. 22.7). The maximal rate

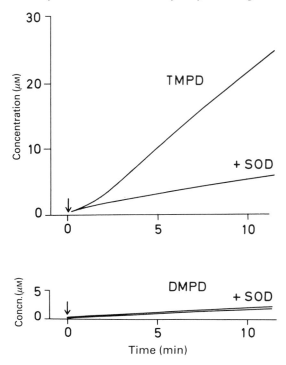

Figure 22.7 Influence of superoxide dismutase on the rate of autoxidation of DMPD and TMPD. DMPD and TMPD (0.1 mM each) were allowed to autoxidize in 0.2 M-sodium phosphate buffer, pH 7.4, at 37°C, under air in the absence or presence of superoxide dismutase (600 U/ml). Formation of the radicals was followed at 550 and 611 nm, respectively.

of autoxidation was about 50 nM/s, i.e. 3% the rate of ferrihaemoglobin formation under the same experimental conditions. Catalase (30 k/ml) had no effect on either the lag phase or the maximal rate, whereas superoxide dismutase (600 U/ml) retarded the maximal rate to one quarter, i.e. the rate observed during the lag phase. Since boiled superoxide dismutase was without effect on the autoxidation kinetics of TMPD, it was assumed that superoxide radicals were involved in the autoxidation of TMPD.

In contrast to TMPD, formation of DMPD radical cations was much slower (Fig. 22.7). The maximal velocity was only 2.5 nM/s, a lag phase was not observed and superoxide dismutase virtually had no effect. Since the observations of radical production does not necessarily reflect the true formation rates but does imply decay reactions as well, the decrease in TMPD and DMPD under the above conditions was also determined.

With TMPD (0.1 mM) the maximal rate of disappearance (spectroscopic determination in ether extracts) was 45 nM/s in the absence and 11 nM/s in the presence of superoxide dismutase (600 U/ml). These values agreed with the formation rates of the TMPD radical cation. With DMPD (0.1 mM) no significant decrease was observed within 10 min. These data indicate that ferrihaemoglobin formation by the above phenylenediamines was faster by two orders of magnitude than the respective autoxidation of the amines. Hence oxyhaemoglobin with its activated oxygen was probably responsible for radical formation. This assumption was confirmed when the rates of ferrihaemoglobin formation were followed at reduced oxygen tension. At 15 Torr (2.2 kPa) oxygen pressure, where haemoglobin was saturated by more than 90%, ferrihaemoglobin formation rates were hardly reduced, whereas autoxidation was diminished to about one-tenth.

22.7 REACTIONS OF THE TMPD RADICAL CATION WITH GSH

22.7.1 Isolation and structural characterization of the reaction products

When TMPD^{+*} (0.5 mM) reacted with GSH (1 mM) at pH 7.4 under anaerobic conditions, the blue solution decolourized in a few minutes. The incubation mixture was exhaustively extracted with ether to determine TMPD in the etheral extract and GSH and GSSG in the aqueous phase. Table 22.2 shows that 0.1 mM-TMPD equivalents and 0.15 mM-glutathione equivalents were missing. Moreover, it appeared that more TMPD^{+*} was reduced to TMPD than GSSG was formed. The u.v. spectrum of the aqueous phase indicated formation of a compound with a bathochromic maximum at 315 nm, compatible with a thioether structure (Eyer and Kiese, 1976; Eckert *et al.*, 1990).

To identify the new compound, 40 μmol TMPD^{+*} perchlorate was reacted with 75 μmol GSH at pH 7.4 under argon. After ether extraction,

Table 22.2 Products of the reaction of TMPD^{+*} with GSH

GSH	GSSG	Products (mM) GSH + 2 GSSG	TMPD^{+*}	TMPD
0.551 ± 0.008	0.143 ± 0.007	0.837 ± 0.012	0	0.406 ± 0.021

TMPD^{+*} (0.5 mM) and GSH (1.0 mM) reacted in 0.2 M-sodium phosphate, pH 7.4, at 20°C under argon for 10 min (means $\pm$ SD, $n = 4$).

the aqueous phase was freeze dried and purified by HPLC. The ninhydrin-positive product was identified by ^{1}H-NMR spectroscopy as 2-(glutathione-S-yl)-N,N,N',N'-tetramethyl-p-phenylenediammonium   salt   (2-GS-TMPD). The signals of the methyl protons were found at 3.27 (6) and 3.35 (6) p.p.m. as expected for dimethylammonium compounds (Hesse *et al.*, 1979). Three aromatic protons exhibited the following signals: H-3, 7.90 p.p.m. d (J = 1.7 Hz); H-5, 7.95 p.p.m. dd (J = 9 and 1.7 Hz); H-6, 7.69 p.p.m. d (J = 9 Hz). The protons of the glutathionyl moiety showed resonances as described by Hinson *et al.* (1982).

22.7.2 Kinetics of the reaction of TMPD^{+*} with GSH

Analysis of the reaction rates over a wide concentration range revealed two different reaction mechanisms. At constant GSH concentration (1 mM) and varying TMPD^{+*} concentrations (10–200 μM), TMPD^{+*} disappeared as expected in a pseudo first order reaction with constant half-life. Unexpectedly, at constant TMPD^{+*} concentration (5 μM) and varying GSH

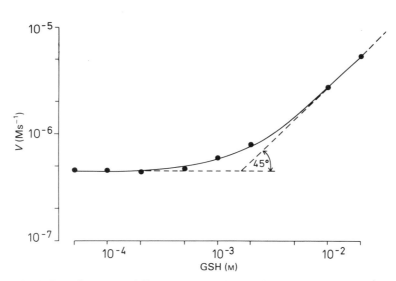

Figure 22.8 Reaction rates of disappearance of the radical cation of TMPD^{+*} in the presence of GSH. TMPD^{+*} (5μM) reacted with various concentrations of GSH in 0.2 M-sodium phosphate, pH 7.4, at 37°C under argon. The disappearance of TMPD^{+*} was followed photometrically. The broken lines are the asymptotes of two different reactions. The curve with a slope of tg α = 1 in the double-logarithmic plot represents the expected second order reaction.

Table 22.3 Rates of decrease of TMPD^{+*} in the presence of GSH and influence of EDTA

GSH (mM)	V_{in} (nM/s)	k_2 (M^{-1}s^{-1})	V_{in} (nM/s) (EDTA)
0.05	46	185	4.6
0.1	46	92	4.8
0.2	44	44	5.2
0.5	47	19	6.0
1.0	60	12	9.0
2.0	80	8	26
10.0	275	5.5	305
20.0	540	4.6	615

TMPD^{+*} (5 μM) reacted with various concentrations of GSH in 0.2 M-sodium phosphate ± EDTA (1 mM) buffer, pH 7.4, at 25°C under argon. Initial velocities of the disappearance of TMPD^{+*} were determined for calculation of the rate constants.

concentrations (0.1–1 mM), the reaction rate of TMPD^{+*} disappearance remained constant, indicating a complex reaction mechanism. At higher GSH concentrations, however, the reaction rates increased, and at GSH concentrations above 10 mM a normal second order reaction was observed (Fig. 22.8). Interestingly, addition of EDTA (1 mM) retarded the first order reaction to about one-tenth, but was without effect on the second order reaction (Table 22.3).

22.7.3 Formation of the thioether from TMPD^{+*} at different GSH concentrations

When TMPD^{+*} (50 μM) reacted with various concentrations of GSH the yield of 2-GS-TMPD remained constant (about 25%) at low GSH concentrations (0.2–1 mM) but decreased gradually at increasing GSH concentration (only traces at 20 mM GSH). These data again indicated two different reaction pathways.

22.8 REACTIONS OF TMPD WITH RED CELLS

Washed human red cells (3 mM Hb) were incubated with TMPD (0.1 mM) in 0.2 M phosphate buffer, pH 7.4, at 37°C under air. Ferrihaemoglobin formation proceeded at an initial rate of 0.5 μM/s, i.e. one-third the velocity observed with purified haemoglobin. When TMPD^{+*} (0.1 mM) was used

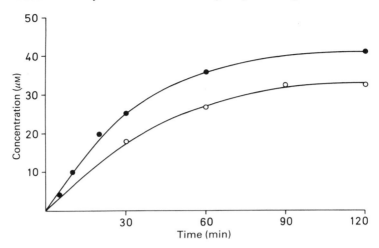

Figure 22.9 Formation of 2-(glutathione-*S*-yl)-*N*,*N*,*N'*,*N'*-tetramethyl-*p*-phenylene-diamine (2-GS-TMPD) during incubation of TMPD with GSH. TMPD (0.5 mM) reacted either with washed human red cells (3 mM Hb) (open circles) or with purified human haemoglobin (3 mM) fortified with GSH (2 mM) (closed circles) in 0.2 M sodium phosphate, pH 7.4, at 37°C under air. After precipitation of aliquots with perchloric acid, 2-GS-TMPD was assayed by HPLC.

instead of TMPD, ferrihaemoglobin formation was virtually identical. After 1 h about 50% of ferrihaemoglobin was formed, indicating that the catalytic mechanism is also operating in red cells.

To investigate whether 2-(glutathione-*S*-yl)-*N*,*N*,*N'*,*N'*-tetramethyl-*p*-phenylenediamine is also formed in red cells, TMPD (0.5 mM) was allowed to react with erythrocytes (3 mM Hb) as described above. After 2 h, about 35 µM thioether was formed. Similarly, reaction of purified human haemoglobin (3 mM) in the presence of GSH (2 mM) yielded the TMPD thioether (45 µM) within 2 h (Fig. 22.9). When TMPD (0.5 mM) was incubated with GSH (2 mM) in phosphate buffer without haemoglobin, only traces of the thioether were detected within 3 h.

22.9 ARYLAMINE-INDUCED FERRIHAEMOGLOBIN FORMATION FROM OXYHAEMOGLOBIN: POSSIBLE REACTION MECHANISMS

All aniline derivatives tested to date promoted ferrihaemoglobin formation from oxyhaemoglobin at rates that exceeded autoxidation distinctly. The most active compounds of this series were the phenylenediamine derivatives which quickly formed many equivalents of ferrihaemoglobin. A linear free-energy relationship was found with those aniline derivatives for which withdrawal of a lone pair electron from the amino nitrogen will result in a radical cation that is stabilized when the positive charge can be delocalized between reaction centre and substituent via 'through resonance' (Johnson, 1973). Hence, a linear correlation of reaction rates with σ_p^+ Hammett constants was expected. In fact, linearity was found only for derivatives with strong positive inductive and resonance effects. For these compounds formation of radical cations as the ultimate ferrihaemoglobin forming species was assumed (see below). Applying the Okamoto–Brown equation

$$\log k_X/k_H = \varrho \times \sigma^+$$

a reaction constant $\varrho = -2.4$ was calculated for compounds with σ^+ values below -0.3.

It should be noted that the oxidation potentials of anilines in non-aqueous media were also found to correlate with σ^+ function, whereas the oxidation of phenols and anilines by horseradish peroxidase compound I showed a linear correlation only with Hammett σ values (Job and Dunford, 1976). This unexpected result was explained by the authors to be due to the fact that the substrate gives an electron to compound I and simultaneously loses a proton. Such a mechanism, however, would be unfavourable for ferrihaemoglobin formation, because a negatively charged aniline anion would result from oxidation of ferrihaemoglobin by a neutral aniline radical.

No conclusions can be drawn from our experiments with regard to the underlying mechanisms of ferrihaemoglobin formation by aniline derivatives with electron-withdrawing substituents, since apparently no Hammett correlation does exist. Because these derivatives produced ferrihaemoglobin very slowly, and poor water-solubility limited an acceleration of the reaction rates compared with autoxidation, we cannot decide whether these compounds produced ferrihaemoglobin in catalytic amounts, too. Golly and Hlavica (1983) detected that 4-chloroaniline is *N*-oxygenated by rabbit haemoglobin in the presence of reduced pyridine nucleotides and cytochrome *c* reductase, and experiments by Sterzl (1984) revealed that ferrihaemoglobin formation of purified human haemoglobin in the presence of aniline and 4-chloroaniline was greatly enhanced by addition of NADPH in the absence of any reductase. These results imply that other than radical mechanisms may be operating during ferri-haemoglobin formation by these latter compounds.

The *N*-alkylated phenylenediamines were the preferred objects of our study, since they offered the opportunity to work with unusually stable radicals that can be obtained even in crystalline form (Wurster's red and Wurster's blue). These radicals with redox potentials between $+0.27$ and $+0.34$ V (Nickel and Jaenicke, 1982; Steenken and Neta, 1982) which thermodynamically favour oxidation of ferrohaemoglobin to ferri-haemoglobin ($+0.13$ V, Antonini *et al.*, 1965) produced ferrihaemoglobin by about three orders of magnitude faster than the parent phenylene-diamines. Hence, steady-state concentrations of these radicals were too low to be followed spectroscopically. Nevertheless, it is reasonable to assume that the radical cations are the ultimate ferrihaemoglobin forming species.

The mechanism of radical formation from DMPD and TMPD deserves separate comment. Since ferrihaemoglobin formation proceeded by orders of magnitude faster than autoxidation of the phenylenediamines, oxidation by the haemoglobin-activated oxygen has to be assumed. Such an oxidative action of haemoglobin has been found in the reaction with aminophenols, *N*-hydroxyarylamines and *N*-hydroxy-*N*-arylacetamides (Eyer *et al.*, 1974, 1975; Lenk and Sterzl, 1984, 1987; Lenk and Riedl, 1989). H-Abstraction from all these compounds by oxyhaemoglobin will initiate these reactions under formation of HbFe(III)OOH with oxygen being in the hydrogen peroxide redox state (compound I complex). Simultaneously, arylaminyl, phenoxyl, phenyl nitroxide, and *N*-phenyl-*N*-acyl nitroxide radicals will be formed. Depending on the type of the xenobiotic, the compound I complex of haemoglobin may decay into ferrihaemoglobin and hydrogen peroxide or may react further with the parent hydrogen donor to give an intermediate compound II complex which then reacts with the third hydrogen donor molecule to give ferrihaemoglobin and water. Such a 'clean' reaction sequence without liberation of free reactive oxygen species has been assumed for 4-dimethylaminophenol (Eyer *et al.*, 1975; Eyer and Lengfelder, 1984), for phenylhydroxylamine (Maples *et al.*, 1990), and for *N*-hydroxy-4-chloroacetanilide (Lenk and Riedl, 1989), where formation of hydrogen peroxide could not be detected.

In case of the phenylenediamines, catalase diminished ferrihaemoglobin formation by 40–50%, an indication for emanation of free hydrogen peroxide. Hence, an alternative reaction sequence may be operating:

$$HbO_2 + TMPD \rightleftharpoons HbFe^{III}-O-O^- + TMPD^{+*} \qquad (22.1)$$
$$HbFe^{II} + TMPD^{+*} \rightleftharpoons HbFe^{III+} + TMPD \qquad (22.2)$$
$$HbFe^{III}-O-O^- + H^+ \rightleftharpoons HbFe^{III}-O-OH \qquad (22.3)$$
$$HbFe^{III}-O-O^- + H_2O \rightleftharpoons HbFe^{III+} + H_2O_2 + OH^- \qquad (22.4)$$
$$2\ HbFe^{II} + H_2O_2 \rightarrow\rightarrow\rightarrow 2\ HbFe^{III+} + 2\ OH^- \qquad (22.5)$$

In this scheme TMPD (or DMPD) acts like a catalyst with formation of two mol of ferrihaemoglobin and 1 mol of hydrogen peroxide which can form another two equivalents of ferrihaemoglobin (besides some green pigments, probably a result of OH radical attack).

The small decrease in the rate of ferrihaemoglobin formation as well as the marked decrease in autoxidation of TMPD by superoxide dismutase is not yet understood. If O_2^{-*} oxidizes TMPD at significant rates, the superoxide dismutase effect on ferrihaemoglobin formation would be explained. However, such a reaction appears rather unlikely. Despite the high positive redox potential of the superoxide anion, it is only a one-electron oxidant, if protons are plentifully available (Ingraham and Meyer, 1985):

$$O_2^{-*} + e^- + 2\,H^+ \rightleftharpoons H_2O_2 \quad (+0.89\;V) \tag{22.6}$$

Hence, at pH 7.4 the oxidation reaction is kinetically unfavoured, and the superoxide anion usually acts as a reductant, e.g. of cytochrome c, of the N,N-dimethylaminophenoxyl radical (Eyer and Lengfelder, 1984) and of the benzosemiquinone radical (Augusto and Cilento, 1975). Since addition of a superoxide radical generating system, hypoxanthine/xanthine oxidase, did not increase the rate of autoxidation of TMPD, but rather reduced TMPD^{+*} to the parent TMPD, the effect of superoxide dismutase cannot currently be explained.

In contrast to the phenolic analogue, 4-dimethylaminophenol that is rapidly bound to haemoglobin, DMPD and TMPD were not readily eliminated from haemoglobin solutions, probably due to the fact that the disproportionation equilibrium is far on the radical side, i.e. at pH 7.4, $k_D/k_S = 0.27$ for DMPD (Nickel *et al.*, 1977) and 0.002 for TMPD (Petterson, 1968).

$$2\;TMPD^{+*} \underset{k_s}{\overset{k_D}{\rightleftharpoons}} TMPD + TMQDI^{++} \tag{22.7}$$

Hence, 1,4-addition reactions of the quinone di-imines are not favoured. Only after prolonged incubations of TMPD with haemoglobin, addition to the reactive SH groups of the β-chains occurred that was responsible for the missing TMPD.

22.10 REACTIONS OF ARYLAMINYL RADICAL CATIONS WITH SULPHYDRYL GROUPS

The reaction of the TMPD radical cation with sulphydryl groups was found to be rather complex. With GSH, formation of 2-GS-TMPD pointed to the quinone di-iminium dication as the ultimate electrophile which undergoes 1,4-addition reactions. This electrophile may result from disproportionation of two radicals. Alternatively, a direct addition of GSH to the radical could not be excluded. The determination of reaction products and reaction rates over a wide range of concentrations has solved this ambiguity. Since the rate of decrease in the radical concentration was constant at GSH concentrations below 1 mM, a direct second order reaction has to be ruled out. Such behaviour, however, is compatible with a preceding, rate-limiting disproportionation reaction followed by a fast reaction of GSH with the quinone di-iminium dication. Only at high GSH concentrations the direct, second order reaction gains weight. In this reaction GSH is oxidized without thioether formation. This direct reaction of GSH with TMPD^{+*} is a rather slow reaction which approaches a second order rate constant of about $5\,M^{-1}s^{-1}$ (pH 7.4, 37°C), whereas the reaction of GSH with the quinone di-iminium dication (TMQDI^{++}) is by orders of magnitude faster. Since $k_D/k_S = 0.002$ (Equation 22.7), the observed 'second order rate constant' found at the lowest GSH concentration studied, $185\,M^{-1}s^{-1}$, has to be multiplied by 1/0.002 to get an impression of the reaction rate of GSH with TMQDI^{++}. Probably the true value is above $10^5\,M^{-1}s^{-1}$ and approaches the value of $2 \times 10^6\,M^{-1}s^{-1}$ as found for the reaction of GSH with benzoquinone (Rossi *et al.*, 1986).

The reaction of GSH with TMQDI^{++} yields about one-quarter thioether (Equation 22.8).

$$TMQDI^{++} + GSH \rightarrow 2\text{-}(GS\text{-})TMPD + 2\,H^+ \qquad (22.8)$$

and three-quarters GSSG according to Equations 22.9 and 22.10.

$$TMQDI^{++} + GSH \rightarrow TMPD^{+*} + GS^* + H^+ \qquad (22.9)$$

$$2\,GS^* \rightarrow\rightarrow\rightarrow\rightarrow GSSG \qquad (22.10)$$

It should be noted that the ultimate formation of GSSG from glutathionyl radicals may involve a cascade of intermediates, particularly in the presence of oxygen (Quintiliani *et al.*, 1977).

The scheme outlined above also explains excess formation of TMPD compared to GSSG (Table 22.2). Formation of 2-GS-TMPD from TMPD^{+*} radicals yields equal amounts of TMPD (Equation 22.7) without simultaneous formation of GSSG. All these data agree with the suggestion that thioether formation does not occur with the radical, but with its disproportionation product.

22.11 CONCLUSIONS

Our knowledge of the enzymic functions of oxyhaemoglobin is increasing: oxyhaemoglobin has an oxidase-like activity also on certain aromatic amines to yield free aminyl radicals that form ferrihaemoglobin and react with thiols. These reactions are also observed in human red cells. Aniline derivatives with electron-donating substituents which allow a high degree of resonance stabilization of the aminyl radical cation are particularly active. Besides, hydrogen peroxide emanates which contributes additionally to ferrihaemoglobin formation.

Although methaemoglobinaemia is the most conspicuous sign after ingestion of aromatic amines, it is not necessarily the most alarming effect. If aromatic amines are activated within red cells and disseminated to sensitive target organs with limited defence systems, the consequences might be worse. Thus the metabolic activity of oxyhaemoglobin which also contributes significantly to detoxification reactions resembles a double-edged sword.

REFERENCES

Antonini, E., Brunori, M. and Wyman, J. (1965) Studies of the oxidation–reduction potentials of heme proteins. IV. The kinetics of the oxidation of hemoglobin and myoglobin by ferricyanide. *Biochemistry*, **4**, 545–51.

Augusto, O. and Cilento, G. (1975) The effect of diphenols upon autoxidation of oxyhemoglobin and oxymyoglobin. *Arch. Biochem. Biophys.*, **168**, 549–56.

Boyer, P.D. (1954) Spectrophotometric study of the reaction of protein sulfhydryl groups with organic mercurials. *J. Am. Chem. Soc.*, **76**, 4331–7.

Eckert, K.-G., Eyer, P., Sonnenbichler, J. and Zetl, I. (1990) Activation and detoxication of aminophenols. II. Synthesis and structural elucidation of various thiol addition products to 1,4-benzoquinoneimine and *N*-acetyl-1,4-benzo-quinoneimine. *Xenobiotica*, **20**, 333–50.

Eling, T.E., Boyd, J.A., Krauss, R.S. and Mason, R.P. (1985) Metabolism of aromatic amines by prostaglandin H synthetase, in *Biological Oxidation of Nitrogen in Organic Molecules: Chemistry, Toxicology and Pharmacology* (eds J.W. Gorrod and L.A. Damani), Ellis Horwood, Chichester, pp. 350–63.

Eyer, P. (1983) The red cell as a sensitive target for activated toxic arylamines. *Arch. Toxicol.*, Suppl. **6**, 3–12.

Eyer, P., Hertle, H., Kiese, M. and Klein, G. (1975) Kinetics of ferrihemoglobin formation by some reducing agents, and the role of hydrogen peroxide. *Mol. Pharmacol.*, **11**, 326–34.

Eyer, P. and Kiese, M. (1976) Biotransformation of 4-dimethylaminophenol: reaction with glutathione, and some properties of the reaction products. *Chem. Biol. Interact.*, **14**, 165–78.

Eyer, P., Kiese, M., Lipowsky, G. and Weger, N. (1974) Reactions of 4-dimethyl-aminophenol with hemoglobin, and autoxidation of 4-dimethylaminophenol. *Chem. Biol. Interact.*, **8**, 41–59.

Eyer, P. and Lengfelder, E. (1984) Radical formation during autoxidation of 4-dimethylaminophenol and some properties of the reaction products. *Biochem. Pharmacol.*, **33**, 1005–13.

Eyer, P., Lierheimer, E. and Strosar, M. (1983) Site and mechanism of covalent binding of 4-dimethylaminophenol to human hemoglobin and its implications to the functional properties. *Mol. Pharmacol.*, **23**, 282–90.

Golly, I. and Hlavica, P. (1983) The role of hemoglobin in the *N*-oxidation of 4-chloroaniline. *Biochim. Biophys. Acta*, **760**, 69–76.

Hesse, M., Meier, H. and Zeeh, B. (1979) *Spektroskopische Methoden in der organischen Chemie*. Thieme, Stuttgart.

Hinson, J.A., Monks, T.J., Hong, M. *et al.* (1982) 3-(Glutathione-*S*-yl)-acetaminophen: a biliary metabolite of acetaminophen. *Drug. Metab. Dispos.*, **10**, 47–50.

Ingraham, L.L. and Meyer, D.L. (1985) *Biochemistry of Dioxygen*. Plenum Press, New York.

Job, D. and Dunford, H.B. (1976) Substituent effect on the oxidation of phenols and aromatic amines by horseradish peroxidase compound I. *Eur. J. Biochem.*, **66**, 607–14.

Johnson, C.D. (1973) *The Hammett Equation*, Cambridge University Press, Cambridge.

Kalyanaraman, B. and Sivarajah, K. (1984) The electron spin resonance study of free radicals formed during the arachidonic acid cascade and cooxidation of xenobiotics by prostaglandin synthase, in *Free Radicals in Biology* (ed. W.A. Pryor), Academic Press, Orlando, FL, vol. VI, pp. 149–98.

Kiese, M. (1966) The biochemical production of ferrihemoglobin-forming derivatives from aromatic amines, and mechanisms of ferrihemoglobin formation. *Pharmacol. Rev.*, **18**, 1091–161.

Kiese, M. (1974) *Methemoglobinemia: A Comprehensive Treatise*, CRC Press, Cleveland, Ohio.

Lenk, W. and Riedl, M. (1989) *N*-Hydroxy-*N*-arylacetamides. V. Differences in the mechanism of haemoglobin oxidation *in vitro* by *N*-hydroxy-4-chloroacetanilide and *N*-hydroxy-4-chloroaniline. *Xenobiotica*, **19**, 453–7.

Lenk, W. and Sterzl, H. (1984) Peroxidase activity of oxyhaemoglobin *in vitro*. *Xenobiotica*, **7**, 581–8.

Lenk, W. and Sterzl, H. (1987) *N*-Hydroxy-*N*-arylacetamides. III. Mechanism of haemoglobin oxidation by *N*-hydroxy-4-chloroacetanilide in erythrocytes *in vitro*. *Xenobiotica*, **17**, 499–512.

Lindquist, T., Hillver, S.-E., Lindeke, B. *et al.* (1985) On the chemistry of the peroxidative oxidation of *p*-phenetidine, in *Biological Oxidation of Nitrogen in Organic Molecules: Chemistry, Toxicology and Pharmacology* (eds J.W. Gorrod and L.A. Damani), Ellis Horwood, Chichester, pp. 350–63.

Maples, K.R., Eyer, P. and Mason, R.P. (1990) Aniline-, phenylhydroxylamine-, nitrosobenzene-, and nitrobenzene-induced hemoglobin thiyl free radical formation *in vivo* and *in vitro*. *Mol. Pharmacol.*, **37**, 311–18.

Mason, R.P. (1982) Free-radical intermediates in the metabolism of toxic chemicals, in *Free Radicals in Biology* (ed. W.A. Pryor), Academic Press, Orlando, FL, vol. V, pp. 161–222.

McGirr, L.G. and O'Brien, P. (1987) Glutathione conjugate formation without *N*-demethylation during peroxidase catalysed *N*-oxidation of *N,N,N',N'*-tetramethylbenzidine. *Chem. Biol. Interact.*, **61**, 61–74.

Michaelis, L. and Granick, S. (1943) The polymerization of the free radicals of the Wurster dye Type: the dimeric resonance bond. *J. Am. Chem. Soc.*, **65**, 1747–55.

Mieyal, J.J., Ackerman, R.S., Blumer, J.L. and Freeman, L.S. (1976) Characterization of enzyme-like activity of human hemoglobin. Properties of the hemoglobin-P-450 reductase-coupled aniline hydroxylase system. *J. Biol. Chem.*, **251**, 3436–41.

Mieyal, J.J. and Blumer, J.L. (1976) Acceleration of the autoxidation of human oxyhemoglobin by aniline and its relation to hemoglobin-catalyzed aniline hydroxylation. *J. Biol. Chem.*, **251**, 3442–6.

Nickel, U., Haase, E. and Jaenicke, W. (1977) Optische Spektren und Gleichgewichtskonstanten der Semichinone und Chinondiimine von substituierten *p*-Phenylendiaminen. *Ber. Bunsenges. Phys. Chem.*, **81**, 849–53.

Nickel, U. and Jaenicke, W. (1982) 1- und 2-Elektronenschritte bei der Oxidation substituierter Paraphenylendiamine mit versch. Oxidationsmitteln. II. Reaktionen mit Hexacyanoferrat. *Ber. Bunsenges. Phys. Chem.*, **86**, 695–701.

O'Brien, P.J. (1984) Multiple mechanisms of metabolic activation of aromatic amine carcinogens, in *Free Radicals in Biology* (ed. W.A. Pryor), Academic Press, Orlando, FL, vol. VI, pp. 289–322.

Petterson, G. (1968) Determination of the equilibrium constant for dismutation of intermediates appearing in redox reactions. *Acta Chem. Scand.*, **22**, 3063–71.

Quintiliani, M., Badiello, R., Tamba, M. *et al.* (1977) Radiolysis of glutathione in oxygen-containing solutions of pH 7. *Int. J. Radiat. Biol.*, **32**, 195–202.

Rossi, L., Moore, G.A., Orrenius, S. and O'Brien, P. (1986) Quinone toxicity in hepatocytes without oxidative stress. *Arch. Biochem. Biophys.*, **251**, 25–35.

Steenken, S. and Neta, P. (1982) One-electron redox potentials of phenols. Hydroxy- and aminophenols and related compounds of biological interest. *J. Phys. Chem.*, **86**, 3661–7.

Sterzl, H. (1984) Exrahepatische Aktivierung aromatischer Amine und ihrer Derivate durch Hämoglobin. Thesis, München.

Störle, Ch. (1989) Bildung und Reaktionen des Radikal-Kations bei der Ferrihämoglobinbildung durch *N,N,N',N'*-Tetramethyl-*p*-phenylendiamin. Diplomarbeit, München.

Wahler, B.E., Schoffa, G. and Thom, H.G. (1959) Nachweis von Radikalzwischenstufen bei der Hämoglobinoxydation nach Einwirkung aromatischer Hydroxylamine. *Arch. Exp. Pathol. Pharmacol.*, **236**, 20–2.

23

Relevance of primary and secondary nitroxide radicals in biological oxidations

W. Lenk*, M. Riedl* and L.-O. Andersson†

*Walther Straub-Institut für Pharmakologie und Toxikologie der LM-Universität München, D-8000 München 2, Nussbaumstrasse 26, Federal Republic of Germany and
†Varian AG, Steinhauser Strasse, CH-6300 Zug, Switzerland

1. Primary aromatic nitroxides are intermediates in the co-oxidation of arylhydroxylamines and oxyhaemoglobin.
2. EPR spectra and decay kinetics of five primary aromatic nitroxides were studied.
3. Secondary aromatic nitroxides are implicated as the catalytically active intermediates in the co-oxidation of N-hydroxy-N-arylacetamides and oxyhaemoglobin.
4. EPR spectra and decay kinetics of seven acetyl arylnitroxides were studied.

23.1 INTRODUCTION

Uehleke (1963) was the first to establish a relationship between the rate of microsomal N-hydroxylation of aromatic amines *in vitro* and the ferrihaemoglobin(HbFe^{3+}) forming activity *in vivo*. Of the five arylamines tested, 4-aminobiphenyl surpassed the others in the rate of N-hydroxylation by rat liver microsomes and in HbFe^{3+} formation in cats. Similarly, Radomski and Brill (1971) found that of the three arylamines tested in the dog, 4-aminobiphenyl surpassed the others in the amount of N-hydroxy derivative determined in urine, in its HbFe^{3+} forming activity, and in its carcinogenicity. Thus, N-hydroxylation of arylamines produces active metabolites which may oxidize oxyhaemoglobin (HbFe$^{2+}O_2$) and initiate tumour growth.

Wahler *et al.* (1959) in studying the mechanism of HbFe$^{2+}O_2$ oxidation in cat erythrocytes by arylhydroxylamines and N-alkylphenylhydroxylamines, claimed to have observed radical intermediates. They postulated the primary and secondary aromatic nitroxide radicals as the catalytically active HbFe^{3+} forming molecules, which were formed from arylhydroxylamines by one-electron oxidation. Consequently, they considered the formation of nitrosoarenes in the coupled oxidation of HbFe$^{2+}O_2$ with arylhydroxyl-amines as a side-reaction. In contrast, Kiese (1959) explained the formation of several equivalents of HbFe^{3+} in erythrocytes *in vivo* and *in vitro* by each μmol of arylhydroxylamine with a repeated hydrogenation of the inactive nitrosoarene (formed from the arylhydroxylamine by co-oxidation with HbFe$^{2+}O_2$ by the removal of two electrons) to the active arylhydroxyl-amine, which entered a new cycle of the coupled HbFe$^{2+}O_2$ oxidation. The cycle is driven by glucose and the NADPH-dependent HbFe^{3+} reductases. In the absence of reducing equivalents, as in the case of solutions of purified human haemoglobin, however, each arylhydroxylamine reacted with HbFe$^{2+}O_2$ only once, i.e., the arylhydroxylamine lacked catalytic activity. Kiese and Reinwein (1950), who studied the mechanism of the coupled oxidation of HbFe$^{2+}O_2$ and phenylhydroxylamine, found that the yield of HbFe^{3+} depended on the ratio of HbFe^{2+}: phenylhydroxylamine. At a ratio of 1, 0.3 equivalents of HbFe^{3+} were formed per μmol of phenylhydroxyl-amine, and at a ratio of 5, 1 equivalent of HbFe^{3+} per μmol of phenyl-hydroxylamine. These results showed that (1) primary nitroxides, if there were any, did not play a role in the coupled oxidation of HbFe$^{2+}O_2$ and arylhydroxylamines, i.e., they lack catalytic activity, and (2) that several reactions or equilibria participate in the coupled oxidation.

That N-acyl-phenylhydroxylamines can also oxidize haemoglobin, was first reported by Heubner *et al.* (1953). They found that N-acetylphenyl-hydroxylamine apparently displayed catalytic activity in the oxidation of HbFe$^{2+}O_2$ in human erythrocytes, but that less HbFe^{3+} was formed much slower than with phenylhydroxylamine. Without further proof, they implicated acetylphenyl nitroxide as the catalytically active molecule, which was formed from *N*-acetyl-phenylhydroxylamine by one-electron oxidation. In contrast, Kiese and Plattig (1959) and Hustedt and Kiese (1959) explained the much slower and weaker reaction of *N*-acyl-phenyl-hydroxylamines with $HbFe^{2+}O_2$ in red cells *in vivo* and *in vitro* as an effect of phenylhydroxylamine, formed by enzymic hydrolysis of the *N*-acyl-phenylhydroxylamines.

This controversial discussion on the mechanism of the coupled oxidation of oxyhaemoglobin and arylhydroxylamines and *N*-acyl-arylhydroxyl-amines stimulated the investigation, which is reported here.

23.2 PRIMARY AROMATIC NITROXIDES

23.2.1 Autoxidation of *N*-hydroxy-4-chloroaniline in buffer, pH 7.4–methanol (9 : 1, v/v)

The coupled oxidation of arylhydroxylamines and $HbFe^{2+}O_2$ can be understood as an interaction of (1) autoxidation of the arylhydroxylamine and (2) co-oxidation of arylhydroxylamine and $HbFe^{2+}O_2$. Autoxidation of *N*-hydroxy-4-chloroaniline, which was rather slow in 0.2 M phosphate buffer, pH 7.4, – methanol (9:1, v/v), accelerated with increasing concentration of $HbFe^{2+}O_2$ and changed into co-oxidation as soon as the molar ratio of $HbFe^{2+}$: *N*-hydroxy-4-chloroaniline became >0.2. Autoxidation of arylhydroxylamines both in the absence and presence of $HbFe^{2+}O_2$ was clearly distinguished from co-oxidation, because autoxidation of *N*-hydroxy-4-chloroaniline was accompanied by a consumption of molecular oxygen, whereas co-oxidation was accompanied by an initial fast phase of oxygen liberation, followed by a slow phase of oxygen consumption.

Autoxidation of *N*-hydroxy-4-chloroaniline (10^{-3} M) in buffer, pH 7.4, – methanol (9:1, v/v) was followed at room temperature for 6 h in a closed Erlenmeyer flask. In a parallel experiment, (1) the same reaction mixture was stirred in the reaction chamber of an oxygen electrode and the changes of oxygen concentration monitored for 6 h, and (2) the time-dependent formation and decay of 4-chlorophenyl nitroxide radical was monitored by EPR spectroscopy. Since actual concentrations of the radical could not be determined, the observed differences in signal height merely reflect differences in its unknown concentration. Figure 23.1 shows the time-dependent changes of the concentrations of *N*-hydroxy-4-chloroaniline, 4,4'-azoxybischlorobenzene, 4-chloronitrosobenzene, 4-chloronitro-benzene, and 4-chlorophenyl nitroxide. 4-Chloronitrosobenzene was the

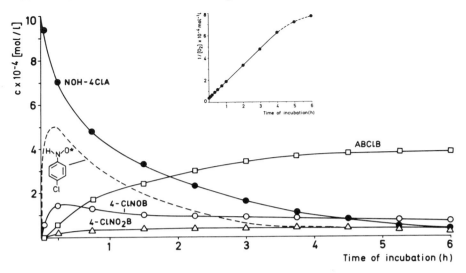

Figure 23.1 Autoxidation of *N*-hydroxy-4-chloroaniline(I) in phosphate buffer, pH 7.4, – methanol. A 10^{-3} M solution of I in 0.2 M phosphate buffer, pH 7, – methanol (9 : 1, v/v) was stirred for 6 h at 25°C in a closed Erlenmeyer flask. At times indicated, 0.5 ml was mixed with 2 ml of methanol and 50 μl samples were injected into the HPLC system. Symbols indicate means of two experiments. ABClB, 4,4'-azoxybischlorobenzene; 4-ClNOB, 4-chloronitrosobenzene; 4-ClNO$_2$B, 4-chloronitrobenzene. As the actual concentration of 4-chlorophenyl nitroxide could not be determined, the dashed line reflects merely differences in its unknown concentration. Inset. The same reaction mixture was stirred at 25°C in the reaction chamber of an oxygen electrode and the changes of oxygen concentration monitored for 6 h. Reciprocal plots of oxygen consumed versus time were linear for 4 h, indicating second order kinetics, $k = 4.3 \times 10^{-8}\,\mathrm{l\,mol^{-1}\,s^{-1}}$.

primary oxidation product which combined with *N*-hydroxy-4-chloroaniline to give 4,4'-azoxybischlorobenzene. The decline of the radical activity paralleled the decline of the concentration of *N*-hydroxy-4-chloroaniline, slow oxygen uptake, and the formation of 4-chloronitrosobenzene during the first minute. But after 15 min the concentration of 4-chloronitroso-benzene and 4-chlorophenyl nitroxide declined, due to the formation of 4,4'-azoxybischlorobenzene. This is an indication that the formation and disappearance of 4-chlorophenyl nitroxide and of 4-chloronitrosobenzene are mutually related according to equilibrium

$$\text{Ar}\!-\!\text{NHOH} + \text{Ar}\!-\!\text{NO} \rightleftharpoons 2\,\text{Ar}\!-\!\text{NH}\!-\!\text{O}^* \tag{23.1}$$

The kinetics of oxygen utilization followed second order for 4 h, when approximately 90% of *N*-hydroxy-4-chloroaniline had disappeared.

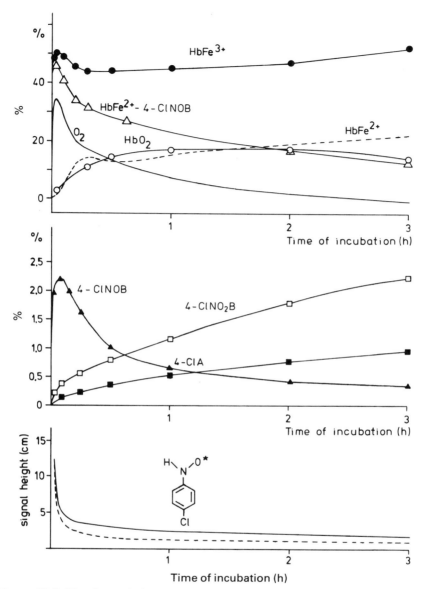

Figure 23.2 Kinetic analysis of the coupled oxidation of *N*-hydroxy-4-chloro-aniline(I) and oxyhaemoglobin. The reaction of purified human haemoglobin (10^{-3}M HbFe^{2+}) with 10^{-3}M (I) in 0.2 M phosphate buffer, pH 7.4, at 25°C was analysed by HPLC, UV and EPR spectroscopy. The curves for HbFe^{3+} and oxygen are based on means of four experiments, the others on means of two experiments. The curve for desoxyhaemoglobin (---) results from the difference of total Hb minus the sum of the directly determined HbFe^{3+}, HbO$_2$ and haemoglobin-4-chloro-nitrosobenzene complex. 4-ClNOB, 4-chloronitrosobenzene; 4-ClNO$_2$B, 4-chloro-nitrobenzene; 4-ClA, 4-chloroaniline.

23.2.2 Co-oxidation of oxyhaemoglobin and N-hydroxy-4-chloroaniline

Pilot experiments on the co-oxidation of $HbFe^{2+}O_2$ and N-hydroxy-4-chloroaniline at different molar ratios of $HbFe^{2+}$: N-hydroxy-4-chloroaniline indicated that several reactions or equilibria accompanied the coupled oxidation, and that 'coupling is not strict' (Kiese and von Ruckteschell, 1951). For example, 1 equiv. of $HbFe^{3+}$ was produced by reaction of 1 equiv. of N-hydroxy-4-chloroaniline with 10 equiv. of $HbFe^{2+}$, but only 0.3 equiv. of $HbFe^{3+}$ arose from a molar ratio of 1 : 1.

The product pattern produced by the reaction of 10^{-3} M-N-hydroxy-4-chloroaniline with equimolar concentrations of $HbFe^{2+}$ in buffer, pH 7.4, at room temperature was determined for 3 h by HPLC and UV spectroscopy. In a parallel experiment, (1) the time-dependent changes in oxygen concentration and (2) time-dependent changes in radical activity were monitored. N-Hydroxy-4-chloroaniline rapidly disappeared during the initial fast phase, since after 1 min only 15% of the initial concentration was left and after 3 min it had completely disappeared. Figure 23.2 shows that after 3 min 46% of the initial concentration of N-hydroxy-4-chloroaniline was present as the ferrohaem-4-chloronitrosobenzene complex and only 2.2% remained as unbound 4-chloronitrosobenzene. Besides traces of 4,4'-azoxybischlorobenzene, 0.3% 4-chloronitrobenzene and 0.1% 4-chloroaniline were determined, in total 48.6% of the applied N-hydroxy-4-chloroaniline. It is most likely, that the major portion of the missing 51.4% were bound to the SH-groups of the globin moieties (Eyer and Ascherl, 1987). Within 3 min 96% of $HbFe^{2+}$ was converted into $HbFe^{3+}$ (50%) and $HbFe^{2+}$–4-chloronitrosobenzene complex (46%), but only 34% of oxygen in $HbFe^{2+}O_2$ was liberated, indicating that a large part of the bound oxygen has served as electron acceptor in the coupled oxidation. The time-dependent decline of 4-chlorophenyl nitroxide activity paralleled that of 4-chloronitrosobenzene, and its unknown concentration was twice as high as in the absence of $HbFe^{2+}O_2$.

23.2.3 Primary aromatic nitroxides, formation and decay

(a) Primary aromatic nitroxides produced by autoxidation of arylhydroxylamines

EPR measurements were performed with a Varian E-109 EPR spectrometer. Arylhydroxylamine (0.1 mmol) was dissolved in 10 ml methanol p.a. and 0.5 ml was added to 4.5 ml 0.066 M phosphate buffer, pH 7.4, to give a final concentration of 10^{-3} M. Whereas clear solutions were obtained with the three monocyclic arylhydroxylamines, turbid solutions with the two polycyclic compounds indicated lower solubility. The solution was poured into a flat aqueous solution cell which was then inserted and centred in the EPR cavity. EPR spectra were obtained at 9.5 GHz using 40 G or 60 G field sweeps and microwave powers between 2 and 10 mW. Field calibration was checked by comparison with the 1.574 G splitting of the tetracyanoethylene anion radical. The g-value of this radical is 2.002777, and this was used as a standard for the g-value measurements.

Figure 23.3 shows the EPR spectra of phenyl nitroxide, 4-ethoxyphenyl nitroxide, 4-chlorophenyl nitroxide, 4-biphenylyl nitroxide and 2-fluorenyl

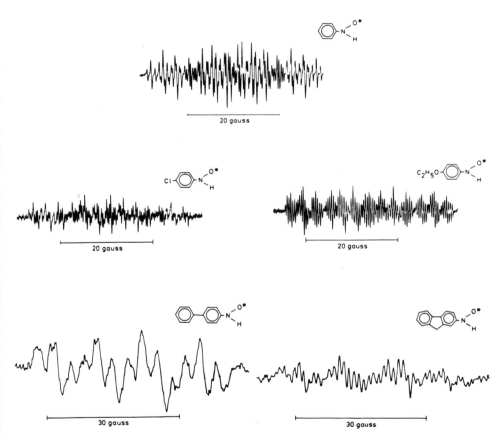

Figure 23.3 EPR spectra of five arylnitroxides in 0.066 M phosphate buffer, pH 7.4, – methanol. The primary aromatic nitroxides were generated by autoxidation of the corresponding arylhydroxylamines (10^{-3}M) in 0.066 M phosphate buffer, pH 7.4, – methanol(9:1, v/v) at room temperature.

Table 23.1 Hyperfine splitting constants in Gauss of five primary aryl nitroxides

Compound	N	H_{amine}	H_{ortho}	H_{meta}	(H, Cl, CH_2)	Accuracy	G-value
Phenyl nitroxide (= PNO)	10.64	13.22	3.31	1.11	3.62	± 0.05	2.00528
4-Ethoxyphenyl nitroxide (= 4-EPNO)	11.01	13.25	3.38	0.97	0.50	±0.05	2.00516
4-Chlorophenyl nitroxide (= 4-ClPNO)	10.41	12.99	3.34	1.14	0.38	±0.05	2.00562
4-Biphenylyl nitroxide (= 4-BNO)	10.14	13.17	3.63	1.0	—	±0.15	—
2-Fluorenyl nitroxide (= 2-FNO)	10.37	13.80	3.45	1.07	—	±0.08	—

structure of the primary nitroxides, which are formed from aryl-hydroxylamines by autoxidation. The hyperfine splitting constants for the five primary nitroxide radicals are listed in Table 23.1.

(b) Kinetics of self-reaction of primary aromatic nitroxides

A prominent line in the EPR spectrum was selected for the kinetic measurements. The E-272 B Field/Frequency-Lock accessory made it possible to remain at the same position of this line during the three hours these experiments lasted. The amplitude of the line was recorded as a function of time, measured from the moment the buffer was added. No data were obtained during the first few minutes while the cell was inserted and the spectrometer adjusted. Figure 23.4 shows the kinetics of formation and decay of the five primary nitroxide radicals. Arbitrary units of amplitude read from the scannings against the base line were plotted (1) on half-logarithmic paper and (2) as reciprocal units versus time on a linear scale to determine whether the order of decay was of first or second order.

(i) PHENYL NITROXIDE Figures 23.4 and 23.5 show that maximal amplitude of a selected line was observed 20 min after initiation of autoxidation. Thereafter half-logarithmic plots of arbitrary units of amplitude were linear, indicating first order decay kinetics, $t_{1/2}$ being 165 min.

(ii) 4-ETHOXYPHENYL NITROXIDE From Fig. 23.4, it can be seen that 4-ethoxyphenyl nitroxide was formed more readily from the corresponding arylhydroxylamine than was phenyl nitroxide from phenylhydroxylamine. The decline of the amplitude of a selected line was monitored for 3 h 18 min, during which time half-logarithmic plots of arbitrary units of amplitude read from Fig. 23.4 were linear, indicating first order decay kinetics, $t_{1/2}$ being 68 min (Fig. 23.5). When the experiment was repeated with 2×10^{-3} M arylhydroxylamine, the same half-life was determined. A comparison with the half-life of 165 min determined for phenyl nitroxide revealed a 2.4-fold faster decay of 4-ethoxyphenyl nitroxide, due to the electron-donating effect of the *p*-substituent.

(iii) 4-CHLOROPHENYL NITROXIDE It is also apparent from Fig. 23.4, that maximal amplitude of a selected line was observed 12 min after initiation of autoxidation of 4-chlorophenylhydroxylamine. The decline of amplitude of a selected line was monitored for 3 h 42 min, during which time half-logarithmic plots of arbitrary units of amplitude read from Fig. 23.4 were curved, excluding first order decay kinetics of 4-chlorophenyl nitroxide.

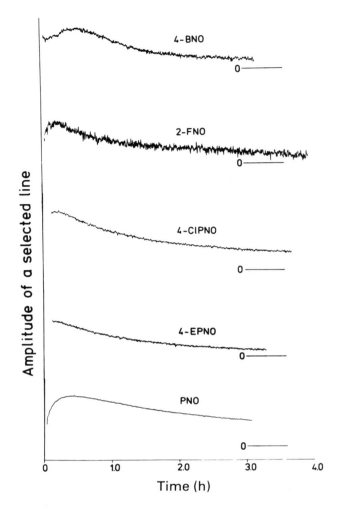

Figure 23.4 Time-dependent formation and decay of five primary aromatic nitroxides. At constant instrument setting, the change of amplitude of a selected signal from the EPR spectrum of 10^{-3} M solutions of phenylhydroxylamine, 4-ethoxyphenylhydroxylamine, 4-chlorophenylhydroxylamine, *N*-4-biphenylyl-hydroxylamine, and *N*-2-fluorenylhydroxylamine in 0.066 M phosphate buffer, pH 7.4, – methanol(9:1, v/v) at room temperature was monitored for more than 3 h. PNO, phenyl nitroxide; 4-EPNO, 4-ethoxyphenyl nitroxide; 4-ClPNO, 4-chlorophenyl nitroxide; 2-FNO, 2-fluorenyl nitroxide; 4-BNO, 4-biphenyl nitroxide.

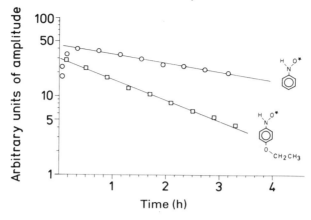

Figure 23.5 Decay kinetics of phenyl nitroxide and 4-ethoxyphenyl nitroxide. The change of the amplitude of a selected line of the EPR spectrum of phenyl nitroxide (○) and 4-ethoxyphenyl nitroxide (□) is plotted in a half-logarithmic scale. The graph shows apparent $t_{1/2}$ values of 165 and 68 min, respectively, for the decay of these nitroxides.

Reciprocal plots of arbitrary units of amplitude were linear from 17 to 227 min after initiation of 4-chlorophenylhydroxylamine autoxidation.

(iv) 2-FLUORENYL NITROXIDE At 12 min after initiation of autoxidation of *N*-2-fluorenylhydroxylamine, maximal amplitude of a selected line was observed and radical activity was monitored for 238 min (see Fig. 23.4). Half-logarithmic plots of arbitrary units of amplitude read from the kinetic curve were curved, excluding first order decay kinetics. Reciprocal plots of arbitrary units became linear after 17 min, indicating that second order decay kinetics apply to the disappearance of 2-fluorenyl nitroxide.

(v) 4-BIPHENYLYL NITROXIDE The decline of the amplitude of a selected line was monitored for 3 h 10 min (Fig. 23.4). Maximal amplitude was observed after 30 min. Reciprocal plots of arbitrary units of amplitude were linear from 1 h to 3 h, indicating that second order decay kinetics apply to the disappearance of 4-biphenylyl nitroxide. Since actual radical concentrations could not be determined, the rate constants could not be evaluated. Only the slopes of the linear plots of reciprocal arbitrary units of amplitude were used for comparison of the different rates of decay: 4-chlorophenyl nitroxide (0.0148 units/h), 4-biphenylyl nitroxide (0.0505 units/h), and 2-fluorenyl nitroxide (0.0312 units/h).

23.3 SECONDARY AROMATIC NITROXIDES

23.3.1 Co-oxidation of *N*-hydroxy-4-chloroacetanilide and oxyhaemoglobin

For a study on the mechanism of co-oxidation of *N*-hydroxy-*N*-aryl-acetamides and $HbFe^{2+}O_2$, *N*-hydroxy-4-chloroacetanilide was chosen, because it surpassed the other five compounds tested in its catalytic activity (Heilmair *et al.*, 1987), and because the product pattern produced by oxidation of *N*-hydroxy-4-chloroacetanilide could be analysed by HPLC. The reaction of 10^{-3} M *N*-hydroxy-4-chloroacetanilide with an equimolar concentration of $HbFe^{2+}$ in 0.2 M phosphate buffer, pH 7.4, at room temperature was followed by injecting aliquots at various times into the HPLC system. $HbFe^{3+}$ and oxygen were determined simultaneously. The results of the product analysis are shown in Fig. 23.6. Parallel with increasing $HbFe^{3+}$ concentration, the concentration of 4-chloronitrosobenzene, 4-chloronitrobenzene, 4-chloroacetanilide, *N*-acetoxy-4-chloroacetanilide, and molecular oxygen increased, indicating that they were formed by co-oxidation. After 1 h, approx. 75% $HbFe^{2+}O_2$ was oxidized, but only 5% of *N*-hydroxy-4-chloroacetanilide was consumed, indicating that 1 equiv. of *N*-hydroxy-4-chloroacetanilide had catalysed the oxidation of 15 equiv. of $HbFe^{2+}$. After 1 h, one-third of the oxygen in the oxidized $HbFe^{2+}O_2$ was liberated and two-thirds reduced, indicating that oxygen in $HbFe^{2+}O_2$ served as acceptor for electrons from $HbFe^{2+}$, transferred by *N*-hydroxy-4-chloroacetanilide. In the subsequent slow phase of the reaction, when $HbFe^{3+}$ formation had slowed down, only the concentration of 4-chloroacetanilide increased further, whereas the concentration of 4-chloronitrobenzene, *N*-acetoxy-4-chloroacetanilide, and oxygen decreased and the concentration of 4-chloronitrosobenzene remained constant for the following 3 h. This is an indication that 4-chloronitro-benzene and *N*-acetoxy-4-chloroacetanilide were consumed by secondary reactions, which became apparent as soon as the disposal of these compounds exceeded their formation. *N*-Acetoxy-4-chloroacetanilide rapidly reacts with nucleophiles and is therefore hydrolysed as well, but the disappearance of 4-chloronitrobenzene cannot be explained at present. After 30 min, approximately 0.3% and after 60 min, 0.1% of the yellow–orange quinoneimine-*N*-oxide was found, which had disappeared completely by 90 min. Within 4 h, the concentration of *N*-hydroxy-4-chloro-acetanilide disappeared by approximately 22%, but the five metabolites accounted for only 5.5%, indicating that the major fraction had escaped detection. However, when the product pattern was analysed by combined gas chromatography-mass spectrometry, additional oxidation products were detected (section 23.3.2).

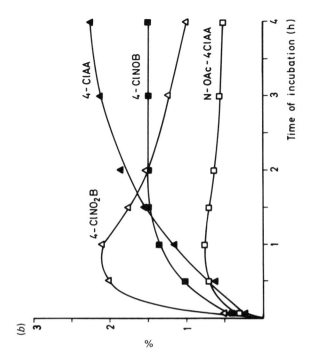

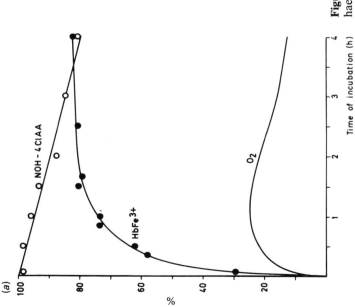

Figure 23.6 Co-oxidation of *N*-hydroxy-4-chloroacetanilide(II) and oxy-haemoglobin. (*a*) The reaction mixture of purified human haemoglobin and (II) each 10^{-3} M in 0.2 M phosphate buffer, pH 7.4, was stirred in the reaction chamber of the oxygen electrode and the changes in the concentrations of O_2, $HbFe^{3+}$ and II were monitored at 25°C. (*b*) At times indicated, aliquots were also analysed by HPLC for 4-chloroacetanilide(II), 4-ClNOB, 4-chloronitrosobenzene; 4-ClNO$_2$B, 4-chloronitrobenzene; 4-ClAA, 4-chloroacetanilide and N-OAc-4ClAA, *N*-acetoxy-4-chloroacetanilide.

As the same products were formed by oxidation of N-hydroxy-4-chloro-acetanilide by lead dioxide in ether or water, it is assumed that a common intermediate in the co-oxidation of N-hydroxy-4-chloroacetanilide(I) and $HbFe^{2+}O_2$ and the chemical oxidation, i.e., acetyl 4-chlorophenyl nitroxide(II), is the primary oxidation product. Radical activity was expected during the initial fast phase of the co-oxidation, but not found. As co-oxidation of N-hydroxy-4-chloroacetanilide and $HbFe^{2+}O_2$ followed second order kinetics (Heilmair *et al.*, 1987), we assume that the one-electron oxidation of N-hydroxy-4-chloroacetanilide by $HbFe^{2+}$ is the rate-limiting step, which yields acetyl 4-chlorophenyl nitroxide (Equation 23.2). The nitroxide in turn oxidizes a second mol of $HbFe^{2+}O_2$ and returns to the hydroxamic acid (Equation 23.3a), which can enter a new cycle of $HbFe^{2+}$ oxidation. Since the radical is not very stable in water, its self-reaction gives rise to two pairs of products (Equations 23.3b), (1) N-acetoxy-4-chloroacetanilide(III) and 4-chloronitrosobenzene(IV), and (2) 4-chloro-nitrobenzene(V) and 4-chloroacetaniline(VII). Such an intermolecular mechanism was first proposed by Forrester *et al.* (1970), who studied the product pattern produced by oxidation of N-hydroxy-acetanilide and N-hydroxy-4-acetylaminobiphenyl in toluene by Ag_2O. They postulated the aminyl(VI) as radical intermediate in the formation of the N-aryl-acetamide(VII), which is not a reduction product, but arises from an oxidation process. The aminyl(VI) also reacted with 4-chloronitroso-benzene in a 'spin-trapping analog' reaction to give the quinoneimine-N-oxide(VIII), which disappeared in the course of the coupled oxidation.

23.3.2 Combined gas chromatography–mass spectrometry (GC–MS) of the product pattern

At 60 min after mixing a 10^{-3}M solution of N-hydroxy-4-chloroacetanilide with an equimolar concentration of $HbFe^{2+}O_2$ (purified human haemo-globin in 0.2 M phosphate buffer, pH 7.4) at room temperature, the reaction mixture was extracted with ether. After drying, the concentrated ether extract was analysed by combined GC–MS using a gas chromatograph 2110 (Carlo Erba, Milano), equipped with a quartz capillary Durabond 1 (30 m × 0.25 mm) and the mass spectrometer Varian CH 7A (Varian-Finnegan), coupled with the system SS 300 MS.

Figure 23.7 shows the gas chromatogram, in which the oxidation products were labelled with the numbers 1 to 12 as they left the column. Among the twelve components numbers 1, 4, 8 and 10 were identified by their mass spectra as 4-chloronitrosobenzene, 4-chloronitrobenzene, 4-chloro-acetanilide, and N-acetoxy-4-chloroacetanilide, respectively. In parallel experiments, mass spectra of the synthetic compounds 1, 4, 8 and 10 gave unequivocal identification.

Table 23.2 shows the retention times and characteristic mass peaks of the twelve products, of which eight remained unidentified. Under the proposition that no fragmentation occurred, M^+ values of these compounds varied from 243 to 129. Since the mass peak of 111 ($= C_6H_4Cl^+$) was found in nine oxidation products it can be deduced that these compounds are metabolites of N-hydroxy-4-chloroacetanilide still containing the intact

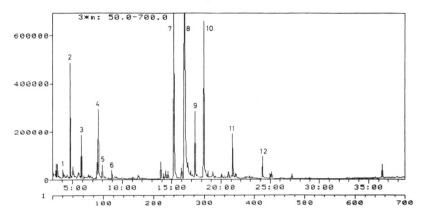

Figure 23.7 Gas chromatogram of the oxidation products of *N*-hydroxy-4-chloro-acetanilide and oxyhaemoglobin. At 60 min after mixing *N*-hydroxy-4-chloro-acetanilide (10^{-3} M) with HbFe^{2+} (10^{-3} M) in 0.2 M phosphate buffer, pH 7.4, at room temperature, the reaction mixture was extracted with ether. Aliquots of the concentrated ether extract were injected into the system. The twelve compounds which left the column were characterized by mass spectrometry (Table 23.2).

chlorobenzene moiety. That twelve products arose from the co-oxidation of *N*-hydroxy-4-chloroacetanilide and HbFe^{2+}O$_2$ is a further indication for a radical intermediate.

23.3.3 Voltammetric determination of the oxidation–reduction potentials of seven biologically relevant *N*-hydroxy-*N*-arylacetamides

Oxidation of *N*-hydroxy-*N*-arylacetamides *in vivo* and *in vitro* has been suggested by several authors in connection with the mechanism of chemical carcinogenesis (Bartsch and Hecker, 1971) and in the catalytic oxidation of haemoglobin (Heilmair *et al.* 1987; Lenk and Riedl, 1989). In order to obtain more information about the probability of the formation and the possible functions of acetyl arylnitroxides, we have determined the oxidation–reduction potentials of *N*-hydroxy-acetanilide, *N*-hydroxy-4-chloroacetanilide, *N*-hydroxy-3,4-dichloroacetanilide, *N*-hydroxy-phenacetin, *N*-hydroxy-4-acetylaminobiphenyl, *N*-hydroxy-2-acetyl-aminofluorene and *N*-hydroxy-2-acetylaminophenanthrene and studied the properties of the corresponding acetyl arylnitroxides, which were produced by one-electron oxidation.

Table 23.2 Mass spectra of the twelve oxidation products eluted from the gas chromatograph

Peak number	Fraction number	Retention time (min)	Mass peaks	Identification (mol. wt.)
1	21	4.1	141, 127, 111	4-chloronitrosobenzene (141.5)
2	42	4.8	129, 114, 99, 43	—
3	57	5.9	129, 113, 101, 43	—
4	90	7.7	157, 141, 127, 111	4-chloronitrobenzene (157.5)
5	98	8.0	171, 156, 140, 127, 111	—
6	117	9.0	143, 141, 140, 126, 111, 43	—
7	240	15.3	199, 157, 141, 126, 111, 43	—
8	268	16.4	169, 153, 149, 137, 127, 111, 43	4-chloroacetanilide (169.6)
9	282	17.3	243, 173, 159, 143, 127, 111, 43	—
10	299	18.3	227, 185, 143, 125, 111, 43	N-acetoxy-4-chloroacetanilide (227.6)
11	356	21.2	227, 207, 185, 168, 143, 125, 111, 43	—
12	416	25.2	229, 207, 183, 169, 142, 125, 111, 43	—

Since the voltammetric investigation has shown a very close relationship in the electrochemical behaviour of the seven *N*-hydroxy-*N*-arylacetamides, it may suffice to explain the characteristics of the obtained voltammograms on one typical example. Figure 23.8 shows the voltammogram of *N*-hydroxy-4-chloroacetanilide obtained with a Polarecord E 506 (deutsche Metrohm GmbH, D-7024 Filderstadt), a silver/silver chloride (3 M-KCl) electrode (EA 441/5) as the reference electrode together with a platinum-helping electrode (EA 282/1) and a glassy carbon electrode (EA 276/2) as working electrode. The interpretation of the three distinct potentials is based on the following observations.

Potential I is not observed in the first run in the anodic direction, but appears as soon as the potential difference of potential II is applied to the working electrode. Therefore, it is obvious that potential I does not arise from the hydroxamic acid itself, but is caused by an oxidation product. Such an assumption could be easily verified by comparison with the voltammogram of 4-chloronitrosobenzene or *N*-hydroxy-4-chloroaniline which showed the same characteristic reversible potential.

Potential II displays the characteristics of a slightly reversible electron transfer process as it may be expected for the formation of quite an unstable oxidation product. Since it is known that *N*-hydroxy-*N*-arylacetamides can be oxidized to the corresponding secondary nitroxides (Aurich and Baer,

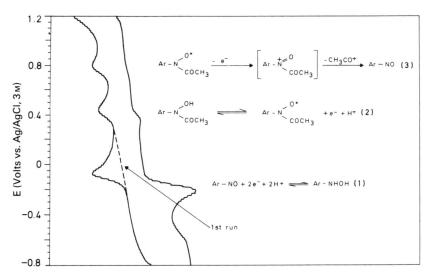

Figure 23.8 Voltammogram of a 10^{-3} M solution of *N*-hydroxy-4-chloroacetanilide in 0.05 M phosphate buffer, pH 7.4 – methanol(9 : 1, v/v) at 22°C. Working electrode, glassy carbon; reference electrode, Ag/AgCl (3 M KCl).

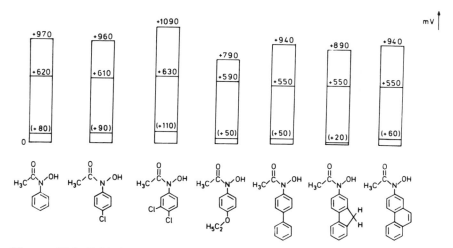

Figure 23.9 Oxidation–reduction potentials (against the normal hydrogen electrode, NHE) of seven *N*-hydroxy-*N*-arylacetamides determined by voltammetric measurements. The lowest potential between + 20 and + 110 mV is not caused by the *N*-arylacetohydroxamic acid itself, but results from the formation of the corresponding nitrosoarenes.

1965; Forrester *et al.*, 1970), we assume that this one-electron transfer is involved in the observed potential II. The value of + 0.61 V for this potential is in agreement with the observation, that only very strong oxidizing agents allow the formation of these nitroxides.

Potential III is characterized by the missing potential wave during the run in the cathodic direction, thus giving the typical shape of an irreversible oxidation process. Therefore, we assume that the *N*-hydroxy-*N*-aryl-acetamide loses a second electron and forms an unstable intermediate that is supposed to decay under *N*-deacetylation.

Differences in the potential height due to different aromatic substitution are apparent from Fig. 23.9. Since we used a silver/silver chloride electrode as a reference electrode, the values given in Fig. 23.9 were obtained by adding + 210 mV to each original potential value.

23.3.4 Chemical oxidation of *N*-hydroxy-4-chloroacetanilide

In order to obtain some information on the stability of the primary oxidation products of *N*-hydroxy-*N*-arylacetamides, i.e., acetyl aryl nitroxides, we have studied the decay of these radicals in toluene and methanol. All the nitroxides produced by oxidation of the corresponding *N*-hydroxy-*N*-arylacetamides with lead dioxide showed absorption bands in the visible region which are presented in Fig. 23.10. Because these nitroxides decay rapidly, we have not determined any extinction coefficient. Figures 23.11 and 23.12 show the kinetics of the radical decay obtained by monitoring the maximum wavelength of each nitroxide in toluene and methanol. The kinetic data of acetyl 4-ethoxyphenyl nitroxide and acetyl 3,4-dichlorophenyl nitroxide are not shown here, because they did not obey second order kinetics, but displayed a formal order of reaction of about 2.5. This was not analysed in detail. Acetyl 4-chlorophenyl nitroxide is also missing from Fig. 23.12 because its decay in methanol followed a first order rate equation. Table 23.3 shows the apparent rate constants determined for the second order decay reaction of several acetyl arylnitroxides in toluene

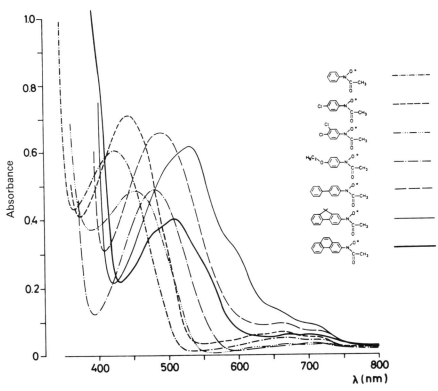

Figure 23.10 Electronic spectra of seven acetyl arylnitroxides in toluene. The relative absorbancies shown here do not reflect the real differences in the extinction coefficients.

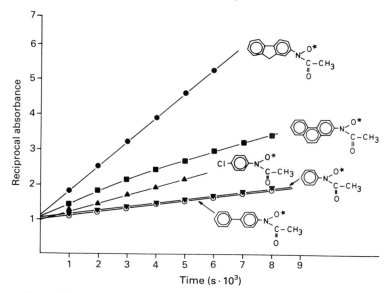

Figure 23.11 Kinetics of the decay of several acetyl arylnitroxides in toluene following second order.

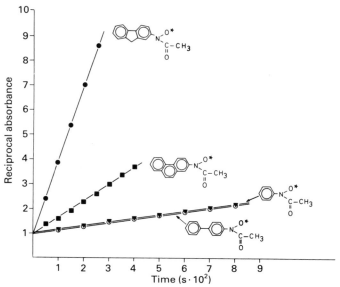

Figure 23.12 Kinetics of the decay of several acetyl arylnitroxides in methanol following second order.

Table 23.3 Decay kinetics of various acetyl arylnitroxides

Compound[a]	Apparent rate constants in	
	Toluene $l\,mol^{-1}\,s^{-1}$	Methanol $l\,mol^{-1}\,s^{-1}$
Acetyl phenyl nitroxide	0.11	1.6
Acetyl 4-chlorophenyl nitroxide[c]	0.22	[b]
Acetyl 4-biphenylyl nitroxide	0.11	1.6
Acetyl 2-fluorenyl nitroxide	0.88	51.3
Acetyl 2-phenanthryl nitroxide	0.34	9.3

[a] The secondary aromatic nitroxides were formed from the corresponding *N*-arylacetohydroxamic acids by oxidation with PbO_2 in either toluene or methanol at room temperature.
[b] The decay of acetyl 4-chlorophenyl nitroxide in methanol apparently followed first order, k being $1.4\,s^{-1}$.
[c] The decay of acetyl 4-chlorophenyl nitroxide in 0.2 M phosphate buffer, pH 7.4, also followed second order, k being $350\,l\,mol^{-1}\,s^{-1}$ (Lenk and Riedl, 1989).

and methanol. Although the rate constants of the decay reactions are from 15 to 58 times higher in methanol than in toluene, the order in which the radicals decayed remained the same for both solvents namely:

acetyl 2-fluorenyl nitroxide > acetyl 2-phenanthryl nitroxide

> acetyl 4-biphenylyl nitroxide = acetyl phenyl nitroxide

We have no explanation as yet for the fact that the last two compounds gave the same rate constants in both solvents. When the second order rate constant for the decay of acetyl 4-chlorophenyl nitroxide in toluene is compared with that in 0.2 M phosphate buffer, pH 7.4, the decay reaction in buffer was even 1600 times faster than in toluene.

23.3.5 EPR spectra of six acetyl arylnitroxides

In order to confirm our assumption about the nature of the primary oxidation products of *N*-hydroxy-*N*-arylacetamides, the EPR spectra of the nitroxides generated in toluene by oxidation with lead dioxide were recorded. The spectra (Fig. 23.13) are in agreement with the structure of acetyl arylnitroxides. The hyperfine splitting constants of acetyl phenyl nitroxide are similar to those reported by Aurich and Baer (1965), and those of acetyl 4-chlorophenyl nitroxide were reported by Lenk and Riedl (1989). In the case of acetyl 3,4-dichlorophenyl nitroxide, acetyl 2-fluorenyl nitroxide, acetyl 4-biphenylyl nitroxide, and acetyl 2-phenanthryl nitroxide the hyperfine splitting constants could not be determined because of the complexity of the spectra.

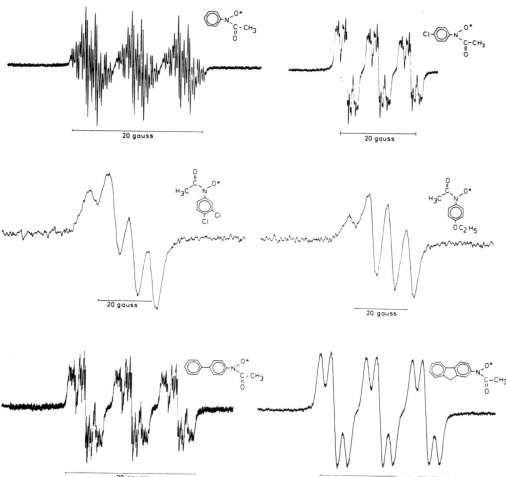

Figure 23.13 EPR spectra of six acetyl arylnitroxides in toluene. The nitroxides were generated by PbO$_2$ oxidation of 10^{-3} M solutions of the corresponding *N*-hydroxy-*N*-arylacetamides. EPR conditions: Varian E 109 spectrometer; magnetic field, 3000 gauss; microwave frequency, 0.15 GHz; microwave power, 2 mW; modulation frequency, 10 KHz; modulation amplitude, 0.05 Gauss.

23.3.6 Product pattern produced by chemical oxidation of *N*-hydroxy-4-chloroacetanilide

N-Hydroxy-4-chloroacetanilide catalytically oxidized HbFe^{2+} and was converted to a small extent into 4-chloronitrosobenzene, 4-chloronitrobenzene, and 4-chloroacetanilide. As acetyl 4-chlorophenyl nitroxide was

Formula 23.1

implicated as the catalytically active intermediate (Heilmair *et al.*, 1987), this radical was produced by chemical oxidation of *N*-hydroxy-4-chloro-acetanilide and the product pattern of its self-reaction studied in various solvents and radical concentrations. To determine whether the same products are produced from acetyl 4-chlorophenyl nitroxide in a similar ratio as those from acetyl phenyl nitroxide or acetyl 4-biphenylyl nitroxide (Forrester *et al.*, 1970), a 10^{-3} M solution of *N*-hydroxy-4-chloroacetanilide in benzene was oxidized with lead dioxide. The product pattern which arose from the decay of the radical was analysed by HPLC and gave 38% 4-chloro-nitrosobenzene, 33% *N*-acetoxy-4-chloroacetanilide, 10% 4-chloro-acetanilide, and 8% 4-chloronitrobenzene, i.e. it was almost identical with that determined by Forrester *et al.* (1970).

Oxidation of 10^{-2} M-*N*-hydroxy-4-chloroacetanilide in water by lead dioxide yielded the same four products and in addition a yellow–orange product, which was subsequently identified as *N*-(2-acetylamino-5-chloro-phenyl)-*p*-benzoquinoneimine-*N*-oxide (Formula VIII, 23.3c). Since this compound was also formed in the reaction of *N*-hydroxy-4-chloro-acetanilide with $HbFe^{2+}O_2$ (section 23.3.1), a further analogy between chemical oxidation and $HbFe^{2+}O_2$ oxidation was established.

As additional proof for this mechanism, we studied the effect of dilution on the product pattern. Since 4-chloronitrosobenzene is a volatile compound, the quantitative determination of the products had to be carried out without any concentrating of the solution. Thus our method of product determination by direct HPLC analysis was limited to concentrations of $\geqslant 10^{-5}$ mol/l (referred to *N*-hydroxy-4-chloroacetanilide before oxidation with lead dioxide). The results given in Table 23.4 show that the decay of acetyl 4-chlorophenyl nitroxide indeed depends on its original concentration. Lowering the radical concentration gradually diminished the concentration of *N*-acetoxy-4-chloroacetanilide but raised the concentration of 4-chloronitrosobenzene. This observation can be explained by the increasing probability of an acetyl transfer to the solvent. Furthermore, the concentration of 4-chloroacetanilide decreased with increasing dilution, as was expected for an intermolecular reaction

Table 23.4 Dependency of the product pattern on concentration of *N*-hydroxy-4-chloroacetanilide

	N-*Hydroxy-4-chloroacetanilide (mol/l)*		
	10^{-3}	*10^{-4}*	*10^{-5}*
4-Chloronitrosobenzene(mol/l)	2.5×10^{-4}	6.3×10^{-5}	7.7×10^{-6}
4-Chloronitrobenzene(mol/l)	3.2×10^{-4}	0.4×10^{-5}	1.1×10^{-6}
N-Acetoxy-4-chloroacetanilide (mol/l)	3.5×10^{-4}	2.5×10^{-5}	0.5×10^{-6}
4-Chloroacetanilide (mol/l)	0.4×10^{-4}	0.2×10^{-5}	0.1×10^{-6}

Different concentrations of *N*-hydroxy-4-chloroacetanilide in water were oxidized by PbO_2 at room temperature and the product pattern determined as described in the text.

mechanism. This finding provides further evidence that 4-chloroacetanilide is not formed from *N*-hydroxy-4-chloroacetanilide by reduction, but rather as the consequence of an oxidative process.

23.4 CONCLUSIONS

23.4.1 Coupled oxidation of *N*-hydroxy-4-chloroaniline and oxyhaemoglobin

Between 3 and 20 min after initiation of *N*-hydroxy-4-chloroaniline-induced $HbFe^{2+}$ oxidation a parallel decline of the concentration of $HbFe^{3+}$, the ferrohaem-4-chloronitrosobenzene complex, 4-chlorophenyl nitroxide, oxygen, and 4-chloronitrosobenzene, and a further increase in the concentration of 4-chloronitrobenzene and 4-chloroaniline was observed. The small but distinct decrease in the $HbFe^{3+}$ concentration of 4% is due to its reduction to $HbFe^{2+}$, which in turn sequestered oxygen for the regeneration of $HbFe^{2+}O_2$. As this experiment was performed under conditions where oxygen supply was limited (in the reaction chamber of the oxygen electrode), the phase of $HbFe^{3+}$ reduction could be observed whereas in a parallel experiment, where oxygen supply was not limited, no such decrease in the $HbFe^{2+}$ concentration was observed. Since by this time, *N*-hydroxy-4-chloroaniline had completely disappeared, 4-chlorophenyl nitroxide was the only possible electron donor in the reaction mixture.

The decay of the ferrohaem-4-chloronitrosobenzene complex does not simply reflect a replacement of 4-chloronitrosobenzene by oxygen from the ferrohaem, but also indicates an intramolecular electron transfer from $HbFe^{2+}$ to 4-chloronitrosobenzene, by which $HbFe^{3+}$ and the primary nitroxide is formed (Equation 23.4)

$$HbFe^{2+}-Ar-N=O \longrightarrow HbFe^{3+} + Ar-\overset{*}{\underline{N}}-\bar{\underline{O}}\text{\scriptsize I} \rightleftharpoons Ar-N\overset{O^*}{\underset{H}{\Big\backslash}} \quad (23.4)$$

From the disproportionation of the latter, 4-chloronitrobenzene and 4-chloroaniline may arise. Alternatively, Equation 23.5 may explain the simultaneous production of 4-chloronitrobenzene and 4-chloroaniline, but not the ratio of 2 : 1.

$$HbFe^{2+}-Ar-N=O + \overset{*}{}O-\underset{H}{\overset{|}{N}}-Ar \rightarrow \left\{ \begin{array}{c} \left[HbFe^{2+}-Ar-\underset{\underset{Ar}{\overset{|}{N}-H}}{\overset{|}{\underset{O}{\overset{|}{N}}}}-\bar{\underline{O}}\text{\scriptsize I}\right] \end{array} \right\} \begin{array}{c} \rightarrow HbFe^{3+} \\ + \\ Ar-NO_2 \\ + \quad (23.5) \\ Ar-\overset{*}{N}-H \\ \downarrow \\ Ar-NH_2 \end{array}$$

As can be seen from Fig. 23.2, the concentration of 4-chlorophenyl nitroxide was twice as high during the reaction of $HbFe^{2+}O_2$ with N-hydroxy-4-chloroaniline as in the experiment without $HbFe^{2+}O_2$. This is an indication that the nitroxide is formed during the coupled oxidation as well as during the decay of the ferrohaem-4-chloronitrosobenzene complex.

23.4.2 Coupled oxidation of *N*-hydroxy-4-chloroacetanilide and oxyhaemoglobin

During the rapid initial phase of the N-hydroxy-4-chloroacetanilide-induced oxidation of $HbFe^{2+}O_2$ a relationship was established between $HbFe^{3+}$ formation and a decrease in the concentration of N-hydroxy-4-chloro-acetanilide, an increase in the concentration of 4-chloronitrobenzene, 4-chloronitrosobenzene, N-acetoxy-4-chloroacetanilide, and 4-chloro-acetanilide, and an increase in oxygen concentration due to its liberation from $HbFe^{2+}O_2$. In addition, N-(2-acetylamino-5-chlorophenyl)-p-benzo-quinoneimine-N-oxide was detected as an intermediate. As the product pattern formed with $HbFe^{2+}O_2$ was formed by chemical oxidation, which produced acetyl 4-chlorophenyl nitroxide as the primary oxidation product, we assume that the nitroxide was also an intermediate in the co-oxidation of $HbFe^{2+}O_2$ and N-hydroxy-4-chloroacetanilide. Attempts to detect radical activity during the rapid initial phase of the reaction have failed. As co-oxidation of $HbFe^{2+}O_2$ and N-hydroxy-4-chloroacetanilide followed second order kinetics (Heilmair *et al.*, 1987), we have assumed that the one-electron oxidation of N-hydroxy-4-chloroacetanilide by $HbFe^{2+}O_2$ is the rate-limiting step, which yields acetyl 4-chlorophenyl nitroxide (Equation 23.2). Therefore, the regeneration of N-hydroxy-4-chloroacetanilide by oxidation of a second $HbFe^{2+}O_2$ (Equation 23.3a) must be faster than the formation of the radical and its stationary concentration therefore must be rather low. This explains our failure to prove directly the presence of the nitroxide by EPR spectroscopy.

One molecule of N-hydroxy-4-chloroacetanilide effected the oxidation of 15 equivalents of $HbFe^{2+}$ in 1 h, indicating its catalytic activity. Although 75% of $HbFe^{3+}$ was formed in 1 h, only one-third of the equivalent oxygen present in $HbFe^{2+}O_2$ was liberated and two-thirds was reduced, indicating that the electrons from $HbFe^{2+}$ were transferred to oxygen in $HbFe^{2+}O_2$ via the catalytically active acetyl 4-chlorophenyl nitroxide. The reaction of molecular oxygen with N-hydroxy-4-chloroacetanilide is slow, as compared with $HbFe^{2+}O_2$, indicating that oxygen in $HbFe^{2+}O_2$ is activated. The reason for this activation is the polarization of the Fe-O bond which can be described by the two resonance structures:

$$HbFe^{2+}-O_2 \leftrightarrow HbFe^{3+}-O_2^{\bar{*}}$$

(Weiss, 1964; Peisach *et al.*, 1968). If the redox potentials of various oxygen species are compared, at pH 7 and 25°C (James, 1978), only the transfer of the first electron to molecular oxygen is an endergonic reaction, whereas the transfer of the other three electrons is exergonic. This explains why molecular oxygen did not efficiently oxidize N-hydroxy-4-chloroacetanilide, whereas oxygen in $HbFe^{2+}O_2$ by way of its superoxide character was much more reactive.

$$
\begin{aligned}
O_2 + e^- &\rightleftharpoons O_2^{\bar{*}} & E_{1/2} &= -400 \text{ mV} \\
O_2^{\bar{*}} + 2 H^+ + e^- &\rightleftharpoons H_2O_2 & E_{1/2} &= +900 \text{ mV} \\
H_2O_2 + 2 H^+ + 2 e^- &\rightleftharpoons 2 H_2O & E_{1/2} &= +1350 \text{ mV} \\
O_2 + 4 H^+ + 4 e^- &\rightleftharpoons 2 H_2O & E_{1/2} &= +800 \text{ mV}
\end{aligned}
$$

The redox potential of the couple $HbFe^{2+}/HbFe^{3+}$ is: $E_{1/2} = +125$ mV (Antonini *et al.*, 1964), that of N-hydroxy-4-chloroacetanilide/acetyl-4-chlorophenyl nitroxide is: $E_{1/2} = +610$ mV, and that of $O_2^{\bar{*}}/H_2O_2$ is: $E_{1/2} = +900$ mV (James, 1978). Therefore, electrons can flow from N-hydroxy-4-chloroacetanilide to oxygen in $HbFe^{2+}O_2$ and from $HbFe^{2+}$ to acetyl 4-chlorophenyl nitroxide. We implicate the $HbFe^{3+}-H_2O_2$ complex (Keilin and Hartree, 1950) as the primary product of one catalytic cycle, because transfer of two electrons to oxygen in $HbFe^{2+}O_2$ is energetically favoured over one-electron transfer. Such an idea is in agreement with the observation that superoxide itself was not detected during the initial phase of the coupled oxidation and that superoxide dismutase had no effect on the kinetics. As no hydrogen peroxide was not found in the reaction mixture and catalase did not affect the kinetics of the coupled oxidation, we think it most likely that the postulated $HbFe^{3+}-H_2O_2$ complex was formed and not free H_2O_2 and that it was decomposed according to the reaction outlined in Equation 23.6.

$$HbFe^{3+}-OOH^- + Ar-N\overset{OH}{\underset{COCH_3}{\diagup}} \longrightarrow Ar-N\overset{O^*}{\underset{COCH_3}{\diagup}} + HbFe^{3+} + OH^- + {}^*OH \quad (23.6)$$

In agreement with Equation 23.6 is the observation that the $HbFe^{3+}-H_2O_2$ complex was rapidly decomposed in the presence of N-hydroxy-4-chloro-acetanilide and gave rise to 4-chloronitrosobenzene and 4-chloro-acetanilide. Furthermore, the electronic spectrum of $HbFe^{2+}O_2$ continuously changed into that of $HbFe^{3+}$ with isosbestic points at 588 and 521 nm when N-hydroxy-4-chloroacetanilide reacted with $HbFe^{2+}O_2$, indicating that no other haemoglobin derivatives were intermediates in the coupled oxidation. Thus, the reactions 23.2, 23.3a and 23.6 can explain the complete reduction of oxygen in $HbFe^{2+}O_2$ and the observation that neither catalase nor superoxide dismutase affected the coupled oxidation.

23.4.3 Comparison of the two mechanisms of haemoglobin oxidation by N-hydroxy-4-chloroaniline and N-hydroxy-4-chloroacetanilide

N-Hydroxy-4-chloroaniline as well as N-hydroxy-4-chloroacetanilide are oxidized by $HbFe^{2+}O_2$ in a coupled oxidation, i.e. under simultaneous formation of $HbFe^{3+}$. In either reaction a large part of the oxygen in $HbFe^{2+}O_2$ was reduced to water, as neither catalase nor superoxide dismutase affected the coupled oxidation. Apparently the initial reduction of oxygen in $HbFe^{2+}O_2$ is the rate-limiting step in either reaction, since both co-oxidations followed second order. N-Hydroxy-4-chloroaniline as well as N-hydroxy-4-chloroacetanilide underwent autoxidation, the former within hours and the latter within months, but they were rapidly oxidized by the reactive oxygen in $HbFe^{2+}O_2$. Both reactions also differed in their activation energy, since it was determined to $E_a = 3.5$ kcal/mol for N-hydroxy-4-chloroaniline and 12.7 kcal/mol for N-hydroxy-4-chloro-acetanilide (Heilmair *et al.*, 1987). Therefore the reaction with N-hydroxy-4-chloroaniline is much faster than that of N-hydroxy-4-chloroacetanilide.

Although nitroxides obviously participate in both reactions, their function is completely different. Whereas the arylhydroxylamine was irreversibly oxidized to the nitrosoarene by a two-electron oxidation without a discrete potential for the one-electron oxidation, N-hydroxy-4-chloroacetanilide was reversibly oxidized by one-electron oxidation to give the secondary nitroxide. N-Hydroxy-4-chloroaniline can produce one equivalent of $HbFe^{3+}$, whereas N-hydroxy-4-chloroacetanilide catalysed the oxidation of 15 or more equivalents of $HbFe^{2+}$, because the catalytically active nitroxide is reduced by $HbFe^{2+}$ before it disappears by self-reaction. The laws of thermodynamics do not allow the reduction of the nitrosoarene by $HbFe^{2+}$, because the redox potential for the couple N-hydroxy-4-chloroaniline/ 4-chloronitrosobenzene, $E_{1/2} = +80\,mV$, is lower than that of $HbFe^{2+}/$ $HbFe^{3+}$, $E_{1/2} = +125\,mV$ (Antonini *et al.*, 1964).

Due to a rapid reaction of 4-chloronitrosobenzene with the SH-groups of globin (Eyer and Ascherl, 1987), the mass balance of N-hydroxy-4-chloroaniline was negative, whereas apparently neither ^{14}C-labelled N-hydroxy-4-chloroacetanilide nor one of its oxidation products was bound to $HbFe^{2+}O_2$ (Heilmair *et al.*, 1987).

REFERENCES

Antonini, E., Wyman, J., Brunori, M. *et al.* (1964) Studies on the oxidation–reduction potentials of heme proteins. I. Human hemoglobin. *J. Biol. Chem.*, **239**, 907–12.

Aurich, H.G. and Baer, F. (1965) Nitroxide II: Die Oxidation von Derivaten des Phenylhydroxylamins. *Tetrahedron Lett.*, **43**, 3879–83.

Bartsch, H. and Hecker, E. (1971) On the metabolic activation of the carcinogen N-hydroxy-N-2-acetylaminofluorene. *Biochim. Biophys. Acta*, **237**, 567–78.

Eyer, P. and Ascherl, M. (1987) Reactions of p-substituted nitrosobenzenes with human hemoglobin. *Biol. Chem. Hoppe Seyler*, **368**, 285–94.

Forrester, A.R., Ogilvy, M.M. and Thomson, R.H. (1970) Mode of action of carcinogenic amines. Part I. Oxidation of N-arylhydroxamic acids. *J. Chem. Soc. (C) (Lond.)*, 1081–3.

Heilmair, R., Lenk, W. and Sterzl, H. (1987) N-Hydroxy-N-arylacetamides. IV. Differences in the mechanism of haemoglobin oxidation *in vitro* between N-hydroxy-N-arylacetamides and arylhydroxylamines. *Biochem. Pharmacol.*, **36**, 2963–72.

Heubner, W., Wahler, B. and Ziegler, C. (1953) Über die Bildung von Hämiglobin durch acylierte Phenylhydroxylamine. *Hoppe-Seyler's Z. Physiol. Chem.*, **295**, 397–403.

Hustedt, G. and Kiese, M. (1959) Umsetzungen von Acetanilid und Acetylphenylhydroxylamin im Organismus. *Naunyn Schmiedebergs Arch. Exp. Pathol. Pharmakol.*, **236**, 435–48.

James, B.R. (1978) Interaction of dioxygen with metalloporphyrins, in *The Porphyrins*, Vol. 5 (ed. D. Dolphin), Academic Press, London, pp. 205–95.

Keilin, D. and Hartree, E.F. (1950) Reaction of methaemoglobin with hydrogen peroxide. *Nature*, **166**, 513–14.

Kiese, M. (1959) in Discussion. *Naunyn Schmiedeberg's Arch. Exp. Pathol. Pharmakol.*, **236**, p. 22.

Kiese, M. and Plattig, K.H. (1959) Hämiglobinbildung durch Benzoylphenylhydroxylamin. *Naunyn Schmiedeberg's Arch. Exp. Pathol. Pharmakol.*, **235**, 373–80.

Kiese, M. and Reinwein, D. (1950) Kinetik der Hämiglobinbildung. VIII. Die Oxydation von Hämoglobin durch Phenylhydroxylamin und Sauerstoff. *Naunyn Schmiedeberg's Arch. Exp. Pathol. Pharmakol.*, **211**, 392–401.

Kiese, M. and von Ruckteschell, A. (1951) Kinetik der Hämiglobinbildung. XI. Die katalytische Wirkung des Hämoglobins auf die Oxidation von Phenylhydroxylamin durch Sauerstoff. *Naunyn Schmiedeberg's Arch. Exp. Pathol. Pharmakol.*, **213**, 128–38.

Lenk, W. and Riedl, M. (1989) *N*-Hydroxy-*N*-arylacetamides. V. Differences in the mechanism of haemoglobin oxidation *in vitro* by *N*-hydroxy-4-chloroacetanilide and *N*-hydroxy-4-chloroaniline. *Xenobiotica*, **19**, 453–75.

Peisach, J., Blumenberg, W.E., Wittenberg, B.A. and Wittenberg, J.B. (1968) The electronic structure of protoheme proteins. III. Configuration of the heme and its ligands. *J. Biol. Chem.*, **243**, 1871–80.

Radomski, J.L. and Brill, E. (1971) The role of *N*-oxidation products of aromatic amines in the induction of bladder cancer in the dog. *Arch. Toxicol.*, **28**, 159–75.

Uehleke, H. (1963) *N*-Hydroxylation of carcinogenic amines *in vivo* and *in vitro* with liver microsomes. *Biochem. Pharmacol.*, **12**, 219–21.

Wahler, B.E., Schoffa, G. and Thom, H.G. (1959) Nachweis von Radikal-Zwischenstufen bei der Hämoglobinoxydation nach Einwirkung aromatischer Hydroxylamine. *Naunyn Schmiedeberg's Arch. Exp. Pathol. Pharmakol.*, **236**, 20–1.

Weiss, J.J. (1964) Nature of iron–oxygen bond in oxyhaemoglobin. *Nature*, **202**, 83–4.

24

Molecular approaches to evaluation of the risk of aromatic amine-induced bladder cancer: smoking-related carcinogen–DNA adducts in biopsied samples of human urinary bladder

G. Talaska* and F.F. Kadlubar

Office of Research (HFT-100), National Center for Toxicological Research, Jefferson, AR 72079, USA

1. Environmental and occupational exposures remain major causes of urinary bladder cancer in the industrialized nations. Cigarette smoking predominates as the leading cause, although occupational exposures can be responsible for high incidence rates in heavily exposed groups.
2. Significant concentrations of aromatic amines are found in tobacco smoke, and as such, these compounds are the common thread linking smoking, occupational exposure, and urinary bladder cancer.
3. Current methods to monitor exposed groups at elevated risk are concerned with detecting the disease at early, non-invasive stages.
4. A ^{32}P-postlabelling method has been developed that allows detection of carcinogen–DNA adducts in small samples of human tissues. We have used this method to monitor a group of smokers, non- and ex-smokers and have found significantly higher levels of several carcinogen–DNA adducts in the urothelium of smokers. These methods enhance our ability to monitor the earliest step in the neoplastic process, identifying individuals at increased risk and thereby preventing the occurrence of irreversible disease.

---

* Current address: Institute of Environmental Health, Kettering Laboratory, University of Cincinnati, Cincinnati, OH 45267, USA.

24.1 INTRODUCTION

Urinary bladder cancer remains a significant health problem in the United States. In 1989, there were an estimated 47 000 cases of bladder cancer diagnosed, with 34 500 of these occurring in males. In the same year, it was estimated that 10 200 persons died from this disease (American Cancer Society, 1989). Epidemiological studies clearly show that the major risk factor in the causation of bladder cancer is tobacco smoking. Between 40% and 85% of all cases of bladder cancer are attributable to smoking cigarettes, and smokers have a 2–10-fold increased relative risk over non-smokers for developing bladder tumours (Mommsen *et al.*, 1985). But although smoking has a tremendous effect upon the rate of bladder cancer in the general population, within certain small groups the effect of occupational exposures can be dramatic. Reports have been published indicating that the prevalence of bladder cancer was as high as 50% in small groups of workers that were heavily exposed to aromatic amines in the earliest part of this century (Zavon *et al.*, 1973). Dye, leather and rubber workers have also been shown to be at increased risk (Cartwright and Glashan, 1984).

The common thread between cigarette smoking, specific occupational exposures, and urinary bladder cancer is a single class of compounds, the aromatic amines (Fig. 24.1). Benzidine, 4-aminobiphenyl and 2-naphthylamine were among the first compounds identified as human bladder carcinogens because of occupational exposure. Accordingly, such exposures to aromatic amines in the workplace have now been severely restricted. However, human exposure to 4-aminobiphenyl and 2-naphthylamine continues as these chemicals are present in significant quantities in mainstream and sidestream cigarette smoke (Patrianakos and Hoffmann, 1979). 4-Aminobiphenyl is now recognized as a liver and bladder carcinogen in rodents (Schieferstein *et al.*, 1985), and a potent bladder carcinogen in lagomorphs (Bonser, 1962) and dogs (Block *et al.*, 1978). Melick *et al.* (1955) found an extremely high incidence of bladder cancer in a group of workers producing 4-aminobiphenyl. Bryant *et al.* (1987) have more recently demonstrated the presence of 4-aminobiphenyl–haemoglobin adducts in peripheral blood samples of cigarette smokers.

This chapter reviews the efforts that have been made to determine if there is a molecular basis for the increased risk of bladder cancer in cigarette smokers. Specifically, we have investigated whether or not it will be feasible to detect carcinogen–DNA adducts in human bladder biopsy samples and to assess whether or not their occurrence can be related to history of cigarette smoking. Furthermore, attempts were made to characterize the specific carcinogen–DNA adducts involved in order to implicate exposure to a particular carcinogenic agent.

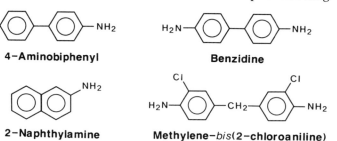

Figure 24.1 Carcinogenic aromatic amines of occupational and environmental importance.

24.2 ^{32}P-POSTLABELLING

The ^{32}P-postlabelling method was developed by Randerath *et al.* (1981) and Gupta *et al.* (1982). DNA thought to contain carcinogen-base adducts is enzymatically hydrolysed to deoxyribonucleoside-3′-phosphates; micrococcal endonuclease and spleen phosphodiesterase are commonly used. The deoxyribonucleoside-3′-phosphates are then 5′-labelled with [^{32}P]ATP using polynucleotide kinase. Labelling each deoxyribo-nucleoside-3′-phosphate with [^{32}P]phosphate is impractical from the viewpoints of safety and economy, as approximately 10 mCi of [^{32}P]ortho-phosphate at a specific radioactivity of 3000 Ci/mmol would be necessary for complete labelling of 1 μg of DNA hydrolysate. Instead, in the standard assay, a given amount of [^{32}P]ATP is added to each sample, usually 150–200 μCi/sample (about 80 pmol), and sufficient non-radioactive ATP is then added to give a modest excess of total ATP over the deoxyribonucleoside-3′-phosphates levels present in the sample. Adducts and normal nucleotides are (^{32}P-phosphate)-labelled to completion under these conditions and the labelling of the adducts relative to the normal deoxyribonucleoside-3′,5′-diphosphates is considered to be an accurate estimate of the adduct levels in the sample. Although quantitation of the standard ^{32}P-post-labelling method is fairly unambiguous, the sensitivity of the assay is limited because only a fraction of the adducts, about one in 70, will be 5′-phos-phorylated with ^{32}P. The limit of sensitivity of the standard ^{32}P-postlabelling method is about one adduct per 10^6 normal nucleotides. This sensitivity is sufficient for most *in vitro* and some *in vivo* applications in animals.

However, the standard ^{32}P-postlabelling assay often lacks sufficient sensitivity to detect carcinogen–DNA adducts in the low levels resulting from human exposure to carcinogens. Several methods have been developed to increase the sensitivity of the assay to the levels necessary to monitor carcinogen–DNA adducts in human samples. In general, if normal nucleotides can be separated from the adducted derivatives prior to labelling, then only the very small number of adducts require ^{32}P. Reddy and Randerath (1986) found that nuclease P_1 can dephosphorylate normal deoxyribonucleoside-3'-phosphates. Since polynucleotide kinase has an absolute requirement for the deoxyribonucleoside-3'-phosphates, the normal nucleosides are not labelled and thus can be effectively removed from the analysis. On the other hand, Reddy and Randerath (1986) reported that many adducted deoxyribonucleoside-3'-phosphates are resistant to nuclease P_1 treatment; therefore, their proportion in the total labelled deoxyribonucleoside-3',5'-diphosphates increases dramatically. These workers reported that several hundredfold intensifications can be obtained with nuclease P_1 for certain carcinogen–base adducts. Obviously, the degree of intensification will vary for each adduct depending on its resistance to the nuclease. Reddy and coworkers reported that the intensification can indeed vary widely depending upon the structure of the compound. Preliminary experiments in our laboratory demonstrated that nuclease P_1, which selectively enhances the recovery of many hydrophobic carcinogen–DNA adducts (Reddy *et al.*, 1986), degrades the C-8-substituted deoxyguanosine adducts that are formed predominantly from the majority of aromatic amines. However, we found that a n-butanol-extraction method was suitable for the enhanced detection of these particular adducts. A more systematic examination of this phenomenon (Gupta and Earley, 1988; Gallagher *et al.*, 1989) corroborated our preliminary findings. Since each method enhances a different class of carcinogen–DNA adducts, it was decided to use both the nuclease P_1 and n-butanol-extraction enhancement methods when analysing human tissue samples. This approach also gave us the opportunity to characterize the adducts seen in a given sample to a rough first approximation.

To summarize, the ^{32}P-postlabelling method offers distinct advantages for the researcher interested in the measurement of carcinogen–DNA adducts. The sensitivity of the method with the modifications described above is such that detection of adducts on the order of 1 carcinogen bound per 10^{10} normal nucleotides can be achieved (Watson, 1987). Moreover, only a small amount of DNA is required for analysis. We routinely analyse 1–10 μg of DNA and have used a modified method which uses only nanogram quantities of DNA (Talaska *et al.*, 1990), allowing the use of very small tissue samples.

24.3 SAMPLING OF HUMAN BLADDER FOR ³²P-POSTLABELLING ANALYSIS

The target for most human urinary bladder carcinogens is the epithelial lining of the bladder (urothelium), resulting in the formation of transitional cell carcinomas. Thus, we sought to estimate the level of carcinogen–DNA adducts in this target cell population of living individuals. Conal bladder biopsy samples were obtained from 32 consenting patients, all males, undergoing cystoscopic examination at the John A. McClellan Memorial Veterans' Hospital in Little Rock, Arkansas, USA. DNA was isolated from the specimens using a solvent extraction method (Gupta and Dighe, 1984). DNA adduct analysis was performed using the nuclease-P_1 and the n-butanol-extraction enrichment procedures to increase the sensitivity of the ^{32}P-postlabelling method (Gupta *et al.*, 1982; Gupta, 1985; Phillips *et al.*, 1988).

The amount of DNA analysed was in the range 2–6 μg depending on the concentration of DNA. Each sample was labelled with 200 μCi [^{32}P]ATP and polynucleotide kinase. Analysis of normal nucleotides indicated that there was excess [^{32}P]ATP in each sample regardless of DNA concentration. This indicates that the available adduct nucleotides are labelled quantitatively. Carcinogen–DNA adduct levels were estimated by visualizing the adduct spots with autoradiography, then excising the corresponding area of the chromatograms, and determining the ^{32}P radioactivity by Cerenkov counting. Discrete adduct spots and diagonal radioactive zones were visualized on the autoradiograms. Where discrete adduct spots lay within a radioactive zone, the adduct spot was excised and quantified separately.

Smoking and occupational histories were obtained from these individuals during a follow-up interview. The group of 32 consisted of nine active smokers, eight persons who never smoked, and 15 ex-smokers who had stopped smoking between 7 and 40 years previously.

24.4 CARCINOGEN–DNA ADDUCTS IN HUMAN URINARY BLADDER

24.4.1 Qualitative results

Several studies have shown that several unknown carcinogen–DNA adducts appear to increase as a function of ageing (Randerath *et al.*, 1986; 1989). An attempt was made to control for this variable in human urinary bladder

(a)   (b)

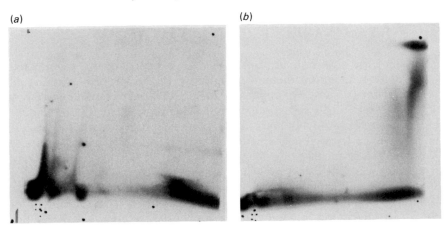

Figure 24.2 Autoradiograms ^{32}P-postlabelled DNA from an ex-smoker. DNA from the bladder biopsy sample from this individual was hydrolysed to the 3′-nucleotides, then either incubated with nuclease P_1 (*a*) or extracted with n-butanol and ^{32}P-postlabelled (*b*). The sample was then spotted at the origin of a polyethyleneimine-cellulose (PEI) thin-layer sheet. After overnight development in 0.65 M sodium phosphate buffer (pH 6.0), the sample was transferred to the origin of another PEI thin-layer sheet (lower left corner) using the magnet-mediated transfer technique (Lu *et al.*, 1986), then developed vertically using 3.6 M lithium formate, 7.7 M urea, pH 3.5. The sheets were rinsed with distilled water, air dried and rotated 90° counterclockwise, then developed in 0.75 M lithium chloride, 0.47 M Tris–HCl buffer, 8.0 M urea, pH 8.0. After this development, the thin-layer sheets were again washed and dried and developed in the same direction with 0.9 M sodium phosphate buffer, pH 6.8. The chromatograms were exposed to photographic film in cassettes with intensifying screens for 60 h.

DNA for the study group of cigarette smokers, ex-smokers and non-smokers. The average age for the entire group was 70.1 years and there was no statistically significant difference between the groups.

Figure 24.2 shows autoradiograms of ^{32}P-postlabelled DNA from a single ex-smoker analysed using butanol-extraction and nuclease-P_1 enhancement. This person quit smoking 20 years previously after a 30 pack-year habit. For comparison, Fig. 24.3 consists of autoradiograms of ^{32}P-postlabelled DNA from one of the smokers; a diagonal zone of radio-activity is seen for both methods. This zone was found to be generally heavier when the samples were analysed by the nuclease-P_1 rather than the butanol-extraction procedure. The diagonal zones in these bladder samples of smokers were less evident than the diagonal zones seen when a series of human peripheral lung samples from smokers were analysed (Talaska *et al.*, unpublished).

(a)   (b)

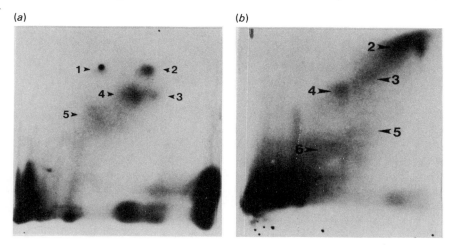

Figure 24.3 Autoradiograms of ^{32}P-postlabelled DNA from a current smoker. Treatment of the sample was as described in the legend to Figure 24.2.

In addition to the diagonal zones, several adduct spots could be readily discerned in autoradiograms of cigarette smokers. These adduct spots were designated numerically as indicated. (The designations for the nuclease P_1 and butanol extraction methods were independent, so that there is no correspondence between the adduct identifiers for each method.) As many as seven distinct adducts were seen using the nuclease P_1 enhancement method, whereas six different adducts were seen using the butanol-extraction method. No single sample contained all the adducts.

24.4.2 Quantitative results

The distribution of the three nuclease P_1-enhanced carcinogen–DNA adducts is indicated in Fig. 24.4. For adduct no. 3, there was no difference between the levels seen in smokers, non-smokers and ex-smokers. The distribution of nuclease P_1-enhanced adduct no. 4 shows that there is a smoking effect. The mean values of adduct no. 4 for smokers, non-smokers and ex-smokers were 12.2, 4.9 and 4.6 adducts per 10^9 nucleotides. A similar distribution is seen for nuclease P_1-enhanced adduct no. 5, where smokers averaged 11.1 adducts per 10^9 nucleotides and non-smokers and ex-smokers had mean values of 4.6 and 2.6, respectively. The differences between smokers and non- and ex-smokers for adducts no. 4 and no. 5 were statistically significant at the 0.05 level using the student's t test.

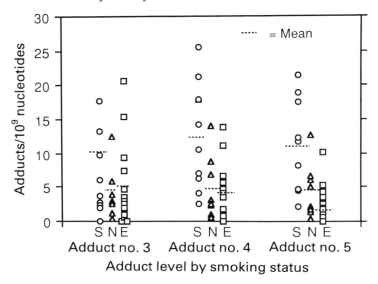

Figure 24.4 The distribution of three nuclease P_1-enhanced, ^{32}P-postlabelled carcinogen–DNA adducts. The values shown are the mean values for an individual from 2–4 independent analyses. Group means are indicated by the dashed lines.

The distribution of the three major adducts enhanced by the butanol-extraction method appears in Fig. 24.5. For adduct no. 3, the mean value for smokers was 7.7 adducts per 10^9 nucleotides, whereas the means for non- and ex-smokers were 3.4 and 2.5, respectively. These values were significantly different at the 0.05 level. For adduct no. 4, which co-chromatographed with the synthetic N-(deoxyguanosin-8-yl)-4-amino-biphenyl 3′,5′-bisphosphate adduct, there was a statistically significant difference in the mean adduct levels between smokers and ex-smokers. The average value for smokers was 11.2 adducts per 10^9 nucleotides, whereas non-smokers averaged 6.4 and ex-smokers had 3.1 adducts per 10^9 nucleotides ($P < 0.05$). A similar situation was seen for the butanol-enhanced adduct no. 6, where the mean value for smokers was 6.1, non-smokers had an average level of 0.6, and ex-smokers averaged 2.9 adducts per 10^9 nucleotides ($P < 0.05$).

These data indicate that the prevalence of carcinogen–DNA adducts in human urinary bladder is smoking-related. Epidemiological studies to date have indicated that the increased risk of bladder cancer in smokers ranges from 2–10 times that of persons who have never smoked (Mommsen *et al.*, 1985). It is quite interesting that the ratios of bladder carcinogen–DNA adducts between smokers and non-smokers are very similar; the range of the

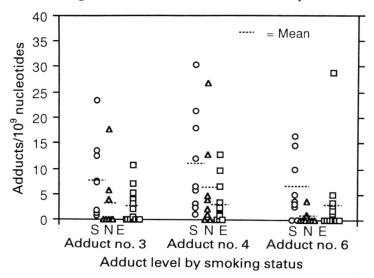

Figure 24.5 The distribution of three n-butanol extraction-enhanced, ^{32}P-post-labelled carcinogen–DNA adducts. See legend to Figure 24.4 for details.

ratios for the five adducts reported here that are related to cigarette smoking was 2.4 to 10. Thus, it seems that monitoring carcinogen–DNA adduct levels may be a good predictor for estimating increased relative risk prior to the occurrence of the disease, at least on a population basis. Of course, it remains to be proven whether or not carcinogen–DNA adduct levels in individuals will predict which persons will develop neoplastic disease.

In an attempt to assess this possibility, the health histories of individuals in this study will be monitored at intervals. Several individual samples obtained from non- or ex-smokers are also of special interest. For butanol-extracted adduct no. 4, a single sample made a major contribution to the relatively high average level in non-smokers. This individual was the only person in the study who reported an occupational exposure to newspaper dyes and, incidentally, also reported that their spouse is a heavy smoker. Reports of heavy smoking by a spouse were also given by the individuals who had the highest values for non- or ex-smokers for each of these adducts. In this regard, some reports have noted that sidestream smoke contains higher levels of some aromatic amines (Patrianakos and Hoffmann, 1979). Again, whether or not these higher adduct levels will be predictive of increased risk for urinary bladder cancer remains to be determined.

We have shown that the level of carcinogen–DNA adducts in ex-smokers with at least seven years abstinence becomes indistinguishable from that of the non-smoking population. A loss of carcinogen–DNA adducts through normal cell turnover is the most plausible explanation for these data as the half-life of the human urothelium is estimated to be between 50 and 200 days (Clayson and Lawson, 1987). The fact that the adduct levels in the ex-smokers are lower on average than in the non-smoking group in this study may be due to a lack of resolution in the assay as the detection limits are approached. However, it may also be that the apparent decrease is real because individuals who have modified their smoking behaviour may also change other factors in their lifestyle that influence carcinogen–DNA adduct formation. For example, their spouses might be less likely to continue smoking or they may no longer frequent taverns or other areas where passive smoking exposure is relatively high. Steineck *et al.* (1988) have reported that the relative risk of bladder cancer in ex-smokers is 1.9 times that of non-smokers, whereas the relative risk in their smoking group was 4.6. This can be interpreted to mean that the carcinogen–DNA adducts formed during the time an individual smokes induce some irreversible event which is responsible for the residual increase in cancer rate. (This identifies a potential limitation in the use of carcinogen–DNA adduct analysis in risk estimation of persons who are no longer exposed to any carcinogen, i.e. that their risk could actually be underestimated.)

The position of adduct no. 4 in the butanol-extracted sample (Fig. 24.3) is notable. This adduct interested us because there was no adduct seen in the corresponding position in the nuclease P_1 maps, indicating that it might possibly be a C-8 deoxyguanosine adduct of an aromatic amine. Moreover, this is the same position to which the major 4-aminobiphenyl–DNA adduct migrated when either the *in vitro*, N-hydroxy-4-aminobiphenyl-reacted DNA standard or DNA obtained from 4-aminobiphenyl-treated dogs was ^{32}P-postlabelled. This matter is currently under investigation.

24.5 CONCLUSIONS

These data indicate that carcinogen–DNA adducts in the human bladder are indeed smoking-related. It is noteworthy that the levels of carcinogen–DNA adducts in ex-smokers are similar to those in non-smokers given at least seven years abstinence. Nuclease P_1 and n-butanol extraction methods enhanced different populations of carcinogen–DNA adducts in human urothelial DNA. It is recommended that both enhancement methods be employed when analysing samples from individuals potentially exposed to complex or unknown mixtures. Some evidence for a passive smoking effect is also shown, although the sample size of this study was too small to demonstrate this conclusively. Finally, data are provided that indicate that a major adduct in the bladder urothelium of human smokers may be N-(deoxyguanosin-8-yl)-4-aminobiphenyl.

REFERENCES

American Cancer Society (1989) *Cancer Facts and Figures—1989*, American Cancer Society, Atlanta, p. 13.

Block, N.L., Sigel, P.M., Lynne, C.M. *et al.* (1978) The initiation, progress and diagnosis of dog bladder cancer induced by 4-aminobiphenyl. *Invest. Urol.*, **16**, 50–4.

Bonser, G.M. (1962) Precancerous changes in the urinary bladder in *The Morphological Precursor of Cancer* (ed. L. Severi), Perugia Press, London, pp. 435–51.

Brooks, T.M. and Dean, B.J. (1981) Mutagenic activity of 42 coded compounds in the *Salmonella*/microsome assay with preincubation, in *Evaluation of Short-term Tests for Carcinogens* (eds F.J. DeSerres and J. Ashby), Elsevier, New York, pp. 327–32.

Bryant, M.S., Skipper, P.L., Tannenbaum, S.R. and Maclure, M. (1987) Hemoglobin adducts of 4-aminobiphenyl in smokers and non-smokers. *Cancer Res.*, **47**, 602–8.

Cartwright, R.A. and Glashan, R.W. (1984) The epidemiology and management of occupational bladder cancer, in *Bladder Cancer* (eds L.T. Smith and D. Prout), Butterworths, London, pp. 125–50.

Clayson, D.B. and Lawson, T.A. (1987) Mechanisms of bladder carcinogenesis, in *Carcinoma of the Bladder* (ed. J.G. Connolly), Raven Press, New York, p. 91.

Gallagher, J.E., Jackson, M.A., George, H.M. *et al.* (1989) Differences in detection of DNA adducts in the ^{32}P-postlabelling assay after either 1-butanol extraction or nuclease P_1 treatment. *Cancer Lett.*, **45**, 7–12.

Gupta, R.C. (1985) Enhanced sensitivity of ^{32}P-postlabeling analysis of aromatic carcinogen:DNA adducts. *Cancer Res.*, **45**, 5656–62.

Gupta, R.C. and Dighe, N.R. (1984) Formation and removal of DNA adducts in rat liver treated with *N*-hydroxy derivatives of 2-acetylaminofluorene, 4-acetyl-aminobiphenyl, and 2-acetylaminophenanthrene. *Carcinogenesis*, **5**, 343–9.

Gupta, R.C. and Earley, K. (1988) ^{32}P-adduct assay: comparative recoveries of structurally diverse DNA adducts in the various enrichment procedures. *Carcinogenesis*, **9**, 1687–93.

Gupta, R.C., Reddy, M.V. and Randerath, K. (1982) ^{32}P-postlabelling analysis of non-radioactive aromatic carcinogen–DNA adducts. *Carcinogenesis*, **3**, 1081–92.

Lu, L-J., Disher, R.M., Reddy, M.V. and Randerath, K. (1986) ^{32}P-postlabeling assay in mice of transplacental DNA damage induced by the environmental carcinogens safrole, 4-aminobiphenyl, and benzo(a)pyrene. *Cancer Res.*, **46**, 3046–54.

Melick, W.F., Escue, H.M., Naryka, J.J. *et al.* (1955) The first reported cases of human bladder tumors due to a new carcinogen-xenylamine. *J. Urol.*, **74**, 760–6.

Mommsen, S., Barfod, N.M. and Aagaard, J. (1985) *N*-Acetyltransferase phenotypes in the urinary bladder carcinogenesis of a low risk population. *Carcinogenesis*, **6**, 199–201.

Patrianakos, C. and Hoffmann, D. (1979) Chemical studies on tobacco smoke, LXIV. On the analysis of aromatic amines in cigarette smoke. *J. Anal. Toxicol.*, **3**, 150–4.

Phillips, D.H., Hemminki, K., Alhonen, A. *et al.* (1988) Monitoring occupational exposure to carcinogens: detection by ^{32}P-postlabelling of aromatic DNA adducts in white blood cells from iron foundry workers. *Mutat. Res.*, **204**, 531–41.

Randerath, K., Reddy, M.V. and Gupta, R.C. (1981) ^{32}P-labeling test for DNA damage. *Proc. Natl. Acad. Sci.*, **78**, 6126–9.

modifications in untreated rats: detection by ^{32}P-postlabelling assay and possible significance for spontaneous tumor induction and aging. *Carcinogenesis*, **7**, 1615–17.

Randerath, K., Liehr, J.G., Gladek, A. and Randerath, E. (1989) Age-dependent covalent DNA alterations (I-compounds) in rodent tissues: species, tissue and sex specificities. *Mutat. Res.*, **219**, 121–33.

Reddy, M.V. and Randerath, K. (1986) Nuclease P_1 mediated enhancement of sensitivity of ^{32}P-postlabeling test for structurally diverse DNA adducts. *Carcinogenesis*, **7**, 1543–51.

Schieferstein, G.J., Littlefield, N.A., Gaylor, D.W. *et al.* (1985) Carcinogenesis of 4-aminobiphenyl in BALB/cStCrlfC3hf/NCTR mice. *Eur. J. Cancer Clin. Oncol.*, **21**, 865–73.

Steineck, G., Norell, S.E. and Feychting, M. (1988) Diet, tobacco and urothelial cancer. *Acta Oncol.*, **27**, 323–7.

Talaska, G., Au, W.W., Ward, J.B. Jr *et al.* (1987) The correlation between DNA adducts and chromosomal aberrations in the target organ of benzidine exposed, partially-hepatectomized mice. *Carcinogenesis*, **8**, 1899–905.

Talaska, G., Dooley, K.L. and Kadlubar, F.F. (1990) Detection and characterization of carcinogen–DNA adducts in exfoliated urothelial cells from 4-aminobiphenyl-treated dogs by ^{32}P-postlabelling and subsequent thin-layer and high-pressure liquid chromatography. *Carcinogenesis*, **11**(4), 639–46.

Watson, W.P. (1987) Post-radiolabelling for detecting DNA damage. *Mutagenesis*, **2**, 319–31.

Zavon, M., Hoegg, U. and Bingham, E. (1973) Benzidine exposure as a cause of bladder tumours. *Arch. Environ. Health*, **27**, 1–7.

25

The role of *N*-oxidation by leukocytes in drug-induced agranulocytosis and other drug hypersensitivity reactions

J. Uetrecht

Faculties of Pharmacy and Medicine, University of Toronto,
Sunnybrook Medical Centre, and the Centre for Drug Safety Research,
19 Russell Street, Toronto, Ontario M5S 2S2, Canada

1. Little is known about the mechanism of drug hypersensitivity reactions. Certain functional groups appear to be associated with a high incidence of such reactions, especially those of arylamines that contain a heteroatom which is easily oxidized.
2. Although the majority of drug metabolism occurs in the liver, many hypersensitivity reactions do not involve the liver. We have found that drugs, such as arylamines, are also metabolized to reactive metabolites by activated neutrophils and monocytes.
3. Such reactive metabolites could be responsible for the agranulocytosis, drug-induced lupus, and generalized hypersensitivity reactions associated with this class of drugs.
4. Furthermore, these reactive metabolites could also be responsible for the anti-inflammatory effects observed with some of these drugs.

25.1 INTRODUCTION

Adverse drug reactions represent a major medical problem. Many of these reactions are due to the pharmacological actions of a drug. Although such reactions can be serious, to a large degree they are predictable, and they can usually be prevented by careful use of the drug. In contrast, other serious drug reactions are idiosyncratic in nature. Although there is disagreement about the definition, the term idiosyncratic drug reaction will be used here to indicate an adverse drug reaction that is unexpected because it does not occur in most patients even if a high dose of the drug is given.

Idiosyncratic drug reactions may be limited to one organ, e.g. skin, liver, bone marrow, kidneys, lungs; however, idiosyncratic reactions can also be systemic with fever, skin rash, and simultaneous involvement of several organs (de Weck and Bundgaard, 1983). Such reactions do not have the characteristics of direct cytotoxicity, and classical toxicology methods have not been successful in elucidating their mechanism. The characteristics of many of these reactions suggest an immune-mediated mechanism (Parker, 1982; Park *et al.*, 1987; Pohl *et al.*, 1988; Uetrecht, 1989). These characteristics include the following:

1. A requirement for either prior exposure to the drug, or a delay of more than a week between starting the drug and the development of toxicity.
2. A lack of delay in toxicity on re-exposure of a patient to the offending drug.
3. An apparent lack of correlation between dose and the risk of toxicity. Although this type of toxicity does not occur in most patients even at high doses, in susceptible individuals, there must be a dose–response relationship. In some cases the dose–toxicity relationship covers a dose range similar to that for the therapeutic effect. In other cases the dose–toxicity relationship can be shifted far to the left of that for the therapeutic effect.
4. An unpredictable nature and usual lack of toxicity in animal species tested.

25.1.1 General mechanism of hypersensitivity reactions

An idiosyncratic adverse drug reaction involving the immune system is usually referred to as a drug hypersensitivity reaction. The term hypersensitivity reaction is also used for immune-mediated reactions caused by other stimuli, and Coombs and Gell (1963) have separated hyper-sensitivity reactions into four types.

Type I: Immediate hypersensitivity reactions – mediated by IgE

Type II: Injury mediated by antibodies against specific tissue antigens
Type III: Injury caused by immune complexes that activate complement
Type IV: Delayed hypersensitivity reactions – injury cell-mediated rather
            than antibody-mediated

The following steps are thought to be involved in the initiation of a hypersensitivity reaction involving antibodies (Unanue and Allen, 1987):

1. The immunogen is internalized by a macrophage or other antigen presenting cell and processed. The processing appears to involve proteolytic hydrolysis of the immunogen. Those fragments of the immunogen that bind to the class II major histocompatibility antigen (MHC II) are saved from further hydrolysis and are transported to the cell membrane bound to MHC II.
2. This fragment is 'presented' on the surface of the cell. Specific clones of T helper cells 'recognize' the processed antigen. The association of the processed immunogen with MHC II is necessary for this recognition.
3. The clones of helper T cells that recognize the processed immunogen stimulate specific clones of B cells that also recognize the immunogen (although B cells probably recognize a different antigenic determinant on the immunogen than the T cells). These B cells proliferate and differentiate into mature antibody-secreting plasma cells.

A similar series of steps is involved in cell-mediated hypersensitivity reactions; the major difference is that other T cells are activated rather than B cells.

Although there may be exceptions, small molecules such as drugs (mol. wt. < 1000) are unable directly to initiate a hypersensitivity reaction (Parker, 1982; Park *et al.*, 1987; Pohl *et al.*, 1988; Uetrecht, 1989). However, small molecules can act as haptens, i.e. bind to a macromolecule and form an adduct which is immunogenic, and such an adduct can lead to a hypersensitivity reaction. The interaction between hapten and macromolecule must be essentially irreversible (i.e. covalent bond). The basis for this requirement is presumably based on a need for the adduct to persist during the several steps of immunogen processing and presentation.

25.1.2 Drug hypersensitivity reactions

With the exception of a few drugs such as penicillin, most drugs do not have sufficient chemical reactivity to form a covalent bond to common macromolecules. However, many drugs are metabolized to chemically reactive species that can act as haptens. Several of the drugs that are associated with drug hypersensitivity reactions are arylamines. A major emphasis in this chapter will be to explore a possible chemical basis for this association.

The antibodies produced during a hypersensitivity reaction are polyclonal, and many of them recognize antigens that have an antigenic determinant or epitope (that part of the molecule recognized by the antibody) similar to that of the immunogen but otherwise have significantly different structures. For example, some antinuclear antibodies against DNA also bind to cardiolipin (Schwartz and Stollar, 1985). This presumably explains how a drug can induce a true autoantibody, e.g. methyldopa induces haemolytic anaemia mediated by an antibody against the red cell membrane (Worlledge, 1973). This antibody does not require the presence of methyldopa on the red cell membrane to bind.

Procainamide is an arylamine the therapeutic use of which is based on its antiarrhythmic activity. However, its use is limited by a high incidence of drug-induced lupus (Henningsen *et al.*, 1975; Woosley *et al.*, 1978) and agranulocytosis (Ellrodt *et al.*, 1984; Meyers *et al.*, 1985; Thompson *et al.* 1988). Drug-induced lupus is an autoimmune disease, i.e. antibodies against an individual's own tissue are observed. The incidence of lupus with chronic procainamide therapy is about 20%. The incidence of agranulocytosis is controversial but is probably about 0.1%. We have demonstrated that procainamide is metabolized to reactive hydroxylamine and nitroso metabolites by human hepatic microsomes (Uetrecht *et al.*, 1984). Generation of these metabolites was associated with covalent binding to microsomal protein (Uetrecht, 1985). The metabolite responsible appeared to be the nitroso metabolite, and ascorbic acid, which reduces the nitroso metabolite back to the hydroxylamine, greatly decreased covalent binding. Binding appeared to involve formation of a sulphinamide bond by reaction of the nitroso metabolite with protein sulphydryl groups, and most of the binding could be hydrolysed by mild acid. Binding to other proteins was also observed, and some of the binding was resistant to acid hydrolysis. The reactive nitroso metabolite could act as a hapten and initiate hypersensitivity reactions such as drug-induced lupus and agranulocytosis. However, it appears that these metabolites are too reactive to escape the liver (Uetrecht *et al.*, 1984). Yet, procainamide rarely causes hepatic toxicity. This suggested that procainamide might be metabolized to reactive metabolites outside the liver.

25.2 METABOLISM OF DRUGS BY LEUKOCYTES AND ADVERSE REACTIONS

After initial failures to detect metabolism of procainamide by peripheral blood mononuclear leukocytes (a mixture of lymphocytes and monocytes), we found that both neutrophils and mononuclear leukocytes oxidize procainamide to reactive hydroxylamine and nitroso metabolites, but only if the cells are activated by a phorbol ester or other stimulus (Uetrecht *et al.*, 1988a). Leukocytes contain little cytochrome *P*-450; however, when neutrophils and monocytes are activated, a membrane enzyme, NADPH oxidase, leads to the generation of superoxide (Weiss, 1989). Superoxide is further converted into hydrogen peroxide, either spontaneously or catalysed by superoxide dismutase. Simultaneously myeloperoxidase (MPO) is released. The combination of H_2O_2 and MPO is used by neutrophils to kill bacteria. In addition, in the presence of chloride ion, hypochlorite ion is also generated, and this is a very effective antibacterial agent as evidenced by its use to kill bacteria in municipal water supplies.

25.2.1 Procainamide

The combination of purified human MPO and H_2O_2 was also found to oxidize procainamide to the same reactive hydroxylamine and nitroso metabolites (Fig. 25.1). In addition, in the presence of chloride ion, a reactive *N*-chloroprocainamide was formed (Uetrecht and Zahid, 1988b). The *N*-chloro metabolite was not observed when procainamide was

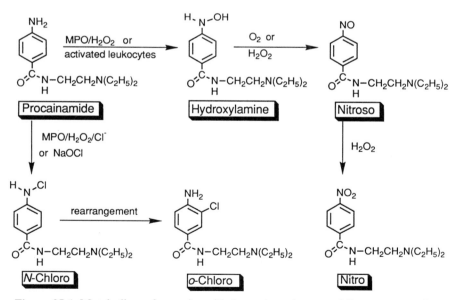

Figure 25.1 Metabolism of procainamide by activated neutrophils or mononuclear cells or catalysed by myeloperoxidase.

incubated with neutrophils, but this presumably reflects its rapid rate of reaction with the neutrophils. Hydrogen peroxide alone did not oxidize procainamide at a significant rate (Uetrecht *et al.*, 1988a). Furthermore, horseradish peroxidase, which is associated with one-electron oxidation rather than two-electron oxidation, was not found to oxidize procainamide. Leukocytes also contain prostaglandin synthetase which might also oxidize an arylamine; however, inhibitors of prostaglandin synthetase, such as aspirin and indomethacin, did not inhibit the formation of the hydroxyl-amine and nitroso metabolites by activated neutrophils.

NADPH oxidase is a membrane protein and the hydrogen peroxide and MPO are released outside the cell, or if the cell is engulfing a particle, they are released into the phagosome that is produced. Therefore, most of the haptenization should occur on the outside of the cell membrane (the inside of the phagosome originates from the outside of the cell membrane). Haptenization of a cell membrane, which would be exposed to the immune system, is more likely to result in an immune response than is haptenization of other proteins (Park *et al.*, 1987). Therefore, it is proposed that the haptenization of neutrophils by reactive metabolites of procainamide formed by MPO/H_2O_2/Cl^- is the initial event that results in agranulocytosis in some patients (Uetrecht, 1989). One of the risk factors for this adverse reaction would be an infection or some other inflammatory condition that leads to activation of cells because resting leukocytes do not metabolize procainamide.

It is also reasonable to postulate that formation of reactive metabolites generated by activated leukocytes is responsible for drug-induced lupus (Uetrecht, 1988). In the case of lupus, the critical cell is likely to be the monocyte. Monocytes also release MPO and generate H_2O_2 when activated. In addition, monocytes are macrophages and process and present immunogens to helper T cells during the initiation of a hypersensitivity reaction. Therefore, it is likely that formation of reactive metabolites by monocytes would have a high probability of leading to a hypersensitivity reaction.

25.2.2 Dapsone

Dapsone is a drug containing an arylamine functional group and is associated with a relatively high incidence of drug hypersensitivity reactions (Ognibene, 1970). The major therapeutic use of dapsone is in the treatment of leprosy. However, in countries where the incidence of leprosy is low, its major use is for the treatment of inflammatory conditions, especially conditions such as dermatitis herpetiformis where the inflammation appears to be due to activated neutrophils (Katz, 1982). It even has therapeutic

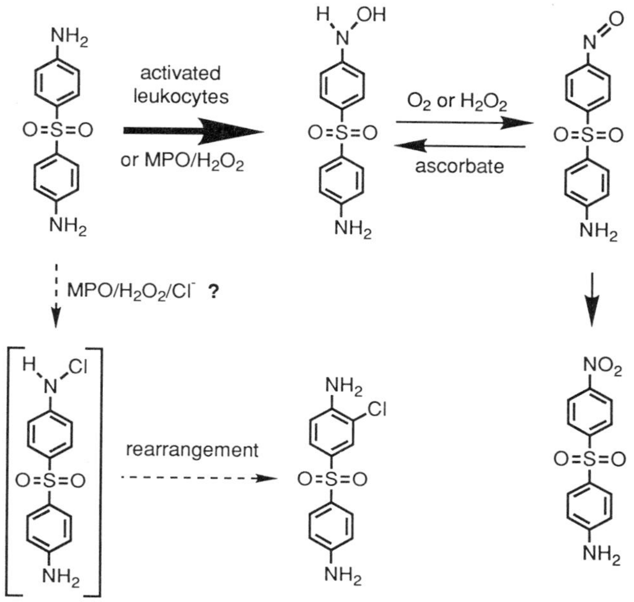

Figure 25.2 Metabolism of dapsone by activated neutrophils or mononuclear cells or catalysed by myeloperoxidase.

efficacy in diseases such as rheumatoid arthritis (Swinson *et al.*, 1981), although it is not the treatment of choice for rheumatoid arthritis.

The metabolism of dapsone by activated neutrophils and mononuclear leukocytes is similar to that of procainamide (Fig. 25.2) (Uetrecht *et al.*, 1988b).

Dapsone is associated with a high incidence of methaemoglobinaemia and haemolytic anaemia. These are not hypersensitivity reactions, but rather are due to redox cycling of the hydroxylamine and nitroso metabolites (Grossman and Jollow, 1988). The major source of these metabolites is presumably the liver, and unlike procainamide, the hydroxylamine of dapsone does escape the liver (Israili *et al.*, 1973; Uehleke and Tabarelli, 1973). This difference probably explains why dapsone is associated with methaemoglobinaemia and oxidative haemolytic anaemia and procainamide is not.

Dapsone is also associated with a high incidence of agranulocytosis (Ognibene, 1970; Firkin and Mariani, 1977) and a generalized hypersensitivity reaction which has clinical characteristics similar to those of mononucleosis (Frey *et al.*, 1981; Kromann *et al.*, 1982; Wille and Morrow,

1988). On the other hand, dapsone is one of the few drugs that is a primary arylamine but does not appear to be associated with a significant risk of drug-induced lupus. It is proposed that, as with procainamide, dapsone-induced agranulocytosis is due to reactive metabolites generated by activated neutrophils, and the mononucleosis-like syndrome is due to reactive metabolites generated by activated monocytes.

The therapeutic activity of dapsone as an anti-inflammatory agent appears to involve inhibition of myeloperoxidase (Stendahl *et al.*, 1978). It is likely that this inhibition of myeloperoxidase is related to the formation of reactive metabolites by myeloperoxidase, either because it acts as a competitive substrate and prevents formation of other reactive species (Niwa *et al.*, 1984), or more likely, because the reactive metabolites formed inactivate the enzyme.

Another peroxidase that can catalyse two electron oxidations is thyroid peroxidase. Dapsone causes thyroid cancer in rats (Griciute and Tomatis, 1980), and this may be due to the formation of reactive metabolites by thyroid peroxidase. Preliminary studies indicate that thyroid peroxidase forms the same reactive metabolites as myeloperoxidase.

25.2.3 Sulphonamides

Sulphonamides are an important class of antibiotics, and they are also arylamines. They are probably associated with the highest incidence of hypersensitivity reactions of any commonly used drug. The most common type of hypersensitivity reaction is a generalized reaction, which can vary in severity from a rash to a life-threatening reaction involving several organs (Shear *et al.*, 1986). Sulphonamides can also cause agranulocytosis (Rinkoff and Spring, 1941; Arneborn and Palmblad, 1978) and drug-induced lupus (Hoffman, 1945; Honey, 1956).

It has been shown that sulphadiazine is oxidized by activated neutrophils to a hydroxylamine (Uetrecht *et al.*, 1986). The metabolism of sulphamethoxazole appears to be more complex, and preliminary studies suggest that the major pathway involves *N*-chlorination. Clearly, leukocytes are capable of oxidizing sulphonamides to reactive metabolites, and it is likely that these metabolites are responsible for the hypersensitivity reactions associated with these drugs (Rieder *et al.*, 1989).

Sulphadiazine also has anti-inflammatory effects similar to those of dapsone (Lang, 1979), and it is likely that these effects are due to metabolism by myeloperoxidase. The sulphonamides may also be metabolized by thyroid peroxidase. They cause hypothyroidism in rats, but this is not clinically significant in man. However, four children have been observed who became hypothyroid while recovering from a severe hypersensitivity reaction to a sulphonamide or an anticonvulsant (Gupta *et al.*, 1988). These patients had antibodies to thyroid peroxidase. Although such antibodies are seen in patients with other forms of autoimmune thyroid disease, this usually occurs in young women, and it would otherwise be a rare finding in children. In addition, this autoimmune hypothyroidism resolved spontaneously after several months, and this is also unusual for idiopathic autoimmune hypothyroidism.

25.2.4 Other primary arylamines

The number of drugs that are primary arylamines is relatively small, presumably because they are associated with a high incidence of adverse drug reactions. Other such drugs include aminoglutethimide (Gez and Sulkes, 1984; Lawrence *et al.*, 1978; McCraken *et al.*, 1980), 4- and 5-amino-salicylic acid (Rab and Alam, 1970), metoclopramide (Harvey and Luzar, 1988; Manoharan, 1988), and nomifensine (Garcia-Morteo and Maldonado-Cocco, 1983; Salama and Mueller-Eckhardt, 1985). These drugs are associated with a similar spectrum of adverse drug reactions which includes agranulocytosis, drug-induced lupus, and generalized hypersensitivity reactions as referenced above. Metoclopramide is very similar in structure to procainamide and is associated with agranulocytosis, but it has not been reported to cause lupus. However, the incidence of procainamide-induced lupus appears to be dose dependent as evidenced by the fact that N-acetylprocainamide can be given to patients who have had procainamide induced lupus even though some of it is hydrolysed to procainamide (Roden *et al.*, 1980; Kluger *et al.*, 1981). Since metoclopramide is given at a much lower dose than procainamide, this may explain its lack of association with drug-induced lupus.

25.2.5 Drugs metabolized to primary arylamines

There are other drugs that are not primary arylamines but are extensively metabolized to arylamines *in vivo*. These drugs include practolol, acebutolol, and sulphasalazine. Practolol is no longer used clinically because it was associated with an unusual oculocutaneous syndrome, and it was also associated with drug-induced lupus (Raftery and Denman, 1973; Milner *et al.*, 1977; Amos, 1979). Acebutolol is the only other β-blocker that is associated with a significant incidence of drug-induced lupus (Cody *et al.*, 1979; Booth *et al.*, 1982). Sulphasalazine is associated with a variety of hypersensitivity reactions including agranulocytosis, drug-induced lupus, and generalized hypersensitivity reactions (Jones and Malone, 1972;

Das *et al.*, 1973; Davies and MacFarlane, 1974; Griffiths and Kane, 1977; Jacobson *et al.*, 1985; Mitrane *et al.*, 1986; Pearl *et al.*, 1986).

Chloramphenicol has an aromatic nitro group. Chloramphenicol is classically associated with aplastic anaemia, and it can also cause agranulocytosis (Wallerstein *et al.*, 1969). It appears that chloramphenicol-induced aplastic anaemia is due to the aromatic nitro group because replacement of this group with a methylsulphone leads to a drug that is still active as an antibiotic but does not appear to cause aplastic anaemia (Yunis, 1988). Anaerobic bacteria in the intestine reduce chloramphenicol to an arylamine (Scheline, 1973), and it has been demonstrated that the liver is capable of oxidizing this amine to a nitroso metabolite which is toxic to bone marrow cells (Gross *et al.*, 1982; Yunis *et al.*, 1980). However, it has also been demonstrated that this nitroso metabolite is too reactive to escape the liver (Ascherl *et al.*, 1985). Attempts to demonstrate metabolism or covalent binding of the amine metabolite by bone marrow cells were unsuccessful (Gross *et al.*, 1982), but in these experiments the bone marrow cells were not activated. We have found that the amine metabolite of chloramphenicol is metabolized by activated neutrophils if they have been activated (Uetrecht, unpublished observation). In addition, Yunis has demonstrated that the nitro group of another bacterial metabolite of chloramphenicol is more readily reduced than that of chloramphenicol itself, and bone marrow cells appear to be capable of reducing it to a different toxic nitroso metabolite (Isildar *et al.*, 1988). Thus, there are at least two possible pathways by which a bacterial metabolite of chloramphenicol may be metabolized to a toxic metabolite by bone marrow cells. The extent of these reactions *in vivo* is unknown.

25.2.6 Tertiary arylamines

The drugs discussed so far are primary arylamines, but other arylamines may also be metabolized to reactive metabolites by activated leukocytes. Aprindine is a tertiary arylamine which is associated with a high incidence of agranulocytosis that has severely limited its use (Danilo, 1979; Opie, 1980). Aminopyrine and dipyrone have a tertiary amine on an aromatic heterocyclic ring. They are both associated with a high incidence of agranulocytosis (Barrett *et al.*, 1976; International Agranulocytosis and Aplastic Anemia Study, 1986). The amine functional group appears to be involved in this adverse reaction because antipyrine, a very similar drug but without the amino group, is rarely associated with agranulocytosis.

A new drug, OPC-8212, which in early clinical trials caused no significant toxicity in 256 Japanese patients, when given to 28 patients in the United States was associated with the development of agranulocytosis in four. Thus there appears to be a striking difference in the incidence of this adverse reaction in these two populations. In addition to the difference in race, there was at least one other important difference in these two populations. In Japan none of the patients received influenza vaccine. By contrast, seven of the US patients received influenza vaccine, one shortly before and the others during treatment with OPC-8212. All of the patients who developed agranulocytosis were among these seven patients.

We found that influenza vaccine was capable of activating neutrophils such that they oxidized procainamide to the reactive hydroxylamine and nitroso metabolites (Uetrecht *et al.*, 1989). Furthermore, using radio-labelled OPC-8212, influenza vaccine-activated neutrophils metabolized the drug and led to covalent binding of a metabolite of OPC-8212 to the neutrophils. The extent of OPC-8212 covalent binding to neutrophils was greater than that to hepatic microsomes. This provides an explanation of how influenza vaccine might predispose patients to OPC-8212-induced agranulocytosis. There may also be racial differences in the risk of OPC-8212-induced agranulocytosis because it would seem likely that, in the Japanese patients, a few of them would have seen some stimulus that led to activation of neutrophils. OPC-8212 also contains a tertiary arylamine, and formation of the reactive metabolite may involve this functional group.

25.2.7 Hydrazines

Another functional group that contains nitrogen and is associated with a high incidence of adverse drug reactions is hydrazine. Hydralazine has a hydrazine functional group, and it is second only to procainamide as a cause of drug-induced lupus (Perry, 1973). It has been reported to cause agranulocytosis, but this is uncommon. Isoniazid is also a hydrazide and is associated with drug-induced lupus (Rothfield *et al.*, 1978). Isoniazid commonly causes a small increase in transaminase levels, but it is also associated with a severe drug-induced hepatitis which has the characteristics of a hypersensitivity reaction (Maddrey and Boitnott, 1973).

The hepatitis associated with isoniazid is thought to be due to the formation of acetylhydrazine which is further metabolized to a reactive metabolite. This is based on careful experiments carried out on rats in which acetylhydrazine was much more hepatotoxic than isoniazid. Furthermore, amidase inhibitors, which decrease the formation of acetylhydrazine, decreased the toxicity of isoniazid but not of acetylhydrazine (Timbrell *et al.*, 1980). However, the predictable hepatotoxicity observed in the rat model is likely to involve a different mechanism than the idiosyncratic reaction that occurs in humans. The idiosyncratic reaction could involve

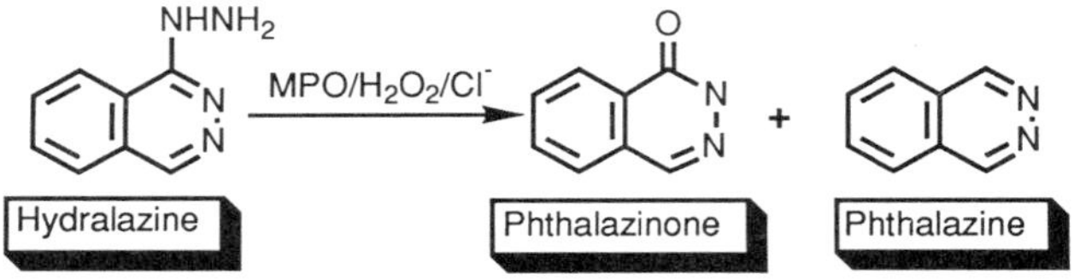

Figure 25.3 Oxidation of hydralazine catalysed by myeloperoxidase.

direct activation of isoniazid. It might be expected that a protein haptenized by an isoniazid metabolite would be more immunogenic than that haptenized by the smaller acetyl group. In fact, antibodies have been detected against the isonicotinyl group in a pharmacist who became hypersensitive to isoniazid after repeated exposure during grinding of the drug, as well as in two patients with tuberculosis who were treated with the drug (Asai *et al.*, 1987).

Hydralazine is metabolized by activated neutrophils to several metabolites (Hofstra and Uetrecht, 1988), among which are phthalazine and phthalazinone (Fig. 25.3). Although neither of these metabolites is chemically reactive, phthalazinone is thought to be produced via a reactive diazine intermediate, and there has been a suggestion that phthalazinone levels are higher in patients who develop hydralazine-induced lupus (Timbrell *et al.*, 1984).

Likewise, isoniazid is metabolized by activated neutrophils to isonicotinic acid, and it is likely that a reactive intermediate is involved in this metabolic pathway (Li and Uetrecht, 1988).

Unlike the other classes of drugs described, hydrazines are very good nucleophiles, and they may also covalently bind to cells without metabolic activation.

25.2.8 Anticonvulsants

Another class of drugs that is associated with a relatively high incidence of hypersensitivity reactions is the anticonvulsants. The most widely used anticonvulsant is phenytoin, which is associated with generalized hypersensitivity reactions and birth defects, and to a lesser degree, with drug-induced lupus and bone marrow toxicity (Beernink and Miller, 1973; Kleckner *et al.*, 1975; Hanson *et al.*, 1976; Spielberg *et al.*, 1981; Brown and Schubert, 1986; Powers and Carson, 1987). It has been suggested that a reactive arene oxide is responsible for these toxicities (Martz *et al.*, 1977). Although the arene oxide has never been observed, there is evidence that it is the precursor to a phenol metabolite (Claesen *et al.*, 1982). Furthermore, inhibitors of epoxide hydrolase increase the toxicity of phenytoin.

Mephenytoin is a closely related drug with a similar spectrum of toxicity except that it is associated with a higher incidence of agranulocytosis and drug-induced fever. Unlike phenytoin, mephenytoin is chiral, and only the *S*-enantiomer is metabolized to a phenol metabolite (Kupfer *et al.*, 1981). Assuming that the mechanism of toxicity is the same and that the phenol metabolite is derived from an arene oxide, the arene oxide hypothesis would predict that only the *S*-enantiomer should be toxic. When this was tested in an animal model of fetal toxicity, it was found that the *R*-enantiomer was actually more toxic than the *S*-enantiomer (Wells *et al.*, 1982). Furthermore, because the *R*-enantiomer is not metabolized to the phenol, more of it is metabolized by *N*-demethylation, and the *R,N*-demethylmephenytoin was more toxic than the parent drug. Because this was an animal model and the toxicity measured was different, this may have no relevance to human hypersensitivity reactions. On the other hand, there are other drugs, such as ethosuximide and trimethadione, that have similar heterocyclic rings and spectrum of hypersensitivity reactions, and yet these drugs do not have a phenyl ring that could be oxidized to an arene oxide (Abbott and Schwab, 1950; Benton *et al.*, 1962; Beernink and Miller, 1973; Dreifuss, 1982).

The observation that the *N*-demethylated mephenytoin was more toxic than the parent drug suggested that the toxicity might involve the heterocyclic nitrogen. We found that phenytoin is metabolized by MPO/H_2O_2/Cl$^-$ to a reactive *N,N'*-dichlorophenytoin (Fig. 25.4) (Uetrecht and Zahid, 1988a). This metabolite was not observed when phenytoin was oxidized by activated neutrophils, but when this metabolite was synthesized it was found to react rapidly with cells. Incubation of radiolabelled phenytoin with activated neutrophils led to a small degree of covalent binding of the radiolabel to the neutrophils, and this binding was dependent on activation of the cells. Thus metabolism of phenytoin by activated neutrophils represents an alternative hypothesis for the initial step in the toxicity of this and similar drugs. In addition, there is a recent report that

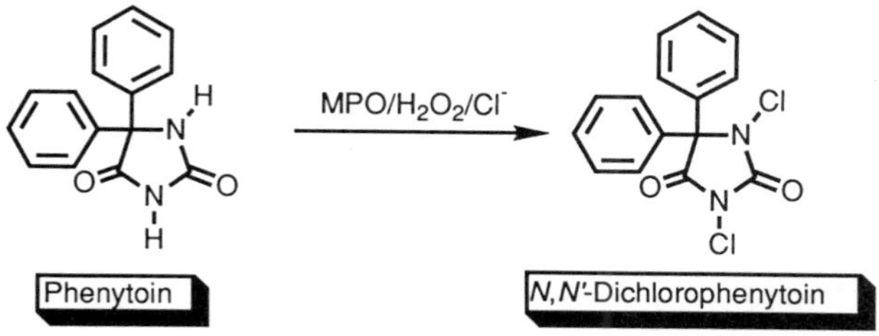

Figure 25.4 Chlorination of phenytoin catalysed by myeloperoxidase.

phenytoin is metabolized by activated neutrophils to a limited extent to the phenol metabolite (Pawluk *et al.*, 1989).

Although the structure of carbamazepine is significantly different from that of the hydantoin anticonvulsants, it is a substituted urea. It is also associated with a relatively high incidence of hypersensitivity reactions and agranulocytosis (Bateman, 1985; Lewis and Rosenbloom, 1982; Livingston *et al.* 1967, 1974; Luchins, 1984). Preliminary experiments have shown that carbamazepine is also metabolized by activated neutrophils or MPO/H_2O_2/Cl$^-$ (Uetrecht, unpublished observation). Most of the metabolism involves the urea moiety. Therefore, the mechanism of carbamazepine hypersensitivity reactions may be similar to that of the hydantoin anticonvulsants.

25.3 CONCLUSIONS

Although little is known about the mechanism of drug hypersensitivity reactions, most investigations favour hypotheses that involve the formation of a chemically reactive metabolite as an initial step. Such reactive metabolites could act as haptens; alternatively, they could functionally alter cells involved in the immune system. Although most drug metabolism occurs in the liver, the ability of neutrophils and monocytes to form chemically reactive metabolites provides a more attractive hypothesis for the initial step in drug hypersensitivity reactions such as agranulocytosis, drug-induced lupus and generalized reactions. Drugs with functional groups that are easily oxidized by peroxidases are usually associated with a high incidence of such reactions, thus lending further credence to this hypothesis. However, direct evidence that such metabolism is responsible for drug hypersensitivity reactions is lacking, and very little is known about subsequent steps in the pathogenesis of such reactions.

In general, the risk factors that predispose specific individuals to hypersensitivity reactions are also unknown. In some cases a decrease in the ability of a patient to metabolize the drug by other pathways such as acetylation appears to be a risk factor (Woosley *et al.*, 1978; Shear *et al.*, 1986). If the leukocyte hypothesis is correct, it would be expected that an infection or some other inflammatory condition that leads to activation of leukocytes would be a risk factor. This is supported by the observations with OPC-8212. However, it is likely that the largest factor in determining risk is the interindividual differences in the immune system.

REFERENCES

Abbott, J.A. and Schwab, R.S. (1950) The serious side effects of the newer antiepileptic drugs: their control and prevention. *N. Engl. J. Med.*, **242**, 943–9.

Amos, H.E. (1979) Immunological aspects of practolol toxicity. *Int. J. Immunopharmacol.*, **1**, 9–16.

Arneborn, P. and Palmblad, J. (1978) Drug-induced neutropenia in the Stockholm region 1973–75. *Acta Med. Scand.*, **204**, 283–6.

Asai, S., Shimoda, T., Hara, K. and Fujiwara, K. (1987) Occupational asthma caused by isonicotinic acid hydrazide (INH) inhalation. *J. Allergy Clin. Immunol.*, **80**, 578–85.

Ascherl, M., Eyer, P. and Kampffmeyer, H. (1985) Formation and disposition of nitrosochloramphenicol in rat liver. *Biochem. Pharmacol.*, **34**, 3755–63.

Barrett, A.J., Weller, E., Rozengurt, N. *et al.* (1976) Amidopyrine agranulocytosis: drug inhibition of granulocyte colonies in the presence of patient's serum. *Br. Med. J.*, **2**, 850–1.

Bateman, D.E. (1985) Carbamazepine induced systemic lupus erythematosus: case report. *Br. Med. J.*, **291**, 632–3.

Beernink, D.H. and Miller, J.J. (1973) Anticonvulsant-induced antinuclear antibodies and lupus-like disease in children. *J. Pediatr.*, **82**, 113–17.

Benton, J.W., Tynes, B., Register, H.B. *et al.* (1962) Systemic lupus erythematosus occurring during anticonvulsive drug therapy. *J. Am. Med. Assoc.*, **180**, 115–18.

Booth, R.J., Wilson, J.D. and Bullock, J.Y. (1982) β-Adrenergic-receptor blockers and antinuclear antibodies in hypertension. *Clin. Pharmacol. Ther.*, **31**, 555–8.

Brown, M. and Schubert, T. (1986) Phenytoin hypersensitivity hepatitis and mononucleosis syndrome. *J. Clin. Gastroenterol.*, **8**, 469–77.

Claesen, M., Moustafa, M.A.A., Adline, J. *et al.* (1982) Evidence for an arene oxide-NIH shift pathway in the metabolic conversion of phenytoin to 5-(4-hydroxyphenyl)-5-phenylhydantoin in the rat and man. *Drug Metab. Dispos.*, **10**, 667–71.

Cody, R.J., Calabrese, L.H., Clough, J.D. *et al.* (1979) Development of antinuclear antibodies during acebutolol therapy. *Clin. Pharmacol. Ther.*, **25**, 800–5.

Coombs, R.R.A. and Gell, P.G.H. (1963) The classification of allergic reactions underlying disease, in *Clinical Aspects of Immunology* (eds P.G.H. Gell and R.R.A. Coombs), Blackwell Scientific Publications, Oxford, pp. 317–37.

Danilo, P. (1979) Aprindine. *Am. Heart J.*, **97**, 119–24.

Das, K.M., Eastwood, M.A., McManus, J.P.A. and Sircus, W. (1973) Adverse reactions during salicylazosulfapyridine therapy and the relation with drug metabolism and acetylator phenotype. *N. Engl. J. Med.*, **289**, 491–5.

Davies, D. and MacFarlane, A. (1974) Fibrosing alveolitis and treatment with sulfasalazine. *Gut*, **15**, 185–8.

de Weck, A.L. and Bundgaard, H. (1983) *Allergic Reactions to Drugs*, Springer, Berlin.

Dreifuss, F.E. (1982) Ethosuximide, in *Antiepileptic Drugs* (eds D.M. Woodbury, J.K. Penry and C.E. Pippenger), Raven Press, New York, pp. 647–53.

Ellrodt, A.G., Murata, G.H., Reidinger, M.S. *et al.* (1984) Severe neutropenia associated with sustained-release procainamide. *Ann. Intern. Med.*, **100**, 197–201.

Firkin, F.C. and Mariani, A.F. (1977) Agranulocytosis due to dapsone. *Med. J. Aust.*, **2**, 247–51.

Frey, H.M., Gershon, A.A., Bordowsky, W. and Bullock, W.E. (1981) Fatal reaction to dapsone during treatment of leprosy. *Ann. Intern. Med.*, **94**, 777–9.

Garcia-Morteo, O. and Maldonado-Cocco, J.A. (1983) Lupus-like syndrome during treatment with nomifensine. *Arthritis Rheum.*, **26**, 936.

Gez, E. and Sulkes, A. (1984) Aminoglutethimide-induced leucopenia: a case report and review of the literature. *Oncology*, **41**, 399–402.

Griciute, L. and Tomatis, L. (1980) Carcinogenicity of dapsone in mice and rats. *Int. J. Cancer*, **25**, 123–9.

Griffiths, I.D. and Kane, S.P. (1977) Sulphasalazine-induced lupus syndrome in ulcerative colitis. *Br. Med. J.*, **2**, 1188–9.

Gross, B.J., Branchflower, R.V., Burke, T.R. *et al.* (1982) Bone marrow toxicity *in vitro* of chloramphenicol and its metabolites. *Toxicol. Appl. Pharmacol.*, **64**, 557–65.

Grossman, S.J. and Jollow, D.J. (1988) Role of dapsone hydroxylamine in dapsone-induced hemolytic anemia. *J. Pharmacol. Exp. Ther.*, **244**, 118–25.

Gupta, A., Waldhauser, L. and Reider, M. (1988) Drug-induced hypothyroidism: the thyroid as a target organ in hypersensitivity reactions. *Soc. Ped. Res.*, **23**, 277A.

Hanson, J.W., Myrianthopoulos, N.C., Sedgwick Harvey, M.A. and Smith, D.W. (1976) Risks to the offspring of women treated with hydantoin anticonvulsants, with emphasis on the fetal hydantoin syndrome. *J. Pediatr.*, **89**, 662–8.

Harvey, R.L. and Luzar, M.J. (1988) Metoclopramide-induced agranulocytosis. *Ann. Intern. Med.*, **108**, 214–15.

Henningsen, N.C., Cederberg, A., Hanson, A. and Johansson, B.W. (1975) Effects of long-term treatment with procaine amide. *Acta Med. Scand.*, **198**, 475–82.

Hoffman, B.J. (1945) Sensitivity to sulfadiazine resembling acute lupus erythematosus. *Arch. Dermatol. Syph.*, **51**, 190–2.

Hofstra, A. and Uetrecht, J. (1988) Oxidation of hydralazine by monocytes as a mechanism of hydralazine-induced lupus. *Pharmacologist*, **30**, A99.

Honey, M. (1956) Systemic lupus erythematosus presenting with sulfonamide hypersensitivity reaction. *Br. Med. J.*, **1**, 1272–5.

International Agranulocytosis and Aplastic Anemia Study (1986) Risks of agranulocytosis and aplastic anemia: a first report with special reference to analgesics. *J. Am. Med. Assoc.*, **256**, 1749–57.

Isildar, M., Abou-Khalil, W.H., Jimenez, J.J. *et al.* (1988) Aerobic nitroreduction of dehydrochloramphenicol by bone marrow. *Toxicol. Appl. Pharmacol.*, **94**, 305–10.

Israili, Z.H., Cucinell, S.A., Vaught, J. *et al.* (1973) Studies of the metabolism of dapsone in man and experimental animals: formation of N-hydroxy metabolites. *J. Pharmacol. Exp. Ther.*, **187**, 138–51.

Jacobson, I.M., Kelsey, P.B., Blyden, G.T. *et al.* (1985) Sulfasalazine-induced agranulocytosis. *Am. J. Gastroenterol.*, **80**, 118–21.

Jones, G.R. and Malone, D.N.S. (1972) Sulphasalazine induced lung disease. *Thorax*, **27**, 713–17.

Katz, S.I. (1982) Commentary: sulfoxone (diasone) in the treatment of dermatitis herpetiformis. *Arch. Dermatol.*, **118**, 809–12.

Kleckner, H.B., Yakulis, V. and Heller, P. (1975) Severe hypersensitivity to diphenylhydantoin with circulating antibodies to the drug. *Ann. Intern. Med.*, **83**, 522–3.

Kluger, J., Drayer, D.E., Reidenberg, M.M. and Lahita, R. (1981) Acetyl-procainamide therapy in patients with previous procainamide-induced lupus syndrome. *Ann. Intern. Med.*, **95**, 18–23.

Kromann, N.P., Vilhelmsen, R. and Stahl, D. (1982) The dapsone syndrome. *Arch. Dermatol.*, **118**, 531–2.

Kupfer, A., Roberts, R.K., Schenker, S. and Branch, R.A. (1981) Sterioselective metabolism of mephenytoin in man. *J. Pharmacol. Exp. Ther.*, **218**, 193–9.

Lang, P.G. (1979) Sulfones and sulfonamides in dermatology today. *J. Am. Acad. Dermatol.*, **1**, 479–92.

Lawrence, B., Santen, R.J., Lipton, A. *et al.* (1978) Pancytopenia induced by aminoglutethimide in the treatment of breast cancer. *Cancer Treat. Rep.*, **62**, 1581–3.

Lewis, I.J. and Rosenbloom, L. (1982) Glandular fever-like syndrome, pulmonary eosinophilia and asthma associated with carbamazepine. *Postgrad. Med. J.*, **58**, 100–1.

Li, A. and Uetrecht, J. (1988) Metabolism of isoniazid by activated neutrophils. *Pharmacologist*, **30**, A99.

Livingston, S., Pauli, L.L. and Berman, W. (1974) Carbamazepine (Tegretol) in epilepsy. *Dis. Nerv. Syst.*, **35**, 103–7.

Livingston, S., Villamater, C., Sakata, Y. and Pauli, L.L. (1967) Use of carbamazepine in epilepsy. *J. Am. Med. Assoc.*, **200**, 204–8.

Luchins, D.J. (1984) Fatal agranulocytosis in a chronic schizophrenic patient treated with carbamazepine. *Am. J. Psychiatry*, **141**, 687–8.

Maddrey, W.C. and Boitnott, J.K. (1973) Isoniazid hepatitis. *Ann. Intern. Med.*, **79**, 1–12.

Manoharan, A. (1988) Metoclopramide-induced agranulocytosis. *Med. J. Aust.*, **149**, 508.

Martz, F., Failinger, C. and Blake, D.A. (1977) Phenytoin teratogenesis: correlation between embryopathic effect and covalent binding of putative arene oxide metabolite in gestational tissue. *J. Pharmacol. Exp. Ther.*, **203**, 231–9.

McCraken, M., Benson, E.A. and Hickling, P. (1980) Systemic lupus erythematosus induced by aminoglutethimide. *Br. Med. J.*, **281**, 1254.

Meyers, D.G., Gonzalez, E.R. and Peters, L.L. (1985) Severe neutropenia associated with procainamide: comparison of sustained and conventional preparations. *Am. Heart J.*, **109**, 1393–5.

Milner, G.R., Holt, P.J.L., Bottomley, J. and Maciver, J.E. (1977) Practolol therapy associated with a systemic lupus erythematosus-like syndrome and an inhibitor to factor XIII. *J. Clin. Pathol.*, **30**, 770–3.

Mitrane, M.P., Singh, A. and Seibold, J.R. (1986) Cholestasis and fatal agranulocytosis complicating sulfasalazine therapy: case report and review of the literature. *J. Rheumatol.*, **13**, 969–72.

Niwa, Y., Sakane, T. and Miyachi, Y. (1984) Dissociation of the inhibitory effect of dapsone on the generation of oxygen intermediates – in comparison with that of colchicine and various scavengers. *Biochem. Pharmacol.*, **33**, 2355–60.

Ognibene, A.J. (1970) Agranulocytosis due to dapsone. *Ann. Intern. Med.*, **72**, 521–4.

Opie, L.H. (1980) Aprindine and agranulocytosis. *Lancet*, **ii**, 689–90.

Park, B.K., Coleman, J.W. and Kitteringham, N.R. (1987) Drug disposition and drug hypersensitivity. *Biochem. Pharmacol.*, **36**, 581–90.

Parker, C.W. (1982) Allergic reactions in man. *Pharmacol. Rev.*, **34**, 85–104.

Pawluk, L., Mays, D., She, Z. *et al.* (1989) Metabolism of phenytoin by activated human neutrophils. *Eur. J. Clin. Pharmacol.*, **36(suppl)**, A251.

Pearl, R.K., Nelson, R.L., Prasad, M.L. *et al.* (1986) Serious complications of sulfasalazine. *Dis. Colon Rectum*, **29**, 201–2.

Perry, H.M. (1973) Late toxicity to hydralazine resembling systemic lupus erythematosus or rheumatoid arthritis. *Am. J. Med.*, **54**, 58–72.

Pohl, L.R., Satoh, H., Christ, D.D. and Kenna, J.G. (1988) The immunologic and metabolic basis of drug hypersensitivities. *Ann. Rev. Pharmacol.*, **28**, 367–87.

Powers, N.G. and Carson, S.H. (1987) Idiosyncratic reactions to phenytoin. *Clin. Pediatr.*, **26**, 120–4.

Rab, S.M. and Alam, M.N. (1970) Severe agranulocytosis during para-aminosalicylic acid therapy. *Br. J. Dis. Chest.*, **64**, 164–8.

Raftery, E.B. and Denman, A.M. (1973) Systemic lupus erythematosus syndrome induced by practolol. *Br. Med. J.*, **2**, 452–5.

Rieder, M.J., Uetrecht, J., Shear, N.H. *et al.* (1989) Diagnosis of sulfonamide hypersensitivity reactions by *in-vitro* 'rechallenge' with hydroxylamine metabolites. *Ann. Intern. Med.*, **110**, 286–9.

Rinkoff, S.S. and Spring, M. (1941) Toxic depression of the myeloid elements following therapy with the sulfonamides: report of 8 cases. *Ann. Intern. Med.*, **15**, 89–107.

Roden, D.M., Reele, S.B. and Higgins, S.B. (1980) Antiarrhythmic efficacy, pharmacokinetics and safety of *N*-acetylprocainamide in human subjects: comparison with procainamide. *Am. J. Cardiol.*, **46**, 463–8.

Rothfield, N.F., Bierer, W.F. and Garfield, J.W. (1978) Isoniazid induction of antinuclear antibodies: a prospective study. *Ann. Intern. Med.*, **88**, 650–2.

Salama, A. and Mueller-Eckhardt, C. (1985) The role of metabolic-specific antibodies in nomifensine-dependent immune hemolytic anemia. *N. Engl. J. Med.*, **313**, 469–74.

Scheline, R.R. (1973) Metabolism of foreign compounds by gastrointestinal microorganisms. *Pharmacol. Rev.*, **25**, 451–523.

Schwartz, R.S. and Stollar, B.D. (1985) Origins of anti-DNA autoantibodies. *J. Clin. Invest.*, **75**, 321–7.

Shear, N.H., Spielberg, S.P., Grant, D.M. *et al.* (1986) Differences in metabolism of sulfonamides predisposing to idiosyncratic toxicity. *Ann. Intern. Med.*, **105**, 179–84.

Spielberg, S.P., Gordon, G.B., Blake, D.A. *et al.* (1981) Predisposition to phenytoin hepatotoxicity assessed *in vitro*. *N. Engl. J. Med.*, **305**, 722–7.

Stendahl, O., Molin, L. and Dahlgren, C. (1978) The inhibition of polymorpho-nuclear leukocyte cytotoxicity by dapsone: a possible mechanism in the treatment of dermatitis herpetiforms. *J. Clin. Invest.*, **62**, 214–20.

Swinson, D.R., Zlosnick, J. and Jackson, L. (1981) Double-blind trial of dapsone against placebo in the treatment of rheumatoid arthritis. *Ann. Rheum. Dis.*, **40**, 235–9.

Thompson, J.F., Robinson, C.A. and Segal, J.L. (1988) Procainamide agranulocytosis: a case report and review of the literature. *Curr. Ther. Res.*, **44**, 872–81.

Timbrell, J.A., Facchini, V., Harland, S.J. and Mansilla-Tinoco, R. (1984) Hydralazine-induced lupus: is there a toxic pathway? *Eur. J. Clin. Pharmacol.*, **27**, 555–9.

Timbrell, J.A., Mitchell, J.R., Snodgrass, W.R. and Nelson, S.D. (1980) Isoniazid hepatotoxicity: the relationship between covalent binding and metabolism *in vivo*. *J. Pharmacol. Exp. Ther.*, **213**, 364–9.

Uehleke, H. and Tabarelli, S. (1973) *N*-Hydroxylation of 4,4'-diaminodiphenyl-sulfone (dapsone) by liver microsomes, and in dogs and humans. *Naunyn-Schmiedeberg's Arch. Pharmakol. Exp. Pathol.*, **278**, 55–68.

Uetrecht, J. and Zahid, N. (1988a) *N*-Chlorination of phenytoin by myeloperoxidase to a reactive metabolite. *Chem. Res. Toxicol.*, **1**, 148–51.

Uetrecht, J. and Zahid, N. (1988b) Procainamide (PA) is *N*-chlorinated by myeloperoxidase (MPO)-implications for toxicity. *Pharmacologist*, **30**, A98.

Uetrecht, J., Zahid, N. and Rubin, R. (1988a) Metabolism of procainamide to a hydroxylamine by human neutrophils and mononuclear leukocytes. *Chem. Res. Toxicol.*, **1**, 74–8.

Uetrecht, J., Zahid, N., Shear, N.H. and Biggar, W.D. (1988b) Metabolism of dapsone to a hydroxylamine by human neutrophils and mononuclear cells. *J. Pharmacol. Exp. Ther.*, **245**, 274–9.

Uetrecht, J.P. (1985) Reactivity and possible significance of hydroxylamine and nitroso metabolites of procainamide. *J. Pharmacol. Exp. Ther.*, **232**, 420–5.

Uetrecht, J.P. (1988) Mechanism of drug-induced lupus. *Chem. Res. Toxicol.*, **1**, 133–43.

Uetrecht, J.P. (1989) Idiosyncratic drug reactions: possible role of reactive metabolites generated by leukocytes. *Pharmacol. Res.*, **6**, 265–73.

Uetrecht, J.P., Shear, N. and Biggar, W. (1986) Dapsone is metabolised by human neutrophils to a hydroxylamine. *Pharmacologist*, **28**, 239.

Uetrecht, J.P., Sweetman, B.J., Woosley, R.L. and Oates, J.A. (1984) Metabolism of procainamide to a hydroxylamine by rat and human hepatic microsomes. *Drug Metab. Dispos.*, **12**, 77–81.

Uetrecht, J.P., Zahid, N. and Spielberg, S.P. (1989) Oxidation of OPC-8212 to a reactive intermediate by influenza vaccine-activated neutrophils: possible relationship to agranulocytosis. *Eur. J. Clin. Pharmacol.*, **36(suppl)**, A53.

Unanue, E.R. and Allen, P.M. (1987) The basis for the immunoregulatory role of macrophages and other accessory cells. *Science*, **236**, 551–7.

Wallerstein, R.O., Condit, P.K., Kasper, C.K. *et al.* (1969) Statewide study of chloramphenicol therapy and fatal aplastic anemia. *J. Am. Med. Assoc.*, **208**, 2045–50.

Weiss, S.J. (1989) Tissue destruction by neutrophils. *N. Engl. J. Med.*, **320**, 365–76.

Wells, P.G., Kupfer, A., Lawson, J.A. and Harbison, R.D. (1982) Relation of *in vivo* drug metabolism to stereoselective hydantoin toxicity in mouse: evaluation of mephenytoin and its metabolite, nirvanol. *J. Pharmacol. Exp. Ther.*, **221**, 228–34.

Wille, R.C. and Morrow, J.D. (1988) Case report: dapsone hypersensitivity syndrome associated with treatment of the bite of a brown recluse spider. *Am. J. Med. Sci.*, **296**, 270–1.

Woosley, R.L., Drayer, D.E., Reidenberg, M.M. *et al.* (1978) Effect of acetylator phenotype on the rate at which procainamide induces antinuclear antibodies and the lupus syndrome. *N. Engl. J. Med.*, **298**, 1157–9.

Worlledge, S.M. (1973) Immune drug-induced hemolytic anemias. *Semin. Hematol.*, **10**, 327–44.

Yunis, A.A. (1988) Chloramphenicol: relation of structure to activity and toxicity. *Ann. Rev. Pharmacol. Toxicol.*, **28**, 83–100.

Yunis, A.A., Miller, A.M., Salem, Z. *et al.* (1980) Nitroso-chloramphenicol: possible mediator in chloramphenicol-induced aplastic anemia. *J. Lab. Clin. Med.*, **96**, 36–46.

26

Phototoxic effects of *N*-oxidized drugs and related compounds

G.M.J. Beijersbergen van Henegouwen

Department of Medicinal Chemistry, Center for Bio-Pharmaceutical Sciences, Leiden University, P.O. Box 9502, 2300 RA Leiden, The Netherlands

1. Biological effects of *N*-oxidized compounds can result from reactive intermediates non-enzymatically formed in (sun)light-exposed parts of the body.
2. Phototoxic effects resulting from photoactivation of *N*-oxidized compounds can concern inner organs as well. Thus, oral or intra-peritoneal administration of nitroarenes or imino-*N*-oxides followed by exposure of the skin to light can produce systemic effects.

26.1 INTRODUCTION

N-Oxidized compounds find extensive application as constituents of drugs, cosmetics, food and agricultural chemicals. Although very useful and almost indispensible, they can also produce adverse biological effects. These toxic (side-) effects include genotoxicity (carcinogenicity), but also immune diseases, such as allergy. Enzymatic activation of N-oxidized compounds as inducers of such processes has received much attention. It is assumed that reactive intermediates or instable products formed during (minor) metabolic processes damage biomacromolecules in the genetic or immune system.

This chapter focuses on the fact that biological effects can also result from reactive intermediates formed non-enzymatically in (sun)light-exposed parts of the body. This is illustrated by some representatives of two classes of compounds: nitroarenes and imino-N-oxides. Introductory remarks on some photobiological phenomena with respect to (sun)light exposure of the skin are given.

26.1.1 Exposure of the skin to (sun)light; some facts

(a) Formation of reactive intermediates

Circumstantial evidence has amassed strongly inferring that reactive intermediates, e.g. free radicals, whether or not coupled to oxygen, and singlet oxygen are highly responsible for deleterious effects of UV-radiation in the skin including accelerated ageing or cancer. These reactive intermediates damage cell membranes by lipid peroxidation and react with proteins, e.g. enzymes and DNA(RNA).

Reactive intermediates not only play an important role in direct effects, as caused by sunlight *per se* (Black, 1987), but also in those resulting from exposure to a xenobiotic in combination with light. Also in the latter case effects may be beneficial or unwanted. For instance, PUVA-therapy is based on the formation of reactive intermediates. Patients suffering from psoriasis are given an oral dose of a psoralen and exposed to long-wave ultraviolet radiation (UVA; 320 – 400 nm). The rationale of PUVA (P = psoralen + UVA) is that the psoralen, complexed to DNA in the skin cells, binds covalently to this biomacromolecule on exposure to UVA (Dall'Acqua and Rodighiero, 1984; Beijersbergen van Henegouwen *et al.*, 1989). Because psoralens preferentially accumulate in diseased parts of the skin, it is just the proliferation of the psoriatic cells which is inhibited. Side effects of PUVA therapy, such as erythema and hyperpigmentation, are not attributed to the photocycloaddition to DNA but to the formation of psoralen radicals and singlet oxygen resulting from energy transfer from the photoexcited drug to oxygen (De Mol and Beijersbergen van Henegouwen, 1981).

Photoallergy is an example of an untoward effect of light-induced decomposition of a xenobiotic to reactive intermediates in the skin. In this respect, the drug chlorpromazine has become notorious. Upon exposure to UVA, chlorpromazine decomposes to radicals and irreversible binding to skin constituents takes place (Schoonderwoerd and Beijersbergen van Henegouwen, 1987; Schoonderwoerd *et al.*, 1988, 1989). Binding to or damage of proteins or cell membranes, especially those of immune cells, can trigger an immune response which may lead to symptoms of photoallergy.

(b) Systemic effects

Besides the fact that reactive intermediates can be formed non-enzymically in the light-exposed skin, it is worth mentioning that biological effects resulting from this process may also occur in other organs (systemic effects). Evidence of this originates from research on some natural photobiological processes in man. Best known is the conversion of 7-dehydrocholesterol into previtamin D_3 upon exposure of the body to UVB (290–320 nm), which constitutes a regular part of the sunlight spectrum at sea level. By careful analysis, Holick (1985) showed that previtamin D_3 in human skin participates in three different photochemical equilibrium reactions which control the level of this biologically very active compound. These interdependent equilibrium reactions, for instance, prevent overproduction of vitamin D_3 during excessive exposure to sunlight. Vitamin D_3 is metabolized in the liver and kidney to the dihydroxyderivative, which, in turn, triggers the formation of a calcium transport factor in the intestine. Vitamin D_3 appears to be absolutely essential for maintaining calcium and phosphorus homeostasis and thus supports regular development of bone structure.

The UVB-induced conversion of 7-dehydrocholesterol is of relevance with respect to research on the photobiological activity of xenobiotics, because it demonstrates that, under the influence of light, compounds can be formed in the skin at extremely low concentrations which become active distant from their site of formation.

Another example of systemic effects arising from light exposure is the phototherapy of neonatal jaundice. In this disease and in the frequently occurring Gilberts syndrome, the blood level of bilirubin is severely increased. Conversion into more water-soluble metabolites is insufficient because of inadequate liver function. As a result of this, excretion of this breakdown product of heme-containing compounds is retarded. Bilirubin is neurotoxic and can cause brain damage. Upon exposure of the body to visible light, bilirubin undergoes a number of processes, such as isomerization and photo-oxidation, leading to products that are more easily excreted than the parent compound (McDonagh, 1985). Because photo-degradation of bilirubin prevents its brain-damaging effects, it can be considered an example of a photobiological process associated with systemic effects.

As already mentioned this chapter aims to highlight the photobiological effects of some nitroarenes and imino-N-oxides. Some of these compounds are drugs or metabolic derivatives, others are frequently used in cosmetics or find extensive application in pig and poultry feed. In each case, the photochemistry, as far as is known, is discussed and special emphasis is put on the possible formation of reactive intermediates. The possibility of a relation between the latter and toxic effects observed is discussed.

26.2 PHOTOBIOLOGICAL REACTIONS WITH NITROARENES

26.2.1 Nifedipine

Nifedipine belongs to an important group of calcium antagonists which have recently become commercialized. The latter are used in the treatment of angina pectoris and arterial hypertension. Nifedipine is extremely sensitive to ultraviolet radiation and to visible light up to 450 nm. The quantum yield for photodegradation is ~0.5 (Thoma and Klimek, 1985a); statistically this means that one out of two photons absorbed causes decomposition of one nifedipine molecule. The main product of photodecomposition is a nitroso compound (Thoma and Klimek, 1985a,b; Fig. 26.1). Nifedipine as well as other 1,4-dihydro-4-(nitrophenyl) pyridines used as drugs are complexed with plasma proteins by ~95%. This, in combination with the formation of reactive photoproducts such as nitroso compounds, implies the possibility of irreversible photobinding to proteins, as has been found (Campbell *et al.*, 1984; Ichida *et al.*, 1989a,b). Since knowledge about the photochemistry

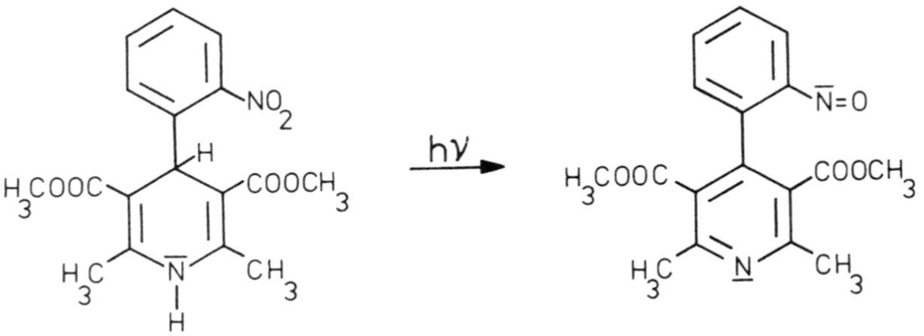

Figure 26.1 Light-induced formation of a nitroso compound from nifedipine.

in vivo is not available, the *in vitro* data may offer only a tentative explanation for several reports on photosensitivity reactions associated with nifedipine intake (Thomas and Wood, 1986).

26.2.2 Musk ambrette (artificial)

Musk ambrette (Fig. 26.2), a widely applied perfume component, is a notorious photoallergen in man (Ford, 1984; Parker *et al.*, 1986). An important route mediating photoallergy is the covalent protein binding of a reactive form of the photosensitizing agent to produce an allergen. Photobinding of musk ambrette to human serum albumin has been demonstrated (Barrett and Brown, 1985). From ESR measurements it was concluded that the first step in the photodecomposition of musk ambrette in the presence of UVA consists in the production of two nitro anion radicals (Motten *et al.*, 1983). Photolysis of a 3% solution in CH_3OH–0.1 M NaOH gave an azoxy derivative (7%) in addition to an azo compound as the main product (42%).

From these data, the mechanism of photodecomposition was derived as follows (Fig. 26.2.): disproportionation of a nitro anion radical formed upon light exposure to a nitroso compound and the original musk ambrette.

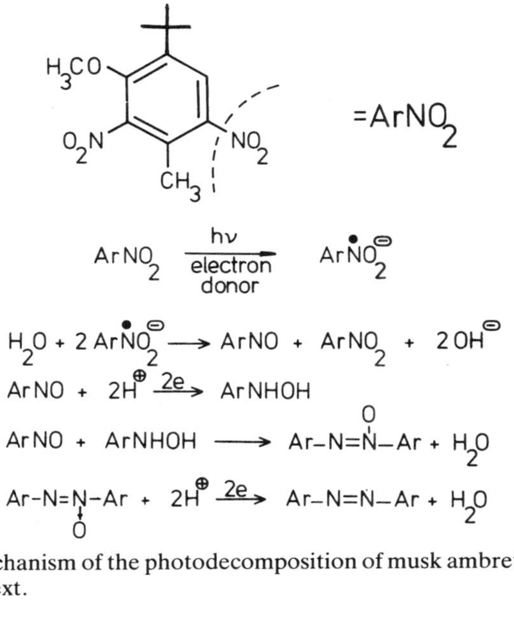

Figure 26.2 Mechanism of the photodecomposition of musk ambrette ($ArNO_2$). For details see the text.

Reduction of the nitroso compound generates the corresponding hydroxylamine derivative, which can condense with another nitroso molecule to yield the azoxy photoproduct. The condensation reaction and the subsequent reduction of the azoxy compound to an azo derivative only occur with relatively high concentrations of the reactants and are not relevant to the *in vivo* situation. More important to the problem of photoallergy is the observation that, along with nitro anion radicals, nitroso and hydroxylamine compounds are formed. The latter two species are considered to be even more reactive towards proteins than the radical itself. In addition to photochemical research *in vitro* some photobiological experiments have been performed with human lymphoid cells (Morison *et al.*, 1982). These showed that musk ambrette is phototoxic to immune cells, as measured by decreased incorporation of [^{3}H]thymidine into nuclear DNA.

From the foregoing it is supposed that, *in vivo*, in the light-exposed skin haptens such as nitro anion radicals or nitroso and hydroxylamine species are formed from musk ambrette which trigger an immune response by reacting with plasma proteins and immune cells. This may eventually lead to the symptoms of photoallergy.

26.2.3 Chloramphenicol

Chloramphenicol (CAP) is an antibiotic effective against a wide range of life-threatening bacteria. It is administered systemically such as in the treatment of central nervous system infections, but also topically such as in the therapy of deep ocular infections, where it represents the drug of choice. CAP is a bone marrow toxicant affecting primarily the erythroid precursors. Although rare, anaemia as a side effect of CAP treatment can be fatal; this has also been observed after topical application.

Metabolic reduction of the nitro group in the liver is supposed to give rise to reactive intermediates. In this regard, *p*-nitrosochloramphenicol (pNOCAP) has been extensively investigated; *in vitro*, it proved to be a very potent bone marrow suppressing agent (Yunis *et al.*, 1980).

De Vries *et al.* (1984) showed that CAP was photodecomposed *in vitro* by ~80% as a result of exposure to sunlight for 45 min (UVA = 14 W/m^2). The initial concentration of CAP (10 mg/l in PBS) was comparable to that found in blood after systemic administration or in the aqueous humour of the eye up to one hour after topical application. Decomposition products were composed as follows: 20% unchanged CAP, 25% *p*-nitrobenzaldehyde (pNB), 36% *p*-nitrosobenzoic acid (pNOBA) and 15% *p*-nitrobenzoic acid (pNBA) (Fig. 26.3). Both the rapid photodecomposition of CAP under conditions relevant to the *in vivo* situation and the reactivity of some of the photoproducts prompted further research on the toxicity of the latter

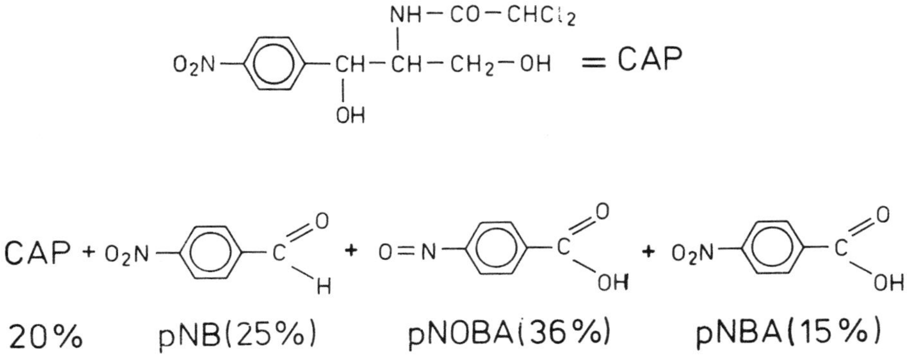

$$NH-CO-CHCl_2$$

$$O_2N-\bigcirc-CH-\underset{\underset{OH}{|}}{CH}-CH_2-OH \;\; = CAP$$

$$CAP + O_2N-\bigcirc-C\overset{\nearrow O}{\underset{\searrow H}{}} + O=N-\bigcirc-C\overset{\nearrow O}{\underset{\searrow OH}{}} + O_2N-\bigcirc-C\overset{\nearrow O}{\underset{\searrow OH}{}}$$

20%   pNB(25%)   pNOBA(36%)   pNBA(15%)

Figure 26.3 Composition (mol%) of CAP in solution (10 mg/l in PBS) after exposure to sunlight (UVA $= 14\,W/m^2$) for 45 min. CAP, chloramphenicol; pNB, *p*-nitro-benzaldehyde; pNOBA, *p*-nitrosobenzoic acid; pNBA, *p*-nitrobenzoic acid.

towards bone marrow cells *in vitro*. For CAP and pNB, the data found corresponded quite well with those reported in the literature (Gross *et al.*, 1982; Table 26.1). As can be seen, *p*-nitrobenzaldehyde (pNB) and *p*-nitrosobenzoic acid (pNOBA) are more potent bone marrow depressants than CAP itself, and their suppressive potency corresponds to that of *p*-nitrosochloramphenicol (*p*-NOCAP). In order to act as a bone marrow toxicant, a reactive intermediate must be stable enough to be transported by the blood stream from its site of formation to the target tissue. Eyer *et al.* (1984) found that pNOCAP was rapidly eliminated from human blood with a half-life of less than 10 s. For this reason it was concluded that pNOCAP, enzymically formed from CAP in the liver, is unlikely to cause bone marrow

Table 26.1 *In vitro* bone marrow cell depressing activity of chloramphenicol and its (photo) metabolites as measured by [^{3}H]-thymidine incorporation. For abbreviations see Fig. 26.3. Each value represent the mean of three experiments

Compound	[^{3}H] Thymidine incorporated (% of control)
CAP	$43.5 \pm 3.9\,(42.6)$[a]
pNB	$1.9 \pm 0.2\,(1.7)$
pNOBA	6.6 ± 1.6
pNBA	57.1 ± 2.0
pNOCAP	ND (1.2)

[a] Figures in parentheses indicate reference data taken from Yunis *et al.* (1980) and Gross *et al.* (1982).

depression. In this respect, the photoproducts pNB and pNOBA seem to be more serious candidates as they have half-lives in human blood of less than 3 and more than 20 min, respectively (De Vries *et al.*, unpublished results). Thus, once formed photochemically in the skin, they are stable enough to reach the bone marrow.

The question whether CAP photodecomposes under *in vivo* conditions was investigated with rats. Six rats (200 g each), whose backs had been shaved, were each given 30 mg [^{3}H]CAP intraperitoneally. Three animals were exposed to UVA (50 W/m^2; 345–410 nm, max. at 370 nm) for 10 h and three others were kept in the dark. This procedure was repeated the next day and then the rats were killed. After dialysis under non-equilibrium conditions, irreversibly bound CAP in the dorsal skin was determined by measuring cutaneous radioactivity which was expressed as dpm/mg of protein. The values thus obtained for UVA-irradiated and non-exposed rats were 463 ($\pm$ 13) and 51 ($\pm$ 15) dpm/mg of protein, respectively. From this it is concluded that CAP photodecomposes in the UVA-exposed skin. Further, irreversible binding suggests that photodecomposition proceeds via the formation of reactive intermediates (De Vries *et al.*, unpublished work).

When [^{3}H]CAP added to human blood *in vitro* was irradiated with UVA, irreversible binding was found to occur not only to plasma proteins, but also to cell constituents. In addition, evidence was obtained that pNB and pNOBA were also formed (De Vries *et al.*, unpublished data).

The *in vitro* and *in vivo* photochemical data described above justify the conclusion that a correlation between sunlight exposure and anaemia as a systemic side effect of CAP treatment should be considered a serious possibility.

26.2.4 Nitrofurantoin

The urinary tract disinfectant nitrofurantoin (NFT) is notorious for its high incidence of serious side effects. Of these, 40% are believed to represent allergic reactions especially concerning the lung and skin. NFT is also known to cause a considerable number of haematological reactions; not less than 20% of the fatalities caused by NFT are due to blood dyscrasia. Blood dyscrasias include methaemoglobinaemia, a situation in which more than 1% of total blood haemoglobin is oxidized to the ferric state. Incomplete metabolic reduction of the nitro group of NFT in the liver, resulting in the formation of reactive intermediates, has been proposed to originate most of the side-effects of this drug. However, the exact mechanism involved is still unknown.

With respect to this and taking into account that NFT is photolabile upon irradiation with UVA, the possible role of light in the activation of NFT was investigated. Upon UVA-exposure of a solution of NFT in PBS (1 mM, pH 7), 60% of the photodecomposed starting material appeared to be converted into 5-nitrofurfural (NFA), which is also photolabile. Under the same conditions, up to 70% of NFA loses nitrite with the formation of 5-hydroxymethylene-2(5H)-furanone (HMF). Extensive research on product formation and the reaction kinetics revealed that HMF is a tautomer of 5-hydroxyfurfural and that conversion of NFA to the latter aldehyde probably proceeds via nucleophilic substitution with a sigma-complex as an intermediate (Fig. 26.4; Busker and Beijersbergen van Henegouwen, 1987).

As nitrite is a photoproduct of NFA and a well-known inducer of methaemoglobin (MetHb) formation, it was investigated whether photo-decomposition of NFT in human blood *in vitro* results in increased MetHb and nitrite levels. Indeed, only irradiation gave a considerable increase in

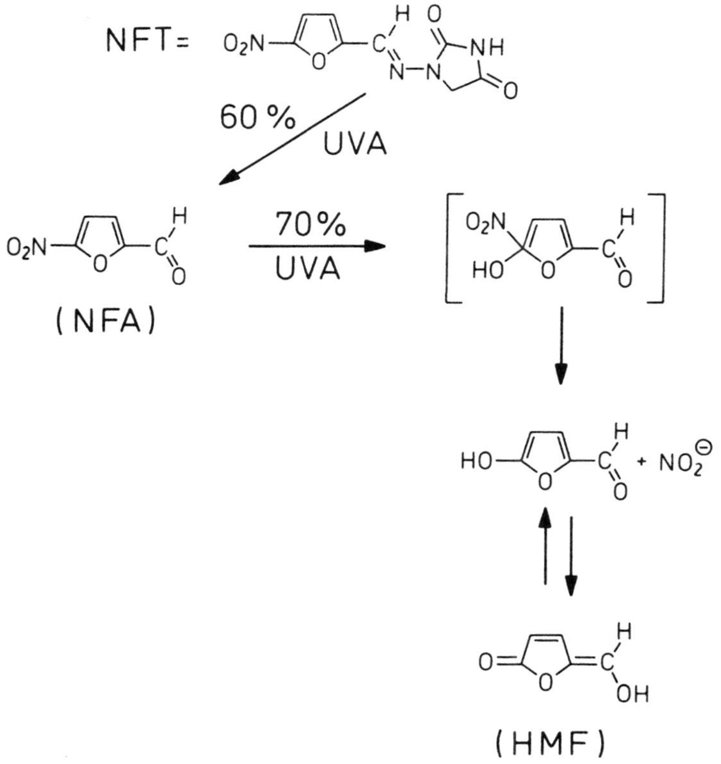

Figure 26.4 Photodecomposition of nitrofurantoin (NFT) to 5-nitrofurfural (NFA) and 5-hydroxymethylene-2(5H)-furanone (HMF). For details see the text.

the concentration of both compounds. The first photoproduct of NFT, NFA, was found to play an important role in the formation of MetHb (Busker *et al.*, 1988a). In order to investigate a possible correlation between exposure to (sun)light and methaemoglobinaemia as a side effect of NFT treatment, *in vivo* experiments were performed with rats (Busker *et al.*, 1988a). The experimental conditions and results are summarized in Table 26.2.

NFA was given intraperitoneally instead of orally and in a lower daily dose as it is a (photo)metabolite of NFT. Exposure to UVA alone (Table 26.2b) or administration of NFT (NFA) without exposure to UVA (Table 26.2c and e) gives MetHb levels not significantly different from those of the controls (Table 26.2a) (for details see Busker *et al.*, 1988a). Although the rat, owing to its ability to compensate rapidly for an increase in MetHb concentration, is not an ideal object for this type of study, the increase in MetHb concentration, which occurs as a result of the simultaneous exposure to NFT (NFA) and UVA, is a clearcut phenomenon (Table 26.2d and f). Together with the *in vitro* photochemical data mentioned above, these *in vivo* experimental results strongly point at a correlation between (sun)light exposure and methaemoglobinaemia as a systemic side effect of NFT administration.

Allergic drug reactions are considered to arise from an immune reponse to adducts formed between the drug and biomacromolecules, e.g. plasma proteins. Adduct formation may result from activation of the drug involved by biotransformation. In view of the photolability of NFT, the effect of UVA irradiation on protein binding was investigated. *In vitro*, efficient photobinding to human serum albumin was demonstrated for both NFT (up

Table 26.2 Methaemoglobin (MetHb) levels in rats after treatment with nitrofurantoin (NFT) or nitrofurfural (NFA) in the absence and presence of UVA. The level of MetHb in untreated rats was 0.5%. UVA exposure ($25\,W/m^2$; 320–410 nm, max. at 370 nm) was 12 h/day. The figures in parentheses indicate the number of animals used. Probabilities for significant differences between (c) and (d) as well as (e) and (f) were < 0.01 (Busker *et al.*, 1988a)

	Compound	Treatment (days)	Dose (mg/day)	Exposure to UVA	MetHb level (%)
a	—	4	—	−	0.5 (10)
b	—	4	—	+	0.6 (10)
c	NFT	4	12[a]	−	0.5 (8)
d	NFT	4	12[a]	+	1.0 (12)
e	NFA	2	3[b]	−	0.4 (5)
f	NFA	2	3[b]	+	1.3 (6)

[a] Oral application.
[b] Intraperitoneal application.

to 50 nmol/mg of protein) and its primary photoproduct NFA. Incubation in the dark of the end-products of photodecomposition of NFT and NFA, such as HMF, with plasma proteins also resulted in irreversible binding. Amino and, to a lesser extent, thiol groups proved to be targets for binding. Furthermore, a significant decrease in the iso-electric point of albumin was observed (Busker *et al.* 1989a).

Whether adducts are also formed *in vivo* was investigated with rats (Busker *et al.* 1988b). Eight rats (150 g each), whose backs had been shaved, were each given an oral dose of 80 mg/kg ^{14}C-labelled NFT. Four animals were exposed to UVA (50 W/m^2; 345–410 nm, max. at 370 nm) for 10 h/day and four others were kept in the dark. This was repeated the next day and the rats were then killed. Irreversibly bound NFT was quantified and expressed as pmol/mg of protein; the latter was taken as a measure of the amount of tissue analysed. The results, presented in Fig. 26.5, show a high extent of irreversible binding in the dorsal skin (both epidermis and dermis), in the ears and in plasma proteins, the spleen and in the tail and eyes (the latter two not shown in Fig. 26.5). A temperature of 32°C instead at 22°C during NFT/UVA treatment, resulted in a higher degree of irreversible binding (Fig. 26.5); this is probably due to an increased dermal blood flow. Under these conditions, irreversibly bound radioactivity was even found in the lungs and in the kidney and liver (the latter two not shown) (Busker *et al.*, 1988b). Other experiments showed that photobinding increased with the dose and light intensity applied. Further, the same extent of irreversible binding was found with rats killed either immediately after the last NFT/ UVA treatment or five days later. The latter observation provides further

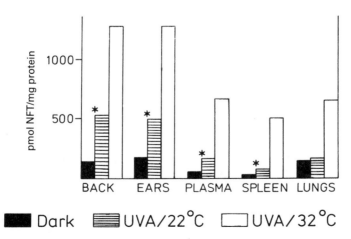

Figure 26.5 Irreversible photobinding of NFT to biomacromolecules in various organs. An asterisk (*) indicates significant difference ($P < 0.001$) between irradiated rats and animals kept in the dark.

evidence that photobinding is irreversible. Irreversible binding found in inner organs can be explained by systemic distribution of plasma proteins to which NFT had been coupled through photoactivation in the skin, but also by transport of reactive photoproducts of NFT, such as HMF, which eventually undergo covalent binding to tissue biomacromolecules distant from their site of formation.

Immunogenic properties of photoadducts formed between NFT and plasma proteins have also been demonstrated (Busker *et al.*, 1989b). Photoadducts, produced *in vitro* with plasma proteins from rabbits, were injected into the same animals from which blood had been drawn. It was found that photoadducts can indeed induce the formation of antibodies, titres ranging from 150–1500 vs. < 60, as found for the controls.

The majority of cutaneous reactions observed during NFT therapy are of the allergic type. This may arise from covalent binding after incomplete enzymatic reduction of NFT in the liver. However, a significantly higher extent of covalent binding was found when rats were exposed to UVA as compared with animals kept in the dark. Therefore, photoactivation of NFT in the skin should be considered a route to immunologically mediated skin damage or rashes. Furthermore, the presence of irreversibly bound NFT in the blood and other organs, especially the spleen, indicates that light-induced adverse reactions are not restricted to the skin, but may also occur systemically.

26.3 PHOTOBIOLOGICAL REACTIONS WITH IMMINO-*N*-OXIDES

26.3.1 Chlordiazepoxide

Extensive research on the photochemistry *in vitro* of the phototoxic chlordiazepoxide (CDZ), Librium® showed that an oxaziridine is the main product. This reactive intermediate was also found on UVA exposure of other imino-*N*-oxides, such as the major metabolites of CDZ (Cornelissen *et al.*, 1980; Fig. 26.6).

In vitro, imino-*N*-oxides irreversibly bind to plasma proteins upon UVA exposure most likely via an oxaziridine intermediate (Bakri *et al.*, 1988). Analogues of CDZ lacking an oxygen atom attached to the nitrogen are photostable, and photobinding to proteins does not occur (Bakri *et al.*, 1988). Based on these results *in vitro*, it was expected that the *N*-oxide functionality in CDZ and its metabolites is responsible for phototoxicity. This was confirmed by using a microbiological test system, plotting the survival of *Salmonella typhimurium* strain TA100 against the concentration of the compound investigated. Compounds lacking an oxygen atom

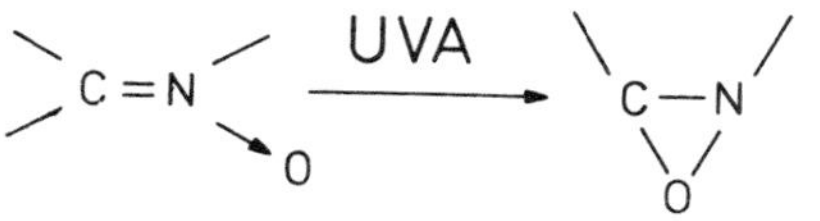

Figure 26.6 Photoisomerization of an imino-*N*-oxide to an oxaziridine.

attached to the nitrogen of the C=N group were not phototoxic at all. Phototoxicity curves for CDZ and some other imino-*N*-oxides corresponded quite well with curves for the toxicity of their oxaziridine derivatives in the absence of irradiation (Cornelissen *et al.*, 1980; De Vries *et al.*, 1983).

UVA-induced effects were also extensively investigated *in vivo* with rats (Bakri *et al.*, 1983, 1985; Beijersbergen van Henegouwen, 1988). Figure 26.7 shows some typical results. As can be seen, UVA has a considerable influence on CDZ metabolism. The percentage of CDZ and *N*-oxy metabolites is only about half that found without UVA irradiation. In contrast to this, the percentage of metabolites lacking an oxygen atom attached to nitrogen, inclusive of R.CDZ, is even about five times higher with UVA-exposed rats. This may be explained by the ability of oxaziridines, formed by photoisomerization in the UVA-exposed skin, to react with compounds containing SH groups. Thus, CDZ, upon UVA irradiation, and its oxaziridine in the absence of light appear to react spon-

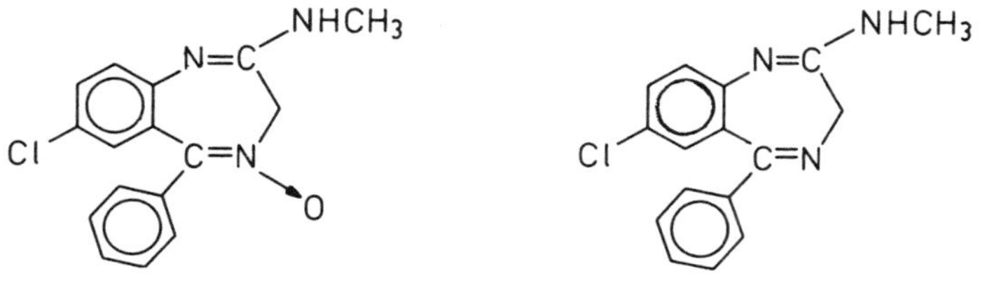

Chlordiazepoxide (CDZ)   Reduced CDZ (R.CDZ).

	UVA	Dark
CDZ+	20%	37%
R.CDZ+	14%	3%
Unknown	11%	5%

Figure 26.7 Urinary excretion (~45% of dose) of CDZ metabolites in rats either exposed to UVA or kept in a dimly lit environment. CDZ+, imino-*N*-oxides (inclusive of CDZ); R.CDZ+, metabolites lacking oxygen attached to the nitrogen (inclusive of R.CDZ). For experimental details see Bakri *et al.* (1983).

taneously with gluthathione (GSH) abundant in the body giving R.CDZ as the ultimate product (Beijersbergen van Henegouwen, 1988). When the reactions with GSH are allowed to proceed in UVA-exposed rats, a lower percentage of *N*-oxides (e.g. CDZ) and a higher percentage of reduced metabolites (e.g. R.CDZ) will be formed. In addition it has observed that the percentage of glucuro-conjugated metabolites of CDZ in UVA-exposed rats is half that in the controls (Bakri *et al.*, 1983).

As was expected from the *in vitro* photochemical data, irreversible binding to tissues was also found with rats exposed to UVA in the presence of CDZ. In a typical experiment (Bakri *et al.*, 1983), four rats (200 g) were each given a single dose of ^{14}C-labelled CDZ (25 mg/kg, i.p.). On 18 consecutive days, two animals were exposed to UVA (14 W/m^2; 320–380 nm, max. at 350 nm) for 10 h each, whereas two others were kept in the dark. Thereafter, the rats were killed and tissues were subjected to dialysis for 26 days under non-equilibrium conditions. Irreversibly bound radioactivity, expressed in cpm/100 mg of tissue, was found in the UVA-exposed rats as follows: dorsal skin, 2440 ($\pm$ 30); ventral skin, 910 ($\pm$ 20); ears, 1940 ($\pm$ 80); liver, 1030 ($\pm$ 90). Both the considerable change in metabolism and the irreversible binding to biomacromolecules in the liver show that the phototoxicity of CDZ is not restricted to the UVA-exposed area, but can affect inner organs as well.

Further information on the role of the *N*-oxide function in phototoxicity was obtained by comparing CDZ with R.CDZ *in vivo*. Investigations with R.CDZ were performed under the same conditions as those with CDZ. UVA exposure of rats did not have any influence on the metabolism of R.CDZ nor did irreversible binding to biomacromolecules occur (Bakri *et al.*, 1985). Similar differences between diazepam-*N*-oxide and diazepam have also been found (Beijersbergen van Henegouwen, 1988). In contrast to diazepam-*N*-oxide, diazepam (Valium®) does not appear to be phototoxic.

26.3.2 Olaquindox

Olaquindox (OLAQ; trade name BAYO-N-OX) belongs to the group of quindoxin (QUIN) derived compounds. Unlike OLAQ, QUIN does not possess a side chain (cf. Fig. 26.8). OLAQ is used extensively in cattle breeding, pig husbandry and poultry farming as a feed additive with antimicrobial and growth-promoting effects. QUIN has been used for similar purposes but has been withdrawn after causing persistent photocontact dermatitis in man. In quite a number of cases, photosensitivity persisted for a period of more than four years (Zaynoun *et al.*, 1976). More recently, OLAQ has also been reported to cause photoallergy in man (Francalanci *et al.*, 1986). Symptoms resemble those of QUIN-induced photoallergy.

Furthermore, severe phototoxicity effects have been observed in pigs treated with OLAQ. QUIN-derived compounds belong to the group of imino-*N*-oxides and appear to form an oxaziridine upon irradiation *in vitro* (cf. Fig. 26.6). Further, they exhibit high photoreactivity towards proteins, suggesting photoallergic properties (De Vries *et al.*, 1990a). Only rats exposed to UVA in the presence of OLAQ suffer from severe dermal erythema, oedema of the feet and necrosis of the ears (De Vries *et al.*, 1990). Moreover a profound influence on OLAQ metabolism was found. Eight rats (140 g each), whose backs had been shaved, were each given OLAQ (60 mg/kg, p.o.) Four rats were exposed to UVA (60 W/m^2; 345–420 nm, max. 370 nm) for 12 h/day, while the other four animals were kept in the dark. Urine was collected every 24 h. This procedure was repeated on each of four consecutive days at 24 h intervals. When rats were kept in the dark, 60% of the daily dose was excreted as unchanged OLAQ. However, after UVA irradiation unchanged OLAQ found in the urine accounted for only 20% of the dose applied (Fig. 26.8).

In the absence of UVA-exposure, reduction of the parent imino-*N*-oxide represents only a minor metabolic route. This changes drastically upon UVA-irradiation: instead of 2%, about 30% of desoxyolaquindox-4-monoxide (desoxyOLAQ-4-M) are formed (Fig. 26.9). This result closely resembles that for chlordiazepoxide (Fig. 26.7) and diazepam-*N*-oxide. Together with the *in vitro* photochemical data, namely formation of an oxaziridine exhibiting photoreactivity towards proteins, the formation of desoxyOLAQ-4-M *in vivo* is an important indication that photo-isomerization of OLAQ to the reactive intermediate mentioned above is responsible for the phototoxic/photoallergic effects observed.

These results with imino-*N*-oxides have been confirmed by Pöhlmann *et al.* (1986) who performed *in vitro* and *in vivo* investigations with

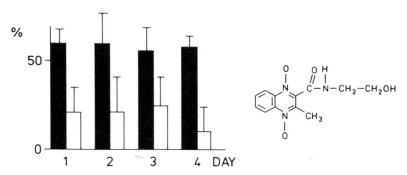

Figure 26.8 Daily urinary excretion (mol% of dose) of unchanged olaquindox in rats exposed to UVA (☐) or kept in the dark (■); ☐ differs significantly from ■ ($P < 0.001$).

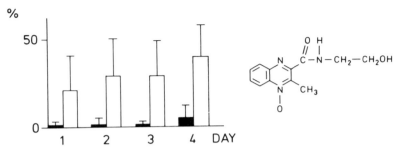

Figure 26.9 Daily urinary excretion (mol% of the dose of olaquindox applied) of the metabolite desoxyolaquindox-4-monoxide in rats exposed to UVA (☐) or kept in the dark (■); ☐ differs significantly from ■ ($P < 0.001$).

methaqualone-N-oxide as a major metabolite of the drug methaqualone. In this case too, a reactive oxaziridine appears to be formed.

26.4 CONCLUSIONS

The results of the investigations summarized in this chapter provide evidence that biological effects of N-oxidized compounds can be caused by reactive intermediates which are formed non-enzymically in (sun)light-exposed parts of the body.

Another important outcome of this research is that biological effects are not restricted to the locus where photoactivation of the N-oxidized compounds takes place. Phototoxic effects can occur in inner organs as well: the simultaneous exposure of the skin to (sun)light and an N-oxidized compound can produce *systemic* effects.

REFERENCES

Bakri, A., Beijersbergen van Henegouwen, G.M.J. and Chanal, J.L. (1983) Photopharmacology of the tranquilizer chlordiazepoxide in relation to its phototoxicity. *Photochem. Photobiol.*, **38**, 177–83.

Bakri, A., Beijersbergen van Henegouwen, G.M.J. and Chanal, J.L. (1985) Involvement of the N_4-oxide group in the phototoxicity of chlordiazepoxide in the rat. *Photodermatology*, **2**, 205–12.

Bakri, A., Beijersbergen van Henegouwen, G.M.J. and De Vries, H. (1988) Photobinding of some 7-chloro-1,4-benzodiazepines to human plasma protein *in vitro* and photopharmacology of diazepam in the rat. *Pharm. Weekbl. (Sci)*, **10**, 122–9.

Barrett, M.D. and Brown, K.R. (1985) Photochemical binding of photoallergens to human serum albumin: a simple *in vitro* method for screening potential photoallergens. *Toxicol. Lett.*, **24**, 1–6.

Beijersbergen van Henegouwen, G.M.J. (1988) *In vitro* and *in vivo* research on phototoxic xenobiotics: structure–reactivity relationships. *Arch. Toxicol. Suppl.*, **12**, 3–9.

Beijersbergen van Henegouwen, G.M.J., Wijn, E.T. and Schoonderwoerd, S.A. (1989) A method for the determination of PUVA-induced *in vivo* irreversible binding of 8-methoxypsoralen (8-MOP) to epidermal lipids, proteins and DNA/RNA. *J. Photochem. Photobiol., B: Biol.*, **4**, 631–5.

Black, H.S. (1987) Potential involvement of free radical reactions in ultraviolet light-mediated cutaneous damage. *Photochem. Photobiol.*, **46**, 213–21.

Busker, R.W. (1989) The formation of antibodies against photoadducts between nitrofurantoin and plasma proteins in rabbits, in *Photoreactivity of Nitrofurantoin as a Possible Cause of Some of its Side-effects*, Ph.D.-thesis, Leiden University, Ch. 7, pp. 78–86.

Busker, R.W. and Beijersbergen van Henegouwen, G.M.J. (1987) The photolysis of nitrofurantoin and of 5-nitrofurfural in aqueous solutions: nucleophilic substitution of the nitro group. *Photochem. Photobiol.*, **45**, 331–5.

Busker, R.W., Beijersbergen van Henegouwen, G.M.J., Menke, R.F. and Vasbinder, G. (1988a) Formation of methemoglobin by photoactivation of nitrofurantoin and of 5-nitrofurfural in rats exposed to UVA light. *Toxicology*, **51**, 255–6.

Busker, R.W., Beijersbergen van Henegouwen, G.M.J., Van Ballegooie, E.P. and Vasbinder, G. (1988b) Irreversible binding of nitrofurantoin to skin and to plasma proteins of UVA exposed rats. *Photochem. Photobiol.*, **48**, 683–8.

Busker, R.W., Beijersbergen van Henegouwen, G.M.J., Vaasen, A.J.H. and Menke, R.F. (1989) Irreversible photobinding of nitrofurantoin and of nitrofurfural to plasma protein *in vitro*. *J. Photochem. Photobiol. B: Biol.*, **4**, 207–19.

Campbell, K.P., Lipshutz, G.M. and Denney, G.H. (1984) Direct photoaffinity labeling of the high affinity nitrendipine-binding site in subcellular membrane fractions isolated from canine myocardium. *J. Biol. Chem.*, **259**, 5384–7.

Cornelissen, P.J.G., Beijersbergen van Henegouwen, G.M.J. and Mohn, G.R. (1980) Structure and photobiological activity of 7-chloro-1,4-benzodiazepines. Studies on the phototoxic effects of chlordiazepoxide, desmethylchlordiazepoxide and demoxepam using a bacterial indicator system. *Photochem. Photobiol.*, **32**, 653–61.

Dall'Acqua, F. and Rodighiero, G. (1984) Biological and medicinal aspects of furocoumarins (psoralens and angelicins), in *Primary Photo-processes in Biology and Medicine* (eds R.V. Bensasson, G. Jori, E.J. Land and T.G. Truscott), NATO ASI Series Life Sciences, Plenum, New York, vol. 85, pp. 277–94.

De Mol, N.J. and Beijersbergen van Henegouwen, G.M.J. (1981) Relation between some photobiological properties of furocoumarins and their extent of singlet oxygen production. *Photochem. Photobiol.*, **33**, 815–19.

De Vries, H., Beijersbergen van Henegouwen, G.M.J. and Wouters, R. (1983) Correlation between phototoxicity of some 7-chloro-1,4-benzodiazepines and their (photo)chemical properties. *Pharm. Weekbl. (Sci)*, **5**, 302–8.

De Vries, H., Beijersbergen van Henegouwen, G.M.J. and Huf, F.A. (1984) Photochemical decomposition of chloramphenicol in a 0.25% eye drop and in a therapeutic intraocular concentration. *Int. J. Pharm.*, **20**, 265–71.

De Vries, H., Beijersbergen van Henegouwen, G.M.J., Kalloe, F. and Berkhuysen, M.H.J. (1990) Phototoxicity of olaquindox in the rat. *Res. Vet. Sci.*, **48**, 240–4.

De Vries, H., Bojarski, J., Donker, A.A., Bakri, A. and Beijersbergen van Henegouwen, G.M.J. (1990a) Photochemical reactions of quindoxin, olaquindox, carbadox and cyadox with oritein, indicating photoallergic properties. *Toxicology*, **63**, 85–95.

Eyer, P., Lierheimer, E. and Schneller, M. (1984) Reactions of nitroschloramphenicol in blood. *Biochem. Pharmacol.*, **33**, 2299–308.

Ford, R.A. (1984) Photoallergenicity testing of fragrance material. A review on photoallergy by musk ambrette in male human patients. *Nippon Koshohin Kagakkaishi*, **8**, 301–4.

Francalanci, S., Gola, M., Giorgini, S. *et al* (1986) Occupational photocontact dermatitis from Olaquindox. *Contact Dermatitis*, **15**, 112–14.

Gross, B.J., Blanchflower, R.V., Burke, T.R. *et al.* (1982) Bone marrow toxicity *in vitro* of chloramphenicol and its metabolites. *Toxicol. Appl. Pharmacol.*, **64**, 557–65.

Holick, M.F. (1985) The photobiology of vitamin D and its consequences for humans, in *The Medical and Biological Effects of Light* (eds R.J. Wurtman, M.J. Baum and J.T. Potts Jr.), *Ann. N.Y. Acad. Sci.* **453**, 1–13.

Ichida, S., Fujisue, T. and Masada, A. (1989a) Characteristics of specific binding of nitrendipine and PN 200-110 to various crude membranes: induction of irreversible binding by UV irradiation. *J. Biochem. (Tokyo)*, **105**, 760–6.

Ichida, S., Fujisue, T. and Masada, A. (1989b) Photoaffinity labeling with dihydropyridine derivatives of crude membranes from rat skeletal, cardiac, ileal and uterine muscles and whole brain. *J. Biochem. (Tokyo)*, **105**, 767–74.

McDonagh, A.F. (1985) Light effects on transport and excretion of bilirubin in newborns, in *The Medical and Biological Effects of Light* (eds R.J. Wurtman, M.J. Baum and J.T. Potts Jr.), *Ann. N.Y. Acad. Sci.* **453**, 65–72.

Morison, W.L., McAuliffe, D.J., Parrish, J.A. and Bloch, K.J. (1982) *In vitro* assay for phototoxic chemicals. *J. Invest. Dermatol.*, **78**, 460–3.

Motten, A.G., Chignell, C.F. and Mason, R.P. (1983) Spectroscopic studies of cutaneous photosensitizing agents – VI. Identification of the free radicals generated during the photolysis of musk ambrette, musk xylene and musk ketone. *Photochem. Photobiol.*, **38**, 671–8.

Parker, R.D., Buehler, E.V. and Newmann, E.A. (1986) Phototoxicity, photoallergy and contact photosensitization of nitro musk perfume raw materials. *Contact Dermatitis*, **14**, 103–9.

Pöhlmann, H., Theil, F.-P., Franke, P. and Pfeifer, S. (1986) *In vitro-* und *in vivo-* Untersuchungen zur fotochemischen Reaktivität von Methaqualon-1-oxid. *Pharmazie*, **41**, 856–8.

Schoonderwoerd, S.A. and Beijersbergen van Henegouwen, G.M.J. (1987) Irreversible photobinding to skin constituents after systemic administration of chlorpromazine to Wistar rats. *Photochem. Photobiol.*, **46**, 501–5.

Schoonderwoerd, S.A., Beijersbergen van Henegouwen, G.M.J. and Luijendijk, J.J. (1988) Photobinding of chlorpromazine and its sulfoxide *in vitro* and *in vivo*. *Photochem. Photobiol.*, **48**, 621–7.

Schoonderwoerd, S.A., Beijersbergen van Henegouwen, G.M.J. and Van Belkum, S. (1989) *In vivo* photodegradation of chlorpromazine. *Photochem. Photobiol.*, **50**, 659–64.

Thoma, K. and Klimek, R. (1985a) Untersuchungen zur Photostabilität von Nifedipin, 1. Mitt.: Zersetzungskinetik und Reaktionsmechanismus. *Pharm. Ind.*, **47**, 207–15.

Thoma, K. and Klimek, R. (1985b) Untersuchungen zur Photostabilität von Nifedipin, 2. Mitt.: Einflusz von Milieubedingungen. *Pharm. Ind.*, **47**, 319–27.

Thomas, S.E. and Wood, M.L. (1986) Photosensitivity reactions associated with nifedipine. *Br. Med. J.*, **292**, 992.

Yunis, A.A., Miller, A.M., Salem, Z. *et al.* (1980) Nitrosochloramphenicol: possible mediator in chloramphenicol-induced aplastic anemia. *J. Lab. Clin. Med.*, **96**, 36–46.

Zaynoun, S., Johnson, B.E. and Frain-Bell, W. (1976) The investigation of Quindoxin photosensitivity. *Contact Dermatitis*, **2**, 343–52.

Index

Page numbers in **bold** refer to figures, and those in *italic* to tables